Acta Physica Austriaca

Supplementum XVIII

Proceedings of the
XVI. Internationale Universitätswochen für Kernphysik 1977
der Karl-Franzens-Universität Graz
at Schladming (Steiermark, Austria)
24th February—5th March 1977

Sponsored by
Bundesministerium für Wissenschaft und Forschung
Steiermärkische Landesregierung
Sektion Industrie der Kammer der
gewerblichen Wirtschaft für Steiermark
International Centre for Theoretical Physics, Triest

1977

Springer-Verlag
Wien New York

Contacts Between High Energy Physics and Other Fields of Physics

Edited by Paul Urban, Graz

With 170 Figures

1977

Springer-Verlag
Wien New York

Organizing Committee

Chairman

Prof. Dr. H. Mitter
Institut für Theoretische Physik
der Universität Graz

Committee Members

Prof. Dr. H. Latal
Dr. A. Mas-Parareda
Dr. L. Mathelitsch
Dr. L. Pittner
Doz. Dr. F. Widder
Dr. H. Zankel

Secretary

M. Krautilik

Softcover reprint of the hardcover 1st edition 1977

Library of Congress Cataloging in Publication Data

Internationale Universitätswochen für Kernphysik der Karl-Franzens-Universität Graz, 16th, Schladming, Austria, 1977.
Contacts between high energy physics and other fields of physics.

(Acta physica Austriaca : Supplementum ; 18)
1. Particles (Nuclear physics)—Congresses.
2. Physics—Congresses. I. Urban, Paul Oskar, 1905- II. Austria. Bundesministerium für Wissenschaft und Forschung. III. Styria. Kammer der Gewerblichen Wirtschaft. Sektion Industrie.
IV. International Centre for Theoretical Physics.
V. Title. VI. Series.
QC793.I57 1977 539.7 77-13624

ISBN-13:978-3-7091-8500-1 e-ISBN-13:978-3-7091-8498-1
DOI: 10.1007/978-3-7091-8498-1

CONTENTS

Acta Physica Austriaca, Suppl. XVIII, 1-62 (1977)

CLASSICAL CONCEPTS IN QUANTUM CONTEXTS+

by

J. R. KLAUDER
Bell Laboratories
Murrary Hill, New Jersey 07974, USA

ABSTRACT

Several recent proposals in quantum field theory that make use of classical soliton (in the wide sense) solutions and semi-classical quantization techniques are reviewed and illustrated with various examples.

+ Lecture given at XVI. Internationale Universitätswochen für Kernphysik, Schladming, Austria, February 24-March 5, 1977.

CONTENTS

I. INTRODUCTION

The equations of quantum mechanics, let alone those of quantum field theory, are sufficiently complex to solve that various approximation schemes are both welcome and desirable. In applications to quantum field theory, renormalized perturbation theory has long been the predominant calculational tool, and it has even tended to shape the entire conceptual outlook on such problems as well. However, the apparently irreconcilable conflict between strong interaction physics and perturbation theoretic calculations has given way in the past few years to some fairly revolutionary (at least for field theory) modes of formulation and calculation. The fires of this revolution have been fed by exact solutions of special classical systems, [1] notably the sine-Gordon equation, and an appreciation of the relevance of classical solutions of nonlinear systems as starting points in quantum analyses [2]. The soliton solutions of the classical sine-Gordon equation exhibit extraordinary stability under scattering, but this feature is not essential for their utility in quantum studies. Consequently, broad studies have been made of nonlinear systems that exhibit localized, persistent solutions, which have also frequently been termed "solitons" (or solitary waves, or kinks) as well [3]. Solitons that have their origin in conventional symmetries and conserved currents (and thus are <u>non</u>topological in origin) have been discussed [4] along with those that arise from asymptotic properties of the solution, as in the sine-Gordon model, (and thus are topological in origin). Commonly, these are all systems for which the classical solution of least energy is nonvanishing, or for which a multitude of classical solutions are degenerate in energy, features which have emphasized the importance of nonperturbative calculational schemes. And several more or less semiclassical schemes, which are analogues of the WKB [2], Hartree-Fock [5], RPA [5], Born-Oppenheimer [6] methods, have been employed with the goal of calculating, at least approximately and often at the expense of explicit covariance, the excitation spectrum and any other accessible property of the system. Attempts to calculate higher-order corrections are often only marginally successful and are beset with problems of incorporating renormalization effects that are important in all but the simplest models. Although much work still needs to be done in the area of higher-order corrections, the initial stage of semi-classical analysis is certainly encouraging.

Semi-classical methods are as old as, if not older than, quantum theory itself, and even the phrase "semi-classical" means different things to different people. The merit of one or another scheme, or even the degree of any metaphysical significance are occasionally matters of debate. However, our aim in these lectures is simply to outline a few of the current trends with a bias toward field theoretic models and methods. For the most part we are reporting on the work of others. Like the authors of these programs we share the hope that one or more of these approaches may be relevant in Nature. But in any case, they make for interesting analyses of field theory and deserve study on their own merits.

Since this is a school of a fairly wide breadth of interest, we content ourselves in the main with conceptual presentations omitting many technical details. Initially we set the stage through a discussion of various field systems that have been proposed and review the salient features leading to soliton solutions. In this review we touch on the work of Lee and collaborators [4], [7], [8], and the SLAC bag [9]. Next we discuss the relevance of these classical analyses in a semi-classical framework, an analysis which, like the SLAC group and others, proposes to use coherent states for the sake of justification. Here we add some remarks of our own drawing on an earlier study [10] of classical-quantum connection and applying it to systems with soliton behavior. In our view this yields a cleaner connection between the classical system and its quantum counterpart and provides a better justification for the classical study. Afterwards, the quantum operator frameworks of Christ and Lee [11] and Goldstone and Jackiw [12] are mentioned, before discussing the path integral methods of Dashen, Hasslacher and Neveu [2], [13], [14]. Finally, as an illustration of some of the general ideas, we discuss a particular model that has been treated by Campbell and Liao [15].

A few brief remarks conclude this section.

Sine-Gordon Model

Models with exact solutions are rare indeed and it is useful to recall a few properties of one such model, namely the two-dimensional sine-Gordon model [6]. This model is characterized by the Lagrangian density

$$L = \frac{1}{2}(\phi_{,\mu})^2 + \frac{m^4}{G^2}\,[1-\cos(\frac{G\phi}{m})]$$

where G denotes a coupling constant and m a mass parameter. The significance of these parameters is most transparent in a development for small G where

$$L = \frac{1}{2}(\phi_{,\mu})^2 + \frac{1}{2}m^2\phi^2 - \frac{1}{4!} G^2\phi^4 + \ldots \quad .$$

For small G this is a ϕ^4 model with the "wrong" sign for m^2. The basic symmetry of this model is invariance under $\phi \to \phi + 2n\pi(m/G)$. The field equation reads

$$(\partial_t^2-\partial_x^2)\phi = \frac{m^3}{G} \sin(\frac{G\phi}{m}) \quad ;$$

and for static (time-independent) solutions, $\phi = \phi(x)$, the field equation reduces to

$$\phi'' = - \frac{m^3}{G} \sin(\frac{G\phi}{m}) \quad .$$

A solution is given by

$$\phi_c(x) = \frac{4m}{G} \tan^{-1} (e^{\pm mx})$$

which has a classical energy given by

$$M_o = 8m^3/G^2 \quad .$$

These solutions represent the soliton (e^{mx}) and anti-soliton (e^{-mx}) solutions. Basic time-dependent solutions are given by

$$\phi_{ss} = \frac{4m}{G} \tan^{-1} [\frac{u \sinh(m\gamma x)}{\cosh(m\gamma t)}]$$

and

$$\phi_{s\bar{s}} = \frac{4m}{G} \tan^{-1} \left[\frac{\sinh(m\gamma ut)}{u \cosh(m\gamma x)}\right]$$

where $\gamma = 1/\sqrt{1-u^2}$. The first solution represents two solitons while the second represents a soliton and anti-soliton. In each case the solitons have velocities $\pm$ u and total momentum zero, and the classical energy of each of these solutions is $2M_o\gamma$. Another solution exists in which $u \to ia$ in the $s\bar{s}$ solution leading to the breather mode

$$\phi_B = \frac{4m}{G} \tan^{-1} \left[\frac{\sin(m\bar{\gamma}at)}{a \cosh(m\bar{\gamma}x)}\right]$$

where $\bar{\gamma} = 1/\sqrt{1+a^2}$. The energy of this solution is $2M_o\bar{\gamma}$, which is less than two isolated solitons and represents a bound state. Note that it is periodic in time with angular frequency $\omega = m\bar{\gamma}a$.

The Grand Design

A fairly simple argument can be advanced for renormalizable field theories that quantum solutions are always associated with classical solitons [4]. Consider a scalar field with Lagrangian density

$$L = \frac{1}{2}(\phi_{,\mu})^2 - G^{-2}U(G\phi)$$

where G denotes a universal coupling constant. In terms of the field $\sigma = G\phi$, $L = G^{-2}L_\sigma$ where L_σ is independent of G. Consequently, every classical solution is of the form $\phi_{c\ell} = G^{-1}\sigma_{c\ell}$, where $\sigma_{c\ell}$ is independent of G, and we restrict our attention to finite energy solutions. Since in a quantum formulation, G^2 enters in the manner of $\hbar$ it follows that an $\hbar$ expansion of the quantum energy reads

$$E_{\text{quantum}} = E_{\text{classical}} \, [1+O(G^2)+ \ldots] \, ,$$

where the correction terms are finite and provide the desired connection between quantum and classical solutions whenever the theory is renormalizable. By construction $E_{\text{classical}} = O(G^{-2})$, and so the classical theory provides a way to calculate the nonanalytic contribution, a term that could not be calculated in perturbation theory. This analysis does not require in any way that ϕ is nonvanishing at spacial infinity, or that this argument applies to only one field at a time. Of course, stability is another question, and this will ultimately determine the significance of the quantum state.

II. MODELS AND CLASSICAL MOTIVATION

A. Nontopological Solitons

Scalar Fields

One interesting development of the past few years has been that suggested by T. D. Lee and collaborators

[4],[7],[8]. This approach exploits a different mechanism than that which leads to solitons in the sine-Gordon equation, and requires the existence of a Noether symmetry leading to a conserved current. Moreover, these models can exist in any number of space-time dimensions. An example consists of a charged scalar field ϕ in four-dimensional space time held together by a scalar glue field χ described by the Lagrangian density

$$L = \phi^{\dagger}{}_{,\mu}\phi_{,\mu} + \frac{1}{2}\chi^{2}{}_{,\mu} - G^{2}\chi^{2}\phi^{\dagger}\phi - U(\chi)$$

where U is a potential with minimum $U(\chi) = 0$ at $\chi=\chi_{vac}\neq 0$. With the restriction to renormalizable theories, $U(\chi)$ is at most a quartic. Let $U(\chi)$ be arranged so that the mass of the χ field is μ. Evidently the mass of the complex ϕ field is $m = G\chi_{vac}$. There is a conserved current

$$j_{\mu} = i\phi^{\dagger}\overleftrightarrow{\partial}_{\mu}\phi$$

that arises from a phase invariance (gauge transformation of the first kind), and a conserved charge

$$Q = i\int(\phi^{\dagger}\dot{\phi} - \dot{\phi}^{\dagger}\phi)d^{3}x \ .$$

Suppose that $\chi(x) \equiv \chi_{vac}$, in which case the Lagrangian reduces to that of a free complex scalar particle of mass $m = G\chi_{vac}$. Each excited particle of momentum p has energy $\sqrt{m^2+p^2}$, and for an excitation of charge Q the energy is

$$E = \sum_{j=1}^{Q} \sqrt{m^2+p_j^2} \geq Qm \ .$$

The question naturally arises whether this is the most favorable energy situation when deviations from $\chi = \chi_{vac}$ are taken into account. To investigate this situation consider the trial functions

$$\phi = \begin{cases} \frac{\sqrt{Q}}{2\pi} \frac{\sin\omega r}{r} e^{-i\omega t}, & r \leq R \\ 0, & r > R \end{cases}$$

where $\omega R = \pi$ and

$$\chi = \begin{cases} 0, & r \leq R \\ \chi_{vac}, & r > R + \varepsilon \end{cases} .$$

For Q and R both large it follows that

$$E = \frac{\pi Q}{R} + \frac{4\pi}{3} R^3 U(0) + O(R^2) ,$$

the second term due simply to the term $U(\chi)$ for $r \leq R$, while the last term represents surface effects. Neglecting the surface effect here, the minimum of E with respect to R occurs when $R = [Q/4U(0)]^{1/4}$ and leads to an energy

$$E = \frac{4\pi}{3} [4U(0)]^{1/4} Q^{3/4}$$

which for sufficiently large Q is less than Qm. When that occurs the state with $\chi = \chi_{vac}$ is not the lowest energy state.

A more complete analysis leads to the picture in Fig. 1 for energy states as a function of charge,

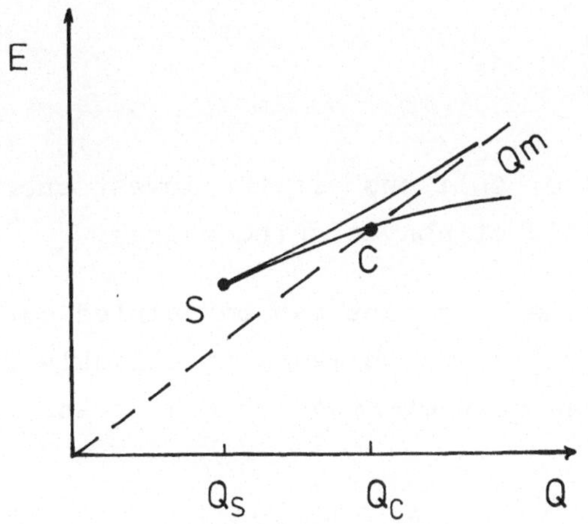

Fig. 1

which is valid for a three-dimensional space, s = 3. If $Q > Q_c$ the soliton states are absolutely stable simply because other states with the same charge have higher energy. For $Q_s < Q < Q_c$ the solutions are classically stable against small perturbations, but unstable quantum mechanically (due to tunneling to lower energy states). The upper branch is not stable. For $Q < Q_s$ there are no soliton solutions.

It is plausible that Q_c depends on the mass m of the ϕ field, and that for large m, Q_c is reduced. In fact, one can show that $Q_c < 1$ provided $m > m_c$, where at least for small G,

$$m_c < \frac{1}{2}\left(\frac{3}{2}\right)^6 \mu \, \frac{4\pi}{G^2} \, .$$

In other words, $Q_c < 1$ whenever

$$\frac{m}{\mu} > \frac{1}{2}\left(\frac{3}{2}\right)^6 \frac{4\pi}{G^2} \, .$$

In such a case, solitons are the lowest energy states for every field of nonvanishing charge.

A parallel argument can be carried out in space dimension s = 1,2, which leads to slightly different conclusions as characterized by Fig. 2 and 3:

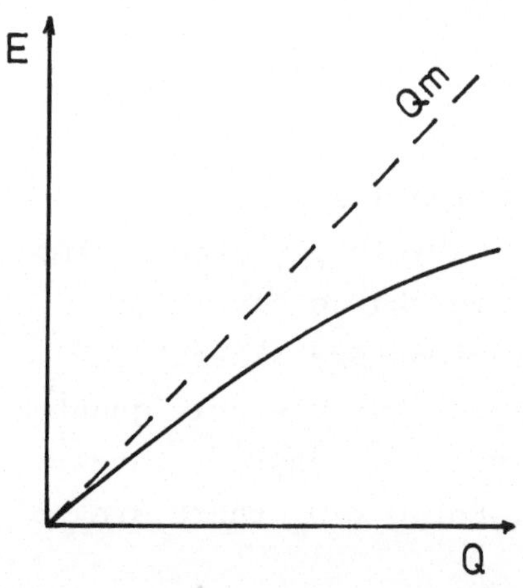

Fig. 2

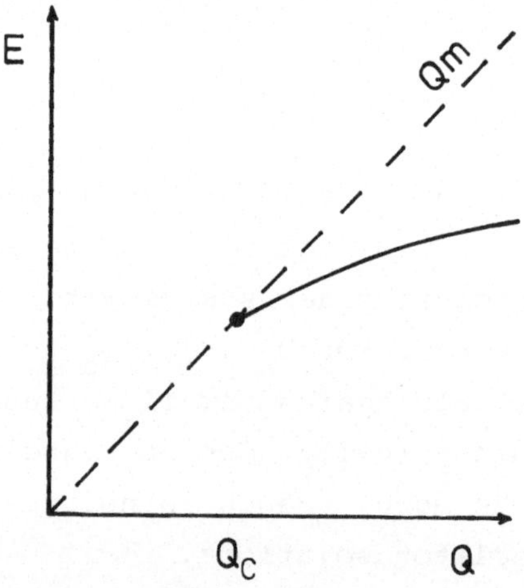

Fig. 3

In one-space dimension the soliton always lies below the free particle states; in two-space dimensions the onset of the soliton is sudden, and there are no classically stable (quantum unstable) solutions as for s = 3.

Vector Fields

The principle illustrated above applies to other systems as well. For a Yang-Mills SU_2 gauge field V_μ coupled to an isospin $\frac{1}{2}$ scalar field $\phi = \phi_1 \oplus \phi_2$ the analogous Lagrangian density to take is given by

$$L = \frac{1}{4}V_{\mu\nu}^2 + (D_\mu\phi)^\dagger (D_\mu\phi) - \frac{1}{2}\lambda(\phi^\dagger\phi - \frac{1}{2}\kappa^2)^2$$

where

$$D_\mu\phi = \partial_\mu\phi - iGV_\mu\phi$$

and

$$V_{\mu\nu} = \partial_\mu V_\nu - \partial_\nu V_\mu + G[V_\mu, V_\nu] \ .$$

There is, by construction, local gauge invariance of this model with regard to the orientation and phase of the field ϕ in isospin space at each point of space. Consequently by a suitable gauge transformation it can be arranged that

$$\phi = 0 \oplus \frac{1}{\sqrt{2}}\chi, \quad \chi = \chi^\dagger ,$$

which is termed the unitary gauge. In such a gauge, the Lagrangian density reduces to

$$L = \frac{1}{4}V_{\mu\nu}^2 + \frac{1}{2}\chi_{,\mu}^2 - \frac{1}{2}G^2\chi^2V_\mu^2 - \frac{\lambda}{8}(\chi^2-\kappa^2)^2$$

and in this form the relation to the charged scalar field studied earlier is apparent. If $\chi = \chi_{vac} = \kappa \neq 0$, then the theory reduces to a model of a massive SU_2 gauge vector bosons with mass $m = G\kappa$. On the other hand, the mass of the field χ is obtained by evaluating the curvature (second derivative) of the last term in the Lagrangian at $\chi = \kappa$. Since this term may be written as

$$\frac{\lambda}{8}(\chi + \kappa)^2 (\chi - \kappa)^2$$

it follows directly that the mass of χ field is $\mu = \sqrt{\lambda}\kappa$.

The conserved quantity in the present case is isospin, T. Just as before conditions for soliton stability may be given; namely, if $m > 2\mu$ and $T > T_c$. Analysis shows a stability curve like Fig. 1 with a spike. Again it can be arranged that $T_c < 1$ provided $m > m_c$.

Fermion Fields

Spinor fields can be incorporated in a similar fashion as well. Consider a typical Yukawa coupling for N massless fermions with a Lagrangian density given by

$$L = \frac{1}{2}\phi_{,\mu}^2 - U(\phi) - \sum_{k=1}^{N} \bar{\psi}_k(i\gamma\cdot\partial+G\phi)\psi_k .$$

The classical Hamiltonian for this system restricted to time-independent ϕ fields reads as

$$H = \int [\frac{1}{2}(\nabla\phi)^2 + U(\phi) + \sum_{k=1}^{N} \psi_k^\dagger(-i\alpha\cdot\nabla + G\beta\phi)\psi_k] d^3x \ .$$

In these expressions N represents the number of fermion species, or colors. The conserved quantity is the fermion number N. Suppose that U vanishes for $\phi = \phi_{vac} \neq 0$. If $\phi \equiv \phi_{vac}$, then the model is that of N fermion species all with equal mass, $m = G\phi_{vac}$. A state of N such fermions has classical energy $E \geq Nm$, equality being restricted by the Pauli principle. However, just as before, it may be energetically favorable if $\phi \neq \phi_{vac}$, in particular $\phi = 0$ for some region $r \leq R$, which then creates a "bag"-like configuration confining the fermions. And indeed this qualitative argument is valid in the sense that a curve of E vs. N is qualitatively similar to that given in Fig. 1, and soliton solutions outlined above exist for $N > N_c$. Again it can be arranged that $N_c < 1$ provided $m > m_c$.

The preceding discussion should indicate that scalar fields can act as glue for scalars, vectors, or fermions giving rise to localized, persistent, and finite energy ($\equiv$ soliton) solutions. An evident feature of such states is their spacial extension and smooth structure, leading to fastly falling form factors, which are features attributed to hadrons. Consequently, one is encouraged by these arguments, as emphasized by Lee, to identify hadrons as bound states of a relativistic field theory, and their large masses are attributed to their large classical energy. On the other hand, Lee argues that leptons satisfy field equations without soliton solutions, leading at least in principle, to their lower mass.

Abnormal Nuclear States

One cannot fail to mention at this point the relation of the preceding discussion to the Lee-Wick theory of abnormal nuclear states [7]. In such a case the Lagrangian density is taken as

$$L = \frac{1}{2}\phi^2_{,\mu} - \bar{U}(\phi) - \sum_{k=1}^{2} \bar{\psi}_k(i\gamma\cdot\partial+\bar{m}+G\phi)\psi_k$$

where the number of species N = 2 representing the proton and neutron and $\bar{U}$ is a new potential term. It is assumed that $\bar{U}$ vanishes at $\phi = 0$, which moreover is the minimum of $\bar{U}$. If we set $\phi \equiv 0$ everywhere we deal with a theory of two kinds of fermions of mass $\bar{m}$, which are identified as the conventional neutron and proton. Standard nuclear forces responsible for holding nuclei together may here be modelled as the attractive Yukawa force about $\phi = 0$ which forms a nucleus of atomic number A and mass $M_A \simeq A\bar{m}$.

Suppose next that $\bar{U}$ has a second minimum situated at $\phi = \bar{\phi} \neq 0$ for which (say) $G\bar{\phi} = -\bar{m}$. Consequently, for large A it may be energetically favorable if $\phi = \bar{\phi}$ in a region $r \leq R$, for then the energy of the fermions (for slowly moving constituents) is much less the value $A\bar{m}$ in the vacuum state of the ϕ field. Thus for $A > A_c$ it may be expected that a stable, abnormal nuclear state exists, and that the value of A_c is small for large values of the ratio $\bar{m}/\bar{\mu}$ where $\bar{\mu}$ is the mass of the ϕ field taken about the second minimum, $\bar{\phi}$.

As may be already anticipated there is a fairly close formal structure with the foregoing fermion soliton state and the SLAC-"bag" [9] . The prototypical example of the SLAC-bag is illustrated by the Hamiltonian

$$H = \int [\tfrac{1}{2}\dot{\phi}^2 + \tfrac{1}{2}(\nabla\phi)^2 + \lambda(\phi^2-\kappa^2)^2 + \psi^\dagger(-i\alpha\cdot\nabla + G\beta\phi)\psi]\, d^3x$$

where $G,\lambda \gg 1$ are large constants. Evidently the minimum energy for the ϕ field alone is attained if $\phi = \pm\kappa$. In such a case the theory describes free fermions of mass $m = G\kappa$, and a state of N fermions has energy $M \geq Nm$. Lower energy states may arise if deviations from $\phi = \pm\kappa$ are taken into account. In particular, if we assume a spherically symmetric distribution in three-dimensional space, then various assumptions can be tried. Let $\phi = 0$, $r \leq R$ and $\phi = \kappa$, $r > R$, with variation of ϕ taking place in a region D. We also assume ψ is contained within the bag, normalized to unity and vanishes for $r > R$, as in Fig. 4.

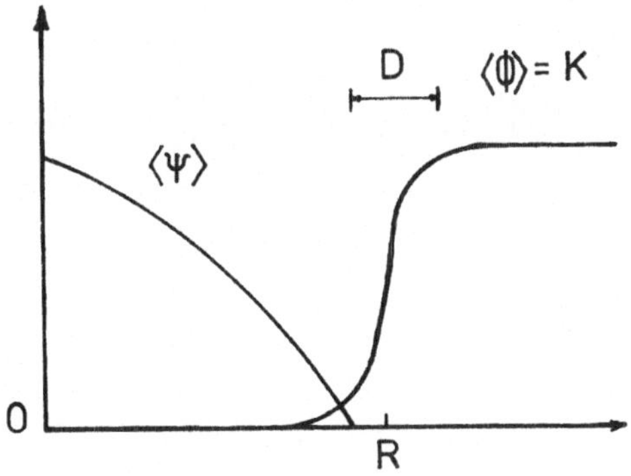

Fig. 4

Then it follows that

$$\int \psi^{\dagger}(-i\underset{\sim}{\alpha}\cdot\underset{\sim}{\nabla})\psi d^3x \sim 1/R,$$

$$\int \frac{1}{2}(\underset{\sim}{\nabla}\phi)^2 d^3x \sim \frac{1}{2}(\frac{\kappa}{D})^2 4\pi R^2 D,$$

$$\int \lambda(\phi^2-\kappa^2)^2 d^3x \sim \lambda\kappa^4(\frac{4}{3}\pi R^3+4\pi R^2 D),$$

where in the last case we have both volume and surface effects (crudely estimated). The total approximate energy

$$E \sim \frac{1}{R} + \frac{2\pi R^2\kappa^2}{D} + \lambda\kappa^4(\frac{4}{3}\pi R^3+4\pi R^2 D)$$

is minimized with respect to D and R to yield

$$\frac{\partial E}{\partial D} = 0 \rightarrow D \sim \frac{1}{\lambda^{1/2}\kappa}$$

and for $\lambda^{1/2}\kappa >> 1/R$ (volume energy dominant over surface energy)

$$\frac{\partial E}{\partial R} = 0 \rightarrow R \sim \frac{1}{\lambda^{1/4}\kappa} .$$

Consequently, the lowest energy reads

$$E = \frac{4}{3R} \sim \kappa\lambda^{1/4}$$

and in addition $D/R \sim \lambda^{-1/4}$ confirming our assumption. Hence if $G >> \lambda^{1/4}$, this single-fermion state is lower in energy than that previously treated, and so a localized bound state is formed. But it is not the best we can do!

For the next case assume that the fields ϕ and ψ are distributed according to the design in Fig. 5.

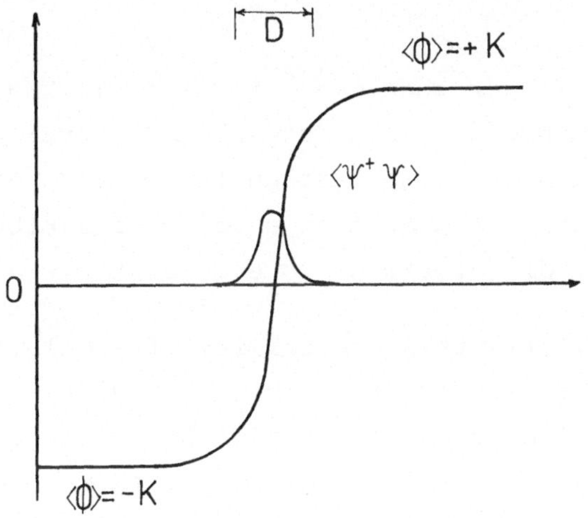

Fig. 5

In this case the volume contribution in the last term of the energy expression is absent, which is favorable, but it would appear that this has been at the expense of raising the energy contribution of the fermion represented by the first term of that expression. On closer inspection this simply turns out not to be the case, i.e., there is no substantial change of fermion energy. Minimization of the so revised energy expression with respect to D and R yields

$$\frac{\partial E}{\partial D} = 0 \rightarrow D \sim \frac{1}{\lambda^{1/2} \kappa}$$

and again for $\lambda^{1/2}\kappa >> 1/R$ and using the estimate for D,

$$\frac{\partial E}{\partial R} = 0 \rightarrow R \sim \frac{1}{\lambda^{1/6}\kappa} \quad .$$

Consequently, the minimum energy is now

$$E = \frac{1}{R} \sim \kappa\lambda^{1/6}$$

which is lower than $\kappa\lambda^{1/4}$ for $\lambda >> 1$. Thus we see that this picture favors a "shell" structure; namely, the localization of the fermion in the form of a surface distribution rather than a volume distribution. (In fact, toroidal distributions do even better.)

Clearly this possibility of shell localization is connected with the fact that the two minima of the potential have equal and vanishing energy. Lift that degeneracy, by going from the solid to the dotted curve as shown in Fig. 6,

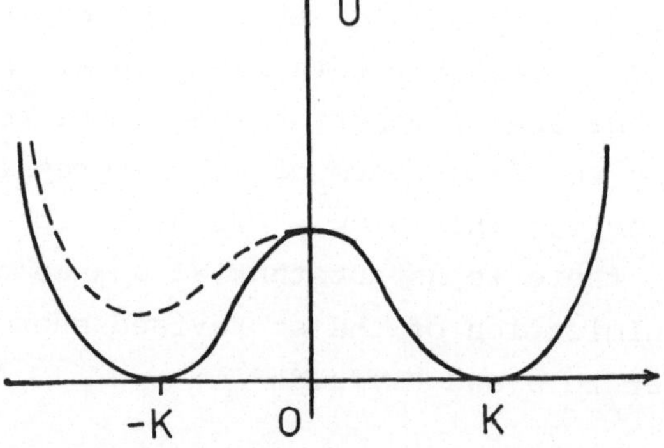

Fig. 6

and the energy cast near $\phi = -\kappa$ is no longer zero. Thus, it may be energetically favorable for the fermion field to obtain some volume component to its distribution. This variation evidently makes the SLAC nonsymmetric model structurally very similar to the Lee-Wick model described earlier. Detailed computations tend to support the qualitative arguments advanced here in that, for suitable model parameters, the fermion field has a nonvanihsing volume density which attains a larger value on the surface simulating thereby a shell-like component as well [16].

We must add that the original SLAC-bag with its shell structure seems to be less successful phenomenologically for hadron physics than models with volume distribution components. Moreover, it should be clear that if these ideas are used to analyze quark confinement, then there is no absolute confinement since given sufficient energy there seems to be no reason not to produce a free fermion.

On the other hand, the MIT-bag [17], which is defined somewhat more phenomenologically and is not taken as a local field theory, attains its confinement by asserting that there exists an externally imposed pressure on the contents of the bag, and by requiring certain boundary conditions at the bag surface that have the effect of containing colored fields (much as the boundary condition of a vanishing tangential electric field - say at the surface of conductors - confines electric charge). In other words, absolute color confinement is built in. The MIT-bag has a better track record from a phenomenological point of view (e.g., g_A/g_V predictions), but a discussion of such questions is outside the scope of these lectures.

III. QUANTUM BASIS BEHIND THE CLASSICAL ANALYSIS

A. Coherent State Justification

Coherent state techniques, by now widely employed in various branches of physics, are frequently used [9], [10], [18] to provide justification of a study of the classical equation in regard to its quantum properties. Let us review the principal point in such arguments. For the sake of illustration let us first examine pure boson systems. We introduce a free-field (mass m) time-zero field,

$$\phi(\underset{\sim}{x}) = \frac{1}{(2\pi)^{3/2}} \int [e^{i\underset{\sim}{k}\cdot\underset{\sim}{x}} a(\underset{\sim}{k}) + e^{-i\underset{\sim}{k}\cdot\underset{\sim}{x}} a^{\dagger}(\underset{\sim}{k})] \frac{d^3k}{\sqrt{2\omega}}$$

where $\omega = \sqrt{k^2+m^2}$, and conjugate momentum

$$\pi(\underset{\sim}{x}) = \frac{-i}{(2\pi)^{3/2}} \int [e^{i\underset{\sim}{k}\cdot\underset{\sim}{x}} a(\underset{\sim}{k}) - e^{-i\underset{\sim}{k}\cdot\underset{\sim}{x}} a^{\dagger}(\underset{\sim}{k})] \sqrt{\frac{\omega}{2}}\, d^3k \,.$$

The creation and destruction operators satisfy

$$[a(\underset{\sim}{k}), a^{\dagger}(\underset{\sim}{k}')] = \delta(\underset{\sim}{k}-\underset{\sim}{k}') \,,$$

and the no-particle state $|0\rangle$ is a unit vector for which, for all $\underset{\sim}{k}$,

$$a(\underset{\sim}{k})|0\rangle = 0.$$

The Hilbert space is spanned by repeated action of the $a^{\dagger}$

on the state $|0\rangle$. Taken together these relations imply that the a and $a^\dagger$ operators constitute an irreducible Fock representation.

The coherent states may be defined according to

$$|f,g\rangle = U[f,g]|0\rangle,$$

where the unitary operators U are given by

$$U[f,g] \equiv \exp\{i\int[f(\underset{\sim}{x})\phi(\underset{\sim}{x})-g(\underset{\sim}{x})\pi(\underset{\sim}{x})]d^3x\}$$

and are labelled by two real, smooth functions f and g. Many properties of these states are, by now, widely known, and suitable reviews of their properties are available [19], [20]. We shall have need of only a few of their important properties in what follows.

These states are overcomplete and not mutually orthogonal - in fact the overlap of two such states, given by

$$\langle f,g|f',g'\rangle = \exp\{-\tfrac{1}{4}\int[(k^2+m^2)^{-1/4}|\tilde{f}-\tilde{f}'|^2 + (k^2+m^2)^{1/4}|\tilde{g}-\tilde{g}'|^2 + 2i(\tilde{f}^*\tilde{g}'-\tilde{g}^*\tilde{f}')]d^3k\},$$

never vanishes, where f denotes the Fourier transform of $\tilde{f}$, etc. These states are so overcomplete, in fact, that diagonal matrix elements

$$G(f,g) \equiv \langle f,g|G|f,g\rangle$$

of an operator <u>uniquely</u> determine the operator. This property is actually rather important in practice as we shall note below.

Without belaboring the point the form for the operator G is easily seen to be

$$G = :G(\pi,\phi):$$

where :: denotes normal ordering with respect to a and $a^\dagger$ (all $a^\dagger$ to the left of all a). For it follows from the relation

$$U[f,g]^\dagger\{\alpha\phi(\underset{\sim}{x})+\beta\pi(\underset{\sim}{x})\}U[f,g]$$

$$= \alpha[\phi(\underset{\sim}{x})+g(\underset{\sim}{x})] + \beta[\pi(\underset{\sim}{x})+f(\underset{\sim}{x})]$$

that

$$\langle f,g|:G(\pi,\phi):|f,g\rangle$$

$$= \langle 0|:G(\pi+f,\phi+g):|0\rangle$$

$$= G(f,g)$$

since any operator other than the identity operator vanishes in virtue of the normal ordering.

The remainder of the argument goes something like the following. Assuming that the Hamiltonian

$$H = \int :\{\tfrac{1}{2}[\pi^2+(\nabla\phi)^2]+\lambda(\phi^2-\kappa^2)^2\}:d^3x$$

it follows that

$$H(f,g) = \langle f,g|H|f,g\rangle$$

$$= \int\{\tfrac{1}{2}[f^2+(\nabla g)^2]+\lambda(g^2-\kappa^2)^2\}d^3x \ .$$

At this point one treats the state $|f,g\rangle$ as a variational state and minimizes $H(f,g)$ with respect to changes in f and g. Along with $f = 0$, this leads to the minimization conditions we have treated earlier, e.g., in regard to the SLAC-bag model - minus, of course, the fermion fields.

Promoting this argument to two scalar fields, it could easily be arranged to generate the scalar model of Lee et al discussed initially. We leave it as an exercise for the interested reader to carry out.

The importance of the fact that diagonal coherent-state matrix elements uniquely determine the operator now becomes clear. For if this were not the case and there was an operator $\mathcal{O} \not\equiv 0$ for which, nevertheless, $\langle f,g|\mathcal{O}|f,g\rangle \equiv 0$, then any amount of $\mathcal{O}$ could be present in the Hamiltonian H influencing the quantum dynamics while its presence in the form

$$H(f,g) \equiv \langle f,g|H|f,g\rangle$$

would go utterly undetected. The evident misrepresentation of the classical dynamics is clear.

B. Fermion Fields

The inclusion of fermion fields can be done in several ways. The most conventional approach when one deals with one fermion (as in the SLAC analysis) [9] is to take a state of the form

$$\int (\tilde{u}b^{\dagger}+\tilde{v}d^{\dagger})\, d^3k\, |0\rangle$$

where $b^{\dagger}$ and $d^{\dagger}$ create fermions and anti-fermions in the normalized state $\tilde{u}$ and $\tilde{v}$, respectively. The functions $\tilde{u}$ and $\tilde{v}$ become variational elements in minimization of the energy along with the arguments of the coherent state for the boson field.

Anticommuting C-Numbers

For larger number of fermions the treatment depends in some measure on the problem. If the number of fermions $n \leq N$, the number of fermion species, then they can each be stacked in the lowest energy state, one for each color, thereby respecting the Pauli principle. For more general situations recourse to different techniques are useful. One such approach exploits the formal analog of coherent states for fermion fields involving anticommuting c-numbers. These are variables that anticommute with themselves with other like quantities and with all fermion operator fields [21]. The square of any such variable vanishes. The usefulness of such entities in formal constructions has been demonstrated repeatedly, and there has even been some recent effort to give them some respectability by including them in phase-space formulations of classical mechanics [22].

In particular, if η and η^* denote anticommuting c-numbers and ψ and $\psi^\dagger$ denote fermion operator fields, let us consider the operators

$$U[\eta,\bar{\eta}] = \exp[-\int(\eta^*\psi-\psi^\dagger\eta)d^3x],$$

where the operator $\psi^\dagger$ represents the Hermitian adjoint to ψ, but η^* is not to be interpreted as the conjugate of η. The fields η and η^* anticommute with everything including themselves and each other (from which it follows that η^* cannot be the conjugate of η) [23]. This business may be made more precise with the aid of Grassmann algebras [24]. The canonical anticommutation relation for ψ and $\psi^\dagger$ translates into the commutation relation

$$[\int\eta^*\psi d^3x,\int\psi^\dagger\eta d^3x] = \int\eta^*\eta d^3x ,$$

and moreover $\int\eta^*\eta d^3x$ commutes with both $\int\eta^*\psi d^3x$ and $\int\psi^\dagger\eta d^3x$, reducing the present problem to one algebraically similar to the boson case. In particular, with (†) suitably defined,

$$U[\eta,\bar{\eta}]^\dagger\{\alpha\psi(\underset{\sim}{x})+\beta\psi^\dagger(\underset{\sim}{x})\}U[\eta,\bar{\eta}]$$

$$= \alpha[\psi(\underset{\sim}{x})+\eta(\underset{\sim}{x})]+\beta[\psi^\dagger(\underset{\sim}{x})+\eta^*(\underset{\sim}{x})]$$

and it then follows that

$$\langle\eta,\bar{\eta}|\int:\psi^\dagger(-i\underset{\sim}{\alpha}\cdot\underset{\sim}{\nabla}+\beta m)\psi:d^3x|\eta,\bar{\eta}\rangle$$

$$= \int\eta^*(-i\underset{\sim}{\alpha}\cdot\underset{\sim}{\nabla}+\beta m)\eta d^3x$$

the remaining operator terms vanishing due to the normal ordering. When combined with the boson terms from a coherent state expression this leads to a variational form for the energy that resembles the classical expressions. In this regard, the "classical" fermion energy involves anticommuting fields η^* and η. The negative energy states of the Dirac equation are rendered positive by switching the order of the η and η^* for those states. Having performed their duty the anti-commuting character of η and η^* is conveniently forgotten and η^* is secretly taken as the complex conjugate to η thus rendering the expectation value real. For, how else is one to add the fermion energy to the boson energy (and the interaction term as well) and minimize the whole sum unless they are entities of the same type?

Fermions Without Anticommuting C-Numbers

There is another approach to justify the fermion analysis that does not make use of anticommuting c-numbers. Here we follow a procedure introduced by Friedrichs [25] that was especially important in the author's thesis [26]. For simplicity in presentation we pretend that the model is nonrelativistic with the property that

$$\psi(\underset{\sim}{x})\,|0\rangle = 0 \ .$$

This is not serious for one could consider that the argument is in reality a multiple label and that $\psi(\underset{\sim}{x})$ represents the positive frequency portions of both the conventional $\psi(\underset{\sim}{x})$ and $\psi^\dagger(\underset{\sim}{x})$. In any case, let us define the basic states

$$|\underset{\sim}{x}_1,\ldots,\underset{\sim}{x}_n\rangle_A \equiv \psi^\dagger(\underset{\sim}{x}_1)\ldots\psi^\dagger(\underset{\sim}{x}_n)|0\rangle,$$

which are clearly antisymmetric in the $\underset{\sim}{x}$ variables (hence the subscript A).

Now, following Friedrichs let us introduce an ordering of points, <, in configuration space (here three dimensional)[+]. We shall say that $\underset{\sim}{x} < \underset{\sim}{y}$ if (1) $x_{(1)} < y_{(1)}$ (these are the first coordinates) or (2) if $x_{(1)} = y_{(1)}$ and $x_{(2)} < y_{(2)}$, or (3) if $x_{(1)} = y_{(1)}$, $x_{(2)} = y_{(2)}$, and $x_{(3)} < y_{(3)}$. Next we introduce the ordering sign function

$$\sigma(\underset{\sim}{x}_1, \underset{\sim}{x}_2,\ldots,\underset{\sim}{x}_n) = \pm 1$$

given by the sign of the permutation P necessary to bring the arguments of σ to the "standard" order $\underset{\sim}{x}_{P_1} < \underset{\sim}{x}_{P_2} < \ldots < \underset{\sim}{x}_{P_n}$; if any two $\underset{\sim}{x}$'s are equal, σ is defined to be zero. With the aid of σ we define the vectors

$$|\underset{\sim}{x}_1,\ldots,\underset{\sim}{x}_n\rangle_S \equiv \sigma(\underset{\sim}{x}_1,\ldots,\underset{\sim}{x}_n)|\underset{\sim}{x}_1,\ldots,\underset{\sim}{x}_n\rangle_A ;$$

the new vectors are <u>symmetric</u> in their arguments, and vanish if any pair of arguments are equal. Clearly, $\sigma^2 = 1$ save at coincident points where both sides vanish already so that the relation

$$|\underset{\sim}{x}_1,\ldots,\underset{\sim}{x}_n\rangle_A = \sigma(\underset{\sim}{x}_1,\ldots,\underset{\sim}{x}_n)|\underset{\sim}{x}_1,\ldots,\underset{\sim}{x}_n\rangle_S$$

[+] We recall that in the classic operator realization of fermion fields by Jordan and Wigner some sort of ordering between variables is needed.

holds as well.

Armed with the S-states we form

$$|\chi\rangle \equiv N \sum_{n=0}^{\infty} (n!)^{-1} \int .. \int d^3x_1 \dots d^3x_n$$

$$\chi(\underset{\sim}{x}_1) \dots \chi(\underset{\sim}{x}_n) |\underset{\sim}{x}_1, \dots, \underset{\sim}{x}_n\rangle_S$$

where N is chosen for normalization and where χ denotes a complex, smooth c-number function. These states play the role of fermion "coherent states". Although these states are not eigenstates of the fermion field operator (there are no longer any anticommuting c-numbers!), they exhibit the essential property that

$$\langle\chi|\psi^\dagger(\underset{\sim}{x})\psi(\underset{\sim}{x})|\chi\rangle = \chi^*(\underset{\sim}{x})\chi(\underset{\sim}{x}) \quad ,$$

which is a prototypical relation needed to derive classical expressions for the energy expression [26].

The argument sketched above can indeed be extended in a fairly straightforward fashion to a multi-component, relativistic, interacting model leading to a Hamiltonian that has the conventional classical form and is expressed entirely in terms of c-number quantities. In establishing this fact, a suitable choice of the ordering "<" introduced above must be exploited. The details appear on pages 161-167 of Ref. 26, particularly pages 165-166. How to deal with the negative energy sea and give positive energies to the classical c-number expression is treated on pages 159-160. All-in-all a perfectly consistent formulation may be given.

Needless to say with such a justification of the strictly c-number classical formalism available, then the variational principle can be invoked to assert the relevance of the minimum value of the expectation value of the Hamiltonian.

C. Improvement on Coherent State Argument

Before leaving the subject of justification of the classical approach we wish to return to the purely boson case for some further remarks. The use of the coherent states in that analysis is implicitly self-contradictory. For it is assumed that the representation of the operators is given by a Fock representation (appropriate to a free-field) and the form of the Hamiltonian is taken as nonfree. Such a choice stands in disagreement with Haag's theorem, which suitably paraphrased, states: Thou shalt not use the Fock representation for interacting fields. Some recognition of these difficulties is apparent in the analysis of the SLAC group. But do not dispare for there is a justification of the classical relations that does not rely on the Fock representation. Since the main points of this procedure have already been published [10], let us here only outline the argument.

For the present we again assume that we deal with a single scalar field ϕ, which obeys canonical commutation relations (CCR). Once again we consider

$$U[f,g] = \exp\{i\int[f(\underset{\sim}{x})\phi(\underset{\sim}{x})-g(\underset{\sim}{x})\pi(\underset{\sim}{x})]d^3x\}$$

and introduce a normalized state $|0\rangle$ about which we assume <u>nothing</u> else at present. The remaining argument is simpler if we assume that the representation of ϕ and π is irreducible. Then every operator is a function of ϕ and π. We introduce the generalized normal ordering implicit in the generating functional

$$ {}^{*}_{*}U[r,s]^{*}_{*} \equiv U[r,s]/\langle 0|U[r,s]|0\rangle $$

where for sufficiently small r and s the denominator is always nonvanishing. (For a Fock representation and $|0\rangle$ the no-particle state this reduces to conventional normal ordering.) Evidently

$$ \langle 0|{}^{*}_{*}U[r,s]^{*}_{*}|0\rangle \equiv 1 $$

which means, on expansion in r and s, for example, that

$$ \langle 0|{}^{*}_{*}\phi^{p}(\underset{\sim}{x})^{*}_{*}|0\rangle = 0 $$

for all $p \geq 1$, etc..

Now let us introduce the states

$$ |f,g\rangle \equiv U[f,g]|0\rangle $$

which for non-Fock representations are simply <u>not</u> coherent states. In the author's general study of such matters the collection of such vectors has been termed an <u>overcomplete family of states</u> (OFS) [27]. It follows simply from the CCR that

$$<f,g|{}^{\star}_{\star}H(\pi,\phi){}^{\star}_{\star}|f,g>$$

$$= <0|{}^{\star}_{\star}H(\pi+f,\phi+g){}^{\star}_{\star}|0>$$

$$= H(f,g) \ .$$

It may be demonstrated that the determination of H (given a representation of ϕ and π) from the diagonal OFS matrix elements is unique provided H is polynomial.

General arguments assert that H(f,g) has the form of the classical Hamiltonian provided the operator here is the Hamiltonian, namely if

$$H = {}^{*}_{*}H(\pi,\phi){}^{*}_{*} \ ;$$

this statement is called the weak correspondence principle [10]. The validity of this identification depends on matching the CCR representation to the dynamics in question, and requires that H is more than just a form and, for example, that $H|f,g>$ is a bona fide vector for sufficiently many f and g. This is the test the CCR representation must pass, and which selects the relevant representation for the model.

The representation of the CCR is of course intimately connected with the choice of the state $|0>$, for in fact the representation is determined by the functional

$$E(f,g) \equiv <0|U[f,g]|0>.$$

Different choices of $|0\rangle$ determine different functionals E(f,g) and potentially inequivalent representations. By their very nature it is presumed that f and g are localized functions. Consequently, it is important to choose $|0\rangle$ in such a way that H(f,g) is <u>finite</u> for localized f and g, which means that $|0\rangle$ is taken to be one or another of the minima.

<u>An Example</u>

As an example take the one-(space) dimensional quartic model normallv written as

$$H = \int[\tfrac{1}{2}(\dot{\phi})^2 + \tfrac{1}{2}(\phi')^2 + \lambda(\phi^2-\kappa^2)^2]dx \ .$$

To make sense of this case we could take $|0\rangle = |0\rangle_+$ meaning that the ϕ field is centered about $\phi = \kappa$. In that case we would have

$${}_+\langle f,g|{}^*_*H_+(\pi,\phi){}^*_*|f,g\rangle_+$$

$$= {}_+\langle 0|{}^*_*H_+(\pi+f,\phi+g){}^*_*|0\rangle_+$$

$$= H_+(f,g)$$

$$\equiv \int[\tfrac{1}{2}f^2 + \tfrac{1}{2}(g')^2 + \lambda(g+2\kappa)^2 g^2]dx \ .$$

In this way a state $|0\rangle$, a CCR representation and a Hamiltonian are associated with the case $\phi = \kappa$.

Although the last expression makes sense for g localized about -2κ as well, the rest of the relations in that equation do not. Instead to focus on that region one must choose another state $|0\rangle = |0\rangle_-$ implying the original field ϕ is centered about $\phi = -\kappa$. Analogously, one finds that

$$_-\langle f,g|{}^*_*H_-(\pi,\phi){}^*_*|f,g\rangle_-$$

$$= {}_-\langle 0|{}^*_*H_-(\pi+f,\phi+g){}^*_*|0\rangle_-$$

$$= H_-(f,g)$$

$$\equiv \int[\tfrac{1}{2}f^2+\tfrac{1}{2}g'^2+\lambda(g-2\kappa)^2g^2]dx \ .$$

In this way one and the same classical Hamiltonian, say simply,

$$\int[\tfrac{1}{2}f^2+\tfrac{1}{2}g'^2+\lambda(g^2-\kappa^2)^2]dx$$

has distinct, inequaivalent quantum implications.

Lastly we examine another type of state $|0\rangle$, namely one that interpolates between $\phi = \pm\kappa$. Let $u(x)$ denote a fixed function such that $u(x) \to -\kappa$ as $x \to -\infty$ and $u(x) \to +\kappa$ as $x \to +\infty$. In this case we seek a state $|0\rangle \equiv |0\rangle_K$ (K for kink) for which

$$_K\langle f,g|{}^*_*H_K(\pi,\phi){}^*_*|f,g\rangle_K$$

$$= {}_K\langle 0|{}^*_*H_K(\pi+f,\phi+g){}^*_*|0\rangle_K$$

$$= H_K(f,g)$$

$$\equiv \int \{\tfrac{1}{2}f^2+\tfrac{1}{2}(g'-u')^2+\lambda[(g-u)^2-\kappa^2]^2\}dx \ .$$

This expression has finite energy for localized f and g since $u^2 \to \kappa^2$ and $u' \to 0$ sufficiently fast by assumption. From a variational point of view the energy is minimized in the state $|0>_K$ for the minimum of

$$H_K(0,0) = \int [\tfrac{1}{2}(u')^2+\lambda(u^2-\kappa^2)^2]dx \ ,$$

which occurs as well known for

$$u(x) = \kappa \tanh [\sqrt{2\lambda}\ \kappa(x-x_o)] \ ,$$

with x_o arbitrary. The value for this minimum energy is

$$H_K(0,0) = \tfrac{4}{3}\sqrt{2\lambda}\ \kappa^3 \ .$$

Thus in no sense is $|0>_K$ a "vacuum" - rather it represents the kink solution at position x_o.

Note in this formulation the value of x_o is not important, and finite changes of x_o - there are no other - can be effected through the function g, and hence lie in the same representation space. This holds since

$$\bar{u}(x) - u(x) \equiv \kappa \tanh[\sqrt{2\lambda}\ \kappa(x-\bar{x}_o)]$$

$$- \kappa \tanh[\sqrt{2\lambda}\ \kappa(x-x_o)]$$

is a localized function. The degeneracy for all different x_o values (i.e., same value for H_K) is the residue of the so-called zero-frequency mode, but it is utterly harmless here.

To summarize, we simply note that different sectors of the quantum theory are connected to distinct regions of the classical Hamiltonian through the identification implicit in the weak correspondence principle for various "rewritings"of the classical Hamiltonian. We feel that the present analysis tends to provide much stronger support for the study of the classical system than is conventionally presented.

D. Role of Classical Solutions in an Operator Formulation

The simplest case in which the question of nonzero classical solutions arises is for scalar fields when the energy minimum is at $\phi = \phi_{vac} = \kappa \neq 0$. To handle this case just set

$$\phi = \phi_\kappa + \kappa,$$

where now the energy minimum occurs at $\phi_\kappa = 0$, and in a quantization ϕ_κ becomes an operator field and κ remains a c-number.

On the other hand, if their are a number of equivalent energy minima labelled by a parameter c, then the field operator, correctly stated, is a direct integral (or a direct sum) over parameters describing the equivalent minima. Specifically,

$$\phi = \int^{\oplus} [\phi_c + u_c] d\mu(c)$$

where u_c denotes the c-number expression that minimizes the energy for parametric value c, and ϕ_c denotes the corresponding expression with energy minimum at $\phi_c = 0$. With quantization, ϕ becomes an operator through promoting ϕ_c to an operator, while u_c remains a c-number. It may happen that c appears explicitly in the Hamiltonian, or it may not. For a Goldstone model Hamiltonian with a charged, massless scalar field ϕ and nonlinear term $\lambda(\phi^\dagger\phi-\kappa^2)^2$, the set of minima are labelled by $u_c = c$ where c is a complex number such that $|c| = \kappa$. For the cases of soliton solutions, c is an integration constant (or set of constants) of the soliton solutions, and c does not appear explicitly in the Hamiltonian. The states of the system are also direct integrals, and various weightings on the states pick out one or another "broken symmetry".

From the point of view of physicists, of course, direct integrals simply correspond to an auxiliary operator variable (with spectrum values c as determined by the support of the measure μ). In application to soliton solutions this approach was introduced by Gervais and Sakita and by Faddeev et al, and has been thoroughly discussed in the literature [11].

An Example

For a one-dimensional ϕ^4 model with kink solution taken, in the spirit of Sec. I, as

$$u(x-x_0) = \frac{\kappa}{G} \tanh[\sqrt{2\lambda}\,\kappa(x-x_0)],$$

the Heisenberg field operator is taken in the form

$$\phi(x,t) = u(x-Q(t)) + \sum_{n=2}^{\infty} q_n(t)\psi_n(x-Q(t))$$

where Q denotes the Heisenberg operator representation of the position variable x_o (the c of the general discussion). The operator Q is just a conventional position operator, and for convenience it appears both in u and in the independent, orthonormal, mode functions ψ_n that describe the remaining degrees of freedom. In the convention of these authors [11], the collective coordinate Q counts as the first variable which explains why the sum starts with n = 2. To ensure independence of the degrees of freedom represented by Q and q_n, it is assumed that

$$\int u'(x)\psi_n(x)dx = 0$$

for all n. Thus the functions u' and ψ_n, $n \geq 2$, taken together constitute a complete, orthogonal set of functions (normalization of u' is dictated by the soliton property of u).

Along with Q, the mode amplitudes q_n are regarded as quantum operators, and there are conjugate momenta associated with all coordinates, which in a Schrödinger representation are given by

$$P = -i\frac{\partial}{\partial Q},\ p_n = -i\frac{\partial}{\partial q_n} .$$

The Hamiltonian is expressed in terms of P, p_n, Q and q_n, and analysis of the quantum theory proceeds along conventional lines. It is clear that the space-translation generator operates on the field ϕ exactly as the variable $P = -i\partial/\partial Q$. Since the particular models under consideration are translation invariant, it follows that the Hamiltonian must commute with P, and so ultimately cannot depend on Q. The hard part, generally speaking, in these treatments is reexpressing the Hamiltonian in terms of the independent set of varialbes P, p_n, and q_n instead of the conventional $\pi(x)$ and $\phi(x)$.

In the previous analysis, Q represents the position operator of the kink, while the q_n represent coordinates of other, conventional ("meson") degrees of freedom. In this sector one and only one kink is always there, as in a first-quantized quantum mechanical treatment, while there may be any number of meson excitations, as in a second-quantized quantum mechanical treatment. Let $|P\rangle$, $|P,k_1\rangle$, $|P,k_1,k_2\rangle,\ldots,$ denote momentum eigenstates for one soliton alone, one soliton plus one meson, one soliton plus two mesons, etc. Then it follows that

$$\langle P|\phi(x)|P'\rangle = \frac{1}{2\pi}\int e^{-iX(P-P')}u(x-X)\,dX + R ,$$

where

$$R = \sum_{n=2}^{\infty} \langle P|q_n|P'\rangle\psi_n(x) .$$

Clearly, the first term in the matrix element of ϕ is $O(G^{-1})$ (since u is) while the rest, R, is $O(G^{0})$. Thus for small G the field operator matrix element above may be approximated by the first term and is thus entirely governed by the soliton solution. This is the ansatz adopted in the analysis of Goldstone and Jackiw [12], and this ansatz directly leads to the approximation that the leading contribution to the operator field equation is given by the classical solution. To calculate correction terms to the field equation, account must be taken of connected matrix elements such as

$$\langle P,k_1,\ldots,k_m|\phi(x)|P',k_1',\ldots,k_n'\rangle_C \quad .$$

For small coupling it is assumed that each meson costs G in amplitude so this expression is $O(G^{(m+n-1)})$, and with that ansatz a consistent approximation scheme can be set up, which, after accounting for the nonanalytic portion through the classical solution, has the general structure of a perturbation expansion.

Other studies of operator equations for soliton problems have been given by Klein and Krejs [28].

IV. SEMI-CLASSICAL APPROXIMATIONS IN PATH INTEGRALS

A. Basics

One of the most common semi-classical quantization conditions is the Bohr-Sommerfeld quantization condition given by

$$\oint p\,dq = 2\pi(n+\tfrac{1}{2}) \quad ,$$

a justification for which can be given in the framework of the WKB approximation. This approximation becomes more valid for large values of n. The term $\frac{1}{2}$ in the factor $n + \frac{1}{2}$ is then a higher-order effect, and its derivation depends on details of the turning formulae. There are analogues of this condition appropriate in field theory as well, and are relevant for periodic classical field solutions. One such periodic solution applies to the breather mode of the sine-Gordon equation presented in Sec. I. Classically, a solution exists for all values of the angular frequency $\omega = ma/\sqrt{1+a^2}$, but that situation changes on quantization. One may evaluate $\oint pdq$ directly for that solution, but there is another way to proceed as well that will be simpler for us [28].

Recall the expression

$$\omega = E_n - E_{n-1} \simeq dE/dn$$

relating the transition frequency to the difference in neighboring energy levels. Inverting this expression leads to

$$n = \int_0^{E_n} \frac{dE}{\omega(E)}$$

where $\omega(E)$ is an expression of the angular frequency of the system for energy E. In Sec. I it was noted that the classical energy of the breather mode was $2M_o/\sqrt{1+a^2}$ (where $M_o \equiv 8m^3/G^2$) and so (on eliminating a)

$$\omega(E) = m(1 - \frac{E^2}{4M_o^2})^{1/2} .$$

It follows directly that

$$n = \frac{2M_o}{m} \sin^{-1} (\frac{E_n}{2M_o})$$

or

$$E_n = 2M_o \sin(\frac{nm}{2M_o})$$

which holds for

$$n = 1,2,... < \frac{\pi M_o}{m} \quad .$$

This semi-classical expression for the quantization states of the breather mode is confirmed by other analyses. When the full quantum treatment is included the only change appears a change of the parameter M_o to M_o', a renormalized soliton mass, where

$$M_o' = \frac{8m^3}{G^2} - \frac{m}{\pi} \quad .$$

As anticipated the correction to the classical energy is $O(G^o)$. This simple change is all that enters the breather mode quanization formula as well. These remarkable results were established by Dashen, Hasslacher and Neveu [13], and by Faddeev and his collaborators [29], and their general significance has been extensively discussed.

Stationary Phase Approximation

The thoroughly classical ingredients that make up the path integral make it ideal for a treatment of semiclassical approximations. Consider the transition amplitude

$$A = \int e^{iI(\phi)} D\phi \ .$$

A common approximation to this integral sets

$$I(\phi) = I(\phi_S) + \frac{1}{2}\int [\delta^2 I/\delta\phi(x)\delta\phi(y)]\Big|_{\phi=\phi_S} [\phi(x)-\phi_S(x)] \times [\phi(y)-\phi_S(y)]dxdy$$

where ϕ_S is a stationary point of I,

$$\delta I/\delta\phi_S(x) = 0 \ .$$

The resultant Gaussian integral can be carried out and leads to the approximation

$$A = \sum_S \det[i\delta^2 I/\delta\phi_S''\delta\phi_S']^{-1/2} e^{iI(\phi_S)}$$

summed over one or more stationary points. Here the Pauli-Morette determinant enters and ϕ' and ϕ'' label initial and final configurations [30]. In the framework of quantum mechanical scattering, Miller has carried through an analysis of such methods [31], and their use in heavy ion scattering has been reviewed by Koeling and Malfliet [32]. In the study of quantum mechanical bound states, Gutzwiller has carefully elaborated such methods [33]. And in the study of bound states in quantum field theory, Dashen, Hasslacher, and Neveu [2] have built on the work of Gutzwiller [33] and of Maslov [34].

Fermion Fields

In forming the trace of the time evolution operator, the Fourier transform of which generates the density of states, one treats the path integral

$$\mathrm{Tr}(e^{-iHT}) = \int e^{iI(\phi)} D\phi$$

subject to periodic conditions on the field,

$$\phi(t+T,x) = \phi(t,x) \quad .$$

The inclusion of fermion fields is less straightforward. Conventionally, fermions in path integrals are regarded as anticommuting c-numbers and so a stationary phase approximation for them is not clear. But for Yukawa coupling (which through the introduction of auxiliary fields can always be arranged) the fermions enter quadratically and can be integrated out using the general rule for anticommuting c-number variables ψ and $\bar{\psi}$ that

$$\int e^{\bar{\psi}A\psi} d\psi d\bar{\psi} = \det(A) \quad .$$

While the result of the fermion integration is all we need, it is interesting to note that it implies anti-periodic boundary conditions on the ψ fields, $\psi(x,t+T) = -\psi(x,t)$ [14]. This elimination of the fermion fields is exact and leads to an effective self-interaction for the boson fields. Stationary phase approximations may then be employed to estimate the remaining integral. This general technique has been applied by Dashen, Hasslacher, and Neveu to study the two-dimensional Gross-Neveu model [14], and, following their method, by Campbell and Liao [15] to study the two-dimensional sigma model with and without a pion field. Since the sigma model (without a pion field) is closest in spirit to the examples treated in Sec. II this is the model we shall discuss further.

B. Two-Dimensional Sigma Model

In notation adapted to this review, the model in question is described by the Lagrangian density

$$L = \frac{1}{2}\phi^2_{,\mu} - \lambda(\phi^2-\kappa^2)^2 - \bar{\psi}(i\gamma\cdot\partial+G\phi)\psi \ .$$

After elimination of the N-component fermion field as discussed above one is faced with the boson path integral

$$\int \exp[i\int\{\tfrac{1}{2}\phi^2_{,\mu}-\lambda(\phi^2-\kappa^2)^2\}d^2x$$

$$+ \operatorname{tr} \ln(i\gamma\cdot\partial+G\phi)]D\phi \ .$$

This expression contains divergences (the Yukawa interaction in two dimensions requires mass renormalization of the boson), and these need to be considered. Before discussing the stationary points of this expression it is expedient to recast it into another form. The subsequent analysis is restricted to time-independent ϕ fields.

Let us focus on eigensolutions of the massless Dirac equation in the presence of a potential,

$$(i\gamma\cdot\partial+G\phi)\psi = 0 \ ,$$

of the form $\psi_j(x,t) = e^{-i\omega_j t}\tilde{\psi}_j(x)$. It is evident that in terms of these modes the formulation of the determinant is "diagonalized". In particular, Dashen, Hasslacher, and Neveu [14] have shown for time independent ϕ that the expression for the determinant takes the form

$$e^{\mathrm{tr}\ \ell n(i\gamma\cdot\partial+G\phi)}$$

$$= \exp(-iTN \sum_{\omega_j<0} \omega_j) \prod_{\omega_j>0} (1+e^{-iT\omega_j})^{2N}$$

$$= \exp(-iTN \sum_{\omega_j<0} \omega_j) \sum_{\{n_i\}} C(2N,\{n_i\})$$

$$\times \exp(-iT \sum_{\omega_i>0} n_i\omega_i)$$

where

$$C(2N,\{n_i\}) \equiv \prod_i \frac{(2N)!}{(2N-n_i)!n_i!}$$

and the sum extends over $0 \leq n_i \leq 2N$. The 2N (rather than N) arises because this model really has O(2N) symmetry [rather than just SU(N)]. For convenience we have assumed the eigenfrequencies are discrete (system in a box) and we have used the symmetry of positive and negative eigenvalues in our expression. The eigenfrequencies are still functionals of the scalar field ϕ, i.e. $\omega_i = \omega_i(\phi)$. This form of the fermion contribution permits one to write

$$\mathrm{tr}(e^{-iHT}) = \sum_{\{n_i\}} C(2N,\{n_i\}) \int e^{iI_{eff}(\phi,\{n_i\})} D\phi$$

where

$$I_{eff}(\phi,\{n_i\}) = T\int[-\tfrac{1}{2}\phi'^2 - \lambda(\phi^2-\kappa^2)^2]dx$$

$$- NT \sum_{\omega_i<0} \omega_i(\phi) - T \sum_{\omega_i>0} n_i\omega_i(\phi) .$$

Here the trace is expressed as a sum of terms, each with degeneracy C, and labelled by the set of indices $\{n_i\}$. One may regard the action as describing a boson field interacting with a filled fermion sea of negative energies and with a number of fermions of positive energies and occupation numbers $\{n_i\}$. Although fluctuations about the extremal points should properly be included, they have not been calculated and as the approximate semi-classical energy for a state labelled by $\{n_i\}$ Campbell and Liao simply take $E = - I_{eff}/T$ evaluated at the point ϕ_S for which $\delta I/\delta\phi_S = 0$.

For purposes of discussion we shall take only one excited mode with n_o fermions and (nonnegative) frequency ω_o. In that case the stationarity condition reads

$$\frac{\delta}{\delta\phi}\{\int[\frac{1}{2}\phi'^2+\lambda(\phi^2-\kappa^2)^2]dx+N\sum_{\omega_i<0}\omega_i(\phi)+n_o\omega_o(\phi)\} = 0 \ .$$

As emphasized by Campbell and Liao, it is pedagogically useful to analyze this relation by first studying the approximate one

$$\frac{\delta}{\delta\phi}\{\int[\frac{1}{2}\phi'^2+\lambda(\phi^2-\kappa^2)^2]dx+n_o\omega_o(\phi)\} = 0 \ .$$

The latter relation is termed "classical" while the former, more accurate one, is termed "semi-classical". The classical case is worth considering independently because it is a common approximation in nuclear matter studies and because it entails no divergences.

Classical Analysis

The classical stationarity condition leads to the coupled equation

$$[\omega_o\gamma_o + i\gamma_1 \frac{d}{dx} - G\phi(x)]\psi_o(x) = 0$$

and

$$\phi''(x) - 4\lambda\phi(\phi^2-\kappa^2) = Gn_o\bar{\psi}_o\psi_o \ .$$

In obtaining this relation, use was made of the identification

$$\delta\omega_o/\delta\phi \equiv - G\bar{\psi}_o\psi_o$$

subject to the normalization condition

$$\int\psi_o^*\psi_o dx = 1 .$$

All n_o fermions are in the same state ψ_o, and distinguished by their color indices. These equations are like a Hartree approximation, and the fermion field here is an ordinary function and not an anticommuting c-number.

One solution to these equations has been previously obtained [9] and is given by

$$\phi(x) = \kappa \tanh(\sqrt{2\lambda}\ \kappa x) ,$$

and in a representation where $\gamma_o = \sigma_z$ and $\gamma_1 = i\sigma_y$,

$$\psi_o(x) = M[\cosh(\sqrt{2\lambda}\ \kappa x)]^{-(\frac{G^2}{2\lambda})^{1/2}} \begin{bmatrix} 1 \\ 1 \end{bmatrix}$$

where M is a normalization factor. This solution requires $\omega_o = 0$, and since

$$\bar{\psi}_o(x)\psi_o(x) = 0$$

there is no feedback of the fermion solution to the boson kink. The classical energy is then just that of the kink itself, namely

$$E_S = \frac{4}{3}\sqrt{2\lambda}\ \ \kappa^3$$

and is independent of n_o. The localization of the fermion field is predominantly in the region where ϕ vanishes, and thus this solution simulates a "SLAC-shell" state, or in the Lee-Wick language an "abnormal nuclei".

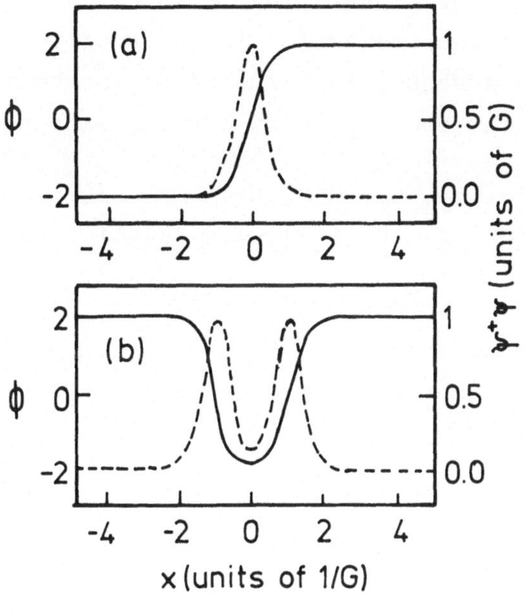

Fig. 7

In Fig. 7a the meson (———) and the fermion (-----) are illustrated for $\kappa = 2$ and $G^2 = 2\lambda$. Note that a solution exists for any value of the ratio $G^2/2\lambda$, but the solution is analytically simple if $G^2 = 2\lambda$.

In order to discuss other solutions of these equations a very different approach is taken. The appearance of the variable ω_o in the action suggests a reparameterization of the action in terms of scattering data - the discrete energy eigenvalues and normalizations, and the reflection coefficient - appropriate to the potential ϕ of the Dirac equation. Variation of the action expressed in terms of scattering data yields algebraic conditions for the scattering variables, and ultimately by invoking inverse-problem solutions the form of ϕ and ϕ_o can be determined. The success of this program hinges on the "trace identities" with which it becomes possible to express the action in terms of the scattering data; and this success, in turn, holds in the present case only for the specific parametric relation $\lambda = G^2/2$ [15].

In particular, the following relation can be established:

$$\int[\tfrac{1}{2}(\phi')^2+\tfrac{1}{2}G^2(\phi^2-\kappa^2)^2]dx$$

$$= \frac{8k_o^3}{3G^2} - \frac{1}{\pi G^2}\int q^2 dq\{\ln[1-|s_{12}^{(+)}(q)|^2]$$

$$+ \ln[1-|s_{12}^{(-)}(q)|^2]\}$$

$$\equiv \frac{8k_o^3}{3G^2} - F(s_{12}) ,$$

where $k_o^2 \equiv G^2\kappa^2 - \omega_o^2$, and $s_{12}(q)$ is the reflection coefficient. Hence, for $\lambda = G^2/2$, it follows that

$$I_{eff}/T = -\frac{8k_o^3}{3G^2} - n_o\omega_o + F(s_{12}) .$$

Variation with respect to s_{12} yields $s_{12}^{(+)} = s_{12}^{(-)} = 0$. Variation with respect to k_o yields the condition

$$\omega_o k_o = n_o G^2/8 ,$$

and in terms of $\omega_o \equiv G\kappa\cos\phi$ and $k_o \equiv G\kappa\sin\phi$, this condition reads

$$\sin 2\phi(n_o) = \frac{n_o}{4\kappa^2} .$$

This is the basis quantization condition; for $n_o/4\kappa^2 > 1$ there are no solutions, while for $n_o/4\kappa^2 < 1$ it exhibits two solutions.

Figure 8a is a plot of the left (——) and right (----) sides of the quantization condition and shows the two possible solutions, which are termed a "shallow bag", for $\phi < \pi/4$, and a "deep bag", for $\phi > \pi/4$. The classical energy for this quantization condition reads

$$E(n_o) = \frac{8}{3} G\kappa^3[\sin\phi(n_o) + \frac{n_o}{4\kappa^2}\cos\phi(n_o)] .$$

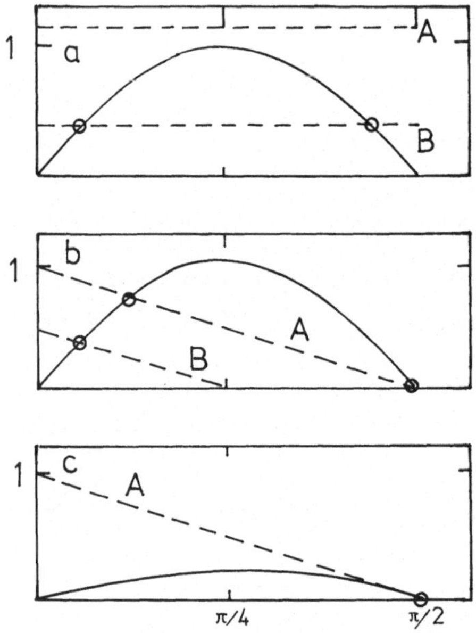

Fig. 8

Inverse scattering techniques determine the form of the "potential", namely

$$\phi(x) = \kappa - \frac{k_o^2}{G\omega_o} \operatorname{sech}[k_o(x+x_o)]\operatorname{sech}[k_o(x-x_o)]$$

and yield for the fermion field

$$\psi_o(x,t) = \left(\frac{k_o}{8}\right)^{1/2} e^{-i\omega_o t} \begin{bmatrix} \operatorname{sech}[k_o(x+x_o)]+\operatorname{sech}[k_o(x-x_o)] \\ -\operatorname{sech}[k_o(x+x_o)]+\operatorname{sech}[k_o(x-x_o)] \end{bmatrix}$$

where the parameter x_o is fixed by the condition $\tanh(k_o x_o) = (G\kappa - \omega_o)/k_o$. It is clear for $\phi(n_o)$ small that k_o is small, and thus the name "shallow bag"; and conversely as well. In the small $\phi(n_o)$ case, it follows for the "shallow bag" that

$$E_{SB}(n_o) \simeq n_o G - G \frac{n_o^3}{384\kappa^3} + \dots ,$$

which are loosely bound "normal nuclei" containing n_o "nucleons". For the "deep bag", on the other hand,

$$E_{DB}(n_o) \simeq \frac{8}{3} G\kappa^3 + \frac{G n_o^2}{16\kappa} + \dots$$

which exceeds the energy of two isolated kinks, namely $8G\kappa^3/3$, valid when $2\lambda = G^2$. Thus this configuration, which is depicted in Fig. 7b for $n_o = 1$, turns out to be unstable to decay into two "shell" states. Actually, a closer analysis of the "shallow bag" states in the range $0.7 \lesssim n_o/4\kappa^2 \leq 1$ shows that these "normal nuclei" are themselves unstable to decay into two "shell" or "abnormal nuclei" states.

Semi-Classical Analysis

In the improved approximation, account is taken of the negative energy sea in deriving the stationary points. It is convenient to once again parameterize the action in terms of scattering data, which again requires that $\lambda = G^2/2$. In so doing a divergent negative sea continuum contribution explicitly enters but it is cancelled by the needed mass renormalization term. We quote only the final result of such a manipulation, namely that

$$I_{eff}/T = -\frac{8k_o^3}{3G^2} + (N-n_o)\omega_o - \frac{2N}{\pi}k_o$$

$$-\frac{2N}{\pi}\omega_o \tan^{-1}\left(\frac{\omega_o}{k_o}\right)$$

where as before $k_o^2 \equiv G^2\kappa^2 - \omega_o^2$ and we have already exploited the stationary criteria $s_{12}^{(+)} = s_{12}^{(-)} = 0$ for the reflection coefficients. Thus it follows from the solution to the inverse problem that the "potential" $\phi(x)$ is the same as that presented in the classical case. Using $\omega_o = G\kappa\cos\phi$ and $k_o = G\kappa\sin\phi$ again, the extremal with respect to k_o yields the new form of the basic quantization condition

$$\sin 2\phi(n_o) = \frac{n_o}{4\kappa^2}\left[1-\left(\frac{N}{n_o}\right)\left(\frac{2\phi(n_o)}{\pi}\right)\right]$$

the left (——) and right (----) sides of which are plotted in Fig. 8b assuming $4\kappa^2/N = 1$. Line A shows two solutions for $n_o = N$, which by analogy with the previous discussion we term a "shallow bag" for $\phi < \pi/4$ and a "deep bag" for $\phi = \pi/2$. Line B shows a unique "shallow bag" solution for $n_o < N$. Thus for $n_o < N$ all "deep bag" configurations disappear. For $n_o = N$, the "deep bag" appears at $\phi = \pi/2$, but at $\phi = \pi/2$, $\omega_o = 0$, $\tanh x_o = 1$

and so $x_o = \infty$. Consequently, this particular form of the "deep bag" configuration is exactly that of a non-interacting, infinitely separated $(S+\bar{S})$ state and does not represent a bound state. In other words, "deep bags" disappear for all $n_o \leq N$. This is also confirmed by the corresponding expression for the field $\phi(x)$.

Along with the quantization condition is one for the energy given by

$$E(n_o) = \frac{8}{3} G\kappa^3[(1 + \frac{3N}{4\pi\kappa^2})\sin\phi(n_o)$$

$$+ 2\sin\phi(n_o)\cos^2\phi(n_o)] \ .$$

We note first that the energy of the fictitious "deep bag" at $\phi = \pi/2$ is given as

$$E_{DB}(n_o) = \frac{8}{3} G\kappa^3 + \frac{2NG}{\pi}$$

which, since it is the sum of two far separated "shell" states, means that the "shell" energy becomes

$$E_S = \frac{4}{3} G\kappa^3 + \frac{NG\kappa}{\pi}$$

when the effects of the negative sea are taken into account. Secondly, one can make a stability analysis of the "shallow bag" states. Recall in the classical analysis it was energetically favorable for "shallow bag" to decay into two "shell" states provided

$n_o/4\kappa^2 \gtrsim 0.7$. When negative-energy sea effects are included the situation changes and all "shallow bag" states are energetically stable. This important result indicates that qualitative changes in the properties can arise if the negative-energy sea contributions are neglected.

ACKNOWLEDGEMENTS

It is a pleasure to thank David Campbell for a discussion of his analysis of the two-dimensional sigma model.

REFERENCES

1. See, e.g., A.C. Scott, F.Y.F. Chu and D.W. McLaughlin, Proc. of the IEEE 61, 1443 (1973); V.E. Zakharov, L.A. Takhtadzhyan, and L.D. Faddeev, Sov. Phys. Dokl. 19, 824 (1975).
2. R.F. Dashen, B.Hasslacher and A.Neveu, Phys. Rev. D 10, 4114, 4130, 4138 (1974).
3. D. Finkelstein and C. W. Misner, Ann. Phys. 6, 230 (1959); J.K. Perring and T.H.R. Skyrme, Nuclear Phys. 31, 550 (1962).
4. T.D. Lee, Phys. Reports 23C, 254 (1976); "Nontopological Solitons", Symposium on Frontier Problems in High Energy Physics, Pisa, 1976 (to be published).

5. F. Cooper, G.S. Guralnik, and S. Kasden, "Collective Phenomena in $\lambda\phi^4$ Field Theory Treated in the Random Phase Approximation" (to be published).

6. R. Jackiw, "Semi-Classical Analysis of Quantum Field Theory", XVI Escuela Latino-Americana, July, 1976 (to be published).

7. T.D. Lee and G.C. Wick, Phys. Rev. D9, 2291 (1974); T.D. Lee and M.Margulies, Phys. Rev. D 11, 1591 (1975); T.D. Lee, Revs. Mod. Phys. 47, 267 (1975).

8. R. Friedberg, T.D. Lee, and A. Sirlin, Phys. Rev. D 13, 2739 (1976); "Gauge-Field Nontopological Solitons in Three Space Dimensions, I, II, III" (to be published); R. Friedberg and T.D. Lee, "Fermion-Field Nontopological Solitons, I" (to be published).

9. W.A. Bardeen, M.S. Chnowitz, S.D. Drell, M. Weinstein, and T.-M. Yan, Phys. Rev. D 11, 1094 (1975); S.D.Drell, "Quark Confinement Schemes in Field Theory", Ettore Majorana Int'l Summer School, Erice, 1975 (to be published). See also P. Vinciarelli, Lett. Nuovo Cimento 4, 905 (1972).

10. J.R. Klauder, J.Math. Phys. 8, 2392 (1967).

11. J.L. Gervais and B. Sakita, Phys. Rev. D 11, 2943 (1975); L.D. Faddeev, P.P. Kulish, and V.E. Korepin, Pizma JETP 21, 302 (1975). See also J.L. Gervais, A. Jevicki, and B. Sakita, Phys. Rev. D 12, 1038 (1975); N. Christ and T.D. Lee, Phys. Rev. D 12, 1606 (1975); E. Tomboulis, Phys. Rev. D 12, 1678 (1975); M. Creutz, Phys. Rev. D 12, 3126 (1975). For a survey, see Phys. Reports 23C (1976).

12. J. Goldstone and R. Jackiw, Phys. Rev. D 11, 1486 (1975).

13. R.F. Dashen, B. Hasslacher, and A. Neveu, Phys. Rev. D 11, 3424 (1975).

14. R.F. Dashen, B. Hasslacher, and A. Neveu, Phys. Rev. D 12, 2443 (1975).

15. D.K. Campbell and Y.-T. Liao, Phys. Rev. D 14, 2093 (1976); D.K. Campbell, Phys. Letts. 64B, 187 (1976).

16. M. Creutz, Phys. Rev. D 10, 1749 (1974); M. Creutz and K.S. Soh, Phys. Rev. D 12, 443 (1975).

17. A. Chodos, R.L. Jaffe, K. Johnson, C.B. Thorn, and V.F. Weisskopf, Phys. Rev. D 9, 3471 (1974); D 11, 2599 (1974).

18. K. Cahill, Phys. Letts. 53B, 174 (1974); P.Vinciarelli, Phys. Letts. 59B, 380 (1975); "Effective Potential Approach to the Quantum Scattering of Solitons" (to be published); R.J. Finkelstein, "Hypothetical Solitons and Hadronic Structure" (to be published); K.Ishikawa, "Coherent State Description of Extended Objects; Stability and Orthonormality Problems" (to be published).

19. J.R. Klauder and E.C.G. Sudarshan, Fundamentals of Quantum Optics (W.A. Benjamin, New York, 1968).

20. R.J. Glauber, Quantum Optics and Electronics, C.DeWitt, A. Blandin, and C. Cohen-Tannoudji, eds. (Gordon and Breach, New York, 1964).

21. J. Schwinger, Proc. Nat. Acad. Sci. 37, 452, 455 (1951); Phys. Rev. 92, 1283 (1953); P.T. Matthews and A. Salam, Nuovo Cimento 2, 120 (1955).

22. F.A. Berezin and M.S. Marinov, "Particle Spin Dynamics as the Grassmann Variant of Classical Mechanics" (to

be published); K. Drühl, "A Theory of Classical Systems Based on Graded Lie Algebras" (to be published); R. Casalbuoni, Phys. Letts. 62B, 49 (1976); Nuovo Cimento 33A, 389 (1976); A.P. Balachandran and P. Salomonson, "Classical Description of Particle Interacting With Nonabelian Gauge Field" (to be published).

23. K. Symanzik, Z. Naturforsch. 9a, 809 (1954).

24. F. A. Berezin, The Method of Second Quantization (Academic Press, New York, 1966); J. Rzewuski, Field Theory II (Iliffe, London, 1969).

25. K. O. Friedrichs, Mathematical Aspects of the Quantum Theory of Fields (Interscience, New York, 1953).

26. J. R. Klauder, Ann. Phys. (NY) 11, 123 (1960).

27. J. R. Klauder and J. McKenna, J. Math. Phys. 6, 68 (1965).

28. A. Klein and F. Krejs, Phys. Rev. D 12, 3112 (1975); D 13, 3282, 3295 (1976).

29. L. Faddeev, "Quantization of Solitons", IAS Preprint (unpublished).

30. Pauli's Lectures on Physics, edited by C. P. Enz (MIT Press, Cambridge, 1973); C. Morette, Phys. Rev. 81, 848 (1951).

31. W. H. Miller, J. Chem. Phys. 53, 1949, 3578 (1970); W. H. Miller and T. F. George, J. Chem. Phys. 56, 5668 (1972); W. H. Miller, Advances in Chemical Physics 25, 69 (1974).

32. T. Koeling and R. A. Malfliet, Phys. Reports 22C, 183 (1975).

33. M. C. Gutzwiller, J. Math. Phys. 8, 1979 (1967); 10, 1004 (1969); 11, 1791 (1970); 12, 343 (1971).

34. V. P. Maslov, Theory of Disturbances and Asymptotic Methods (Moscow Univ. Press, 1965); Teor. Mat. Fiz. 2, 21 (1970).

Acta Physica Austriaca, Suppl. XVIII, 63-109 (1977)

TOPOLOGY OF MAGNETIC FIELDS IN PARTICLE PHYSICS, IMPLICATIONS ON THE QUARK MODEL+

by

H. JEHLE ++

Univ. of Amsterdam, Uppsala University
and University of Munich

ABSTRACT

The flux loop model of quarks attributes simple topological field structures to quarks. It assumes that the quark may be represented by a Seifert-Threlfall fibration [1] of ordinary three-dimensional space, i.e. by a continuous manifold of closed field lines ("loopforms" which a loop of quantized magnetic flux may adopt). Seifert and Threlfall's analysis of fibration showed that, because of continuity conditions, these field lines should be considered as representing a manifold of similar torus loopforms; the field thus may be constructed by superposing these "alternative loopforms" of the manifold with complex probability amplitudes [2].

+ Seminar given at XVI.Internationale Universitätswochen für Kernphysik, Schladming, Austria, February 24-March 5, 1977.
++ On leave from University of Maryland.

An N, P, λ quark corresponds to a loop of winding numbers (2,1), (3,1), (3,2) respectively (one circuit of a loop leading 2, 3, 3 times respectively around the "dough" of a doughnut torus while turning 1, 1, 2 times around the hole of the doughnut). These quarkloops thereby generate the familiar fractional electric charge fields if they spin with Zitterbewegung angular velocity [3]. It is then assumed that a meson is constituted from two interlinked coaxial loops, i.e. two toroidal fields, one essentially located in the inside of the dough of that doughnut, the other in the outside region [4]. They are able to perform their spinning motion independently, and have independent magnetic moment orientation too, coaxial of course. The sign of the electric charge depends on the product of these two. - Similarly, three interlinked toroidal fields represent a baryon. A basic assumption is that any "crosscutting" (a loop cutting across itself, thereby changing its topological structure, or across a loop with which it is interacting or with which it is interlinked)is a slow, i.e. a weak process. Therefore the loopforms, not only the alternative loopforms of one quark, but those of all quarks of a meson, or of a baryon, hang together. This concept essentially implies quark confinement. The interlinkage, of course, makes quarks localized objects, and permits the customary symmetric spin-isospin functions, or other spin-isospin functions, to be not in violation of the Pauli principle. - With this interpretation of the quark model, the quark's contribution to the magnetic moment of a particle should, as usually, be determined by that particle's spin-isospin function and, instead of the quark's magnetic moment being assumed as exactly proportional to its electric charge, it depends on the number of times which the respective quark loop goes through the "core" of the particle, which is the same as the number of "wings" of the quark loop, and which is indicated by its first winding number.

The principal difficulties of the quark model arise exactly when a quark or an antiquark are treated as if they were particles. The loop model does not have the aforementioned difficulties, the quark and the antiquark of a meson having the properties of coaxial toroidal fibrated fields. L-excited states of mesons should therefore not be treated as a quark and an antiquark orbiting about each other (as the charmonium model assumes). One should rather consider a $\bar{q}q$ orbiting about another, perhaps similar $\bar{q}q$, or several $\bar{q}q$ orbiting crudely like balls of a ball-bearing.

The usual trend of physics is in a direction from the more concrete to the more abstract. The present project seems, however, to be a reversal of that direction: At the onset of a new development in which topology is to play a major role, it is evident that elementary model aspects have to be developed, and one may well be advised to start with a conservative approach and check carefully on its consistency [5].

INTRODUCTORY REMARKS

Already before the advent of quantum mechanics it was recognized that a dual formulation of the Maxwell-Lorentz theory permits the introduction of magnetic charges (i.e. the monopoles) in analogy to electric charges. This results in a beautifully symmetric formulation of electromagnetism [6]. There are certain difficulties inherent in the monopole theories and in the particle physics models based on magnetic monopoles; we shall report on those difficulties further below.

There exists also a more conservative extension of electromagnetic theory which avoids the introduction of that hypothetical feature of magnetic monopoles, and which studies, however, the implications of the definition of electromagnetic fields in terms of distributions of "alternative forms" of closed quantized flux loops [7]. Such a distribution may be defined by means of probability amplitudes for the alternative forms which one flux loop, characterizing a lepton, may adopt, the distribution resulting in a smooth magnetic (and electric) field. For mesons and baryons, the distribution of two and of three interlinked loops (instead of one leptonic loop) are considered [8]. It is these simple concepts of quantized flux loops and of the probability amplitude distribution for alternative flux loopforms (somewhat analogous to alternative path history superposition in Feynman's space-time approach to quantum mechanics) [9] which are considered in the present model.

These manifolds of flux loopforms will be seen to represent the electromagnetic field of a particle; the electromagnetic field playing the role of an observable whereas the probability amplitudes (which characterize the distribution) represent the quantum state.

ELECTROMAGNETIC GAUGE INVARIANCE

Let us discuss the familiar basis on which magnetic flux quantization (or also quantization of magnetic charge) arises. The word gauge invariance is used in very different contexts.

It is used in the widest sense for gauge transformations which make no reference to conserved quantities at all. It is used, with Nambu strings, for the group of transformations of the space and time-like variables σ and τ. It is used when gauge transformations correspond to some approximately conserved quantities.

It was originally, however, used as "electromagnetic gauge", to correspond to an absolutely rigorous conservation law, i.e. that of electromagnetic charge. The physical significance and necessity of this electromagnetic gauge invariance was understood by Hermann Weyl [10]. It was recognized to be not merely a kind of accidental freedom admitted by the formulation of the laws of electromagnetism, but as an invariance of the electromagnetic equations, an invariance as important as their Lorentz invariance: Invariance with respect to the gauge variable ϕ is expressed by

$$0=\delta L=\delta\int j^k A_k\, d^4r = -\int j^k (\hbar c/e)\partial_k\, \phi d^4r = \int (\partial_k j^k)\phi d^4r \qquad (1)$$

$$\rightarrow \partial j^k/\partial r^k = 0 \qquad (2)$$

(cf. the notation ϕ used below).

Electromagnetic gauge invariance was then recognized by Infeld and van der Waerden [11] as entering into the formalism of affine connections of spinors by necessity, in the same way as invariance with respect to general point transformations enters into the formalism of affine connections of vectors, a very important result (the papers of Infeld-van der Waerden and of Laporte-Uhlenbeck [12] have been presented in shortened form by Bade-Jehle [13] and by Parke-Jehle [14],[15]. In this way a formulation of the laws of classical electrodynamics and dynamics was achieved which properly combines covariance with respect to gauge transformations and point transformations, a formulation which incorporates electric charge conservation law and dynamical conservation laws in a unified theory.

FLUX QUANTIZATION

Now, the electromagnetic gauge permits to introduce the concept of magneticflux quantization. Let us describe this quantum phenomenon in some detail. Consider a field lepton, described by a ψ function which satisfies a wave equation, interacting with a source, described by A_k, i.e. a four potential which satisfies the Maxwell-Lorentz equations. An electromagnetic gauge transformation with the gauge variable ϕ, applied on

$$[\partial_k - i(e/\hbar c)A_k]\psi\ , \quad A_k = V, -\vec{A}, \quad k = 0,1,2,3, \tag{3}$$

transforms to primed quantities:

$$\psi = \psi'\exp(i\phi),\ A_k = A'_k + (\hbar c/e)\partial_k\phi = A'_k - \delta A_k\,; \tag{4}$$

ϕ is considered to be a single valued function of t, x, y, z. Such a gauge transformation is not changing the physical situation at all.

Without violating the condition that ψ be a single valued function of time and space, one may introduce, however, instead of the single valued gauge variable ϕ, a pseudogauge function θ which is only singlevalued modulo 2π and which thereby brings about a change of the physical situation (we refer to the very useful terminology of pseudogauge, in J.M. Blatt, Superconductivity) [16]:

$$\psi = \psi' \exp(i\theta) \text{ or } \psi \exp(-i\theta) = \psi' \exp(-iO), \ \theta' = O \ . \quad (5)$$

$\nabla\theta$, as function of the space variables x, y, z has singular lines around which θ changes by an integer multiple of 2π, usually $\pm 2\pi$.

What is the physical significance of these singular lines? Starting with a field lepton whose wave function ψ' is without singularities in the above sense, starting e.g. with $\theta' = O$, a field lepton which moves in a field $A'_k = O$,

$$O = A'_k - (\hbar c/e)\partial_k\theta' = A_k - (\hbar c/e)\partial_k\theta \equiv A_k \ . \quad (6)$$

From a solution of the wave equation with nonsingular $\partial_k \psi'$ in the nullfield $A'_k = O$, one may obtain another solution ψ with singular $\partial_k\theta$; the corresponding A_k field has a quantized amount of flux $\Phi_q = \hbar c/e$ in this singular line, because

$$-2\pi = \oint\nabla\theta\cdot d\vec{r} = (e/\hbar c)\oint\sum_1^3 A_k \, dr^k$$

$$= -(e/\hbar c)\iint \text{curl}\,\vec{A}\cdot d^2\vec{r} = -(e/\hbar c)\Phi_q\ . \qquad (7)$$

Let us consider now the source to which A_k is attached by the Maxwell-Lorentz equations, i.e. the electromagnetic charges and currents which, by Maxwell-Lorentz, are connected with that A_k field. To specify that statement, we may be reminded that (e.g. in case of a leptonic source) the one quantized flux line is, with probability amplitudes, continuously distributed over alternative forms so as to correspond to a smooth A_k field which is the field of the lepton under consideration. We shall again return to this issue further below, but, before we do that, digress on an issue concerning the physical character of an individual flux line.

Dirac considers a singular $\partial_k\theta$ line to start and to end at a positive and a negative magnetic monopole. One of his important aims was to explain that great puzzle: Why are the total electric charges of any one particle always <u>integer</u> multiples of the electronic charge? That indeed seemed to be a major achievement of Dirac's monopole theory. The trouble about it is this: In principle, such explanation of integrity is nice, but electric charges ±2e, ±3e come up so far only in hadrons whose complex internal structures can scarcely be expected to be represented by one magnetic monopole of charge ±e, ±2e, ±3e.

There have been other difficulties. One concerns the complexity of a scheme which introduces twice as many types of four-currents and four poentials as the conventional Maxwell-Lorentz theory, a scheme which is considerably more difficult to quantize than quantum electrodynamics. Such an argument in itself might not be a real argument against such a theory. But it certainly means that the more conservative way of investigating what may be achieved with closed quantized flux loops without monopoles, should first of all rigorously be pursued in an effort of applying the well established electromagnetic laws to particle physics.

A rather drastic criticism against magnetic monopoles (not against Barut's magnetic dipoles [17] with a poleseparation of the order of $\hbar/mc$ where m is the mass of the respective particle) comes from experiment: Because of their large coupling constant, free magnetic monopoles would, for a long time already, have shown up in experiment if they existed.

Quite independent of the question of admissibility of the magnetic monopole concept in particle physics is the issue of whether quarks may be described as particles, i.e. by field- and other theories based on the particle concept of quarks. Naturally it is convenient to try which results may be obtained from the quark-particle model, because the mathematical tools have already been developed, and because the quark parton model has indeed achieved great success. But at important other points it is exactly through the assumption that quarks might be handled as particles, that trouble arises. We are reminded of the statistics problem in low lying baryons which led to the hypothesis of parastatistics (O.W. Greenberg) [18] and color. And we are reminded of the quark confinement problem which may only be appeared with the help of ad hoc concepts of bags, and ad hoc particles, gluons and color gluons among them.

That there is an alternative to the quark-particle hypothesis, is a most welcome property of the loop model, i.e. the topological model which takes advantage of the concept of fibration to which we return after some historical remarks about the origin of the ideas developed in this note.

Fritz London considered closed lines of quantized flux as a basis for the formulation of a phenomenological theory of superconductivity [19]. Considering the quantized flux Φ_q of Eq. (7) as due (by Maxwell-Lorentz) to a new current of density $\vec{j}$, one interprets

$$(2e/\hbar c)\oint\vec{A}\cdot d\vec{r} = -(2e/\hbar c)\Phi_q = -(2e/\hbar c)(4\pi\lambda^2/c)\oint\vec{j}\cdot d\vec{r} \qquad (8)$$

as a relationship between this $\vec{j}$ and $\vec{A}$, so that

$$-\Box^2\vec{A} = 4\pi\vec{j}/c = \lambda^{-2}\vec{A} \quad . \qquad (9)$$

It is of course the pairing of electrons which, in superconductivity, brings in the factor 2e instead of the e in the context of quantized flux in particle physics.

Although the pairing of electrons was of course not possible to understand in the 1930's, London's theory of superconductivity and of flux quantization was indeed very much ahead of his time. It was not only til two decades later that Deaver and Fairbank, Doll and Näbauer experimentally discovered flux quantization, predicted by London. But also another element in London's understanding of superconductivity was far ahead of his time:

In the hole of a superconductor, the magnetic field is a smooth magnetic field. How could such be reconciled with the discrete character of a line singularity of quantized flux? The concept of superposition of alternative forms which one quantized flux loop may adopt (i.e. the forms of magnetic field lines in the space surrounding the superconductor), this concept was evidently in London's picture of the phenomenon. Whether he had the concept of superposition of alternative forms with probability amplitudes already developed, I am, unfortunately unable to say. In any case, such superpositions were, at that time, not considered as a permissible interpretation. Looking at it now, the lowest (or any other) quantum state of a superconductor should, as a quantum state, be considered as a probability amplitude, not a probability, superposition of alternative forms of a quantized flux loop, somewhat analogous to the path history superposition permitting to formulate a quantum path.

Not only is a superposition with complex probability amplitudes mandatory for the construction of a quantum state, it is also a necessity when we regard the formation of the Bohr magneton of an electron from quantized flux [20].

The probability amplitude superposition is altogether a necessary condition for obtaining a consistent formulation of leptons, as it was a necessary step for a consistent formulation of the magnetic field around a superconductor.

The lepton issues which had been discussed in those quoted papers [21] are not repeated in the present note; we want to turn here to indicate a start on the problems of quark models of hadrons in the present context.

TOPOLOGICAL CONDITIONS FOR THE MAGNETIC FIELD [22]

The manifold of alternative loopforms was defined as the manifold of singular lines of any field lepton's ψ function. That manifold, through Eq. (6), describes the field A_k in which such a field lepton moves. While this characterizes the relationship of A_k with any field lepton, we should now state how that A_k field, i.e. the manifold of singular loops (the field lines), is connected with a source. As already pointed out, that should be through the Maxwell-Lorentz equations, i.e. we want to consider the statistical manifold of these field lines to correspond, by Maxwell Lorentz, to a charge-current distribution of the source which distribution may be calculated if the loopform manifold is given. If that is requested to result e.g. in a magnetic dipole source of the strength of a Bohr magneton with an electronic Coulomb field, this request represents conditions imposed on the distribution of probability amplitudes of the loopform manifold.

As these loopform amplitudes are, by that request, not yet uniquely determined, important further conditions are imposed, i.e. conditions as regards the distribution of the loopform manifold over the Euler angle parameters (Fig. 3), over the "winding numbers" (below), in addition to the distribution implied by the SU(6) functions in the case of hadrons. We may shortly illustrate this over-completeness of the manifold of flux loopforms in regard to the winding numbers which among further topological properties, register whether we have righthanded toroidal field lines or lefthanded ones. A superposition of the magnetic field from equal shares of right-and of left-handed fields my result in an ordinary dipole field [23].

We note that both these fields have, with the resultant dipole field, the property in common that they are smooth. What is, topologically speaking, this concept of smoothness? H. Seifert and H. Threlfall have analyzed this concept in their investigations on "fibration of three-dimensional space", and given a mathematical definition of fibration which permits a complete discussion of the ensuing topological structures [24]. Suppose that, at every point of ordinary three dimensional space, a direction, a line element, is given (like a vector field). Suppose that, following these directions, (i.e. drawing field lines), one obtains always closed loops. Suppose furthermore that two loops which are neighbors, stay neighbors all along these loops and that this holds for any loop's neighbors. This defines, in simplified terminology, a Seifert-Threlfall fibration. These assumptions imply that the lines are in a local sense, somewhat parallel, related to each other like the muscle fibres of a single muscle, not like neighboring fibres of two differently directed adjacent muscles.

The Seifert-Threlfall fibrations are then shown to be manifolds of torus loops, i.e. "loopforms"; their forms may be visualized by drawing a closed line on a torus (heavily drawn line in Fig. 2, and Fig. 1c), these figures refer to a line (2.1) and a line (3.2) respectively; (3.2) denotes a line which goes three times around the dough of a doughnut, while going twice around the hole of that doughnut. Clearly, one may fill out the entire threedimensional space by drawing the twoparametric manifold of those lines (their parameters: azimuth and size of the torus loopforms; the azimuth parametrization of a (2.1) loopforms is illustrated with the manifold of loopforms drawn on the torus).

We mentioned a magnetic dipole field, and also right-and left-handed toroidal fields in connection with a resultant dipole field, above. Each one fulfills the Seifert-Threlfall conditions. But, as long as those refer to a point dipole source, the field structure (and field intensity) become quite singular at the origin. To avoid an unnecessary discourse of that kind of singularity, we may anticipate already the point made below, i.e. that we are interested in the discussion of extended sources and therefore may have to deal with toroidal field structures corresponding to a finite size torus, corresponding, by Maxwell-Lorentz, to the magnetic field of an extended source distribution of equivalent current (and electric charge).

Let us return then to the discussion of the Seifert-Threlfall fibration. We may talk of a "loop (3,2)" simply to define a fibration, covering the entire three-space, with a manifold of (3,2) loop-forms, and we may distinguish a left-hand screw (3,+2) fibration from a right-hand screw (3,-2) fibration by the signature of the second "winding number" 2. Consider now loopforms which are very close to the circular middle line ("axis") of the torus (Fig. 1). One has to go twice around the hole to make a full circuit along the loopform to complete the circuit. If the loopform coincides, however, in the limit, with that circle, one has to go around only once to be back. The topologists therefore call that circle a "singularity" of the fibration, we may call it the circular axis of the fibration. Of course, "circular" only in a topological, not in a metrical sense. It is evident that this concept of singularity is not to be mixed up with the concept of singular flux line. - Consider now also the (3,2) loopforms of exceedingly large size, they go almost parallel to the central straight axis of the torus (Fig. 1), and go in a big loop back from almost $+\infty$ to almost $-\infty$, three times to complete a circuit along the loopform. In the limit, when the loopform coincides with the straight central axis, only one traverse is needed to fulfill the circuital ride. Therefore again, that axis is called a singularity of the fibrated field [25].

So far, we described the torus loop fibrations by winding numbers and handedness, to which orientation of flux and orientations of motion may be added. To this essentially topological definition which is so important when the interactions of loops of one or of different particles are to be formulated, further characterizations, as already mentioned earlier, are to be added: A detailed differential description would characterize the manifold of loopforms by amplitudes which are functionals of the forms of the loop. Instead of a formalism with functionals, a simplified parametrized description was given in this project. In order to set up a model of particle structure, in order to formulate interactions, and in order to test the consistency of the model, the parametrized description turned out to be most efficient and convenient. In addition to the aforementioned topological parameters, we characterize the loopforms by 3 Euler angles (direction of the central axis and angle about it), and by 1 size parameter (one of the Euler angles, i.e. azimuth, is only represented in the distribution of Fig. 2). For dipole field lines, as an example, that parametrization is trivial; Fig. 3 indicates a distribution of directions of the central axis. The Maxwell-Lorentz equations are then again to connect the probability amplitude distribution of a loopform parameter manifold with the source currents which correspond to such loopform manifold.

A word about the history of the Seifert Threlfall fibration, only a causal remark, of relevance to us here. As a quantized flux loop is a "knot", it is of interest to remark that a knot, i.e. a closed loop of general type cannot be associated with a manifold of knots of the same type which fill out the entire three-dimensional space, without complicated discontinuities and singularities arising. Furthermore, general knots cannot unambiguously classified.

In order at mathematically well definable structures of manifolds of lines or of loops in three-dimensional space, E. Artin [26] investigated braids and closed braids, i.e. wreaths. A beautifully simple and general grouptheoretical classification was achieved by him.

As regards the problem of smooth coverage of the entire threedimensional space by closed loops, the above given specifications of Seifert-Threlfall fibrations permitted to arrive at a complete description of the types and properties of those fibrations. At the same time, these S.T. specifications are necessary conditions, expected to be satisfied by magnetic fields of leptons and of quarks.

EXTENDED SOURCE MODEL

Relativistic quantum mechanics has recognized a stationary, single spin 1/2 lepton to perform a Zitterbewegung of an amplitude of the order of magnitude $\hbar/mc$. Such a lepton, when placed in a small external potential A_k, has been shown by quantumelectrodynamics to have a <u>local</u> interaction with that A_k field, with minimal coupling $\sum_k j^k A_k$. The Zitterbewegung lepton thus behaves, in appropriate time averaging, as an extended particle of that size $\hbar/mc$. One may, of course, also describe it by the meanposition X which is sharp for that stationary single lepton,

$$\vec{X} = \vec{x} + (2|E|)^{-1}\, i\hbar c\beta\vec{\alpha}$$

$$-[2|E|^2(E+mc^2)]^{-1}\, \hbar c^2[i\beta(\vec{\alpha}\cdot\vec{p})\vec{p}+(\vec{\sigma}\times\vec{p})|E|/c] + \ldots \tag{10}$$

$$\vec{V} = d\vec{X}/dt = \pm\, c^2\, \vec{p}/|E| \tag{11}$$

by the Pryce-Foldy-Wouthuysen-Tani transformation [27].

$$\psi' = (\exp iS)\psi,\ S = -i(\beta\vec{\alpha}\cdot\vec{p}/2p)\ \mathrm{arc\ tg}\ (p/mc)\ . \tag{12}$$

A further analysis of the Zitterbewegung, by K.Huang [28], has shown it to be interpretable as a spinning motion with angular velocity

$$\Omega = 2\, mc^2/\hbar \ . \tag{13}$$

We make the assumption that for a stationary single lepton, a model may be employed which shows that "quasi-extension" as an extension of the source, i.e. in our terminology, shows a distribution of the magnetic moment like the magnetization of a spherically symmetric "core" (of the size of the "circular axis" of the fibration), and which shows that spinning angular velocity (13) as an angular velocity of each "sheaf" of quantized flux loopforms. (A "sheaf" is the manifold of loopforms of Fig. 2 when that distribution over the azimuth parameter is supplemented by a distribution over the size parameter to fill out all of space. The complete manifold of parametrized loopforms still implies superposition of sheaves of the 2 parametric manifold of orientation, indicated in Fig. 3).

Such a hypothesis of taking Zitterbewegung extention and Zitterbewegung motion realistically into account avoids the singularities which a (non-physical) stationary point-source model would have. Furthermore, as indicated below, the spinning angular velocity of the model implies the proper connection between Bohr magneton and Coulomb field (in analogy to the connection between Coulomb field and the Bohr magneton of the Dirac electron).

It should be noted that the first adequate model of the electron is that due to J.C. Slater who in 1926

[29] showed that $\hbar/mc$, where m = the mass of the lepton in question, would have to characterize the size of the extended model. And it may readily be seen that only this choice makes it possible to formulate a model whose electromagnetic energy is compatible with mc^2 and whose electromagnetic angular momentum is compatible with $\hbar/2$. [30].

ELECTRIC FIELD

The very same equations (6) which define quantized flux through Eq. (7), also define the electric potential due to the motion of the flux loopform manifold

$$-\vec{A} = (\hbar c/e)\nabla\theta, \qquad V = (\hbar c/e)\partial_o\theta \ . \tag{14}$$

If the distribution of flux loopforms corresponds to a Bohr magneton dipolefield (let us start with a point dipolefield) and if the loopforms, or sheaves of loopforms spin about their magnetic axis with angular velocity $2mc^2/\hbar$ (Eq.(13)), then V represents a Coulomb-type potential [31]. More precisely: if the distribution of flux loopforms (i.e. of sheaves of flux loopforms) corresponding to a quantum state "magnetic moment up" is superposed from sheaves with a proper statistical distribution of (fluctuating) inclination (Fig. 3), having a resultant magnetic moment equal to a Bohr magneton, and if every sheaf spins about its axis with angular velocity $2mc^2/\hbar$, the electric potential (14) is isotropic and of magnitude

$$V = \pm\ 2e/3r \ . \tag{15}$$

It is not proper to repeat the proof of this statement here because it was shown in previous papers (cf.l.c. 1975). But two comments should be made here: First, for an extended source the averaged potential is then simply obtained by "smearing out" the potential (15).

Second, it is important to be clear about the pseudogauge function θ, function of t, x, y, z. Instead of plotting θ as a multivalued function of x, y, z, by plotting for example the values of θ in an equatorial plane z = 0 on some kind of Riemannian sheat (as we did e.g. in l.c.1975), we may simply consider the principal value of $\theta(x, y, z)$ and indicate where θ jumps by $+2\pi$ or -2π. That jump occurs at a 2-dimensional "cut surface" (a kind of analogue to the cut line of a Riemannian sheet over a complex plane. This cut surface is of course bounded by the quantized flux loopform under consideration and may be obtained by dipping a wire-or rubbertube model of the flux loopform into a soap solution; that soap film is very instructive. (Apart from the boundary, that cut surface is not determined as the soap fim would be, because an ordinary gauge transformation with a single valued gauge function $\phi(x,y,z)$ is still open). When we consider the loopform with its soapfilm spinning and then consider a space-fixed point, the electric potential (14) is then given by the frequency with which the soapfilm passes across that point.

The immediate corrolary to that is the soapfilm of a (3,2) loop (i.e. a trefoil, Fig. 1c). If that slides through space like a coasting tree-bladed propeller, the electric field it produces is seen to be exactly equal to zero.

This is, as it should be for this to be the neutrino loop. - A Lorentz transformation to arbitarily high velocity along the direction of motion of the neutrino (i.e. along the central axis), does not change its handedness. Accordingly there is no rest system for that neutrino and thus its mass is zero [32].

LINKAGE OF LOOPFORMS

Let us first consider the loopforms corresponding to one sheaf of a lepton. Would these loopforms be of the type (1,0) as ordinary dipole field lines are, these would not be interlinked, and it might be plausible to argue that such a manifold of alternative loopforms would have difficulties in staying together. After all, the flux loop is the basic entity of that leptonic structure, and, in consequence of the Maxwellian stress tensor (sideways: pressure, along the lines: tension) the alternative forms of that type of loop would not hold together. (That such argumentation might be applicable to flux loopforms, is an assumption, though a plausible one, which needs to be substantiated, however). - On the other hand, loopforms of type (2,+1) are interlinked (Figs. 1a and 2) (similarly also loopforms (2,-1)), and they go all around the center of the core; one may therefore assume that such interlinkage and such tension along the loopforms would hold the loopforms together, so that they may not fall apart.

Finally: a superposition of a lefthanded manifold (2,+1) and a righthanded (2,-1) will result in the effective dipole field (1,0) [33].

TOPOLOGY AND MOTION OF FLUX LOOPFORMS

We mentioned so far the spinning angular velocity $2\ mc^2/\hbar$ of the flux loopform manifold about the central straight (torus symmetry) axis. Topologically, the two axes, the central straight axis and the circular torus midline axis, are equivalent in many respects. It would therefore be absurd in a topological model to assume "spinning" about the one axis without a corresponding velocity, simultaneously.

The spinning angular velocity and the whirling angular velocity may form a left-handed or a right--handed motion. Let us consider a situation in which the center of the fibrated structure stays at rest (unlike the neutrino); a superposition of loopform manifolds (to be discussed in a later note) is responsible for the possibility of a standstill of the center.

If the set of loopforms is lefthanded, i.e.(2,+1), and the combined spinning-whirling motion is righthanded, the loopforms are readily seen to spin with the effective spinning angular velocity [34]

$$(1 + \frac{1}{2})\ 2mc^2/\hbar = 3mc^2/\hbar \ ; \tag{16}$$

if the combined motion is left-handed, the effective spinning angular velocity is

$$(1 - \frac{1}{2})\ 2mc^2/\hbar = mc^2/\hbar \ . \tag{17}$$

That is so because a pure whirling turn by 2π corresponds to a half spinning revolution when a (2,1) loop is under consideration. The angular velocities $3mc^2/\hbar$

and $mc^2/\hbar$ are the effective angular spinning velocities due to both motions combined. They lead to

$$V = \pm\, e/r \qquad \text{and to} \qquad V = \pm\, e/3r \qquad (18)$$

respectively because of (15), and correspond to electron or muon charge, and to the charge of an N quark [35].

Again we may refer to previous papers to remind that the ± sign corresponds to parallelism or anti-parallelism of effective spinning angular velocity and magnetic moment. And we may be reminded that the mass of the particle (e.g. in the leptonic case, the mass of electron or of muon) cancels out rigorously, not accidentally, in the calculation of the electric field (15) [36].

Immediately the suggestion is obvious that whereas for leptons one has cooperative effects between spinning and whirling motion, for quarks it is the noncooperative, let us say "sliding" motion which applies.

The simplest loopform manifolds of interlinked types are (2,1), (3,1), (3,2); a simple count of the number of fluxwings which pass, per unit time over a spacefixed point, i.e. the frequency with which the θ discontinuity surface (soap-bubble) passes over a space fixed point, measures the effective electric charges of quarks, resulting in

$$\pm\, e/3r, \qquad \pm\, 2e/3r, \qquad \pm\, e/3r \qquad (19)$$

respectively.

One finds also (cf. l.c. 1975) that loops (1,2), (1,3) and (2,3) show the same electric charges as the (2,1), (3,1) and (3,2), which circumstance simplifies the discussions considerably.

A word of caution is needed here: The total charge of the resulting particle is to be integer, that is a rule not explained by the present model either. This integrity means of course that the mesons are built from $\bar{q}q$ and the baryons from qqq, and it will bring important conditions on the hadrons built with or without admixture of the higher quarks, e.g. (4,3) or the like, cf. the tables I and II of l.c.1975.

Of quite drastic importance becomes this integrity rule in the case of leptons: Of the loops with low winding numbers it is only the (2,1) or (1,2), not the (3,1), (1,3) or the (3,2), (2,3) loops which permit integrity of electric charge, except the helically moving neutrinos (3,2) (and also (2,3)) which we already discussed.

It is nice to find the simplest types of loops exactly corresponding to the simplest leptons and quarks.

COAXIAL LOOPS OF HADRONS. WEAK INTERACTIONS

A meson is assumed as being constituted of a q loopform manifold which mainly occupies the toroidal region adjacent to the circular axis ("inside the dough of the doughnut") but still reaches throughout all of 3 space, and a $\bar{q}$ loopform manifold which mainly occupies the "toroidal" region outside the dough of the doughnut, adjacent to the central straight axis [37].

Similar definitions hold with q and $\bar{q}$ interchanged, and for qqq baryons where we consider three coaxial toroidal regions.

The spinning angular velocities and the magnetic moments and the handednesses of those quarkloop fibrations are independent which permits the independent assignments of quarks and antiquarks to form SU(6) hadrons.

Let us try to relate electromagnetic interactions with weak [38] and strong interactions. To that effect we may introduce a basic assumption regarding weak interactions.

We shall assume that the "crosscutting" of a loopform manifold over itself (i.e. a change of its own topological structure), a crosscutting over another manifold which is coaxial with the former manifold or a crosscutting over a manifold with which the former one interacts, is an improbable, i.e. a weak process. Although that seems to be a mere qualitative statment, it may be shown below (cf. "strong interactions") that this permits a sensible quantitative formulation, l.c.1977.

With this assumption we may now point to some important features of the quark loop model.

First, the loop (3,2) has the correct electric charge $\pm e/3$ which a λ quark should have. It has been identified with the λ quark not only because of that value of electric charge, but also, as it is more massive than the (2,1)quark(obviously, because the (3,2) is more complex in structure than the (2,1)), therefore (2,1) is assigned to the N quark and (3,2) to the λ quark, both having the same electric charge. But, most of all, the (3,2) loop has the "unknotting number" ±1, i.e. it would have to undergo a loopcrosscutting over itself if it were to be transformed into an unknotted loop. Such crosscutting being improbable, i.e. representing a weak process, it is plausible to associate (3,2) with the strange λ quark. With such an assignment, a strangeness changing process should eo ipso become a weak process. Associated production of loops of opposite strangeness is again topologically plausible as a short look at Fig. 4 indicates. One should, however, caution, that it is only a complete discussion of all of those kinds of processes, in detail, which will give us the means of knowing the definitive assignments of internal quantum numbers and of correct superpositions of loops. Level one of that discussion is indicated in this Fig. 4 which illustrates possible mergings of a quark with a corresponding antiquark, on the level of one q loopform and one $\bar{q}$ loopform, disregarding interlinkage of these with other loopforms. Level two considers a typical valence-quark interaction, and quark transfer from meson to baryon with consideration of the presence of the other (spectator) quarks (cf. Fig.5 and the discussion of strong interactions below). Level three of the discussion considers such processes, but in consideration of the detailed quark loop superpositions in such a process, superpositions which e.g. represent SU(6) states.

Second, the issue of quark confinement has a very different aspect from the customary one. Interlinkage of two or of three loops (i.e. of 2 or of 3 loopform manifolds) now is the main point of quark confinement, i.e. that the quarks hold together [39]. Again, as above, the Maxwell stress tension inherent in fluxlines may at least qualitatively make it plausible, in addition, that the entire structure remains to keep its size (corresponding to a core of extension $\hbar/mc$). The introduction of gluons or even color gluons is here altogether besides the point.

Third, the so described quarks are localized objects; therefore no conflict arises between the Pauli principle and a symmetric quark function for baryons or perhaps even some otherquark function. The possibilities of symmetries of wave functions, the partition functions and other functions, are entirely different for localizable objects, [40] in contrast with those for nonlocalizable objects such as are quark "particles" in a bag. Color is accordingly not needed to satisfy the Pauli principle.

MAGNETIC MOMENTS OF HADRONS [41]

A short comment about magnetic moments may be in order. As quarks are localizable objects, one may still have to decide about what functions, e.g. symmetric or other with respect to exchange of positions (inner, middle, outer), should represent the hadrons. The antisymmetric ones are incompatible with experimental data, the symmetric ones, as long as not much is known about magnetic moments of hyperons, are tolerably in agreement with experiment. In any case, the proportionality between electric charge and magnetic moment of quarks is purely hypothetical, and it is not the hypothesis which one should choose in the flux loop model.

Instead, it should be reasonable here to assume the magnetic moments to be proportional to the number of "coretraverses", i.e. the number of intersections of the totality of all two or three fluxloops of the hadron in question, with the core (Fig. 6). That number is then to be multiplied with the size of the core which is of the order $\hbar/m_{hadron}c$ --, actually, to be precise, one should multiply the number of coretraverses with $\frac{1}{2}\, e\hbar/2m_{hadron}c$, in analogy to the electrons $\frac{1}{2}e\hbar/2m_{electron}c$ which has two coretraverses. In l.c.1975 the magnetic moments had been calculated on the basis of symmetric state functions, but these numbers may change if other state functions are required, in particular when the (1,2), (1,3), (2,3) loops are included. The experimental knowledge of magnetic moments of hadrons is of great importance for a correct understanding of the quark model.

STRONG INTERACTIONS [42]

We should like to report (cf. Figs. 5) on a prototype of strong interactions, i.e. of a meson getting absorbed by a baryon without implying a weak process. We consider flux loop interactions when one coaxial pair of $\bar{q}_2 q_1$ approaches a coaxial triplet $q_2 q_3 q_4$ by sharing their central straight symmetry axes. We consider first the "merging" of the valence quarks $\bar{q}_2$ and q_2 of meson and baryon respectively, and thereafter the transfer of the second quark q_1 of the meson to the baryon. q_3 and q_4 are spectator quarks. To illustrate, we may picture the meson absorption

$$\bar{\lambda}\uparrow\ P\downarrow\ +\ \lambda\downarrow\ q_3\ q_4\ \rightarrow\ P\uparrow\ q_3\ q_4 \tag{20}$$

by considering only the single terms shown in this equation (not the full superpositions which represent the hadrons). The valence quarks $\bar{\lambda}$, λ occupy the outside toroidal regions, i.e. the regions adjacent to the central straight symmetry axes, of the meson torus and of the baryon torus respectively. The first set of pictures (a), (b), (e), show only the valence quarks; the other quarks are indicated in (a), (b), (c), (e), (f) by the toruses to which they are essentially confined. The arrows in Eq. (20) indicate spin, the arrows on the loops in the pictures give the direction of magnetic flux. The handedness of spinning and whirling are "sliding", in relation to the handedness of the respective quark loop (i.e. motion and structure have same handedness), for every one of these loops. At this moment, we do not yet want to make assertions about whether a λ valence quark is right handed or left handed; for the present discussion it suffices to use the relative handedness assignments shown in the pictures.

The upper torus may always represent the meson, the lower one the baryon. Effective spinning angular velocity (parallel to spin), and magnetic moment, of this $\bar{\lambda}$ are shown to be parallel, accordingly the electric charge is positive, as it should be for a $\bar{\lambda}$. To the baryon's λ valence quark we give opposite handedness of that of the meson's $\bar{\lambda}$.

The purpose of the sequence of pictures is to indicate how, without crosscutting, i.e. without involving a weak process, first of all, $\bar{\lambda}$ and λ merge to form three non-interlinked "link loops" (f) which chain the meson torus (i.e. the P↓) to the baryon torus (i.e. the q_3q_4). That proceeds as follows: In (a) the straight central axes of meson and of baryon are shown to become aligned. Meson and baryon then approach and touch each other along the two fictional circular lines indicated in (c). With the touch-down we imagine the two torus loops $\bar{\lambda}$ and λ to unroll themselves into two planar structures (d) lying directly side by side, as indicated by the two parallel flat layers in the center of picture (d); this picture shows two toroidal rubber tubes in the process of being flattened out after application of two circular cuts. While this unrolling occurs, as indicated by the sequence of three positions in picture (d), the two flux loops (which had been drawn on these two toroidal rubber tubes) will of course remain closed loop lines, i.e. they will bridge over the gaps of the cut-open rubber tubes. These six bridging sections loop over the meson and baryon toruses, as shown in picture (e). The flattened spiral sections of the $\bar{\lambda}$ and λ loops match exactly and have opposite magnetic flux, they may therefore merge, leaving only the three link loops. The result is shown in pictures (e), (f); the link loops have spin zero.

It is now readily seen that these three link loops may in their upper parts merge with the inner, remaining quark P↓ of the meson (cf.Pict.(g)). Thereupon, as again illustrated by toroidal rubber tubes (h), (i), this P↓ loop is simply transfered as a P↑ to the outside of the baryon torus which represents the spectator quarks q_3q_4. The effective spinning angular velocity (or spin) of P gets reversed in the process because the charge of P stays invariant whereas its magnetic moment (and handedness) gets reversed. (Angular momentum conservation might proceed through involvement of orbital angular momentum of the two initial hadrons, or through flip of the entire final hadron; we are not elaborating on that here). The final pictures (j) and (k) still show the meson "ghost", i.e. nothing is left of that former meson. - The entire set of processes relates to what topologists call cobordism, it represents a prototype of a strong process.

SOME REMARKS ABOUT STRING MODELS [43]

Whereas the present effort is directed at a representation of the topological character of electromagnetic fields in particle physics, centering on a parametrized description of those fields and on a discussion of conservation laws, the more abstract string models yield the interesting connections to dual amplitudes.

It seems to me of special importance that one may achieve that result by using only closed strings (cf. Arfei's preprint, from Mandelstam's group). It is therefore to be hoped that the string model, in which a string represents a valence quark, may get closely related to the present flux loop model. On the other hand, the choice of open strings implies the unphysical hypothetical massive monopoles at the ends of the strings, the boundary conditions of the strings having the purpose to accomodate the internal quantum numbers. - With the internal quantum numbers being, in the present model, mapped on the topological properties of quark loops, i.e. of fibrations representing the electromagnetic fields, the need for monopoles and their inherent problems disappears, and the closed electromagnetic strings (loops) may hopefully be formulated to yield the beautiful results obtained from the string models.

IMPLICATIONS OF THE TOPOLOGICAL QUARK MODEL ON THE GROUND- AND EXCITED STATES OF MESONS

We pointed out that the assignment of a manifold of forms of one loop, i.e. a Seifert-Threlfall torus loop fibration corresponds to a lepton or to a quark, depending on whether the motion has a handedness opposite to that of the structure (cooperative motion), or the same handedness as that of the structure (sliding motion); accordingly the leptons and the ordinary SU(3) quarks have integer and fractional electric charges respectively.

Mesons consist of two coaxial interlinked torusloop fibrations, each one filling out all of the ordinary three space but one, q_1, essentially confined to the inner region (close to the circular axis), the other, $\bar{q}_2$, essentially confined to the outer region, i.e. adjacent to the central straight axis of the torus, or, for antimesons, the roles of $\bar{q}_1$ and q_2 interchanged. For baryons, similarly, three interlinked loops are to be considered. When we write the state function of a baryon, $q_3q_1q_2+\ldots$ $+\ldots$, the first entry, q_3, may always refer to the outermost, the "valence quark", the second, q_1, to the middle, and the third, q_2, to the innermost quark; q_1 and q_2 being spectators as regards this term $q_3q_1q_2$. The above meson would be denoted $\bar{q}_2q_1$ + .. + ...

The interlinkage of hadronic quark loops, together

with the assumption that the "crosscutting" of loops is an improbable process, has several interesting consequences:

Quark loops being thus localized objects, are no longer in conflict with the Pauli principle (as quark particles are) if their state functions are symmetric (or perhaps even other functions) with respect to exchange of position. (One may be reminded of the distinction of partition functions of localized molecules in a crystal, and of nonlocalizable molecules in the gaseous state, this distinction being responsible for the sublimation energy). Color is therefore not needed to reconcile quarks with the Pauli principle in this topological model.

The interlinkage of quark loops drastically changes the problem of confinement. Gluons are not needed as tools to explain confinement, nor are bags part of the model.

Even the usefulnes of a W boson becomes questionable as the weakness of a process is considered as due to a change of topology: A loop, if it has to crosscut over itself or over another loop which is coaxial with it, or over a loop with which it is interacting, is considered to perform a weak process, and its rates may be estimated (cf. l.c 1977).

The two interlinked torusloop fibrations of a meson are, of course, never performing orbital motion about each other, such as quark particles might do. The question of orbital (L excited) states will have to be considered in a different context which we shall refer to at the end of this section.

It might be mentioned that the (2,1), (2,1), (3,2) loops are the simplest leptonic loops (refering to electron, muon, neutrino resp.), and the (2,1), (3,1), (3,2) the simplest quark loops (refering to N quark, P quark and λ quark resp.), if one requires, as we shall do, that the conditions of interlinkage even of the loopforms belonging to every single fibration be satisfied, and that, in addition, the electric charges of the resulting $\bar{q}q$ or qqq hadrons be integer.

That leads us to the question for the next complicated and therefore more massive loops which had been listed in the U.Md. report Spring 1974 (cf.l.c.1976). The loops (4,1) or (4,3) have been seen to be the next ones to come up in the list of quarks. The (4,1), interestingly enough, being of integer electric charge, cannot occur together with ordinary fractionally charged quarks in a $\bar{q}_2 q_1$ or a $q_2 q_1 q_3$ combination. As the (4,1) has zero unknotting number (as do the (2,1) N and the (3,1) P quarks), there is not strangeness (nor "charm") connected with it.

On the other hand, the (4,3) fibration is interesting as a SU(4) quark; it is impeded very much in any effort to "unknot"; three crosscuttings would be necessary to unknot - which makes such a process very weak. In other words it has strangeness to a higher degree (three) than the λ quark (one). An associated production of (4,+3) and (4,-3), like the ordinary (3,+2) and (3,-2) may be expected to be not subject to such delay. Because of that property, the (4,+3), (4,-3) are most likely candidates involved in a J/ψ particle. It is important to note that the electric charge of (4,3) quarks is ±e/3, not ± 2e/3. Not only because of this circumstance, it is different from charm but, more important still, the property of the introduction of "charm" on the basis of an analogy of

$$\begin{matrix} e & \nu_e \\ \mu & \nu_\mu \end{matrix} \quad \text{with} \quad \begin{matrix} N & P \\ \lambda & c \end{matrix}$$

is inconsistent with the topological model of quarks.

The excited states ψ',ψ'' etc. again should not be constructed on the basis of a quark-particle model in analogy to positronium. As pointed out above, the motion of $\bar{q}$ with respect to q is a coaxial motion of two fibrations, not an orbital motion. Besides, the coaxial fibrations $\bar{q}q$ or qqq may move like a top. "Orbital excitation" may arise when one $\bar{q}q$ orbits about another $\bar{q}q$, or several $\bar{q}q$ orbiting in a fashion resembling the balls ($\bar{q}q$) of a ball bearing. If the motion in that somewhat circular orbit should occur with a standard velocity, and if, what is very plausible, the distances between neighboring $\bar{q}q$ is standard, such a motion implies (as was pointed out to me by H.B.Nielsen) a possibility of understanding the linearity of Regge trajectories. The total mass is then proportional to the number of $\bar{q}q$, and the total orbital angular momentum is that mass times the radius, i.e. prop. to the number squared, i.e. prop. to the mass squared.

Figs. 1. Lefthanded torus loops (2,1), (3,1) and (3,2). Considering first the (2,1) loop: A two parametric (azimuth and size) manifold of similar coaxial

(2,1) loopforms (as shown for different azimuths in Figs. 2) fill the three-dimensional space in a continuous smooth manner and thus represent a Seifert-Threlfall fibration. These loopforms are interlinked with each other. - The same statements hold for the two other, the (3,1) and the (3,2) loops. Furthermore, the (3,2) is called "knotted" because a (3,2) loop can only be unknotted into a circle if the loop is cutting across itself. These (2,1), (3,1), (3,2) fibrations are the simplest ones with the aforementioned properties. Electrons, muons and neutrinos are characterized by (2,1), (2,1), (3,2) fibrations, N, P, λ quarks by (2,1), (3,1), (3,2) fibrations. The leptons and quarks differ drastically from each other: for leptons the handedness of the motion is opposite to that of the structure, for quarks handedness of motion and of structure are the same.

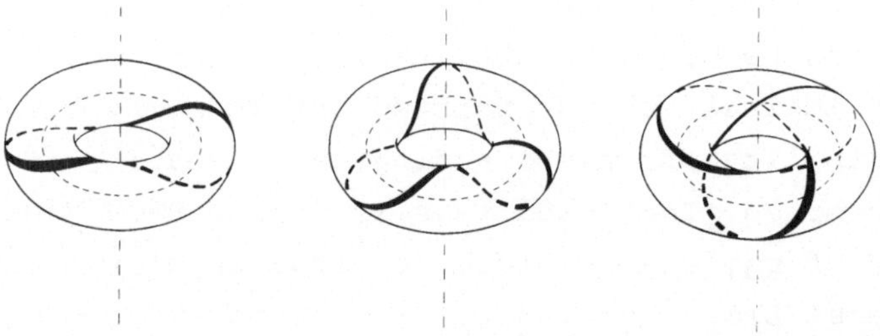

Fig. 2. A(2,1) loop. Its forms for different values of the azimuth parameter are shown by those torus loopforms. The interlinkage is readily recognized from the respective drawing of two loopforms drawn on the torus surface .

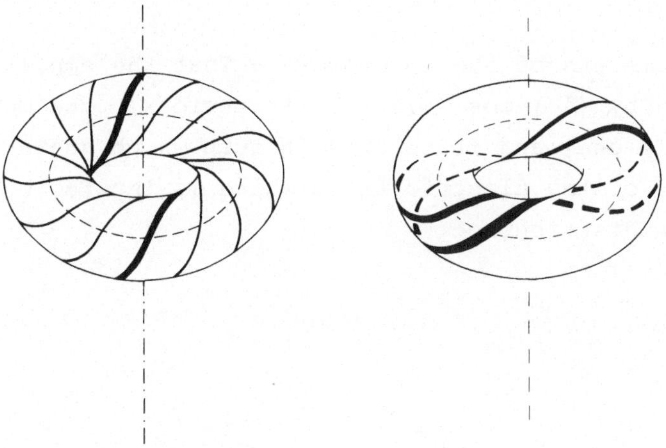

Figs. 3. A magnetic dipolefield (Fig.D) whose axis is named the +z axis) may be considered as resulting from a superposition of "sheaves" (Fig. B) whose axes $\vec{\zeta} = (\beta, \theta)$ are preferentially oriented towards the +z direction. Each sheaf should be considered as a two-parametric manifold of loopforms of different sizes σ and azimuths α. Fig. A shows the manifold of sizes σ of loopforms, indicated by three values of the size parameter σ. These loopforms, by rotation with the azimuth α, generate the sheaves Fig. B. The distribution of sheaves over orientations $\vec{\zeta}$ is indicated by showing a distribution of the generators Fig. A over the Euler angle θ. The strongest contributions in that superposition over $\vec{\zeta}$ are indicated by the thickest lines in Fig. C; the probability amplitudes are proportional to $(1 + \cos(\vec{\zeta},\vec{z}))^{1/2}$. Obviously, such a symmetric distribution of sheaves orientations about the $+\vec{z}$ axis results in the field Fig.D. - The distribution over a <u>manifold</u> of $\vec{\zeta}$ orientations is obviously needed when a probability (amplitudes) superposition of loopforms is to be constructed, because a rigidly oriented distribution $\vec{\zeta} = \vec{z}$ would contradict the concept of a statistical distribution with consideration of quantum mechanical zero point motion which affects

all parameters of the loopforms. - That the amplitude be given by $(1 + \cos(\vec{\zeta},\vec{z}))^{1/2}$ is obviously the most plausible one, and it is this in turn which guarantees isotropy of the electric field which is implied by the loopform distribution.

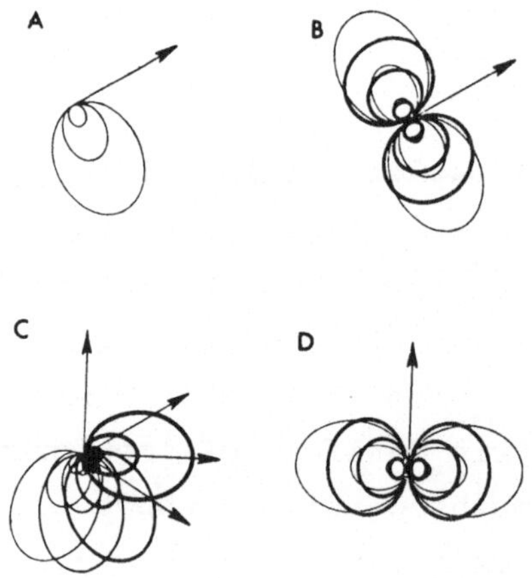

Fig. 4. A lefthanded trefoil and a righthanded trefoil loop with opposite flux orientation may be placed on top of each other (sharing their central straight axes). These two loops may "merge", their magnetic fields cancelling each other; that process may proceed without any of the loops crosscutting over themselves or over the partner loop, because the two loops have opposite unknotting numbers.

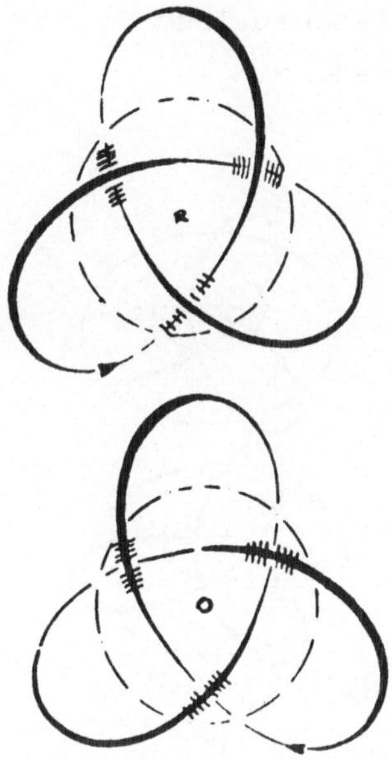

Fig. 6. Loop-contributions to a kaon. The (2,1) loop (N quark) is, in this picture, mainly confined to the inner region, and the other, (3,2) loop (λ quark) mainly to the outside region. Their spins and their magnetic moments are independent for these two quarks except for the coaxiality condition which means that a quantum state spin up for the N quark implies spin up <u>or</u> spin down for the λ quark.

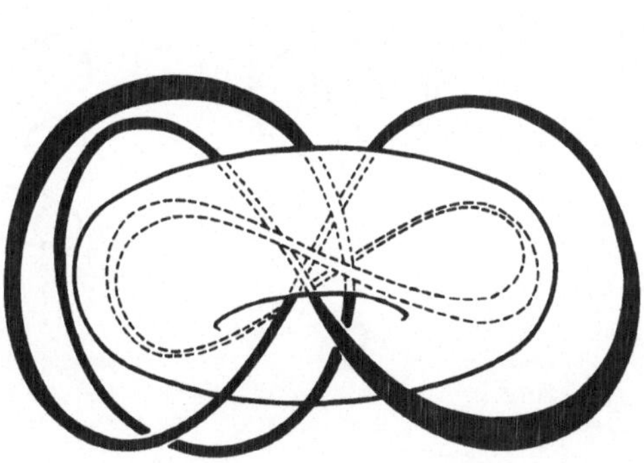

Fig. 5. Absorption of a kaon by a baryon. The sequence of steps is described in the section on "strong interactions".

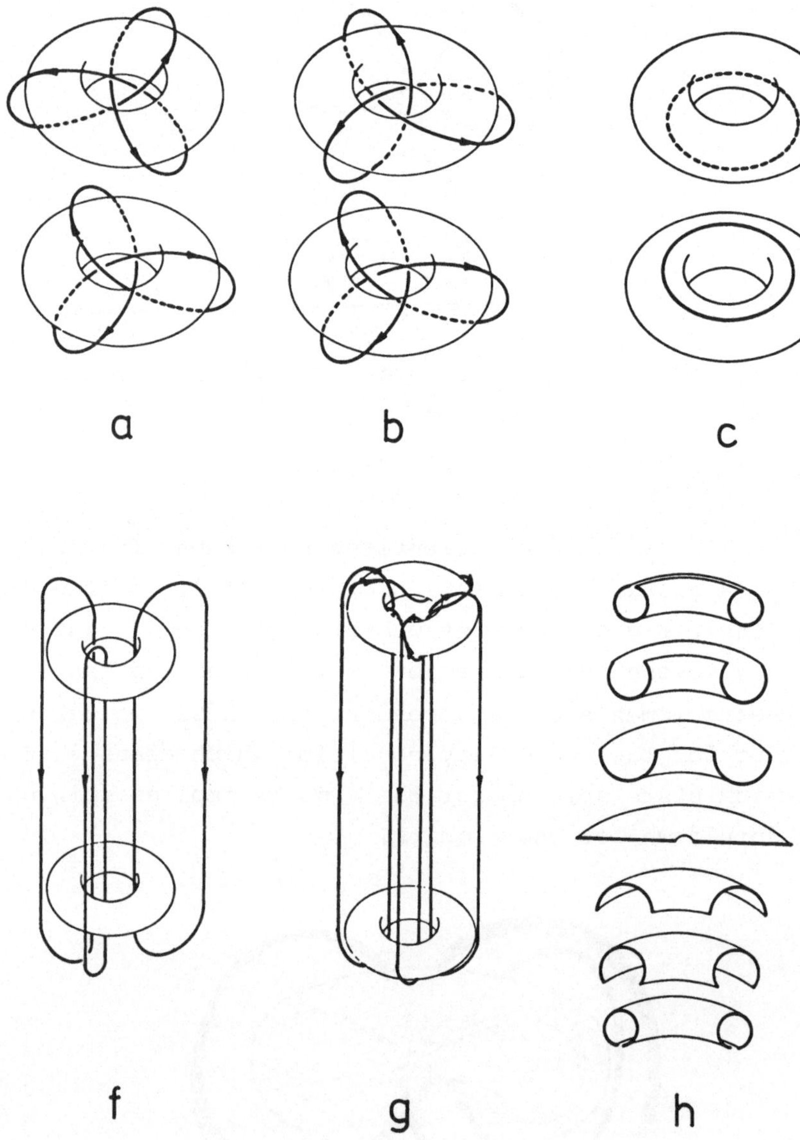

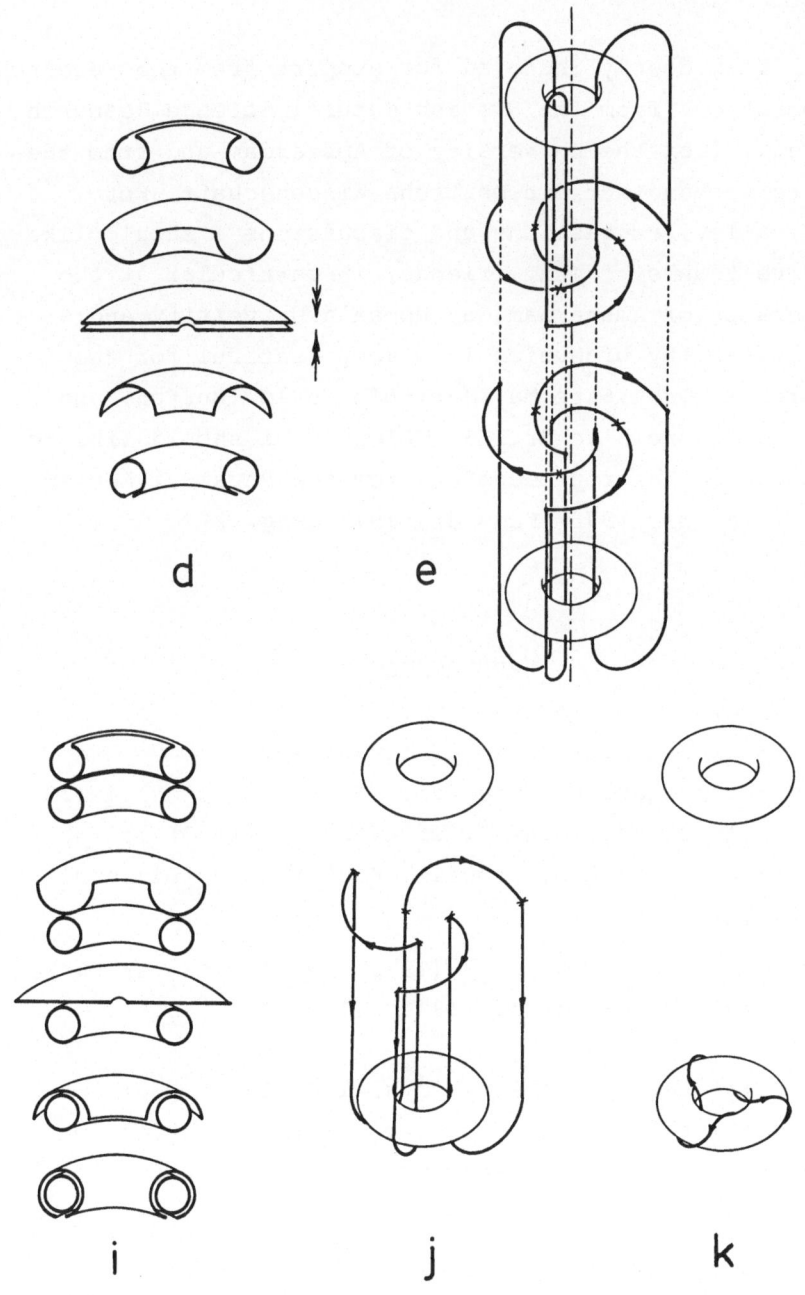

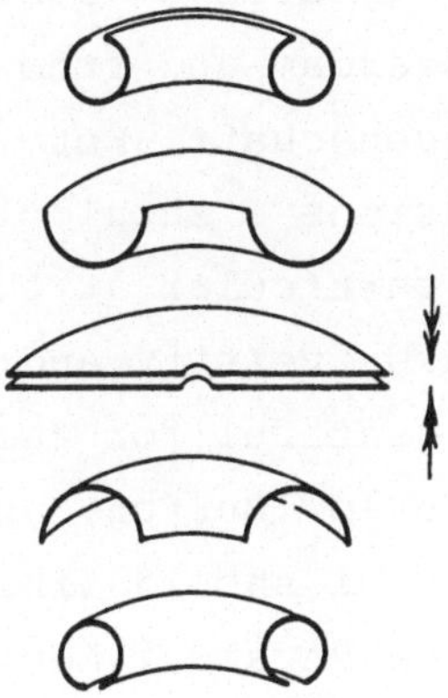

d

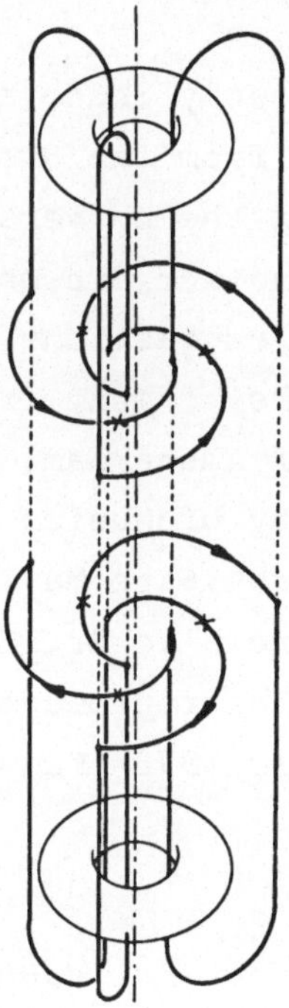

e

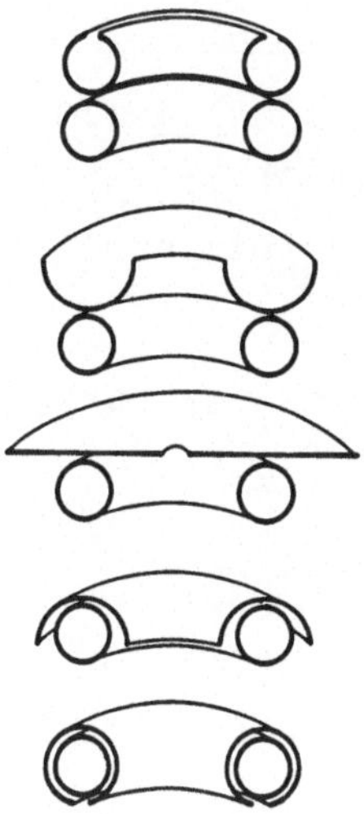

i

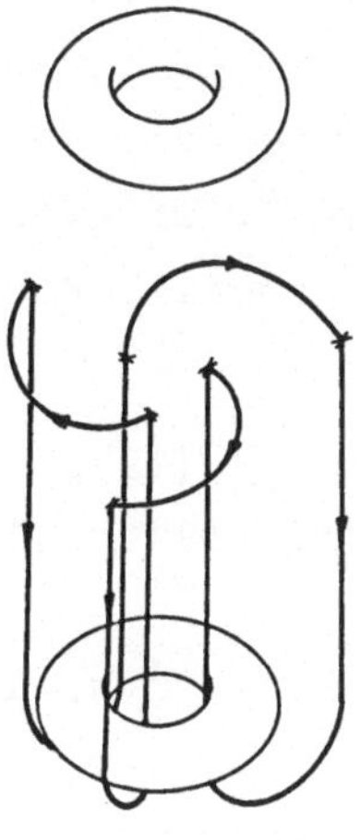

j

k

ACKNOWLEDGEMENTS

I am deeply indebted for support from the Research Corporation, from the Swedish Natural Science Research Council, from the University of Amsterdam and from the Stifterverband für die Deutsche Wissenschaft. For hospitality, suggestions and discussions I should like to give thanks to many friends, in particular at the University of Amsterdam, at Uppsala University and at the University of Munich I am very grateful for the figures 1,2,3,4,6 to Ms. Jeri-Lin Furlow Burton, and for the figure 5 to Mr. Jan Veldhoen. I should like to acknowledge the reproduction from the Physical Review 1971 (Fig. 3), 1975 (Fig. 4), 1977 (Fig. 2).

REFERENCES

1. H. Seifert, Acta Math. 60 (1933) 147;
   H. Seifert and W. Threlfall, Math. Annalen 104 (1931) 1; 107 (1932) 543; Can. J. of Math. 2 (1950) 1;
   R. H. Fox and R.H. Crowell, Knot Theory (Blaisdell-Ginn, New York 1963).

2. H. Jehle, Phys. Rev. D3 (1971) 306 (quoted as "A"); D6 (1972) 441 ("B"); D11 (1975) 2147 ("C"); D15 (1977) ("D").
   Boulder Lect. in Theor. Phys. 10B (1967) 673;
   Special reference is made to the introductions of "C" and of "D". A more extensive literature reference list is to be found in these papaers.

3. cf. l.c. "C".

4. cf. Fig. 6.

5. For comments to these and related basic issues, I should like to refer to
J. R. Klauder, Schladming Lectures 1977;
R. Haag, Schladming Lectures 1975;
H. Rollnik, Teilchenphysik II p. 139 (Bibliogr. Inst. Mannheim-Wien-Zürich 1971);
A.E. Ruark, pers. comm.;
G. Süssmann and N. Fiebiger, ed., Atome, Kerne, Elementarteilchen (Umschau, Frankfurt 1968).
A. O. Barut, Phys. Blätter 31 (1975) 352;

6. P.A.M. Dirac, Proc. Roy. Soc. A133 (1931) 60;
A.O. Barut and G.L. Bornzin, Nucl.Phys. B81 (1974) 477;
A.O. Barut, Int. Conf. on Math. Phys., Warzawa 1974;
A.O. Barut and R. Raczka, preprint 1974.

7. cf. l.c. "C".

8. cf. Fig. 6.

9. R.P. Feynman, Rev. Mod. Phys. 20 (1948) 367.

10. H. Weyl, Raum, Zeit, Materie, Springer 1921, Dover 1922;

11. L. Infeld and B.L. van der Waerden, Sitzb. Preuss. Akad. Wiss. p. 380 (1933);

12. O. Laporte and G.E. Uhlenbeck, Phys. Rev. 37, (1931) 1380 and 1552;

13. W.L. Bade and H. Jehle, Rev. Mod. Phys. 25 (1953) 714;

14. W.C. Parke and H. Jehle, Boulder Lect. Theor. Phys. 7a (1964) 297, p. 360.

15. W. Kofink, Ann. der Physik 38 (1940) 421, 436, 565, 583; 1 (1947) 133.

16. J.M. Blatt, Theory of Superconductivity, Ac. Pr. 1964.

17. A.O. Barut, preprint (Univ. of Munich and of Colorado, Internat. Center Theor. Phys., Trieste 1975/76).

18. O.W. Greenberg, Phys. Rev. Lett. 13 (1964) 598;
M. Gell-Mann, Schladming Lect. 1972 p. 733;
H. Rollnik, Phys. Blätter 32 (1976) 704; 33 (1977) 70.

19. F. London, Superfluids (Wiley - Dover, N.Y. 1960).

20. cf. l.c. 2 above.

21. cf. l.c. "D", cf. also
P. Roman, Theory of Elementary Particles (North Holland, Amsterdam (1964));
L.H. Thomas, In Quant. Theory of Atoms, Molecules and Solids, Slater Festschrift, ed. by P.O. Löwdin 93 (Ac. Press 1966);
V. Bargmann, Z. f. Physik 99 (1936); Rev. Mod. Phys. 34 (1962) 829;
S.S. Schweber, Relat. Quant. Field Theory (Row Peterson, N.Y. 1961);
V. Fock, Z. f. Phys. 98 (1935) 145;
R.J. Finkelstein, J. Math. Phys. 8 (1967) 443.
F. Bopp and R. Haag, Z. f. Naturf. 5a (1950) 644.
R. Hagedorn and I. Montway, Nucl. Phys. B59 (1973) 45;
R. Hagedorn, Lectures, CERN 71-12 (1971);
M. Chaichian, R. Hagedorn and M. Hayashi, CERN report 1975;
M. Bander, Ann. N.Y. Ar. Sci. 229 (1974) 78;
R. Hagedorn and J. Ranft, Nucl. Ph. B48 (1972) 157;
E. Fermi, Progr. Th. Phys. 5 (1950) 570.

22. In the present case space-time is flat, the fluxloops show torsion. In a complementary investigation, space has torsion:
R.J. Finkelstein, J. Math. Phys. 1 (1960) 440; 7 (1966) 1632; Ann. of Phys. 15 (1961) 223.

cf. also R.J. Finkelstein, Nonrelativistic Mechanics (Benjamin, Reading 1973);
R.J. Finkelstein and W. Ramsay, Ann. of Phys. 17 (1962) 379; 21 (1963) 408;
R. Penrose, J. Math. Phys. 8 (1967) 345;
Ch. W. Misner, pers. comm.;
Ch. W. Misner, K.S. Thorne, J.A. Wheeler, Gravity (Freeman, San Francisco 1973).

23. cf. l.c. "C".

24. cf. l.c. 1 above, and
W. Drechsler, Fortschr. d. Phys. 23 (1975) 607;
Max Planck Institute preprint MPI-PAE/P Th 8/76,
and Int. Symp. on Math. Phys., Mexico City (1976).

25. cf. l.c. "B".

26. E. Artin, Coll. Papers, ed. by S. Lang and J.T.Tate, Add. Wesley 1965.

27. M.H.L. Pryce, Proc. Roy. Soc. London A195 (1948) 62;
L.L. Foldy and S.A. Wouthuysen, Phys. Rev. 78 (1950) 29;
A.S. Wightman and S.S. Schweber, Phys.Rev. 98 (1955) 812;
W.C. Parke, thesis George Washington Univ., (Univ. Microfilms, Ann. Arbor Mich. 1967);
K.M. Case, Phys. Rev. 95 (1954) 1323;
H. Salecker, Z. f. Phys. 160 (1961) 385; 164 (1961) 463; Z. f. Natf. 10A (1955) 349;
M. Magg, K. Ringhofer and H. Salecker, Nucl. Ph. B40 (1972) 367;
J. Kraus, Nucl. Ph. B89 (1975) 133.
S. Tani, Progr. Theor. Phys. 6 (1951) 267;
K.M. Case, Phys. Rev. 95 (1954) 1323.

28. K. Huang, Am. J. of Phys. 20 (1952) 479.

29. J.C. Slater, Nature 117 (1926) 587, cf. footnote in Phys. Rev. D3 (1971) 331;
D. Finkelstein, J. Math. Phys. 7 (1966) 1218;
D. Finkelstein and J.Rubinstein, J. Math. Phys. 9 (1968) 1762;
R.J. Finkelstein, Phys. Rev. 75 (1949) 1079: cf. in particular.
H.C. Corben, Class. and Quant. Theories of Spinning Particles (Holden-Day, San Francisco 1968).

30. l.c. "C" App. B.

31. l.c. "C" or l.c. "A".

32. cf. also K.M. Case, Phys. Rev. 107 (1957) 307.

33. cf.l.c. "C".

34. cf. l.c. "C".

35. cf. l.c. "C".

36. cf. l.c. "A".

37. cf. Fig. 6.

38. cf. l.c. "D" App. D.

39. cf. Fig. 6.

40. cf. l.c. "B", "C", "D".

41. cf. l.c. "C".

42. The process described in this Fig. 5 is completely equivalent to the somwhat different pictures given elsewhere.

43. M. Rasetti and T. Regge, Physica 80A (1975) 217;
F. Lund and T. Regge, Inst. Adv. St. Preprint (1976);
H.B. Nielsen and P. Oleson, Nucl. Phys. B61 (1973) 45; B57 (1973) 367;

H. Arfaei, Nucl. Phys. B112 (1976) 256;
S. Mandelstam, Phys. Reports 13 (1974) 254;
J. Scherk, Rev. Mod. Phys. 47 (1975) 123;
S. Fubini and G. Veneziano, N. Cim. 67 (1970) 29;
A.O. Barut and G. Bornzin, Nucl. Ph. B81 (1974) 477;
Phys. Rev. D7 (1973) 3017; Phys. Blätter 31 (1975) 352.

Acta Physica Austriaca, Suppl. XVIII, 111–151 (1977)

CLASSICAL LIMIT OF QUANTUM ELECTRODYNAMICS[+]

by

I. BIALYNICKI-BIRULA
Inst. of Theoretical Physics
Warsaw, Poland
and
Department of Physics
Univ. of Pittsburgh, Pittsburgh, PA.

CONTENTS

[+] Lecture given at XVI. Internationale Universitätswochen für Kernphysik, Schladming, Austria, February 24-March 5, 1977.

1. INTRODUCTION

The relationship between quantum and classical electrodynamics is a complex subject, with many aspects, not all of which are at present well understood. Clearly, it can not be a simple problem, since we do not even know what is the true content of the two theories which we are supposed to compare. Quantum electrodynamics exists now as a perturbation theory, with strong indications that the perturbation series is at best only asymptotic. Classical electrodynamics is not in a much better shape, since to my knowledge, there is no systematic approximation scheme that would be guaranteed to give finite results in each step and which would have at the same time the properties of relativistic invariance, causality, positive definitness of energy and indubitable physical interpretation. A very good historical review of the fundamental problems of classical electrodynamics can be found in a recent article by Rohrlich [1].

These statements apply to quantum electrodynamics and classical electrodynamics understood as complete theories of the electromagnetic field and interacting charged particles. In various truncated versions of electrodynamics, the relationship between the quantum and the classical theory is easier to uncover. There are two extreme cases in which the relationship can be fully understood both physically and mathematically. One is the nonrelativistic mechanics of charged particles moving in an external electromagnetic field and the other is the Maxwell theory of the electromagnetic field interacting with external sources. I believe that these two greatly simplified models do help us to understand some aspects of the relationship between the fullfledged quantum and classical theories. It seems, however, that there is much more to that relationship that can be learned from these two simple models and it is quite possible that we are still far from understanding the very nature of this relationship.

One may question the need for this understanding arguing that after all it is QED, not its classical counterpart, which is the correct theory and that lack of a direct link between the quantum and the classical description only emphasizes this fact. Such a pragmatic approach is easily refuted, however, by observing that QED like any other quantum theory, in the words of Bohm [2], "presupposes a classical level and the correctness of classical concepts describing this level". Accepting this proposition, we should intensify rather than abandon the search for the classical level of QED, especially in view of all the successes of this theory.

It was the fusion of quanta and relativity that led to the emergence of QED and it is precisely the combination of quantum effects (controlled by the Planck constant $\hbar$) and relativistic effects (controlled by the velocity of light c) that makes the classical limit of QED so difficult to reach. The only available calculational tool - perturbation theory - means an expansion in powers of the dimensionless fine structure constant $\alpha = e^2/\hbar c$. In this perturbative framework the simple "$\hbar \to 0$" rule does not work since it implies $\alpha \to \infty$. Such statements as: "It is well known that if one makes $\hbar$ tend to zero in any formula of the quantum theory, one obtains the corresponding classical theory formula" [3] or "In the limit, when $\hbar \to 0$, the laws of quantum mechanics must reduce to those of classical mechanics" [4] are just not applicable to QED, because it is very unlikely that QED in its present form can be extended to arbitrarily large values of α.

Not all the problems with the classical limit of QED arise in connection with higher order terms in α. The electron spin is also a source of complications in the classical limit. Contrary to the widespread opinion, which originated with Pauli [5], the effects of spin do not necessarilly disappear in the classical limit. This problem was first discussed by de Broglie [6] and later was further clarified by Rubinov and Keller [7], who have shown that the spin may affect the trajectory in the classical limit if the motion of a particle is studied over large distances (of the order of $1/\hbar$). This is precisely what we are dealing with in all scattering problems.

In the absence of exact solutions of the relativistic scattering problems we can not prove that the spin influences the cross section in the classical limit, but there are strong indications that this indeed happens. For example, the ratio of the differential cross section $d\sigma(1/2)$ for a spin 1/2 particle to that of a spinless particle $d\sigma(0)$ in the Born approximation is

$$\frac{d\sigma(1/2)}{d\sigma(0)} = 1 - \left(\frac{v}{c} \sin\frac{\theta}{2}\right)^2 \quad . \tag{1}$$

The decrease in the differential cross section in the backward direction, predicted by this formula for spin 1/2 particles, is clearly seen in experiments on the scattering of relativistic electrons on nuclei. However, we can not draw a definite conclusion from here that the spin influences the scattering in the classical limit, because in wave mechanics the regions of validity of the Born approximation and the classical approximation in general do not overlap.

We will return to these problems after a systematic analysis of the classical limit in the nonrelativistic theory.

2. CLASSICAL LIMIT OF NONRELATIVISTIC QUANTUM MECHANICS

Historically, the first example of the classical limit of a quantum theory was, of course, the transition from the Planck law for the black body radiation to the classical Raleigh-Jeans formula. In the words of Planck himself [8] "for a vanishingly small value of the quantum of action, $\hbar$, the general formula

$$E_\lambda = \frac{c^2\hbar}{\lambda^5}\frac{1}{\exp(\frac{c\hbar}{k\lambda T})-1} \tag{2}$$

degenerates into Raleigh's formula"

$$E_\lambda = \frac{ckT}{\lambda^4}\,. \tag{3}$$

This first example exhibits very clearly one important feature of the classical limit. Namely, we should be comparing the same physical quantity (the spectral energy density) calculated according to two different theories. It is only in this restricted sense that we will use the term classical limit, always having in mind a connection between the same concept described by the two theories.

The second example of the classical limit is the correspondence principle of Bohr, which played a crucial role in the discovery of quantum mechanics. According to this principle [9] "in the region of high quantum numbers the intensities of spectral lines become asymptotically equal to the intensity of radiation computed classically from the corresponding Fourier amplitude". Again we have the same situation as before that the same physical characteristic (light intensity) derived from the two theories are compared.

After this historical introduction let us turn to the study of the classical limit of quantum mechanics.

Several approaches have been developed to investigate this limit, which differ in the choice of the physical characteristics that are to be compared between the quantum and the classical theory. One natural possibility is to choose as those characteristics the trajectories of particles. In view of the statistical nature of quantum mechanics, we must compare average trajectories evaluated in this theory with classical trajectories. This problem was investigated for the first time by Ehrenfest [10] in the early days of quantum mechanics. Starting from the Schrödinger equation he derived what is now called the Ehrenfest theorem. We will discuss the Ehrenfest theorem here with the help of the hydrodynamic formulation, which by itself represents a convenient tool to explain some aspects of the classical limit.

2.1. Hydrodynamic formulation

In the hydrodynamic formulation, which was discovered by Madelung [11], we replace the complex wave function by the set of four real hydrodynamic-like variables: the probability density ρ and the velocity $\vec{v}$ defined through the probability current, $\vec{j} = \rho\vec{v}$. For a charged particle moving in an external EM field we obtain the following relation between the real amplitude R and the phase $S/\hbar$ of the wave function and the hydrodynamic variables

$$\rho = R^2 , \tag{4}$$

$$\vec{v} = m^{-1} (\nabla S - \frac{e}{c} \vec{A}) . \tag{5}$$

The Schrödinger equation rewritten in terms of ρ and $\vec{v}$ reads

$$\partial_t \rho = -\nabla (\rho \vec{v}) , \tag{6a}$$

$$m\partial_t (\rho v_k) = e\rho (\vec{E} + \frac{\vec{v}}{c} \times \vec{B})_k - \nabla_i T_{ik} , \tag{6b}$$

where the stress tensor T_{ik} is

$$T_{ik} = - \rho \frac{\hbar^2}{4m} \nabla_i \nabla_k \ln\rho + m\rho v_i v_k . \tag{7}$$

There are more solutions to Eqs. (6) than to the Schrödinger equation, since we have replaced two real functions by four. Only those solutions to Eqs. (6) correspond to solutions of the original Schrödinger equation which satisfy the following quantization condition

$$\int_{\Sigma} d\vec{\sigma}\,(m\nabla \times \vec{v} + \frac{e}{c}\vec{B}) = 2\pi n\hbar \qquad (8)$$

for every surface Σ. If this does not hold we can not construct ψ from ρ and $\vec{v}$. It is sufficient to impose the condition (8) on initial values of $\vec{v}$ to have it satisfied at all times [12].

The hydrodynamic variables satisfying Eqs. (6), unlike the wave functions obeying the Schrödinger equation, can be directly compared with the corresponding statistical classical concepts. It is worthwhile noticing what becomes of the quantization condition (8) in the classical limit. The correct way of taking this limit comes with the recognition that Eq. (8) is equivalent to one of the Bohr-Sommerfeld quantization conditions of the old quantum theory. In the limit, when $\hbar \to 0$, the r.h.s. of (8) can be made equal to anything, because the separation between the allowed values gets smaller with the decrease of $\hbar$. In the classical limit there is no restriction on the velocity $\vec{v}$.

The Ehrenfest theorem can be obtained from Eqs.(6) by first multiplying Eq. (6a) by $\vec{r}$ and integrating it over the whole space and next integrating Eq. (6b). The results can be written in the form

$$\frac{d}{dt}\langle\vec{r}\rangle = \langle\vec{v}\rangle , \qquad (9a)$$

$$m\,\frac{d}{dt}\langle\vec{v}\rangle = e\langle(\vec{E} + \frac{\vec{v}}{c} \times \vec{B})\rangle , \qquad (9b)$$

where the brackets denote the standard quantum-mechanical averages (i.e. integrals with ρ over the whole space).

There is no direct contribution to the Ehrenfest equation (9b) from the stress tensor because only the divergence of this tensor enters Eq. (6b). This is why the Ehrenfest equations have a purely classical look; the only place where the Planck constant appears in the hydrodynamic equation is in the first term of this tensor. We may note here in passing that the Ehrenfest equations are remarkably stable; they do not change under nonlinear modifications of the Schrödinger equation [13].

The Ehrenfest equations do not have the same content as the classical equations of motion. The average position $\langle\vec{r}\rangle$ and the average velocity $\langle\vec{v}\rangle$ vary with time in a different fashion than their classical counterparts because the average force is, in general, not equal to the force evaluated at the average position and average velocity

$$\langle(\vec{E} + \frac{\vec{v}}{c} \times \vec{B})\rangle \neq \vec{E}(\langle\vec{r}\rangle) + \frac{\langle\vec{v}\rangle}{c} \times \vec{B}(\langle\vec{r}\rangle) \ .$$

The above inequality becomes less and less pronounced when the probability density shrinks to the Dirac δ function, but this can not happen uniformly for all times due to the spreading of wave packets. This spreading is caused by the so called quantum pressure term i.e. the $\hbar$-dependent term in the stress tensor. In the classical limit, when $\hbar \to 0$, the spreading disappears and the hydrodynamic equations (6) assume the form of the equations of motion for a statistical ensemble of classical charged particles (charged dust) moving in an external EM field, each particle following its classical trajectory.

2.2. Weyl - Wigner - Moyal transforms

One can go one step further in imitating classical statistical mechanics and introduce in quantum mechanics the counterpart of the distribution function $f(\vec{r},\vec{v},t)$. This method goes back to Weyl [14] and Wigner [15] and was fully developed by Moyal [16].

The classical distribution function $f(\vec{r},\vec{v},t)$ generalizes the probability density $\rho(\vec{r},t)$. It describes the statistical distribution of both the position and the velocity. In quantum mechanics the uncertainty principle precludes the possibility of having authentic distribution function of $\vec{r}$ and $\vec{v}$, but the Weyl - Wigner - Moyal transform (WWM transform) is as close an approximation to the distribution function as we can get under the circumstances.

In the simplest case of a system with only one degree of freedom, the WWM transform $\mathcal{O}(q,p)$ of a operator $\mathcal{O}$ is defined by the formula

$$\mathcal{O}(q',p') = \int d\Gamma(q,p)\ \mathrm{Tr}\{W(q,p)\,\mathcal{O}\}\, e^{-\frac{i}{\hbar}(qp'-pq')} , \qquad (10)$$

where

$$W(q,p) = e^{\frac{i}{\hbar}(qP-pQ)} , \qquad (11)$$

$d\Gamma(q,p) = (2\pi\hbar)^{-1}\, dq\, dp$, and Q and P are the position and momentum operators. The definition (10) may be applied to dynamical operators and to the density operators.

Calculations involving the family of unitary

operators W(q,p) are often made simpler with the help of the following orthogonality and completeness relations

$$\mathrm{Tr}\{W(q,p)W^{\dagger}(q',p')\} = 2\pi\hbar\delta(q-q')\delta(p-p') , \tag{12a}$$

$$\int d\Gamma(q,p)(\psi_1|W^{\dagger}(q,p)\psi_2)(\psi_3|W(q,p)\psi_4)=(\psi_1|\psi_4)(\psi_3|\psi_2) . \tag{12b}$$

In proving these relations one repetitiously uses the Baker-Hausdorff formula

$$W(q,p) = e^{\frac{i}{\hbar}qP - \frac{i}{\hbar}pQ}\, e^{\frac{i\hbar}{2}qp} , \tag{13}$$

or its equivalents and the completeness relations for the (generalized) eigenfunctions of q and p, for which we adopt the following normalization conventions

$$\int dq|q\rangle\langle q| = 1 = \int\frac{dp}{2\pi\hbar}|p\rangle\langle p| , \tag{14a}$$

$$\langle q|p\rangle = e^{\frac{i}{\hbar}qp} . \tag{14b}$$

With the help of Eq.(12b) we can invert the formula (10) and obtain a representation for $\mathcal{O}$ in terms of its WWM transform

$$\mathcal{O} = \int d\Gamma(q,p)\int d\Gamma(q',p')W(q,p)e^{-\frac{i}{\hbar}(qp'-pq')}\mathcal{O}(q',p') . \tag{15}$$

The WWM transforms of dynamical operators (also known as

Weyl representations of these operators) coincide with the expressions for the corresponding classical dynamical variables, provided we always fully symmetrize products of Q and P when we construct quantum operators.

The WWM transforms of density matrices (known as Wigner functions) are the counterparts of classical distribution functions. In order to justify this claim we will express in terms of WWM transforms the average value $\langle A \rangle$ of the dynamical operator A evaluated in the state described by the density operator ρ,

$$\langle A \rangle = \mathrm{Tr}\,\{A\rho\} = \int d\Gamma(q,p)A(q,p)\rho(q,p) \quad . \tag{16}$$

This formula is obtained from (15) and (12a). It has the same form as in the classical theory. Despite the general validity of (16), the function $\rho(q,p)$ is not a genuine distribution function: even though it is real and normalized it is nonnegative only for a very restricted class of states.

To study the motion of charged particles in an EM field, we will introduce a gauge invariant version of the WWM transform [17] which employs the kinetic momentum $\vec{K}$ rather than the canonical momentum $\vec{P}$,

$$\vec{K} = m\vec{v} = \vec{P} - \frac{e}{c}\vec{A} \quad . \tag{17}$$

In this case

$$\rho(\vec{r}\,',\vec{k}\,')=\int d\Gamma(\vec{r},\vec{k})\,\mathrm{Tr}\{W(\vec{r},\vec{k})\rho\}e^{-\frac{i}{\hbar}(\vec{r}.\vec{k}\,'-\vec{k}\cdot\vec{r}\,')} \tag{18}$$

and

$$W(\vec{r},\vec{k}) = e^{\frac{i}{\hbar}(\vec{r}\cdot\vec{K}-\vec{k}\cdot\vec{R})} \quad . \tag{19}$$

In particular, for a pure state we obtain

$$\rho(\vec{r},\vec{k}) = \int d^3\ell\,\psi(\vec{r}+\tfrac{\vec{\ell}}{2})\psi^*(\vec{r}-\tfrac{\vec{\ell}}{2})\exp\{-\tfrac{i}{\hbar}\vec{\ell}\cdot[\vec{k}+\tfrac{e}{c}\int_0^1 d\alpha\,\vec{A}(\vec{r}+\tfrac{\vec{\ell}}{2}-\alpha\vec{\ell})]\} . \tag{20}$$

In the derivation of this formula the following generalization of the Baker-Hausdorff formula (13) was used

$$W(\vec{r},\vec{k}) = e^{\frac{i}{\hbar}\vec{r}\cdot\vec{P}}\, e^{\frac{i}{2\hbar}\vec{r}\cdot\vec{k}} \exp\{-\tfrac{i}{\hbar}[\vec{k}\cdot\vec{R}+\tfrac{e}{c}\int_0^1 d\alpha\;\vec{r}\cdot\vec{A}(\vec{R}-\alpha\vec{r})]\} \quad . \tag{21}$$

The line integral in Eq.(20) guarantees the invariance of $\rho(\vec{r},\vec{k})$ under the simultaneous gauge transformations of the wave function and of the vector potential.

The classical limit can not be directly taken on the formula (20), because of the presence of the wave functions in it. However, as we shall show now, the equation of motion for $\rho(\vec{r},\vec{k},t)$ does have a simple limit, when $\hbar \to 0$. To this end we calculate the time derivative of $\rho(\vec{r},\vec{k},t)$, assuming that the time evolution of the density operator is determined by the von Neumann-Liouville equation,

$$\partial_t \,\mathrm{Tr}\{W\rho(t)\} = (i\hbar)^{-1}\,\mathrm{Tr}\{(WH - HW)\rho(t)\} \quad . \tag{22}$$

For simplicity, we have assumed that the EM field is

time independent. Otherwise we would have to include an additional term in Eq. (22): the time derivative of W.

The calculation of the r.h.s. of (22) is made simple with the help of the following identities:

$$(-\frac{\hbar}{i}\nabla_k + \frac{1}{2}\vec{r})\, W = W\vec{R}\ , \tag{23a}$$

$$(-\frac{\hbar}{i}\nabla_k - \frac{1}{2}\vec{r})\, W = \vec{R}W\ , \tag{23b}$$

$$[\frac{\hbar}{i}\nabla_r + \frac{1}{2}\vec{k} + \frac{e}{c}\int_0^1 d\alpha\,\alpha\,\vec{r}\times\vec{B}(-\frac{\hbar}{i}\nabla_k + \frac{1}{2}\vec{r} - \alpha\vec{r})]W = W\vec{K}\ , \tag{24a}$$

$$[\frac{\hbar}{i}\nabla_r - \frac{1}{2}\vec{k} - \frac{e}{c}\int_0^1 d\alpha\,(1-\alpha)\,\vec{r}\times\vec{B}(-\frac{\hbar}{i}\nabla_k + \frac{1}{2}\vec{r} - \alpha\vec{r})]W = \vec{K}W\ , \tag{24b}$$

which can be derived with the help of (21). We also notice that under the integral in formula (18) we can make the replacements

$$-\frac{\hbar}{i}\nabla_k \rightarrow \vec{r}\,'\ , \tag{25a}$$

$$\frac{\hbar}{i}\nabla_r \rightarrow \vec{k}\,'\ , \tag{25b}$$

$$\vec{r} \rightarrow -\frac{\hbar}{i}\nabla_{k'}\ , \tag{26a}$$

$$\vec{k} \rightarrow \frac{\hbar}{i}\nabla_{r'}\ . \tag{26b}$$

From Eqs. (18) and (22-26) we obtain in the limit, when $\hbar \rightarrow 0$,

$$[\partial_t + \vec{v}\cdot\nabla_r + e(\vec{E}(\vec{r}) + \frac{\vec{v}}{c}\times\vec{B}(\vec{r}))\cdot\nabla_v]\rho(\vec{r},\vec{v},t) = 0 , \qquad (27)$$

where we have used the velocity instead of kinetic momentum $\vec{k}$. All quantum corrections can be obtained by expanding relations (23-24) in which the substitutions (25-26) are made, into powers of $\hbar$. These corrections are small for slowly varying fields and they disappear completely for homogeneous fields.

Eq. (22) determines the time evolution of the states of charged particles. Similar equations can be written for all dynamical variables:

$$\partial_t \, \mathrm{Tr}\,\{WA(t)\} = -(i\hbar)^{-1}\, \mathrm{Tr}\,\{(WH - HW)A(t)\} . \qquad (28)$$

Since this differs from Eq. (22) only in the sign of the time derivative, we can immediately write down from (27) the classical limit of Eq. (28),

$$\partial_t A_{CL}(\vec{r},\vec{v},t) = [\vec{v}\cdot\nabla_r + \vec{F}(\vec{r})\cdot\nabla_v]A_{CL}(\vec{r},\vec{v},t) , \qquad (29)$$

where $\vec{F}(\vec{r})$ is the Lorentz force. The differential operator in the square bracket is the Liouville operator i.e. the Poisson bracket with the Hamiltonian:

$$\partial_t A_{CL} = \{A_{CL}, H\} . \qquad (30)$$

Since the Hamiltonian is time independent and the Poisson bracket is invariant under canonical transformation, in particular under the time translation, we can write

Eq. (29) in the form

$$\partial_t A_{CL}(\vec{r},\vec{v},t) = [\vec{v}(t)\cdot\nabla_{r(t)} + \vec{F}(\vec{r}(t))\cdot\nabla_{v(t)}]\, A_{CL}(\vec{r},\vec{v},t)\,, \quad (31)$$

provided $\vec{r}(t)$ and $\vec{v}(t)$ obey the equations of classical dynamics. The solution of this equation, therefore, can be written in the form

$$A_{CL}(\vec{r},\vec{v},t) = A_{CL}(\vec{r}(t),\ \vec{v}(t),t = 0)\ . \quad (32)$$

In order to obtain the quantum analogy of the particle trajectory we must specify $A(\vec{r},\vec{v},t)$ assuming it to be the WWM transform of the position operator $\vec{R}(t)$,

$$\vec{R}(\vec{r}\,',\vec{v}\,',t) = \int d\Gamma(\vec{r},\vec{k})\,\mathrm{Tr}\{W(\vec{r},\vec{k})\vec{R}(t)\}\, e^{-\frac{i}{\hbar}(m\vec{r}\cdot\vec{v}\,' - \vec{k}\cdot\vec{r}\,')}\ . \quad (33)$$

According to Eq. (31), this function together with its time derivative (velocity) obeys in the classical limit the classical equations of motion.

One can generalize the description in terms of WWM transforms to the case of many particles moving in an arbitrary (also time dependent) external field or even interacting mutually, provided the interaction is described by a potential term in the Hamiltonian. Thus, in the nonrelativistic quantum mechanics we have a complete understanding of the classical limit. Quantum mechanical counterparts of classical dynamical variables exist and they tend to their classical limits, when $\hbar \to 0$.

Difficulties are encountered, when we try to generalize this approach to a relativistic theory and to include field quantization.

3. CLASSICAL LIMIT OF RELATIVISTIC QUANTUM MECHANICS

Relativistic quantum mechanics of a single particle moving in an external field is not a selfconsistent theory. Creation and annihilation of particles caused by external fields force us to abandon the quantum mechanical description, whose main characteristic is the fixed number of particles. To deal with this new situation we must introduce a collection of wave functions, each describing one component of the state vector and this is usually done in the field-theoretic framework. It is only in the special case of static fields that we may consider one particle at a time, since in most static fields creation and annihilation of particles do not take place. At first one may think that these processes can never occur in static fields because of the energy conservation. However, as was shown by Schwinger [18], even in a very simple case of a uniform electric field one finds continuous pair creation from the vacuum.

We will begin our discussion with the simpler case of spinless particles. Assuming that the creation and annihilation of particles do not occur we can consider only those solutions of the Klein-Gordon equation

$$[(i\hbar\partial_\mu - \frac{e}{c}A_\mu)(i\hbar\partial^\mu - \frac{e}{c}A^\mu) - m^2c^2]\Phi(x) = 0 \ , \tag{34}$$

which describe particles. The time dependence of such solutions contains only positive frequencies ($e^{-i\omega t}$, $\omega > 0$). It turns out that we can easily generalize the method of WWM transforms to the relativistic situation by taking the "square root" of the Klein-Gordon equation

$$i\hbar\partial_t \Phi(x) = \pm\, [c\sqrt{m^2c^2 + (\frac{\hbar}{i}\nabla - \frac{e}{c}\vec{A})^2} + e\phi]\Phi(x) \ , \qquad (35)$$

and hence by considering only one particle or one antiparticle at a time. Following the same procedure as that described previously for static fields we can obtain in the classical limit the relativistic analog of Eq.(27). What is lacking in the relativistic case even for the static fields as compared to the nonrelativistic case is the clear physical interpretation of basic dynamical variables. As was shown by Newton and Wigner [19], the position operator must be modified in the relativistic theory and it ceases to be identical with the argument $\vec{r}$ of the external field. In other words, the center of mass can not be identified with the center of charge. That is why the WWM does not have an unambiguous physical interpretation, even though it can be formally applied to the theory based on Eq. (35). We can shed some light on these problems by considering again the hydrodynamic formulation and deriving from it the relativistic analog of the Ehrenfest theorem.

The hydrodynamic formulation of the relativistic theory based on the Klein-Gordon equation is obtained in the same manner as in the nonrelativistic case, except that we introduce the velocity in the form of a four-vector v^{μ} and scale the density ρ by a factor of 2. To obtain the hydrodynamic formulation there is no need to exclude time dependent EM fields.

The equations of motion for the hydrodynamic variables read

$$\partial_{\mu}(\rho v^{\mu}) = 0 \ , \tag{36a}$$

$$\partial_{\mu} T^{\mu\nu} = \frac{e}{c} \rho f^{\nu\lambda} v_{\lambda} \ , \tag{36b}$$

where

$$\rho = 2R^2 \ , \tag{37a}$$

$$v_{\mu} = - m^{-1} (\partial_{\mu} S + \frac{e}{c} A_{\mu}) \ , \tag{37b}$$

and

$$T^{\mu\nu} = - \rho \frac{\hbar^2}{4m} \partial^{\mu} \partial^{\nu} \ell n\rho + m\rho \, v^{\mu} v^{\nu} \ . \tag{37c}$$

Eqs. (36) can also be written as the equations for the propagation of ρ and v^{μ} along the stream lines,

$$v^{\mu} \partial_{\mu} \rho = -\rho \partial_{\mu} v^{\mu} \ , \tag{38a}$$

$$v^{\mu} \partial_{\mu} v_{\lambda} = \frac{e}{mc} f_{\lambda\rho} v^{\rho} + \frac{\hbar^2}{2m^2} \partial_{\lambda} (\rho^{-1/2} \Box \rho^{1/2}) \ . \tag{38b}$$

The quantization condition for the velocity takes on now also a relativistic form [20]:

$$\int_{\Sigma} d\sigma^{\mu\nu}[m(\partial_\mu v_\nu - \partial_\nu v_\mu) + \frac{e}{c}f_{\mu\nu}] = 2\pi\hbar n . \tag{39}$$

In the classical limit the equations of motion (36) or (38) become identical with those for a charged dust moving in an external field and the quantization con-condition disappears.

From Eqs. (36) we can obtain by integration the relativistic analog of the Ehrenfest theorem:

$$\partial_t \int d^3r\, \vec{r}\, \rho\, v^o/c = \int d^3r\, \rho\vec{v} , \tag{40a}$$

$$\partial_t\, m\int d^3r\, \rho\vec{v}\, v^o/c = \int d^3r\, e\rho(\vec{E} + \frac{1}{c}\vec{v}\times\vec{B}) . \tag{40b}$$

As in the nonrelativistic case, the Planck constant does not appear explicitely in these equations. However, the appearance of factors v^o/c on the l.h.s. is a clear indication of the fact that the center of charge does not coincide with the center of mass.

Things become more complicated for the Dirac equation. The difference between the center of charge and the center of mass is even more pronounced in this case owing to the presence of Zitterbewegung.

It is true that we can rewritte the Dirac equation

$$[\gamma^\mu(i\hbar\partial_\mu - \frac{e}{c}A_\mu) + mc]\psi = 0 \tag{41}$$

in the "squared" form

$$[(i\hbar\partial_\mu - \frac{e}{c}A_\mu)(i\hbar\partial^\mu - \frac{e}{c}A^\mu) - m^2c^2 - \frac{e\hbar}{2c}\sigma^{\mu\nu}f_{\mu\nu}]\psi = 0 \tag{42}$$

and apply to this equation the methods developed for the Klein-Gordon equation. In particular, we may decompose into the phase factor $\exp(iS/\hbar)$ and the remaining spinor field u and obtain in the limit, when $\hbar \to 0$, the classical equations:

$$\partial_\mu (\bar{u}\, u\, v^\mu) = 0 \ , \tag{43a}$$

$$v^\mu \partial_\mu v_\lambda = \frac{e}{mc} f_{\lambda\rho} v^\rho \ . \tag{43b}$$

However, since the classical limit resulting from the decomposition

$$\psi = e^{(i/\hbar)S} u \tag{44}$$

is essentially the same as obtained by Pauli, the objections raised by de Broglie [6,7] are valid here.

The problem is basically that of uniqueness of the classical limit. We may obtain different limits, making different assumptions about the asymptotic behaviour, when $\hbar \to 0$, of various physical variables. For example, if we hold the size of the spatial domain, to which the particle is confined, fixed while taking the limit, then we obtain the result of Pauli. If this size increases as $1/\hbar$, then small corrections to particle trajectories of the order of $\hbar$, which are due to the effects of the magnetic moment, get multiplied by $1/\hbar$ and contribute in the classical limit.

I believe that the pair creation processes may also contribute in the classical limit if we only knew how to calculate contributions from pair creation summed over all orders of perturbation theory with respect to the external field.

In the absence of the exact solutions of the Klein-Gordon and Dirac equations that would exhibit pair creation in time-varying external field, we must leave this problem unresolved.

4. CLASSICAL LIMIT OF THE QUANTUM THEORY OF THE EM FIELD

The principal difference between the classical limit of the quantum theory of the EM field and of the quantum mechanics of particles was explained very clearly by Sakurai and we will begin by quoting from his book [21].

"The classical limit of the quantum theory of radiation is achieved when the number of photons becomes so large that the occupation number may as well be regarded as a continuous variable. The space-time development of the classical electromagnetic wave approximates the dynamical behavior of trillions of photons. In contrast, the classical limit of Schrödinger's wave mechanics is the mechanics of a single mass point obeying Newton's equation of motion. Thus it was no coincidence that in the very beginning only the wave nature of light and the particle nature of the electron were appearent".

This description of the classical limit for the EM field is also indirectly contained in Bohr's correspondence principle. After all, photons are the excitations of the EM field and hence the number of photons can be viewed as a quantum number, characterizing the state of a certain oscillator of the EM field. Classical limit means large quantum numbers which in turn imply high intensity of the field.

We may give a more formal argument confirming this view by noticing that $\sqrt{\hbar}$ multiplies all creation and annihilation operators in the expansion of the vector potential of the EM field into a Fourier series,

$$\vec{A}(\vec{r}) = \sum_{\vec{k},\lambda} \sqrt{\frac{\hbar c^2}{2V\omega}}\, \vec{\varepsilon}(\vec{k},\lambda)\, a(\vec{k},\lambda)\, e^{i\vec{k}\cdot\vec{r}} + \text{h.c.} \quad . \tag{45}$$

In the classical limit the average values of the EM field must approach their classical values and this can happen only if matrix elements of annihilation and creation operators go to infinity as $1/\sqrt{\hbar}$ so that the average photon numbers go to infinity as $1/\hbar$.

To describe the transition to the classical limit for the EM field systematically let us consider an arbitrary system of charged particles, described by nonrelativistic quantum mechanics, interacting with the quantized EM field.

The Hamiltonian operator of the system can be split into three parts

$$H = H_F + H_A + H_I \ , \tag{46}$$

where

$$H_F = \frac{1}{2} \int d^3r \ : \left(\frac{1}{c^2}\left(\frac{\partial \vec{A}}{\partial t}\right)^2 + (\nabla\times\vec{A})^2\right) : \ , \tag{47a}$$

$$H_A = \sum_i \frac{1}{2m_i} \vec{p}_i^{\,2} + \frac{1}{2} \sum_{i \neq j} \frac{e_i e_j}{|\vec{r}_i - \vec{r}_j|} \,, \tag{47b}$$

$$H_I = - \sum_i \frac{e_i}{m_i c} \vec{p}_i \cdot \vec{A}(\vec{r}_i) + \sum_i \frac{e_i^2}{2m_i^2 c^2} : \vec{A}^2(\vec{r}_i) : \,, \tag{47c}$$

where the colons denote normal ordering of creation and annihilation operators.

We will study the time evolution of this system in the interaction picture, in which the time evolution governed by the Hamiltonian $H_O = H_F + H_A$ is separated out

$$e^{-\frac{i}{\hbar}Ht} = e^{-\frac{i}{\hbar}H_O t} \, V(t) \,, \tag{48}$$

where

$$V(t) = T \exp^{-\frac{i}{\hbar} \int_0^t d\tau \, H_I(\tau)} \tag{49}$$

and

$$H_I(\tau) = e^{\frac{i}{\hbar}H_O t} \, H_I \, e^{-\frac{i}{\hbar}H_O t} \,. \tag{50}$$

The EM field evolves in the interaction picture according to the free Maxwell equations:

$$\vec{A}(\vec{r},t) = \sum_{\vec{k},\lambda} \sqrt{\frac{\hbar c^2}{2V\omega}} \, \vec{\varepsilon}(\vec{k},\lambda) a(\vec{k},\lambda) e^{-i\omega t + i\vec{k}\cdot\vec{r}} + \text{h.c.} \tag{51}$$

4.1. Classical limit in terms of photon number states

We will first calculate matrix elements of the evolution operator V(t) between photon number states containing a large number of photons. We will follow the method described in our recent papers [22]. The calculation will first be performed for a single mode of electromagnetic radiation i.e. only one term in the sum (51) survives the classical limit. Generalization to many modes is straightforward.

Let us represent photon number states $|n+m\rangle$ by harmonic wave functions of the phase ϕ

$$|n + m\rangle \rightarrow e^{im\phi} , \tag{52}$$

where n is a fixed reference point and m varies from -n to ∞. An arbitrary state vector can be represented as a function of the phase ϕ

$$\psi = \sum_k c_k \; |k\rangle \rightarrow \psi(\phi) = \sum_{m=-n}^{\infty} c_{n+m} \; e^{im\phi} \tag{53}$$

and the scalar product takes on the form

$$(\Phi|\Psi) = \int_0^{2\pi} \frac{d\phi}{2\pi} \; \Phi^*(\phi)\Psi(\phi) \; . \tag{54}$$

Annihilation and creation operators act on these wave functions in the following way,

$$a \rightarrow e^{-i\phi} (n + \frac{1}{i}\frac{\partial}{\partial\phi})^{1/2} , \tag{55a}$$

$$a^{\dagger} \rightarrow \left(n + \frac{1}{i}\frac{\partial}{\partial\phi}\right)^{1/2} e^{i\phi} \; . \tag{55b}$$

We also obtain

$$a^{\dagger}a \rightarrow n + \frac{1}{i}\frac{\partial}{\partial\phi} \; . \tag{56}$$

If the reference point is large and if we restrict ourselves to values of m much smaller than n, we can approximate formulas (55) by expanding them into powers of $1/\sqrt{n}$ and keeping only two lowest terms

$$a \rightarrow e^{-i\phi}\sqrt{n}\left(1 + \frac{1}{2ni}\frac{\partial}{\partial\phi}\right) , \tag{57a}$$

$$a^{\dagger} \rightarrow \sqrt{n}\left(1 + \frac{1}{2ni}\frac{\partial}{\partial\phi}\right) e^{i\phi} \; . \tag{57b}$$

In the classical limit, when $\hbar \rightarrow 0$ and $n \rightarrow \infty$ in such a way that $n\hbar$ is kept constant we may keep only the first term in the expressions (57). Since the representation (52) and (57) can be used for every mode, we obtain the following formula for the vector potential operator in the classical limit

$$\vec{A}_{CL}(\vec{r},t;\phi) = \sum_{\vec{k},\lambda} \sqrt{\frac{\hbar n(\vec{k},\lambda)c^2}{2V\omega}}\,\vec{\varepsilon}(\vec{k},\lambda)\,e^{-i(\omega t - \vec{k}\cdot\vec{r} + \phi(\vec{k},\lambda))} + \text{h.c.}, \tag{58}$$

where ϕ on the l.h.s. stands for the collection of all phases. Owing to its dependence on the phases, A_{CL} is still an operator, even though it has the same appearance as the corresponding classical object. Upon introducing this operator into the formula for V(t) we obtain

$$V_{CL}(t;\phi) = T \exp\{-\frac{i}{\hbar}\int_0^t d\tau\, H_I^{CL}(\tau)\}\,, \tag{59}$$

where

$$H_I^{CL}(t) = -\sum_i \frac{e_i}{m_i c}\vec{p}_i(t)\cdot\vec{A}_{CL}(\vec{r}_i,t;\phi) + \sum_i \frac{e_i^2}{2m_i^2c^2}\vec{A}_{CL}^2(\vec{r}_i,t;\phi)\,. \tag{60}$$

Formulas (59) and (60) for the evolution operator are the same as in the case of a classical external field, but owing to the operatorial character of $\vec{A}_{CL}$, we can still obtain from them the transition amplitudes between photon number states. Assuming again that only one mode of the EM field is important, we obtain in the classical limit

$$\langle n+m|V(t)|n\rangle = \int_0^{2\pi} \frac{d\phi}{2\pi}\, e^{-im\phi} V_{CL}(t;\phi) \tag{61}$$

or inversly

$$V_{CL}(t;\phi) = \sum_m e^{im\phi}\langle n+m|V(t)|n\rangle\,. \tag{62}$$

Thus, by Fourier analysing the evolution operator in an external field with respect to the phase of the field we obtain photon transition amplitudes in the limit when $n\to\infty$.

Having obtained the classical limit of the transition amplitudes, we may proceed to calculate the time evolution of the density operator in this limit, assuming that initially the photon field was in the n photon state (one mode only). In the interaction picture we obtain the formula

$$\rho(t) = V(t)|n\rangle\rho_A\langle n|V^{\dagger}(t) \quad , \tag{63}$$

where ρ_A is the initial density operator of the atomic system. The atomic density matrix $\rho_A(t)$ at time t is equal to the trace of $\rho(t)$ with respect to the photon states

$$\rho_A(t) = \sum_m \langle n+m|V(t)|n\rangle\rho_A\langle n|V^{\dagger}(t)|n+m\rangle \quad . \tag{64}$$

Substituting (61) for the matrix elements of V and $V^{\dagger}$ and using the identity

$$\sum_m e^{im(\phi-\phi')} = 2\pi\ \delta(\phi-\phi') \tag{65}$$

we arrive at

$$\rho_A(t) = \int_0^{2\pi} \frac{d\phi}{2\pi}\ V_{CL}(t;\phi)\rho_A\ V_{CL}^{\dagger}(t;\phi) \quad . \tag{66}$$

Therefore, the density operator of the atomic system in the limit of large n is related to the density operator in the presence of an external EM field through the phase averaging operation. Such averaging is carried out in quantum optics whenever one uses the so-called semi-classical approach in the description of the EM field, but it is introduced then as an ad hoc time averaging procedure (for harmonic oscillations time and phase averaging are equivalent).

All the results of this section were obtained

under the assumption that the number m of emitted or absorbed photons is much smaller than the number of photons in the beam n, so that the reference point practically did not change during the interaction. In other words, the depletion of the photon beam by the atomic system is negligible. This is true only when we are dealing with not too many atomic systems at a time which in addition do not possess the ability to emit copiously photons. This is the case for bound systems with discrete levels, but scattered particles may radiate many photons. Such difference in behavior is due to the presence of the factor $1/\sqrt{\omega}$ in the expansion (45) of the vector potential. For bound-bound transitions, ω can not be very small due to energy conservation, but for the transition between scattering states ω can be small and the effective coupling constant between the atomic system and the EM field increases making the emission of photons more likely.

If the number of photons emitted (or absorbed) by the atomic system is large, it is natural to assume that a single act of emission (or absorption) does not influence significantly the state of the emitters. This assumption leads us directly to the concept of coherent states of radiation.

4.2. Classical limit in terms of coherent states

Coherent states appear in a natural manner in QED when we assume that the quantized EM field is coupled to a classical, external current. These states made their appearance in QED for the first time in connection with infrared divergencies. In their fundamental paper on this subject, Bloch and Nordsieck [23] made the crucial assumption that in the description of electron interactions with low energy photons" in first approximation the motion of electron is treated as given". This assumption leads to the appearance of coherent states of the EM field. The term coherent states for eigenstates of annihilation operators was introduced quarter of a century after the work of Bloch and Nordsieck by Glauber [24], who recognized the importance of these states for quantum optics.

The description of the sources of radiation in terms of a given external current is justified when these sources are so strong that they are practically not affected by the emission of photons. This is precisely what happens in the analysis of infrared divergencies, as can be seen from formula (51). As we observed before, the effective coupling constant increases with the decrease of ω, so that the current coupled to field oscillators with low frequencies is strong and this is the physical reason why the emission of soft (low frequency) photons can be described in terms of an external current.

There is more to the similarity between the problems of the classical limit and the infrared divergencies than only the appearance of classical currents in the limit, when $\omega \to 0$. One may indeed give a physical argument that when $\omega \to 0$ we must obtain exactly the classical limit. For low frequencies i.e. long wavelengths only the macroscopic features of the radiating sources are relevant and they can always be described by classical physics. Moreover for low frequencies the number of photons must be very large if the energy stored in the field is to be finite.

The usefulness of coherent states in the analysis of various aspects of the classical limit has been shown in a large number of publications. In a broader context of continuous representations, coherent states were used to investigate the relation between Quantum and Classical Dynamics by Klauder [25], who later formulated the weak correspondence principle [26]. According to this principle the expectation values of the dynamical variables in coherent states have the same form as in the classical theory. Equations of motion were studied with the use of coherent states by the present author [27]. It was shown that in the limit, when $\hbar \to 0$, the expectation values of quantum operators satisfy the classical equations of motion. Recently, Hepp [28] proved this result with full mathematical rigor using unitary operators $W(q,p)$ (cf. formula (11)) rather than (unbounded) position and momentum operators. All these works were incomplete in one respect: there was no discussion of the classical limit of the theory of particles interacting with a field. That is why the problem of the classical limit is still unresolved.

We will not discuss the classical limit of quantum mechanics in terms of coherent states referring readers to Refs.[25-28],because when used in quantum mechanics these states are defective in one important respect. With the only exception of the harmonic oscillator, the choice of coherent states in quantum mechanics is to a large extent arbitrary [25,27]. It is only in the theory of fields, where coherent states appear in a very natural way.

There is a vast textbook literature, where one can find information about the properties of coherent states (cf. for example [29,30,31]). We hope that the reader has at least a rudimentary knowledge of this subject.

Let us begin with a brief discussion of the interaction of a single mode of radiation with an external current. The interaction Hamiltonian and the evolution operator for this system in the interaction picture can be written in the form (cf. Refs. [29,30] or [32])

$$H_I(t) = \alpha^*(t)\ a\ e^{-i\omega t} + \alpha(t) a^\dagger\ e^{i\omega t} , \tag{67}$$

$$V(t) = e^{i\psi}\ e^{i(\eta a^\dagger + \eta^* a)} = e^{i\psi}\ e^{-|\eta|^2/2}\ e^{i\eta a^\dagger}\ e^{i\eta^* a}, \tag{68}$$

where

$$\eta = -\frac{1}{\hbar} \int_0^t d\tau\ \alpha(\tau)\ e^{i\omega\tau} \tag{69}$$

and $e^{i\psi}$ is a phase factor which is unobservable. Acting on the vacuum, the evolution operator produces a coherent state

$$V(t)|0\rangle = e^{i\psi}\ e^{-|\eta|^2/2}\ e^{i\eta a^\dagger}|0\rangle = e^{i\psi}\ |i\eta\rangle\ . \tag{70}$$

A coherent state is also obtained when V(t) acts on another coherent state. Therefore, in the world of classical currents it suffices to consider only coherent states.

Disregarding the phase factor in the formula (70), we can write the coherent state in the canonical form

$$|z\rangle = e^{za^{\dagger} - z^{*}a}|0\rangle \ . \tag{71}$$

There are various heuristic arguments which indicate that coherent states of the radiation field are the quantum analogs of pure classical states (i.e. the classical states which are characterized by given values of field variables and not by statistical distributions). For a single radiation mode, the classical field is characterized by the intensity and the phase. It is believed that the coherent state $|z\rangle$ is the analog of the classical state, whose intensity is $|z|^2$ and the phase is the phase of z. This belief is based on the following facts. First, the average value of the field operator $a + a^{\dagger}$ in the coherent state $|z\rangle$ is equal to $z + z^{*}$. Second, the product of dispersions of the hermitian and antihermitian parts of the operator a attains its minimal value of coherent states (gaussian functions saturate the uncertainty relation). Third, the values of z(t) for a mode driven by an external current $\alpha(t)$ are identical in the classical case and in the quantum case,

$$z(t) = \frac{1}{i\hbar}\int_0^t d\tau \ \alpha(\tau) \ e^{i\omega\tau} \ . \tag{72}$$

The analogy between coherent states and pure classical states can be extended to statistical mixtures of such states.

Statistical mixtures of coherent states formed

according to the rules of quantum mechanics (the density matrix of a mixed state is equal to the superposition of the projectors on the pure states contained in the mixture)

$$\rho = \int \frac{d^2z}{\pi} \, |z\rangle \, P(z) \langle z| \tag{73}$$

should correspond to classical statistical mixtures of pure classical states. The representation (73) of density matrices is known as the P-representation. The integration in the formula (73) is extended over the whole complex plane of the variable z.

The role of the P-representation in the theory of the EM field is very similar to the role of WWM transforms in quantum mechanics and these two concepts, as a matter of fact, are mathematically closely related. Both concepts are used to formulate the quantum theory in terms of statistical distributions which are in turn amenable to classical interpretation.

The possibility to interpret P(z) as a classical distribution function was stressed by Sudarshan in the context of the so-called optical equivalence theorem (cf. Ref. 29 p. 192).

We are now ready to state the following classical correspondence property valid for systems driven by external currents (for simplicity, we give it for one mode):

The time evolution of the density matrix, when expressed as the change in time of the weight function P(z), coincides with the change of the distribution function determined by the classical theory.

This statement follows from formula (68) and from the fact that the operator exp $(z_0 a^\dagger + z_0^* a)$ displaces the variable z of the coherent state by z_0. In the interaction picture we obtain

$$\rho_I(t) = V(t) \int \frac{d^2z}{\pi} |z\rangle P(z) \langle z| V^\dagger(t)$$

$$= \int \frac{d^2z}{\pi} |z+i\eta(t)\rangle P(z) \langle z+i\eta(t)| = \int \frac{d^2z}{\pi} |z\rangle P(z-i\eta(t)) \langle z| . \quad (74)$$

This formula translated to the Schrödinger picture reads

$$\rho(t) = e^{-\frac{i}{\hbar}H_0 t} \rho_I(t) e^{\frac{i}{\hbar}H_0 t}$$

$$= e^{-i\omega a^\dagger a} \int \frac{d^2z}{\pi} |z\rangle P(z-i\eta(t)) \langle z|$$

$$= \int \frac{d^2z}{\pi} |z\rangle P(ze^{i\omega t} - i\eta(t)e^{i\omega t}) \langle z| . \quad (75)$$

The change in time described by the formula

$$P_t(z) = P(ze^{i\omega t} - i\eta(t)e^{i\omega t}) \quad (76)$$

is the same as the change of the distribution function predicted by the classical theory. For example, we would have obtained the same result, equating the function P(z) with $\rho(q,p,t)$ and solving Eq. (27) for the harmonic oscillator. All the results obtained in this subsection can be generalized to include many modes of the EM fields. In most instances this generalization will just involve inserting sums over all the modes of the radiation field and it will not lead to any clarification of the one very important problem. Under what precise conditions we can use external, classical currents to describe the sources and whether this can be done in the classical limit. From the existing literature on the infrared problems starting with the paper by Bloch and Nordsieck [23] we learn that the use of classical currents is justified for very soft photons. Hence, if the classical limit is understood not only as limiting case of large photon numbers but also as a transition to very low frequencies and long wavelenghts, we can then obtain the equivalence of the quantum and the classical descriptions.

4.3. Classical limit and the low-frequency limit

The low-frequency limits of QED transition amplitudes have been extensively studied after it was shown by Low [33] that such soft-photon amplitudes are uniquely determined by the requirements of Lorentz invariance, current conservation and some simple analytic properties. With the help of these low-frequency or low-energy theorems, one can establish also the connection between classical and quantum results in several simple cases. For example, from the Klein-Nishina formula describing the Compton scattering

$$\frac{d\sigma}{d\Omega} = \frac{r_o^2}{2} \frac{1 + \cos^2\theta}{[1+(1-\cos\theta)\frac{\hbar\omega}{mc^2}]^2}\{1+\frac{\frac{\hbar\omega}{mc^2}(1-\cos\theta)^2}{(1+\cos\theta)^2[1+(1-\cos\theta)\frac{\hbar\omega}{mc^2}]}\}, \quad (77)$$

one obtains the classica Thompson formula

$$\frac{d\sigma}{d\Omega} = \frac{r_o^2}{2} (1 + \cos^2\theta) \tag{78}$$

by either taking the limit $\hbar \to 0$ or the limit $\omega \to 0$. This equivalence of the two limits is simply a result of the energy-momentum conservation at each vertex: $p_\mu + \hbar k_\mu = q_\mu$. If the electron four-momenta are kept fixed when we take the limit $\hbar \to 0$, then even virtual electrons must have the same momenta as the initial electron, as if the photons had the frequencies and the wave vectors equal to zero. What seems to be difficult to achieve is the simultaneous transition with $\hbar$ to zero and with the intensity of the photon beam to infinity, keeping their product finite. Such a simultaneous transition is essential if we are to account for finite amounts of energy being transfered in electromagnetic interactions. In this respect some results obtained by Brown and Goble [34] are encouraging. They have been able to perform an infinite summation over soft-photon contributions and even to include, to some extent, a change in the electron four-momentum due to the interaction with the soft photons. However, the main part of the change in four-momentum was attributed to some unspecified nonelectromagnetic interactions and it was not shown that the change attributed to photons is introduced in a self-consistent way.

Complete calculation still remains an unfulfilled program. Even in a very simple case of the Coulomb scattering of an electron, the classical formula for the bremsstrahlung spectrum [35],

$$I(\omega) = \frac{16}{3} \frac{Z^2 e^2}{c} \left(\frac{e^2}{mc^2}\right)^2 \left(\frac{c}{v}\right)^2 \ln \left(\frac{\lambda mv^2}{Ze^2 \omega}\right) \tag{79}$$

has not been obtained from full QED in the classical limit.

Our review of the classical limit of QED would not be complete without mentioning recent papers by Moniz and Sharp [36]. They have addressed themselves to the problems of runaway solutions and preaccelerations which emerge in classical electrodynamics, but seem not to appear in QED. First they showed classically that preacceleration and runaways disappear when one assumes an extended electron whose radius L is larger than the classical electron radius [37]. Next they summed up a certain class of terms in the nonrelativistic approximation to quantum theory, showing that the Compton wave length plays the role of the radius L. Thus, in their approach the electrostatic self-energy of the electron remains finite even in the limit of the point electron.

However, as was shown by Weisskopf [38], the electrostatic term is just one part of the self-energy the remaining two being the part due to the magnetic moment and the part due to the interaction with zero-point fluctuations of the EM field. It is a very important question whether the results of Moniz and Sharp also hold in the relativistic theory, but even their nonrelativistic approximation strongly indicates that the inconsistencies of the classical theory will disappear if we give the electron its quantum extension $\hbar/mc$. Complete answer to this question must also contain an explanation how is the vilation of causality avoided for classical extended electrons.

I wish to end here my review of various topics connected with the classical limit of QED, hoping to have convinced you that the first sentence of the Introduction is indeed true.

REFERENCES

1. F. Rohrlich, The electron: Development of the first elementary particle theory, in The Physicist's Conception of Nature, Ed. J. Mehra (Reidel, Dordrecht, 1973).

2. D. Bohm, Quantum Theory (Prentice-Hall, New York, 1951), Ch. 23.

3. N.F.Mott and H.S.W. Massey, Theory of Atomic Collisions (Oxford University Press, London, 1965) Ch. 5 § 5.

4. A. Messiah, Méchanique Quantique (Dunod, Paris, 1959) Ch. 6 §1.

5. W. Pauli, Helv. Phys. Acta 5 (1966) 179.

6. L.de Broglie, La Théorie des Particules de Spin 1/2 (Gauthier-Villars, Paris, 1952) Ch. 10.

7. S.I. Rubinov and J.B. Keller, Phys. Rev. 131 (1963) 2789.

8. M.Planck, The theory of Heat Radiation (Dover, New York, 1959) p. 170.

9. E.C. Kemble, Principles of Quantum Mechanics (Dover, New York, 1958) p. 376.

10. P.Ehrenfest, Z. Physik 45 (1927) 455.

11. E. Madelung, Z. Physik 40 (1926) 322.

12. I.Bialynicki-Birula and Z. Bialynicka-Birula, Phys. Rev. D3 (1971) 2410.

13. I.Bialynicki-Birula and J.Mycielski, Ann. of Phys. 100 (1976) 62.

14. H. Weyl, The Theory of Groups and Quantum Mechanics (Dover, New York, 1931) §14.

15. E.Wigner, Phys. Rev. 40 (1932) 749.

16. J.E. Moyal, Proc. Cambr. Phil. Soc. 45 (1949) 99.

17. S. Fujita, Introduction to Non-Equilibrium Quantum Statistical Mechanics (Saunders, Philadelphia, 1966) Ch. 5.

18. J. Schwinger, Phys. Rev. 82 (1951) 664.

19. R. Newton and E.P. Wigner, Rev. Mod. Phys. 21 (1949) 400.

20. I.Bialynicki-Birula, Phys. Rev. D3 (1971) 2413.

21. J.J. Sakurai, Advanced Quantum Mechanics (Addison-Wesley, Reading 1967) Ch. 2 §3.

22. I.Bialynicki-Birula and Z.Bialynicka-Birula, Phys. Rev. A14 (1976) 1101.

23. F.Bloch and Nordsieck, Phys. Rev. 52 (1937) 54.

24. R.J.Glauber, Phys. Rev. 130 (1963) 2529.

25. J.R. Klauder, J.Math. Phys. 4 (1963) 1058; 5 (1964) 177.

26. J.R. Klauder, J. Math. Phys. 8 (1967) 2392.

27. I.Bialynicki-Birula, Ann. of Phys. 67 (1971) 252.

28. K. Hepp, Comm. Math. Phys. 35 (1974) 265.

29. J.R. Klauder and E.C.G. Sudarshan, Fundamentals of Quantum Optics (Benjamin, New York, 1968) Ch. 7.

30. R. Loudon, The Quantum Theory of Light (Clarendon Press, Oxford, 1973) Ch. 7.

31. I. Bialynicki-Birula and Z. Bialynicka-Birula, Quantum Electrodynamics (Pergamon Press, London, 1975) Ch. 4 § 11.

32. I.Bialynicki-Birula and Z.Bialynicka-Birula, Phys. Rev. A8 (1973) 3146.

33. F.E. Low, Phys. Rev. 110 (1958) 974.

34. L.S. Brown and R.L.Goble, Phys. Rev. 173 (1968) 1505.

35. L.D. Landau and E.M.Lifshitz, Classical Theory of Fields (Addison-Wesley, Reading, 1951) Ch.8 and 9.

36. E.J. Moniz and D.H. Sharp, Phys. Rev. D10 (1974) 1133 and to be publ.

37. H. Levine, E.J.Moniz and D.H.Sharp, Am. J. Phys. to be published.

38. V. Weisskopf, Phys. Rev. 56 (1939) 72.

Acta Physica Austriaca, Suppl. XVIII, 153-384 (1977)

THE PHYSICS OF STRONG FIELDS IN QUANTUM ELECTRODYNAMICS AND GENERAL RELATIVITY[+]

by

B. MÜLLER and W. GREINER
Institut für Theoretische Physik
der Johann Wolfgang Goethe Universität
Frankfurt am Main, W-Germany

CONTENTS

[+]Lecture given at XVI.Internationale Universitätswochen für Kernphysik,Schladming,Austria,February 24-March 5, 1977.

INTRODUCTION

Quantum electrodynamics, i.e. the theory of electrons, positrons, and their interaction with the radiation field has been one of the most successful disciplines of physics. Its formalism which essentially dates back to the early works of Dirac, Heisenberg and Weisskopf in the 1930's and was completed in the 1940's by Feynman, Schwinger, Dyson and others, allows the calculation of atomic properties with virtually arbitrary precision. In spite of the somewhat unsatisfying divergencies in the renormalization scheme, QED may be regarded as a completed theory. There is, however, one phenomenon in QED which only recently has been fully understood and which leads to a qualitatively new concept: The charged vacuum in strong electrostatic fields.

Starting point for a discussion of this idea is the following question: What happens to the atomic electrons if the charge of the nucleus is considerably increased? As we shall discuss below in detail, relativistic effects will become dominant and qualitatively change the level spectrum (leading e.g. to a large fine structure splitting). A first attempt to account for them is the famous Sommerfeld fine structure formula (we mostly set $\hbar = c = 1$)

$$E_{nj} = m_e \left[1 + \left(\frac{Z\alpha}{n-|\kappa|+\sqrt{\kappa^2-(Z\alpha)^2}}\right)^2\right]^{-1/2} \tag{1}$$

with $\kappa = \pm 1, \pm 2, \ldots$ and $n = 1, 2, \ldots$. It describes the spectrum of electronic bound states in the external Coulomb potential $A_o(r) = \frac{-Z\alpha}{r}$ of a point charge. In this case the appropriate wave equation - the Dirac equation - can be solved analytically. Due to the term $\sqrt{\kappa^2 - (Z\alpha)^2}$ Eq. (1) obviously breaks down at $Z\alpha > |\kappa|$. For example, the curves of energy as function of Z, for the $1s_{1/2}$ state with $E_{1s} = m_e \sqrt{1 - Z^2\alpha^2}$ and all other states with $j = 1/2$ cease to exist at $Z = 1/\alpha \sim 137$ as shown in Fig. 1. The corresponding wave functions become non-normalizable at the origin. This, however, does not imply that the Dirac equation has no solution at high Z as was first believed. Taking into account the finite extension of the nucleus one can trace any level E_{nj} down to a binding energy of twice the electronic restmass if the nuclear charge is increased as a parameter. At the corresponding charge number, which we will call critical (Z_{cr}), the state reaches the negative-energy continuum of the Dirac equation ("Dirac sea") which according to the hole-hypothesis is totally occupied by electrons. If the strength of the external field is further increased, the bound state "dives" into the continuum. The overcritical state obtains a width and is spread over the continuum. Still, the electron charge distribution does remain localized.

The related phenomena have been analyzed and investigated very carefully. At this place, we will only stress the most important aspect: The overcritical vacuum state is charged. If an empty atomic state dives into the continuum it will be filled spontaneously by the emission of a free positron moving to infinity. The remaining electron cloud of the supercritical atom necessarily is negatively charged. While in ordinary undercritical physics we can define a vacuum state $|0\rangle$ without charges or currents by choosing the Fermi surface (up to which the levels are occupied) below the lowest bound state, this is not possible in the overcritical case. Thus we are led to the concept of charged vacuum (Müller, Rafelski, Greiner [1,2]). Note that this is a fundamentally new process leading to a new understanding of the vacuum. Everytime when at increasing external field strength a bound state joints the continuum, the vacuum undergoes a new phase transition and becomes successively higher charged.

In the first part of this review article we extensively discuss the results on overcritical atomic phenomena which have been obtained since the year 1968 essentially by two groups in Frankfurt and Moscow. After some general considerations we introduce a model which treats the decay of the neutral vacuum as an autoionization process of positrons. Chap. 3 quantitatively justifies this picture, giving the exact single particle solutions of the Dirac equation with external Coulomb potential for all Z. The important task of quantization of the theory is accomplished in Chap. 4, still maintaining the concept of overcritical states.

A typical and important feature of the quantized theory is vacuum polarization. Its behaviour in strong fields is shown according to the work of various authors in Chap. 5. The consequences of certain phenomenologically introduced nonlinear extensions of the theory are scrutinized in the next chapter, namely nonlinear electrodynamics of Born-Infeld type and higher order nonlinear terms in the spinor field. The results show, that diving of electronic levels cannot be prevented by any known causes. Finally, Chap. 7 gives a short statistical description of the charged vacuum which is interesting in connection with hypothetical very highly charged nuclei.

Though an interesting problem in itself, the physical relevance of QED of strong fields has yet to be demonstrated (Sec. II). Historically the first detailed investigations were inspired by the possible existence of superheavy nuclei. The deformed shell model of theoretical nuclear physics tells us that nuclei near the magic proton numbers Z = 114 and 164 should have strongly enhanced lifetimes. The production of long-lived superheavy elements by fusion of ordinary nuclei would allow the extension of the exact measurements of atomic spectroscopy to a new region of binding energy. The overcritical region (Z > 172) cannot be reached. Luckily, however, it is at least possible to assemble a supercritical charge for a short period of time in the collision of very heavy ions. For example in U-U-collisions near the Coulomb barrier the lowest bound state joints the contiunuum for some 10^{-20} sec. This should be long enough to observe the decay of the neutral vacuum by detecting emitted positrons. We deal with the somewhat delicate dynamical description of this process. It requires the formation of a K-hole during the collision which then can autoionize.

Lecture I

in collaboration with J. Reinhardt

1. GENERAL DISCUSSION OF THE DIRAC EQUATION WITH COULOMB POTENTIAL

AT the beginning of relativistic quantum theory and especially QED stands the Dirac equation [3]

$$H_D\psi(\vec{r}) \equiv [\vec{\alpha}\cdot\vec{p} + \beta m]\psi(\vec{r}) = E\psi(\vec{r}) \tag{2}$$

with the 4x4 matrices

$$\vec{\alpha} = \begin{pmatrix} 0 & \vec{\sigma} \\ \vec{\sigma} & 0 \end{pmatrix} \quad \text{and} \quad \beta = \begin{pmatrix} I & 0 \\ 0 & -I \end{pmatrix} . \tag{3}$$

The four-component wave function exhibits the properties of a bi-spinor and describes fermions with spin 1/2. The spectrum of the free Dirac equation consists of a gap between -m and +m and two continua of free particles with the dispersion relation

$$E = \hbar\omega = \pm [(\hbar\vec{k})^2 + (m)^2]^{1/2} . \tag{4}$$

To avoid the decay of electronic states under emission of an infinite amount of energy when turning on the coupling to the radiation field, Dirac [4] postulated that all states in the negative continuum are occupied, i.e. he defined a Fermi surface lying in the gap. The so constructed vacuum can be excited by e.g. photons to produce particle-hole-pairs at a threshold energy of 2m. The hole states then, having the properties of a particle with electron mass, positive energy and positive charge are identified with positrons. In the language of second quantization (Chap. 4) Dirac's prescription takes the form of suitable anticommutation relations between electron and positron operator (for an extensive treatment of the formalism of QED refer to Bjorken and Drell [5], Jauch and Rohrlich [6]).

The electromagnetic field A_μ is introduced into the Dirac equation by minimal coupling

$$[\gamma_\mu (p_\mu - e A_\mu) + m] \psi(\vec{r},t) = 0 . \qquad (5)$$

For stationary states in a static Coulomb potential $A_0(r) = V(r)$ this reads

$$[\vec{\alpha}\cdot\vec{p} + \beta m + V(\vec{r})]\psi(\vec{r}) = E\psi(\vec{r}) . \qquad (6)$$

In low Z atoms it is sufficient to take the potential of a pointlike nuclear charge $V(\vec{r}) = -\frac{Z\alpha}{r}$. $V(\vec{r})$ may also contain electron-electron interaction terms.

Turning on the binding potential $V(\vec{r})$ has two effects on the solutions of the Dirac equation. At the edge of the positive continuum bound states emerge, which shift down into the gap with increasing strength of the potential. It is a peculiarity of the long range 1/r potential that at once an infinite number of bound states is created unlike the situation of e.g. a square well or Yukawa potential, where a few states are truly bound and the remaining ones only show up as resonances[+]. Also the continuum states are influenced and deformed by the central Coulomb potential, which attracts electronic and repels positronic wave functions. This gives rise to a vacuum displacement charge and is the origin of virtual vacuum polarization.

An explicit derivation of the exact solution of (6) for various cases will be given in the chapter after next. Here let us look at its qualitative behaviour. To this end it is advantageous to use the squared form of the Dirac equation, obtained by eliminating one of the components of the bi-spinor. Assuming a spherically symmetric potential the spin orbit angular momentum operator

$$K = \beta(\vec{\sigma}\cdot\vec{L} + 1) \tag{7}$$

commutes with the Hamiltonian and the total angular momentum operator, producing a good quantum number κ. We separate angular and radial part by the ansatz

$$\psi(\vec{r}) = \frac{1}{r}\begin{bmatrix} u_1(r) & \chi_\kappa^\mu \\ i\, u_2(r) & \chi_{-\kappa}^\mu \end{bmatrix} \tag{8}$$

with the spherical spinors χ_κ^μ,

$$\chi_\kappa^\mu = \sum_{m=\pm\frac{1}{2}} (\ell\tfrac{1}{2}j\,|\,\mu-m\ m\ \mu)Y_\ell^{\mu-m}(\theta,\phi)\ \chi^m \ . \tag{9}$$

[+]Rose [7] speaks of an additional bound state appearing at $Z \geq 109$. This solution has to be ruled out, since the expectation values of kinetic and potential energy do not exist separately.

(For detailed treatments of the Dirac equation in central potentials see the textbook of Rose [7] and Akhiezer and Berestetzki [8]). We are led to the system of first order differential equations

$$\frac{d}{dr} u_1 = - \frac{\kappa}{r} u_1 + (E + m-V)\, u_2$$

$$\frac{d}{dr} u_2 = - (E-m-V) u_1 + \frac{\kappa}{r} u_2 \quad . \tag{10}$$

Eliminating u_2 leads to a second order wave equation

$$\frac{d^2u_1}{dr^2} + \frac{dV/dr}{m+E-V} \frac{du_1}{dr} +$$

$$+ \left[(E-V)^2-m^2- \frac{\kappa(\kappa+1)}{r^2} + \frac{\kappa}{r} \frac{dV/dr}{m+E-V} \right] u_1 = 0 \; . \tag{11}$$

Using the transformation $u_1(r) = \sqrt{m+E-V(r)}\, \chi(r)$ it is further possible to reduce (11) to the form of a self-adjoint Schrödinger equation (cf. Popov [9], Zeldovich and Popov [10]):

$$\chi'' + k^2(r)\chi = 0 \tag{12}$$

with

$$k^2(r) = E^2-m^2 + \left(V^2-2EV - \frac{\kappa(\kappa+1)}{r^2}\right) -$$

$$- \left[\frac{1/2\, V''}{m+E-V} + \frac{3}{4} \left(\frac{V'}{m+E-V}\right)^2 - \frac{\kappa}{r} \left(\frac{V'}{m+E-V}\right)\right] \; . \tag{13}$$

The effective potential in this equation is quite a complicated expression.

Looking at the asymptotic solution of (11) one can easily distinguish between bound and free states. Assuming that V(r) and its derivatives vanish at $r \to \infty$ we obtain

$$u_1(r) \sim (m+E-V)^{1/2} \exp[-r(m^2-(E-V)^2)^{1/2}]. \tag{14}$$

Wave functions inside the gap ($-m < E < m$) therefore are localized and vanish exponentially

$$u_1(r) \sim \exp[-r(m^2-E^2)^{1/2}] \to 0 \tag{15a}$$

while continuum solutions ($|E| > m$) oscillate

$$u_1(r) \sim \exp[\pm ir(E^2 - m^2)^{1/2}] . \tag{15b}$$

Particularly interesting is the limit $|E| = m$. The wave functions behave like

$$u_1(r) \sim \begin{cases} \exp[\pm ir(2m|V|)^{1/2}] & \text{for} \quad E = +m & (16a) \\ \exp[-r\,(2m|V|)^{1/2}] & \phantom{\text{for}} \quad E = -m\,. & (16b) \end{cases}$$

Thus the bound states with negative energy do not display any irregular behaviour and remain localized when reaching the negative continuum.

Now let us turn to the special potential $V=-\frac{Z\alpha}{r}$. The relativistic problem with this potential was first solved by Darwin and by Gordon [28], leading to the Sommerfeld formula (1) already mentioned in the introduction. The obtained solutions break down at $Z\alpha > |\kappa|$. This can be understood by looking at the behaviour of the wave functions.

The form of the wave function at small distances r is deduced from (11) or (12)

$$u_1 \sim r^{\sqrt{\kappa^2-(Z\alpha)^2}} \quad . \tag{17}$$

When $Z\alpha > |\kappa|$ the wave function suddenly begins to oscillate near the origin like $\cos(\sqrt{(Z\alpha)^2 - \kappa^2}\,\ln r)$ and thus loses its physical meaning. We can explain the reason for the breakdown of the solution by looking at the effective potential in Eq. (13) which at small r takes the form

$$V_{eff} \sim - 2E\frac{Z\alpha}{r} - \frac{(Z\alpha)^2 - \kappa^2 + 1/4}{r^2} \quad . \tag{18}$$

For large $Z\alpha$ large enough the $1/r^2$ term becomes attractive and leads to the collapse of the motion. This is a general feature of the relativistic motion in a Coulomb potential. It can already be concluded from the relativistic energy momentum relation

$$p^2 + m^2 - (E-V)^2 = 0 \tag{19}$$

with the effective potential+

$$V_{eff}(r) \sim 2\ EV + \frac{(j+1/2)^2}{r^2} - V^2 \qquad (20)$$

and occurs also in classical mechanics.

We want to stress, however, that the collapse of the wave function is no physically real effect. Mathematically it means, that the Hamiltonian loses its hermitecity. The solution of the wave equation then has to be specified by an additional parameter as discussed by Case [12] in a somewhat different context. Cf. also the work of Alliluev [13].

The wave function is stabilized in a very natural way by recognizing the finite extension of the potential source. In normal nuclei the potential is practically cut off at the nuclear charge radius R of a few fm. We can write

$$V_i(r) = \begin{cases} -\frac{Z\alpha}{r} & \text{for } r \geq R \\ -f_i(r)\,\frac{Z\alpha}{R} & \text{for } r \leq R \end{cases} \qquad (21)$$

with

$$f_I(r) = 1 \qquad \text{or} \qquad f_{II}(r) = -\frac{1}{2}\left(\frac{r}{R}\right)^2 + \frac{3}{2} \qquad (22)$$

corresponding to the models of a spherical shell or a

+The difference between (18) and (19) is the correction of Langer [11].

homogeneously charged sphere, respectively. The regular solution of the Dirac equation with potential (21) in the interior region $r \leq R$ is easily obtained. Its logarithmic derivative then has to be matched to the exterior solution on the sphere $r = R$. In this way a boundary condition is specified for the exterior wave function. Therefore we can construct meaningful solutions also at $Z\alpha > |\kappa|$.

A simple argument (Zeldovich and Popov [10]) shows that the atomic levels can be traced continuously down to the negative continuum. Let $|\psi(Z)\rangle$ be any bound state wave function in the gap with energy $E(Z)$ and let $Z \cdot U(r)$ be the potential with a strength determined multiplicatively by Z

$$(H_o + Z\, U\,(r))|\psi\rangle = E(Z)|\psi\rangle . \qquad (23)$$

Then one can show that

$$\frac{dE}{dZ} = \langle\psi|U(r)|\psi\rangle \; . \qquad (24)$$

Assuming $U(r)$ to be an integrable, bounded negative function like (21) we can deduce from (24) that dE/dZ is finite and negative, so that the energy $E(Z)$ continuously decreases with Z. Note that this reasoning depends on the positive definiteness of the norm $\langle\psi|\psi\rangle$ in the Dirac theory and would not be valid for Klein-Gordon particles.

Let us at length discuss the exact energy levels for superheavy nuclei. The first authors to treat the cut-off potential were Pomeranchuk and Smorodinsky [14] whose solution, however, was numerically inadequate. Later calculations were performed by Werner and Wheeler [15] in connection with their speculation on superheavy elements and by Voronkov and Koleznikov [16]. They had no influence at all on the modern development and understanding of QED of strong fields, however.

The first precise calculation of the eigenenergies of the Dirac equation with extended nuclear potential for all elements up to $Z = 170$ was performed by Pieper and Greiner [17]. These results are shown in Fig. 1, exhibiting a number of interesting properties. Substantial deviations from the Sommerfeld energies begin to occur only in the vicinity of $Z = 1/\alpha \approx 137$. The levels show two kinds of splitting which attain extremely large values. For example the spin-orbit-splitting between the $2p_{1/2}$ and $2p_{3/2}$ reaches 800 keV. Also the exact degeneracy of states which differ only by the sign of κ, e.g. $2p_{1/2}$ and $2s_{1/2}$ is broken.

Most interesting, however, is the fact, that several of the shown levels do reach the negative energy continuum. The critical charge numbers for three lowest levels $1s_{1/2}$, $2p_{1/2}$, $3s_{1/2}$ are approximately $Z_{cr} = 170, 185, 245$. The magnitude of the first of these numbers is vital for the idea of testing the decay of the neutral vacuum in heavy ion collisions, where one can reach $Z_1 + Z_2 = 184$.

Beyond the critical charge the Dirac equation no longer supports the bound state. It was expressed by Pieper and Greiner [17] and later by Gershtein and Zeldovich [18], that an empty atomic level reaching $-mc^2$ leads to the spontaneous energyless emission of a positron. The physics and mathematics of this "diving" into the continuum and of the overcritical state will be clarified in the next three sections.

To end our general discussion, we remark, that electron screening (Fricke and Greiner [19]) slightly modifies the energies of Fig. 1. Results of Hartree-Fock-Slater calculations for all elements between Z = 100 and 173 were published by Fricke and Soff [20]. Although electron-electron interaction is important for the chemistry and spectroscopy of superheavy elements, it does not influence the inner shells significantly. The most elaborate calculation (Soff, Müller and Rafelski [21]) for the critical charge leads to Z = 172 for the $1s_{1/2}$ orbital.

2. THE AUTOIONIZATION MODEL

When the nuclear charge exceeds the critical value Z_{cr} a bound state vanishes from the gap and becomes degenerate with the negative energy continuum. To describe this process we note, that it is formally equivalent to the autoionization of an excited state in atomic or nuclear physics. For the reader familiar with nuclear physics, Fig. 2 schematically shows a situation where a 2p-2h state is degenerate with a 1p-1h-continuum state. Both configurations mix because of the residual interaction which leads to the decay of the 2p-2h bound state. One of the excited nucleons jumps back below the Fermi surface while the other one is ejected into the continuum.

Using the mathematical method of Fano [22] a model was

developed (Müller, Peitz, Rafelski, Greiner [23]; Müller, Rafelski, Greiner [1,2]) describing the overcritical behaviour in terms of the single particle Dirac equation. The field theoretical extension will be treated in Chap. 4.

We are interested in the eigenstates of the overcritical Dirac Hamiltonian

$$H(Z) = H_o + V(Z) = H(Z_{cr}) + V(Z') \tag{25}$$

with the free Dirac Hamiltonian $H_o = i\gamma_\mu \frac{\partial}{\partial x_\mu} + m$, the total electromagnetic potential $V(Z)$ and the overcritical extrapotential defined by $V(Z') = V(Z) - V(Z_{cr})$. As the basis for a diagonalization procedure let $|\phi\rangle$ and $|\psi_E\rangle$ denote the bound and scattering states of the atom just reaching the critical charge

$$H(Z_{cr})|\phi\rangle = E_o|\phi\rangle = -m|\phi\rangle \tag{26a}$$

$$H(Z_{cr})|\psi_E\rangle = E|\psi_E\rangle \tag{26b}$$

with

$$\langle\phi|\phi\rangle = 1, \ \langle\psi_E|\psi_{E'}\rangle = \delta(E-E') \ . \tag{27}$$

Without introducing a serious error, we have restricted our basis to one discrete state since the energetic distance to the next level with the same spherical symmetry (necessary for the interaction) is very large.

For convenience we neglect the weak dependence of the nuclear size on Z and factorize the additional potential

$$V(Z') = Z' \cdot U(r) \quad . \tag{28}$$

Next we need the matrix elements of the overcritical Hamiltonian with our undercritical basis. They can be written

$$\langle\phi|H(Z_{cr} + Z')|\phi\rangle = E_o + Z' \ \langle\phi|U|\phi\rangle$$

$$\equiv - m + \Delta E \tag{29}$$

$$\langle\psi_{E'}|H(Z_{cr}+Z')|\phi\rangle = \langle\psi_{E'}|H_{cr}|\phi\rangle + \langle\psi_{E'}|Z'U|\phi\rangle$$

$$= Z'\langle\psi_{E'}|U|\phi\rangle \equiv V_{E'} \tag{30}$$

$$\langle\psi_{E''}|H(Z_{cr}+Z')|\psi_{E'}\rangle = \langle\psi_{E''}|H_{cr}|\psi_{E'}\rangle + Z'\langle\psi_{E''}|U|\psi_{E'}\rangle$$

$$\equiv E'\delta(E'' - E') + Z'U_{E''E'} \quad . \tag{31}$$

Using these matrix elements the supercritical state

$$H(Z_{cr} + Z')|\Psi_E\rangle = E|\Psi_E\rangle \tag{32}$$

can be obtained analytically. Let us expand

$$|\Psi_E\rangle = a(E)|\phi\rangle + \int_{E'<-m} dE' \ b_{E'}(E)|\psi_{E'}\rangle \ . \tag{33}$$

Inserting this wave function and projecting with the basis states $\langle\phi|$, $\langle\psi_{E'}|$ the wave equation (32) is equivalent to a system of equations for the coefficients

$$(E_o + \Delta E)a(E) + \int dE' \ b_{E'}(E) \ V^*_{E'} = E \ a(E)$$

$$V_{E'}a(E) + b_{E'}(E)E' + \int dE'' \, b_{E''}(E) \, Z'U_{E'E''} = E \, b_{E'}(E). \quad (34)$$

The explicit solution of these equations is not possible due to the integral in (34b). In the following, we neglect the term, $U_{E'E''}$ what can be justified by two arguments: The matrix element $Z'U_{E'E''}$ describes a re-arrangement of continuum states due to the extra potential which should be not very significant for Z' not too large. Furthermore, one can think of the continuum as being prediagonalized in which case our treatment would be exact, implying only a change in the values of V_E.

Let us impose upon the wave function (33) the normalization

$$\langle \Psi_{E'} | \Psi_E \rangle = \delta(E' - E) \; . \quad (35)$$

Then the solution of (34a,b) neglecting $U_{E'E''}$ takes the form

$$a(E) = \frac{V_E^{*}}{E-E_o-\Delta E-F(E) + i\pi |V_E|^2} \quad (36a)$$

$$b_{E'}(E) = \delta(E-E') + \frac{a(E) \, V_{E'}}{E-E'-i\varepsilon} \quad (36b)$$

with the principal value integral

$$F(E) = P \int_{E'<-m} dE' \, \frac{|V_{E'}|^2}{E-E'} \; . \quad (37)$$

ε is a small positive quantity. The convergence of the integral at $E \to -m$ will be proven below (Eq. (42)).

Eqs. (35a,b) are the complete solution to our problem of the diving of a level in terms of the reduced basis set $|\phi\rangle, |\psi_E\rangle$. Generalizations of the result to include more than one bound state and more than one continuum are laborious but straight-forward.

We are mainly interested in the fate of the previously bound state $|\phi\rangle$ when the interaction $Z'U(r)$ is turned on. Since the overcritical wave function $|\psi_E\rangle$ is normalized to unity (Eq. (35)), the probability with which $|\phi\rangle$ is promoted to the energy $E < -m$ under the action of the overcritical additional potential, is given by $|a(E)|^2$.

We obtain

$$|a(E)|^2 = \frac{|V_E|^2}{[E-(E_0+\Delta E+F(E))]^2 + \pi^2 |V_E|^4} \; . \tag{38}$$

This expression obviously shows resonance behaviour. To further simplify Eq. (38), we neglect the principal value integral $F(E)$. That is justified by the relatively weak energy dependence of V_E and will be supported by the numerical phase shift analysis in the next section. Writing $2\pi |V_E|^2 = \Gamma$

$$|a(E)|^2 = \frac{1}{2\pi} \frac{\Gamma}{[E-(E_0+\Delta E)]^2 + \frac{1}{4}\Gamma^2} \tag{39}$$

takes the form of a symmetric Breit-Wigner curve centered

around $E = E_o + \Delta E < -m$ with the width Γ.

The position of the resonance is determined by

$$\Delta E = Z'\langle\phi|U(r)|\phi\rangle \equiv - Z'\delta \quad . \tag{40}$$

Therefore the level dives approximately linearly with respect to Z into the continuum. The width

$$\Gamma = 2\pi|Z' \langle\psi_E|U(r)|\phi\rangle|^2 \equiv Z'^2\gamma \tag{41}$$

roughly increases quadratically with the overcritical charge.

To be specific, the following values of the two parameters δ,γ have been calculated [24],

$$1s_{1/2} : \delta \approx 29 \text{ keV}, \ \gamma \approx 0.04 \text{ keV} \ , \ \text{and}$$

$$2p_{1/2} : \delta \approx 38 \text{ keV}, \ \gamma \approx 0.08 \text{ keV} \ . \tag{42}$$

To avoid misapprehensions, however, Eq. (41) has to be interpreted carefully. The matrix element $\langle\psi_E|U(r)|\phi\rangle$, although fairly constant at large $|E|$, vanishes exponentially with $E \to -m$. This clearly is caused by the repulsion of the negative energy continuum wave functions by a positively charged nucleus. The amplitude of the positron wave function with momentum p near the origin is

$$|\psi_E(0)|^2 \sim \exp\left(- \frac{\pi Z\alpha m}{p}\right) \tag{43}$$

so that the simple quadratic scaling now does not hold near Z_{cr}. This result also ensures the existence of the integral (37) near $E \sim -m$, where only one branch of the pole lies in the integration region.

Let us now once more look at the physical meaning of the described diving process. At first, we stress the important fact, that the charge distribution of a filled K-shell remains localized when it enters the continuum. This is illustrated in Fig. 8 where the $1s_{1/2}$ density distributions $|r\psi(r)|^2$ for different nuclei are compared. For $Z > Z_{cr}$ the bound state shows up as a strong distortion of the continuum wave functions. The collective effects of the continuum state producing a negatively charged cloud still bound to the nucleus has been termed <u>real vacuum polarization</u> in contrast to the normal virtual vacuum polarization where only positive and negative charge densities are displaced in their relative position.

In a normal, undercritical system there are many stationary stable states corresponding to various electron occupations. Especially one can define the electron vacuum state (in the presence of an external field) where all bound states are empty. This is no longer possible in the overcritical situation. Since here the diving state loses its identity and is spread over the continuum (Fig. 3 show this schematically in the case of a discretized continuum) it can no longer be defined empty. If a previously empty bound state becomes overcritical, it will be filled by spontaneous emission of a positron (Pieper and Greiner [17], Gershtein and Zeldovich [18]).The resulting new ground state is doubly (due to spin degenerycy) charged: The neutral vacuum decays into a charged vacuum [1,2]. Of course, charge conservation is observed and the obsolute positive charge is carried to infinity by the emitted positron.

The charged vacuum is a basically new physical concept. It is derived solely from the properties of the relativistic wave equation and applies generally to deeply bound fermionic systems. The overcritical atom is only the most transparent example, where a well-known external potential is specified. It is, however, quite interesting to note that in other areas of physics analogous phenomena have been found. For example, Bastard and Nozières [25] in the context of solid state physics investigate the behaviour of a donator level which moves through a band edge into the conduction band+. (More precisely under the influence of increasing external pressure a gap is formed and enlargened while the hole state stays in its position.) As in the case of the overcritical atom the bound state becomes quasistationary but a localized charge distribution is retained.

Our treatment of the overcritical phenomena in the autoionization model has been physically very elucidating. It was, however, not complete in two ways. Most important, the formalism has to be extended and justified by a quantum field theoretical treatment (Sect.4). Before doing this we will take in the next section a closer look at the solutions of the Dirac equation in a central Coulomb potential and verify our statements on energies and wave functions.

+We thank Prof.H.Thomas, Zürich, for drawing our attention to this work.

3. EXACT SOLUTION OF THE SINGLE-PARTICLE DIRAC EQUATION

In this section we will investigate the exact solutions of the Dirac equation with Coulomb potential generated by a point - or extended nuclear charge - distribution. The special case of the wave function at E = -m and the determination of the critical radius (or charge, resp.) will be illustrated explicitly. It will turn out in our discussion, that the motion of an overcritical point-like Coulomb potential is physically not admissible since the limit $R \to 0$ cannot be defined in a unique manner.

We follow the elaborate papers [17,24,26,27] and the already quoted standard reference books. Throughout we will use the abbreviations

$$\lambda = [m^2 - E^2]^{1/2} \tag{44}$$

$$\gamma = [\kappa^2 - (Z\alpha)^2]^{1/2} \tag{45}$$

$$\mu = \frac{Z\alpha E}{\lambda} \quad . \tag{46}$$

To begin with, the bound state problem for $|Z\alpha| < \kappa$ has a wellknown solution due to [28,29]. The radial Dirac equation (10) with Coulomb potential $V(r) = -Z\alpha/r$ can be transformed in the shape of Kummer's differential equation, leading to confluent hypergeometric functions. In terms of the Whittaker function

$$M_{\mu,\gamma}(x) = x^{1/2+\gamma}\, e^{-x/2}\; {}_1F_1(\tfrac{1}{2}-\mu+\gamma,\; 1+2\gamma, x\,) \; . \tag{47}$$

The solution of (10) which satisfies the asymptotic boundary condition $u_1(r)$, $u_2(r) \to 0$ for $r \to \infty$ can be written

$$u_1(r) = c_o (m+E)^{1/2} (2\lambda r)^{-1/2} [\phi_1(r) - \phi_2(r)]$$

$$u_2(r) = -c_o (m-E)^{1/2} (2\lambda r)^{-1/2} [\phi_1(r) + \phi_2(r)] \tag{48}$$

with

$$\phi_1(r) = \Gamma(1-2\gamma)\Gamma(-\mu+\gamma)\, M_{\mu+\frac{1}{2},\gamma}(2\lambda r) + \Gamma(1+2\gamma)\Gamma(-\mu-\gamma)\, M_{\mu+\frac{1}{2},-\gamma}(2\lambda r)$$

$$\phi_2(r) = \frac{E}{\kappa E-\mu m}[\Gamma(1-2\gamma)\Gamma(1-\mu+\gamma)\, M_{\mu-\frac{1}{2},\gamma}(2\lambda r) + \Gamma(1+2\gamma)\Gamma(1-\mu-\gamma)\, M_{\mu-\frac{1}{2},-\gamma}(2\lambda r)]. \tag{49}$$

c_o is a normalization constant determined by the condition

$$\int_0^\infty (u_1^2 + u_2^2)\, dr = 1 \quad . \tag{50}$$

The energy E entering the parameters λ and μ will be determined by the second boundary condition, imposed on the wave function at small distances. In the case of a 1/r-potential the solution (48), (49) has to be valid at all $r \to 0$. Since the functions are irregular at small distances their coefficients in (5) have to vanish compared to the regular solution. This condition yields

$$-\mu + \gamma = -n \tag{51}$$

which is just the Sommerfeld formula (Eq.(1)). The wave functions then become finite polynomials multiplied by $(2\lambda r)^{\gamma}\, e^{-\lambda r}$. To take into account the finite nuclear dimensions one has to solve the Dirac equation inside the nucleus separately and match the solutions u_k and $u_k^{(i)}$ at the nuclear radius $R(\rho(r) = 0$ for $r \geq R)$ by

$$\frac{u_1(R)}{u_2(R)} = \frac{u_1^{(i)}(R)}{u_2^{(i)}(R)} \quad . \tag{52}$$

Taking the potential of type II defined in (21) the matching condition (52) was solved numerically for the bound state energies [17]. As shown in Fig. 1 each level can be traced down to -m and $Z\alpha = \kappa$ has no special significance. Although γ becomes imaginary it is immediately shown from the relations

$$\Gamma(\bar{Z}) = \overline{\Gamma(Z)} \qquad \text{and} \qquad M_{\bar{\alpha},\bar{\beta}}(\bar{x}) = \overline{M_{\alpha,\beta}(x)} \tag{53}$$

that the functions (49) remain real.

For the sake of transparency we will discuss the explicit determination of the critical charge Z_{cr}. Treating this problem it is sufficient to use the simplified wave function (48) at the energy E = -m or

$\lambda \to 0$, $\mu \to -\infty$. The solution of the accordingly specialized differential equation can be obtained directly. Instead, one can also take the appropriate limit in Eq. (50). E.g. the limit formula

$$\lim_{a\to\infty} {}_1F_1(a,b,\tfrac{Z}{a}) = Z^{\frac{1-b}{2}} \Gamma(b) I_{b-1}(2\sqrt{Z}) \tag{54}$$

can be employed. The Bessel function of the second kind $I_\alpha(x)$ may be transformed to the regular solution $K_\alpha(x)$. Not bothering about the normalization constant we obtain

$$u_1(r) = \bar{c}_o\, K_{i\nu}\,(\sqrt{8Z\alpha r}) \tag{55}$$

and

$$u_2(r) = \frac{\bar{c}_o}{Z\alpha}\,[\sqrt{2Z\alpha r}\; K'_{i\nu}\,(\sqrt{8Z\alpha r}) + \kappa\, K_{i\nu}\,(\sqrt{8Z\alpha r})]$$

with

$$\nu = 2\,\sqrt{(Z\alpha)^2 - \kappa^2}\;.$$

$K_{i\nu}(x)$ and $K'_{i\nu}(x)$ are the MacDonald function and its derivative with respect to the argument. Closer inspection of its asymptotic properties proves that the exact electron wave function at the diving point is localized (exponential decrease at $r \to \infty$) and exhibits an infinite number of oscillations when approaching the origin ($r \to 0$). Based on general arguments we had discussed this already in Eqs. (15) to (17). To obtain physically meaningful solutions the wave function inside the nucleus with finite extension R has to be specified.

Taking the square well potential V_I of Eq.(21) for simplicity the solutions at $r < R$ are determined by spherical Bessel functions

$$u_1^{(i)}(r) = c_1 r\, j_{\ell_1}(pr)$$

$$u_2^{(i)}(r) = c_1\, \mathrm{sgn}(\kappa)\, \frac{pr\, j_{\ell_2}(pr)}{E+Z\alpha/R+m} \tag{56}$$

where $\ell_1 = \kappa + \frac{1}{2}(\mathrm{sgn}\,\kappa - 1)$, $\ell_2 = \ell_1 - \mathrm{sgn}(\kappa)$, $p^2 = (E + Z\alpha/R)^2 - m^2$. For the $1s_{1/2}$-level at $E = -m$ the interior solution is

$$u_1^{(i)} = \frac{G}{p} \sin pr \qquad u_2^{(i)} = -\frac{GR}{Z\alpha}\left(\frac{\sin pr}{pr} - \cos pr\right). \tag{57}$$

Since the nuclear charge radius happens to be small compared to the electron Compton-wave length, $R \ll 1/m$, we have $p = \frac{Z\alpha}{R}$; the matching condition then reads

$$\cot \frac{Z\alpha}{R} = \left(\frac{2R}{Z\alpha}\right)^{1/2} \frac{K'_{i\nu}(\sqrt{8Z\alpha R})}{K_{i\nu}(\sqrt{8Z\alpha R})}. \tag{58}$$

This simple transcendental equation allows the calculation of the $1s_{1/2}$ critical charge (at fixed radius R). Eq.(58) and several other formulas explicitly treating the level motion and wave functions with the approximation $\Lambda = \ln \frac{1}{R} \gg 1$ are discussed in [26,27].

Let us now turn to the continuum solutions which are essential for the understanding of the overcritical state [24]. It is useful to transform the radial equations (10) by

$$u_1 = \sqrt{E+m}\ (\phi_1 + \phi_2)$$

for energies $E > m$ (59)

$$u_2 = i\sqrt{E-m}(\phi_1 - \phi_2)$$

and

$$u_1 = \sqrt{-E-m}\ (\phi_1 + \phi_2)$$

for energies $E < -m$. (60)

$$u_2 = -i\sqrt{-E+m}\ (\phi_1 - \phi_2)$$

For a pure $-Z\alpha/r$-potential we obtain the system of coupled differential equations

$$\frac{d}{dx}\,\phi_1 = \left(\frac{1}{2} + \frac{iZ\alpha E}{px}\right)\phi_1 - \left(\frac{\kappa}{x} - \frac{iZ\alpha m}{px}\right)\phi_2$$

$$\frac{d}{dx}\,\phi_2 = \left(\frac{\kappa}{x} + \frac{iZ\alpha m}{px}\right)\phi_1 - \left(\frac{1}{2} + \frac{iZ\alpha E}{px}\right)\phi_2 \qquad (61)$$

which may be transformed to the second order equation

$$\frac{d^2\phi_1}{dx^2} + \frac{1}{x}\frac{d\phi_1}{dx} - \left[\frac{1}{4} + \left(\frac{1}{2} + iy\right)\frac{1}{x} + \frac{\gamma^2}{x^2}\right]\phi_1 = 0 \quad . \qquad (62)$$

We use the abbreviations

$$p^2 = E^2 - m^2, \qquad x = 2ipr\ , \qquad y = \frac{Z\alpha E}{p}\ . \qquad (63)$$

The differential equation (62), as in the discrete case, has a fundamental system of solutions consisting of Whittaker functions

$$\phi_1^{(\pm)}(x) = x^{\pm\gamma} e^{-x/2} \; {}_1F_1(\pm\gamma+1+iy, \pm 2\gamma+1, x)$$

$$\equiv x^{-1/2} M_{-(iy+\frac{1}{2}), \pm\gamma}(x) , \qquad (64)$$

where $\phi^{(+)}$ is the regular and $\phi^{(-)}$ the irregular solution at the origin. Since $\gamma = \sqrt{\kappa^2 - (Z\alpha)^2}$ becomes imaginary at $Z\alpha > \kappa$ the continuum functions $\phi^{(\pm)}$ again obtain the already encountered infinite oscillations (essential singularity at $r = 0$).

At first we treat the case $Z\alpha < \kappa$, i.e. γ real. To construct the most general continuum solution one has to adopt a linear combination

$$\phi_1 = [a_+ M_{-(iy+\frac{1}{2}), \gamma}(x) + a_- M_{-(iy+\frac{1}{2}), -\gamma}(x)] x^{-1/2} \qquad (65)$$

or, defining new constants η and N,

$$\phi_1 = N x^{-1/2} [\cos\eta \, e^{i\alpha_+} M_{-(iy+\frac{1}{2}), \gamma}(x) +$$

$$+ \sin\eta \, e^{i\alpha_-} M_{-(iy+\frac{1}{2}), -\gamma}(x)] \qquad (66)$$

with

$$e^{2i\alpha_\pm} = \mp\ (\gamma \pm iy)\ \frac{e^{\mp im\gamma}}{\kappa + iym/E}\ . \qquad (67)$$

ϕ_2 is obtained immediately from ϕ_1 by use of (61b). In (66) together with (67) the reality condition for u_1 and u_2, namely $\phi_1 = \phi_2^*$ has been incorporated. The wave functions u_1, u_2 must be real in order to satisfy real boundary conditions on the nuclear surface and to approach a spherical wave at infinity.

If we choose the continuum normalization

$$\int d^3x\ \psi^+_{E''}(x)\psi_{E'}(x) = \delta(E'-E") \qquad (68)$$

the normalization constant N is fixed by comparing the asymptotic behaviour of our solution with

$$\phi_1 \rightarrow \frac{e^{i(pr+\Delta)}}{2(\pi p)^{1/2}}\ , \qquad (69)$$

where Δ is the Coulomb phase shift.

Finally, the quantity η which determines the relative contribution of the regular $(+\gamma)$ and irregular $(-\gamma)$ Whittaker function is obtained from the matching condition at the nuclear surface. One immediately obtains

$$\tan\eta = -\frac{u_2^{(i)}(R)\,\mathrm{Re}[B_+(2ipR)]\pm[(E-m)/(E+m)]^{1/2}.}{u_2^{(i)}(R)\,\mathrm{Re}[B_-(2ipR)]\pm[(E-m)/(E+m)]^{1/2}.}\ \cdots$$

$$\ldots \frac{u_1^{(i)}(R)\ \mathrm{Im}[B_+(2ipR)]}{u_1^{(i)}(R)\ \mathrm{Im}[B_-(2ipR)]} \tag{70}$$

with the abbreviation

$$B_{\pm}(x) = x^{-1/2}\, e^{i\alpha_\pm}\, M_{-(iy+\frac{1}{2}),\pm\gamma}(x) \ . \tag{71}$$

The upper and lower sign apply for $E > m$ and $E < -m$, respectively.

Now we approach the most important part of our analysis: the continuum solution in the case $Z\alpha > \kappa$. It can be treated in a similar manner. Defining

$$\bar{\gamma} = [(Z\alpha)^2 - \kappa^2]^{1/2} = -i\gamma \tag{72}$$

where now γ is imaginary, the general solution takes the form

$$\phi_1 = [a_+ M_{-(iy+\frac{1}{2}),i\bar{\gamma}}(x) + a_- M_{-(iy+\frac{1}{2}),-i\bar{\gamma}}(x)]x^{-1/2} . \tag{73}$$

Now the reality condition $\phi_1 = \phi_2^*$ takes a general form, finally leading to the wave function

$$\phi_1 = Nx^{-1/2}[e^{i\eta} M_{-(iy+\frac{1}{2}),i\bar{\gamma}}(x) + ie^{-i\eta-\pi\bar{\gamma}} \ .$$

$$\cdot \frac{\bar{\gamma}-y}{\kappa+iym/E} M_{-(iy+\frac{1}{2}),-i\bar{\gamma}}(x)] \ . \tag{74}$$

The matching phase η is again determined by a condition similar to the one obtained above. Inserting

$$B_{\pm}(x) = M_{-(iy+\frac{1}{2}),i\bar{\gamma}}(x) \pm ie^{-\pi\bar{\gamma}} \frac{\bar{\gamma}-y}{\kappa+iym/E} M_{-(iy+\frac{1}{2}),-i\bar{\gamma}}(x) \tag{75}$$

it reads

$$\tan \eta = \frac{u_2^{(i)}(R)\mathrm{Re}[B_+(2ipR)]\pm[(E-m)/(E+m)]^{1/2} \cdot}{u_2^{(i)}(R)\mathrm{Im}[B_-(2ipR)]\mp[(E-m)/(E+m)]^{1/2} \cdot} \cdots$$

$$\cdots \frac{\cdot u_1^{(i)}(R)\mathrm{Im}[B_+(2ipR)]}{\cdot u_1^{(i)}(R)\mathrm{Re}[B_-(2ipR)]} \ . \tag{76}$$

Since it will be essential for the following discussion let us also give the scattering phase [24]

$$\Delta = y \ \ln\,(2pr) + \arg\,[e^{i\eta} \frac{\Gamma(2i\bar{\gamma}+1)}{\Gamma(i\bar{\gamma}+1+iy)} +$$

$$+ ie^{-i\eta+\pi\bar{\gamma}} \frac{\bar{\gamma}-y}{\kappa-iym/E} \cdot \frac{\Gamma(-2i\bar{\gamma}+1)}{\Gamma(-i\bar{\gamma}+1+iy)}] \ . \tag{77}$$

The so obtained Δ does not vanish in the limit of low energy and has to be renormalized. It turns out that the properly defined physical scattering phase shifts δ which satisfy the required limit

$$\lim_{|E|\to m} \delta = 0 \tag{78}$$

are obtained by subtracting an asymptotic phase, constant for all multipoles

$$\delta = \Delta - \delta_{\log} \equiv \Delta - y\left(\ln \frac{2pr}{|y|} + 1\right) + \frac{\pi}{4} . \tag{79}$$

By the way it is worthwhile to note that the physical phase shift in the limit of very high energy of the scattering particle does not vanish. Indeed, taking for example a point nucleus with $Z\alpha < \kappa$ we have $\gamma \to \pm Z\alpha$ and the phase shift approaches

$$\lim_{E\to\pm\infty} \delta = \arg \Gamma(\gamma+1\mp iZ\alpha) \pm Z\alpha(\ln Z\alpha - 1) - \\ - \frac{\pi}{2}\left[\gamma\mp(1 + \operatorname{sgn}\kappa)\right] - \frac{\pi}{4} . \tag{80}$$

This behaviour is contrary to the results of nonrelativistic scattering theory where $\gamma = \frac{Z\alpha}{r} \to 0$ and $\delta \to 0$. The discrepancy is easily explained since in the correct relativistic treatment the velocity remains limited and therefore the particle spends a finite time in the interaction region.

Having constructed the exact single-particle solutions of the Dirac equation with a realistic nuclear Coulomb potential we now turn again to the overcritical phenomenon. To this end the phase shift δ has been calculated numerically for nuclei with very large Z. As anticipated by our earlier results the phase shift undergoes a resonance: $\delta(E_{res}) = \pi/2$ at a certain energy depending on Z and R. Fig. 4 shows $\sin^2\delta$ as a function of energy for the nucleus Z = 184 yielding a narrow Breit-Wigner-shaped curve with $\Gamma = 4.8$ keV. In Fig.5 the dependence of the location E_{res} of the resonance on the nuclear charge is illustrated. The dashed curve shown for comparison is the "linear diving" result taken from $\Delta E = \langle\psi|Z'U(r)|\psi\rangle$ of the autoionization model.

Although this model is logically independent from the phase shift analysis, both methods can be linked by the relation [22,24]

$$\tan \Delta_E = \frac{-\frac{1}{2}\Gamma_E}{E-(E_o+\Delta E)-F(E)} \tag{81}$$

or

$$|a(E)|^2 = \frac{2}{\pi\Gamma_E} \sin^2 \Delta_E \,. \tag{82}$$

Employing the parametrization $E_{res} \sim Z'\delta - Z'^2\tau$, $\Gamma_E \sim Z'^2\gamma$. the following precise numerical results have been obtained for the resonance parameters :

	$1s_{1/2}$	$2p_{1/2}$
Z_{cr}	172	185
δ (keV)	29	37.8
τ (keV)	.33	.22
γ (keV)	.04	.08

To end this section we take a look at the over-critical point nucleus problem. Continuum solutions at $Z\alpha > \kappa$ in the potential $V = -Z\alpha/r$ cannot be constructed. It can be shown that also the solution with cut-off potential has no unambiguous limit at $R \to 0$. The proof starts form Eq. (76).

The matching phase can be written

$$\mathrm{tg}\,\eta = -\mathrm{tg}(\bar{\gamma}\,\ln 2pR + \varepsilon(R)) \qquad (83)$$

where ε turns out (for $pR << 1$) to be independent of R, leaving the divergent logarithmic term $\bar{\gamma}\ln 2pR$. Therefore no point charge limit can be defined. If we require, that the resonance condition is satisfied we find

$$Rp_{res} \to \text{const}$$

which implies

$$E_{res} \to -\left(m^2 + \frac{\text{const.}}{R^2}\right)^{1/2} \sim \frac{\text{const.}}{R}\ . \qquad (84)$$

Hence all bound states with $|\kappa| < Z\alpha$ obtain an infinite binding energy when $R \to 0$. This, by the way, explains the vertical tangent of e.g. the 1s-state in the Sommerfeld formula at $Z\alpha = 1$. Fig. 6 shows quantitatively the dependence of the critical charge Z_{cr} on the extension R_o of the nucleus, demonstrating that even for a very small radius of about 1 fm Z_{cr} is significantly greater than 137.

Physically, of course, even in the case of an extremely small overcritical nucleus there will not be infinitely many "diving" states. Screening of the nuclear charge, i.e. electron-electron-interaction will stop this process and produce a certain effective reduced charge of the system (see Sect.7). This is very similar to the case of overcritical Boson fields, where the energyless pair-production is only limited by the mutual particle repulsion. (Contrary to the Fermionic situation the Pauli principle is not at work and a single overcritical state is sufficient to produce many particles.)

In Fig. 7 two wave functions of the negative energy continuum, computed from the above derived exact solution of the Dirac equation with extended-charge Coulomb potential are compared. While the ordinary continuum wave function ($E = -1.5$ m) decreases at small distances (Coulomb repulsion) the curve for $E = -1.7455$ m clearly exhibits a localized bound state resonance (the dived 1s level). At larger distance the huge 1s-bump joins the oscillating spherical wave solution. This behaviour can be interpreted in terms of the effective potential (cf. Eq.(13)) which has a maximum at intermediate distances. The continuum and bound state parts of the wave function therefore do not join directly but they are connected by a tunneling process through a potential barrier. For further mathematical details of the effective potential cf! the works of Popov and collaborators quoted in the references.

At this point we take the opportunity to make a remark which is important both from the theoretical and historical point of view. The ideas on the charged vacuum and on spontaneous positron production in strong fields which now seem to be fully understood and which have become liable to experimental examination, in fact are based on a number of earlier observations. Soon after the emergence of Dirac's theory, O. Klein [30] in 1929 investigated the reflection of an electron at a one-dimensional potential barrier of heigt V. Instead of the usual exponential decay inside the barrier he observed oscillating waves. They emerge in the energy interval $m < E < V - m$ when the potential V is larger than 2m. This phenomenon first was considered as a puzzling inconsistency of relativistic quantum mechanics and was christened "Klein's paradox". Later Sauter [31] showed that the transmission into the non-relativistically forbidden region is not caused by the discontinuity of V but happens as well for smoothly rising potential barriers (Nikolsky [32] came to the same conclusion for an oscillator potential). The penetrability, however, decreases exponentially if the extension of the barrier wing becomes larger than a few Compton wave lengths. Therefore the effect cannot be observed on a laboratory scale with macroscopic electrostatic potentials.

In view of the hole theory Klein's paradox can be ascribed to the possibility of pair production. The "negative energy" solution (compared to the position of the gap) describes a positron. To get rid of the unphysical infinitely extended barrier let us look at a potential well. If one draws the potential V(r) and the gap with the boundaries V + m and V - m one immediately realizes that the overcritical atomic phenomena are closely connected with Klein's paradox. A bound state inside the well can aquire an energy value E = -m which belongs to free waves in the negative energy continuum at large distance r. Both regions are separated by a classical forbidden area where only exponentially decreasing wave functions are possible (hence the damping factor of Sauter). Quantum mechanically tunneling through this barrier is possible as we have seen in Fig. 7. This leads to the spontaneous filling of unoccupied levels with $E < -m$ and to the emission of positrons producing a charged vacuum state. The arguments presented in the case of the supercritical atom (especially the quantized treatment in Sect. 4) apply also for more general potentials and are, in fact, a rigorous description of our present understanding of Klein's "paradox". For earlier references related to this subject cf. [33-35] and the quite recent and illuminating paper [36].

4. QUANTIZED DESCRIPTION OF OVERCRITICAL STATES

A consistent and reliable description of strong field phenomena certainly must be based on a formally correct quantized theory [37]. In the present section the decay of the neutral vacuum will be treated field theoretically. Without loss of generality one can restrict considerations on the problem of one deeply bound state interacting with the negative energy continuum. Many-particle effects and virtual excitations (for example vacuum polarization) will be discussed in the next section.

Let us start from the Dirac state vector in a weak external potential. One usually writes

$$\hat{\psi}(\vec{r},t) = \sum_{E_p>0} \hat{b}_p \phi_p(\vec{r}) e^{-iE_p t} + \sum_{E_p<0} \hat{d}_p^+ \phi_p(\vec{r}) e^{-iE_p t} \qquad (85)$$

with the anticommutation relations for Fermion operators

$$\{\hat{b}_p^+, \hat{b}_{p'}\} = \delta_{pp'} \qquad \{\hat{d}_p^+, \hat{d}_{p'}\} = \delta_{pp'} \ . \tag{86}$$

All other anticommutators vanish. The wave functions ϕ_p are eigensolutions of the classical Dirac Hamiltonian

$$H_D = \vec{\alpha}\cdot\vec{p} + \beta m + V(r) \tag{87}$$

with energy E_p. The vacuum state is defined by

$$\hat{b}_p|0\rangle = \hat{d}_p|0\rangle = 0 \ . \tag{88}$$

Since in (85) the negative energy solutions are introduced by a creation operator $\hat{d}_p^+$ this quantization corresponds to the familiar concept of the completely filled Dirac sea. $\hat{b}_p$ and $\hat{b}_p^+$ are annihilation and creation operators for electrons, $\hat{d}_p$ and $\hat{d}_p^+$ those for positrons. If the spectrum of H_D acquires deeply bound states the expansion (85) has to be rethought. The partition of the sum at energy 0 which goes back to the bound state interaction picture of Furry (1951) is not mandatory when a bound state in the region $-m < E_p < 0$ emerges. Introducing an adjustable Fermi surface [37], which divides occupied and empty states we may generalize (85)

$$\hat{\psi}(\vec{r},t) = \sum_{E_p>E_F} \hat{b}_p\phi_p(\vec{r})e^{-iE_pt} + \sum_{E_p<E_F} \hat{d}_p^+\phi_p(\vec{r})e^{-iE_pt} \ . \tag{89}$$

If E_F is chosen slightly above the negative energy continuum at -m, the ψ-operator (89) describes a bare nucleus with an empty electron shell). Although energy might be gained by filling the deeply bound level this state is stable since charge conservation prohibits the creation of a single electron. Thus we face a neutral and stable ground state. Generally, the vacuum will be defined as the state with lowest energy which is stable under the given interactions. If the external potential becomes overcritical, i.e. a bound state joins the lower continuum as a resonance, the so defined vacuum will be charged.

To proceed with the proof of this statement let us consider two different sets of classical wave functions, $|q,Z_{cr}\rangle$ and $|q,Z_{cr}+Z'\rangle$, which are eigenstates of the Hamiltonian H_D with near critical (Z_{cr}) or overcritical ($Z_{cr}+Z'$). Both sets are complete since H_D is a hermitean operator. (This would not be true for an overcritical point charge, since here the wave function is not integrable at the origin.) Analogously to the treatment in Sect. 2 we can expand

$$|E,Z_{cr} + Z'\rangle = a(E)\,|1s,Z_{cr}\rangle + \int_{E'<-m} dE'\,h_{E'}(E)\,|E,Z_{cr}\rangle \tag{90}$$

leading to

$$a(E) = V_{E^*}\left[E-E_{1s} - F(E) + i\pi|V_E|^2\right]^{-1} \tag{91}$$

$$h_{E'}(E) = \frac{V_{E'}}{E-E'+i\varepsilon}\,a(E) + \delta(E-E') \tag{92}$$

where V_E and $F(E)$ have been defined in Eqs. (30), (37). Coupling to higher bound or continuum states and the rearrangement of the negative energy continuum have been neglected.

Let us now analyze three different descriptions of the overcritical state.

(i) The atom is initially prepared with a filled K-shell. Then the ψ-operator in the undercritical basis reads

$$\hat{\psi}(\vec{r},t) = \sum_{E'>E_{F_0}} \hat{b}_{E'}|E',Z_{cr}> + \hat{d}^+_{1s\uparrow}|1s\uparrow,Z_{cr}> + \hat{d}^+_{1s\downarrow}|1s\downarrow,Z_{cr}>$$

$$+ \sum_{E'<-m} \hat{d}^+_{E'}|E',Z_{cr}> \tag{93}$$

and E_{F_0} lies slightly above E_{1s}. The sums are understood to include spin and angular momentum quantum numbers.

The ground state $|0>$ corresponding to (93) is stable and normalized $<0|0> = 1$. It carries two negative units of charge. This holds as well in the overcritical region. Here, however, the K-Hole states $\hat{d}^+_{1s\uparrow}|0>$, $\hat{d}^+_{1s\downarrow}|0>$ will become unstable.

Adopting the Hamiltonian

$$\hat{H}(Z) = : \hat{\psi}^+ H_D(Z)\hat{\psi} : \tag{94}$$

(where normal ordering was introduced to eliminate the infinite energy of the occupied levels below the Fermi surface) the energy of the 1s-state if given by

$$E_{1s} = \langle 0|\hat{d}_{1s}\ \hat{H}(Z_{cr}+Z')\ \hat{d}^+_{1s}|0\rangle =$$

$$= \langle 1s,Z_{cr}|H_D\ (Z_{cr}+Z')|1s,Z_{cr}\rangle = -m-Z'\delta, \qquad (95)$$

with $\delta = \langle 1s,Z_{cr}|\ U(r)\ |1s,Z_{cr}\rangle$.

(ii) If the atom is prepared with an empty K-shell, it is appropriate to define the Fermi surface below the 1s-level. Again expanding in the undercritical basis the ψ-operator now takes the form

$$\hat{\psi}'(\vec{r},t) = \sum_{E'>E_{F_0}} \hat{b}_{E'}|E',Z_{cr}\rangle + \hat{b}_{1s\uparrow}|1s\uparrow,Z_{cr}\rangle + \hat{b}_{1s\downarrow}|1s\downarrow,Z_{cr}\rangle$$

$$+ \sum_{E'<-m} \hat{d}^+_{E'}|E',Z_{cr}\rangle\ . \qquad (96)$$

The so defined vacuum $|0'\rangle$ (with $b_{1s}|0'\rangle = 0$, $d_E|0'\rangle = 0$), though neutral and normalized, is not stable for overcritical external fields. It will mix with the one electron - one positron state $\hat{b}_{1s}\hat{d}^+_E|0'\rangle$. The resulting eigenstates will be

$$|\psi'(E)\rangle = a(E)|0'\rangle + \int dE' h_{E'}(E)\ \hat{b}^+_{1s}\ \hat{d}^+_{E'}|0'\rangle\ . \qquad (97)$$

Following [37], we show that the coefficients $a(E)$, $h_{E'}(E)$ are those of Eqs. (36). Let us consider, for convenience, the Hamiltonian

$$\hat{H}' = :\hat{\psi}'^+ H_D \hat{\psi}' : + E_{1s} \qquad (98)$$

incorporating the energy E_{1s} (this is permissible since a constant term in the Hamiltonian leads only to an unobservable overall energy shift). By construction the state (97) obeys the equation

$$\hat{H}'|\psi'(E)\rangle = E|\psi'(E)\rangle . \tag{99}$$

To solve (97), (99) for the expansion coefficients we need a number of matrix elements of H' with the vacuum. We employ the explicit form of H', namely (discarding the states $E > E_{F_o}$)

$$\hat{H}' = (1+\hat{b}^+_{1s}\hat{b}_{1s})E_{1s} + \sum_{E'<E_{F_o}} (\hat{d}_{E'}\hat{b}_{1s}V_{E'} + \hat{b}^+_{1s}\hat{d}^+_{E'}V^*_{E'}) -$$

$$- \sum_{E,E'>E_{F_o}} \hat{d}^+_E\, \hat{d}^+_{E'}\, U_{E'E} \quad . \tag{100}$$

This leads to the matrix elements

$$\langle O'|\hat{H}'|O'\rangle = E_{1s} \tag{101}$$

$$\langle O'|\hat{H}'\ \hat{b}^+_{1s}\ \hat{d}^+_E|O'\rangle = V_E \tag{102}$$

$$\langle O'|\hat{d}_{E'}\hat{b}_{1s}\hat{H}'\hat{b}^+_{1s}\hat{d}^+_E|O'\rangle = E\delta(E-E') + U_{E'E} \ . \tag{103}$$

These matrix elements are identical with Eqs. (29) - (31). Therefore the expansion coefficients $a(E), h_{E'}(E)$ of the classical autoionization model (7), (8) hold also in the quantized treatment. The interpretation too remains the same. If the inital state was $|O'\rangle$ it will be spread over the negative energy continuum. The probability to find it in the final state $\psi'(E)$ is

$$p(E) = |\langle 0'|\psi'(E)\rangle|^2 = |a(E)|^2 \quad . \tag{104}$$

Since $\psi'(E)$ asymptotically describes a continuum state the bound positron of the undercritical case becomes a free positron moving to infinity.

The state (97) was written for only one spin orientation. The spin degeneracy, however, results only in a minor change. We include the two electron - two positron states (87) and obtain

$$|\psi_{E'}\rangle = a(E)|0'\rangle + \int dE'\,(h_{E'}(E)\hat{b}^+_{1s\uparrow}\hat{d}^+_{E'\uparrow}|0'\rangle + g_{E'}(E)\hat{b}^+_{1s\downarrow}\hat{d}^+_{E'\downarrow}|0'\rangle)$$

$$+\int dE'dE''f_{E'E''}(E)\hat{b}^+_{1s\uparrow}\hat{b}^+_{1s\downarrow}\hat{d}^+_{E'\uparrow}\hat{d}^+_{E''\downarrow}|0'\rangle \quad . \tag{105}$$

If we consider only the overcritical potential there are no finite matrix elements from the ground state to the two particle - two hole state since $Z'U(r)$ is a one body operator. The Electron-electron-interaction on the other hand - which is a two body-force - changes smoothly as a function of Z and therefore does not lead to a qualitative modification of our results. Thus we can neglect the last term in Eq.(105). Then the solution of (105) reduces to that of Eq.(97) up to the trivial factor of two. This reflects the fact that both spin states decay independently and two positrons are emitted, yielding $p(E) = 2|a(E)|^2$.

(iii) To complete our discussion of the overcritical state we describe the dynamics of the change between neutral and charged vacuum. Consider the occupied 1s-state, embedded in the negative energy continuum. The field operator is

$$\hat{\psi}(\vec{r},t) = \sum_{E'>-m} \hat{b}_{E'}|E',Z_{cr}+Z'>+ \sum_{E'<-m} \hat{d}^{+}_{E'}|E',Z_{cr}+Z'>. \quad (106)$$

Projecting the overcritical states (which contain the 1s-resonance)on the undercritical basis my means of

$$|q,Z_{cr}+Z'>= \sum_{n} <n,Z_{cr}|q,Z_{cr}+Z'>|u,Z_{cr}>=\sum_{n} a(n,q)|n,Z_{cr}> \quad (107)$$

we obtain in first order

$$\hat{\psi}(\vec{r},t) = \sum_{E'>-m} \hat{b}_{E'}|E',Z_{cr}> + [\int dE'\hat{d}^{+}_{E'}\ a(E)]|1s,Z_{cr}>$$

$$+ \sum_{E',E<-m} \hat{d}^{+}_{E'}\ h_{E'}(E)|E,Z_{cr}> . \quad (108)$$

We have employed the approximations $a(1s,E) \approx a(E)$, $a(E',E) \approx h_{E'}(E)$ for $E,E' < -m$ and $a(E',E) = \frac{U_{E',E}}{E-E'} + \delta(E-E') \approx \delta(E-E')$ for $E,E' > -m$.

Eq.(108) shows, that we can define a generalized ("collective") operator for vacancies in the embedded 1s-state

$$\hat{\tilde{d}}^{+}_{1s} = \int dE'\ \hat{d}^{+}_{E'}\ a(E') \quad . \quad (109)$$

This operator, however, is not linearly independent in the overcritical case (for undercritical fields it just reduces to the 1s-hole-operator). If we prepare a bound state vacancy in this way at a certain time t = 0, the true time-dependent state can be written

$$|\psi(t)\rangle = y(t)\,\hat{\tilde{d}}^{+}_{1s}|0\rangle + \int dE'\, W_{E'}(t)\hat{d}^{+}_{E'}|0\rangle . \qquad (110)$$

$|0\rangle$ is the charged vacuum defined above. The initial state $\hat{\tilde{d}}^{+}_{1s}|0\rangle$ plays the role of a collective vacuum excitation and corresponds to a 1s-hole. The time development of $|\psi(t)\rangle$ will be described by the Schrödinger equation

$$\hat{H}|\psi(t)\rangle = i\frac{\partial}{\partial t}|\psi(t)\rangle = i\frac{dy}{dt}\,\hat{\tilde{d}}^{+}_{1s}|0\rangle + i\int_{E'<-m} dE'\frac{dW_{E'}}{dt}\,\hat{d}^{+}_{E'}|0\rangle . \qquad (111)$$

Projecting on $\langle 0|\hat{\tilde{d}}_{1s}$ and $\langle 0|\hat{d}_{E}$ we obtain two coupled differential equations for the coefficients y(t) and $W_E(t)$. The required matrix elements are

$$\langle 0|\hat{\tilde{d}}_{1s}\,\hat{H}\,\hat{\tilde{d}}^{+}_{1s}|0\rangle = \int E dE|a(E)|^2 \equiv \tilde{E}_{1s}$$

$$\langle 0|\hat{\tilde{d}}_{1s}\,\hat{H}\,\hat{d}_{E}|0\rangle = E\, a^{*}(E)$$

$$\langle 0|\hat{d}_{e}\,\hat{H}\,\hat{d}_{E'}|0\rangle = E\,\delta(E-E')$$

$$\langle 0|\hat{\tilde{d}}_{1s}\,\hat{\tilde{d}}^{+}_{1s}|0\rangle = \int dE|a(E)|^2 \sim 1$$

$$\langle 0|\hat{\tilde{d}}_{1s}\,\hat{d}^{+}_{E}|0\rangle = a^{*}(E) . \qquad (112)$$

The system of coupled differential equations is therefore

$$y(t)\,\tilde{E}_{1s} + \int_{E'<-m} W_{E'}(t)\; a^*(E')E'\; dE' =$$

$$= i\dot{y}(t) + i \int_{E'<-m} dE'\; \dot{W}_{E'}(t)\; a^*(E') \tag{113}$$

$$y(t)\; a(E)E + W_E(t)E = i\dot{y}(t)a(E) + i\dot{W}_E(t)\; . \tag{114}$$

Additionally the solution must obey the normalization condition

$$\langle\psi(t)|\psi(t)\rangle = 1 = |y(t)|^2 + \int dE'\; |W_{E'}(t)|^2\; . \tag{115}$$

After some manipulations the solution of (113) - (115) satisfying the initial conditions $y(0) = 1$, $W_{E'}(0) = 0$ reads

$$y(t) = e^{-i\tilde{E}_{1s}t}\; e^{-\frac{1}{2}\Gamma|t|} \tag{116}$$

$$W_E(t) = a(E)\; [e^{-iEt} - e^{-i\tilde{E}_{1s}t}\; e^{-\frac{1}{2}\Gamma|t|}]. \tag{117}$$

$\Gamma = 2\pi|V_{\tilde{E}_{1s}}|^2$ is the width at the resonance energy $\tilde{E}_{1s} = E_{1s} + F(\tilde{E}_{1s})$. Eqs. (116), (117) evidently describe an exponential decay of the prepared 1s-hole. At given time t the probability to find a localized vacancy is

$$p(t) = |y(t)|^2 = e^{-\Gamma|t|} \quad . \tag{118}$$

The positron spectrum at time $t >> 1/\Gamma$ approaches

$$|W_E(\infty)|^2 = |a(E)|^2 \tag{119}$$

which, as we know, describes a Breit-Wigner Shape.

This completes our discussion of the overcritical state. We were able to corroborate the conceptually clear result of the autoionization of positrons. Eq. (116) proves that the neutral vacuum $|0'\rangle = \hat{\tilde{d}}^{+}_{1s\uparrow}\hat{\tilde{d}}^{+}_{1s\downarrow}|0\rangle$ becomes unstable under the action of an overcritical potential and decays into the vacuum $|0\rangle$ bearing two negative units of charge [10,23]. The charge distribution in the new vacuum remains localized and closely resembles that of a normal filled electron shell. It may be derived taking the properly understood vacuum expectation value of the operator

$$\hat{\rho}(\vec{r}) = -\frac{e}{2}[\hat{\bar{\psi}}(\vec{r}),\gamma^{0}\hat{\psi}(\vec{r})]_- \tag{120}$$

$$\rho_{inel} = \frac{e}{2}\int_{-\infty}^{-m} dE\psi_E^{+}(\vec{r})\psi_E(\vec{r}) - \frac{e}{2}\int_{m}^{\infty}\psi_E^{+}(\vec{r})\psi_E(\vec{r})dE - \frac{e}{2}\sum_{BS}\rho_{BS} \tag{121}$$

where the sum includes all empty bound states. Renormalization problems are avoided if one subtracts the undercritical vacuum charge and integrates over an energy interval in the vicinity of the resonance [1]:

$$\tilde{\rho}_{1s}(\vec{r}) = e \int_{\tilde{E}_{1s}-\Delta E}^{\tilde{E}_{1s}+\Delta E} dE[\psi_E^+(\vec{r})\psi_E(\vec{r}) - \psi_{-E}^+(\vec{r})\psi_{-E}(\vec{r})]. \qquad (122)$$

Fig. 8 shows the charge density for several (collective) supercritical bound state resonances.

The distribution ρr^2 has a shape completely analogous in the undercritical and overcritical case. With increasing charge the radial extension of the bound states shrinks to about 20 fm (maximum of ρr^2) and below. This smoothly continues the "collapse" of the spatial extension of the wave function with increasing Z which is illustrated in Fig. 9 for the 1s-level [21]. Fig.8 also demonstrates the drastic difference between the real vacuum charge and the virtual vacuum polarization. The latter (dashed line) is about two orders of magnitude smaller and, of course, has a vanishing space integral (this is not evident from the figure since it shows only the contribution from a finite energy interval).

The discussion of the charged vacuum presented in this section is due to [37]. Some of the results have been found independently by Fulcher and Klein [38]. Fulcher and Klein [39] and Klein and Rafelski [40] present a treatment equivalent to the one described above.

They investigate the reduced Hamiltonian

$$\hat{H}_r = -\varepsilon \sum_\sigma \hat{b}_\sigma^+ \hat{b}_\sigma + \sum_{p,\sigma} \varepsilon_p \hat{b}_{p\sigma}^+ \hat{b}_{p\sigma} - \sum_{p\sigma} U_p \hat{b}_{p\sigma}^+ \hat{b}_\sigma - \sum_{p\sigma} U_p^* \hat{b}_\sigma^+ \hat{b}_{p\sigma} \qquad (123)$$

where the summation runs over energy states $-\infty<-m-\varepsilon_p<-m$ and spin orientation $\sigma = \pm 1/2$. $\hat{b}^+_\sigma$ and $\hat{b}^+_{p\sigma}$ are positron creation operators in the bound and continuous region. The energies are measured relative to -m.

Since the Hamiltonian (123) is quadratic it can be diagonalized by a linear transformation

$$\hat{\beta}_\sigma = A_o \hat{b}_\sigma - \sum_p A_p \hat{b}_{p\sigma}$$

$$\hat{\beta}_{q\sigma} = - A^{(t)}_{oq} \hat{b}_o + \sum_p A^{(t)}_{pq} \hat{b}_{p\sigma} ; \qquad (124)$$

we obtain the form

$$\hat{H}_r = \sum_\sigma \omega \hat{\beta}^+_\sigma \hat{\beta}_\sigma + \sum_{q\sigma} \varepsilon(q) \hat{\beta}^+_{q\sigma} \hat{\beta}_{q\sigma} . \qquad (125)$$

This leads to an eigenvalue condition for the bound state

$$\chi = F(\chi - \varepsilon) \qquad (126)$$

with

$$\chi = \omega + \varepsilon \quad \text{and} \quad F(x) = \sum_p \frac{|U_p|^2}{x-\varepsilon(p)} \equiv \int_0^\infty dE \frac{|V_E|^2}{x-E} . \qquad (127)$$

The integral equation (126) has real solutions only for $\varepsilon > 0$ as illustrated in the scetch below once again demonstrating the disappearance of the discrete bound state in a supercritical potential. The solutions for the coefficients A in (40) are identical with those obtained earlier.

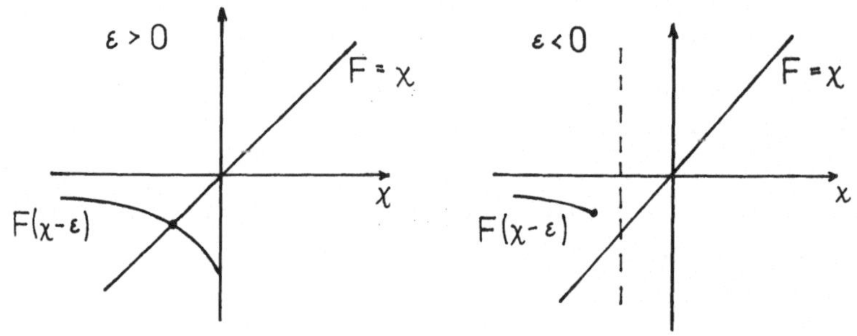

5. VACUUM POLARIZATION IN STRONG EXTERNAL FIELDS

Quantum electrodynamics of strong fields shows interesting features which are worthwhile investigating also apart from the overcritical phenomena. For all atoms except hydrogen one has to deal with two different parameters defining the coupling strength, namely α and $Z\alpha$. The latter quantity reaches about 0.7 in the heaviest stable elements and can even exceed unity in superheavy systems. Thus the coupling is neither weak ($g^2 << 1$) nor strong ($g^2 >> 1$) and the usual series expansion in $(Z\alpha)^n\alpha^m$ becomes questionable. We will therefore describe a method which employs the exact Dirac propagator in the external Coulomb field and includes all orders $(Z\alpha)^n$. The coupling to the radiation field characterized by the small constant α will be treated in perturbation theory. We will obtain many-particle equations unifying the Hartree-Fock method of atomic physics with the treatment of vacuum polarization and selfenergy of usual QED [37,41,42]. At the end of this section we will discuss the explicit results on vacuum polarization recently obtained by several authors.

Following Schwinger [43] we start from the Lagrangian

$$L = L_{oe} + L_{oem} + \frac{e}{2}\{j_\mu, A^\mu\} + ej_{ex_\mu} A^\mu . \tag{128}$$

L_{oe} and L_{oem} are the standard Lagrangians for the free electron (e) and photon field. A_μ and j_μ are the electromagnetic potential and electronic current while j_{ex_μ} describes an external current, e.g. the electrostatic potential due to the nuclear charge distribution. The four current j_μ is expressed by the bilinear covariant

$$j_\mu = -\frac{1}{2}(\bar{\psi}\gamma_\mu\psi - \bar{\psi}^c\gamma_\mu\psi^c) = -\frac{1}{2}[\bar{\psi},\gamma_\mu\psi] . \tag{129}$$

The equations of motion derived from (128) are well known

$$(\gamma\cdot p - m)\psi - \frac{e}{2}\gamma_\mu\{A^\mu,\psi\} = 0 \tag{130}$$

$$\partial^\mu F_{\mu\nu} = e\,(j + j_{ex})_\nu \tag{131}$$

with the electromagnetic field tensor $F_{\mu\nu} = \partial_\mu A_\nu - \partial_\nu A_\mu$. In the Lorentz gauge $\partial_\mu A^\mu = 0$ (131) is a Poisson equation with the formal solution

$$A_\mu(x) = e \int d^4y\, D_o(x-y)(j + j_{ex})_\mu(y) . \tag{132}$$

We adopt the Feynman propagator

$$D_o(y-x) = \int \frac{d^4q}{(2\pi)^4} \frac{e^{-iq(y-x)}}{q^2 - i\varepsilon}$$

$$= \frac{1}{4\pi} (\delta((y-x)^2) - \frac{i}{\pi} P\ (y-x)^{-2})$$

$$= \frac{i}{4\pi^2} [(y-x)^2 + i\varepsilon]^{-1} . \qquad (133)$$

Insertion of (132) in the Dirac equation (130) leads to

$$(\gamma \cdot p - e\gamma \cdot A_{ex} - m)\psi - \frac{e^2}{2}\{\gamma^\mu \psi(x), \int d^4y\ D_o(x-y) j_\mu(y)\}. \qquad (134)$$

To perform actual calculations we expand the field operator in the basis of exact classical solutions of an effective single particle equation

$$\psi_e(x) = \sum_{E_q > E_F} b_q \phi_q(\vec{x}, E_q)\ e^{-iE_q t} +$$

$$+ \sum_{E_q < E_F} d_q^+ \phi_q(\vec{x}, E_q)\ e^{-iE_q t} , \qquad (135)$$

and a similar equation for $\bar{\psi}_e(x)$. The operators b_q and d_q satisfy Fermion anticommutator relations. A typical choice of the Fermi surface E_F is illustrated in Fig.10.

The Feynman Dirac propagator connected with Eq. (125) is defined as the time ordered product of field operators

$$i\ S_{F\alpha\beta}(x,y) = \langle 0|T(\psi_\alpha(x), \bar{\psi}_\beta\ (y))|0\rangle =$$

$$= \begin{cases} \langle 0|\psi_\alpha(x)\bar{\psi}_\beta(y)|0\rangle & (t_x > t_y) \\ -\langle 0|\bar{\psi}_\beta(y)\psi_\alpha(x)|0\rangle & (t_x < t_y) \end{cases} \tag{136}$$

and therefore

$$i\, S_{F\alpha\beta}(x-y) = \begin{cases} \sum_{E_q > E_F} \phi_{q\alpha}(\vec{x},E_q)\bar{\phi}_{q\beta}(\vec{y},E_q)\, e^{i(t_y-t_x)E_q} & (t_x>t_y) \\ -\sum_{E_q < E_F} \phi_{q\alpha}(\vec{x},E_q)\bar{\phi}_{q\beta}(\vec{y},E_q)\, e^{i(t_y-t_x)E_q} & (t_x<t_y) . \end{cases} \tag{137}$$

Obviously the propagator is only uniquely defined by prescribing an additional condition, namely the position of the Fermi surface.

The ground state expectation value of the current is (see for example [5])

$$\langle 0|j_\mu|0\rangle = \mathrm{Tr}\,(i\, S_F(x,x)\gamma_\mu) , \tag{138}$$

where the propagator at the point x = y is defined by the prescription

$$S_F(x,x) = \frac{1}{2} \lim_{\varepsilon \to 0} [S_F(x,x+\varepsilon) + S_F(x,x-\varepsilon)] . \tag{139}$$

Here $\varepsilon^2 > 0$ and in a particular Lorentz frame $\varepsilon_\mu = (\varepsilon',0,0,0)$. We thus obtain

$$\langle 0|j_\mu|0\rangle = \frac{1}{2} \sum_{E_q<E_F} \bar{\phi}_q(\vec{x},E_q)\gamma_\mu\phi_q(\vec{x},E_q) -$$

$$- \frac{1}{2} \sum_{E_q>E_F} \bar{\phi}_q(\vec{x},E_q)\gamma_\mu\phi_q(\vec{x},E_q)$$

$$\equiv \tilde{\sum_q} \bar{\phi}_q\gamma_\mu\phi_q \quad . \tag{140}$$

The "tilde sum" is defined by

$$\tilde{\sum_q} = \frac{1}{2} \left(\sum_{E_q<E_F} - \sum_{E_q>E_F} \right) = \sum_q^{HF} + \frac{1}{2} \left(\sum_{E_q<-m} - \sum_{E_q>-m} \right) . \tag{141}$$

It equals the Hartree-Fock sum $\sum_q^{HF} = \sum_{-m<q<E_F}$ over all occupied bound states plus the "Uehling sum". The latter usually accounts for the effects of vacuum polarization since it describes the induced current due to the presence of the external source [44,45].

We wish to derive a system of classical self-consistent one particle equations for the set ϕ_q. Let us consider the matrix elements of the equation of motion between the vacuum and the single particle (or hole) state $b_q^+|0\rangle$, $d_q^+|0\rangle$. Eq. (134) reduces to

$$(\gamma\cdot p - e\gamma\cdot A_{ex} - m)\phi_q(x) = \langle 0|\frac{e^2}{2}\int d^4y D_o(x-y)\{j_\mu(y),\gamma_\mu\psi(x)\}b_q^+|0\rangle . \tag{142}$$

Using the definition of the current operator (129) and the commutation relations of the field operators the equation may be transformed

$$(\gamma\cdot p-e\gamma\cdot A_{ex}-m)\phi_q(x)=e^2\gamma_\alpha\int d^4yD_o(x-y)\tilde{\sum_p}\bar{\phi}_p(y)\gamma^\alpha\phi_p(y)\phi_q(x)$$

$$-e^2\gamma_\alpha\int d^4yD_o(y-x)\tilde{\sum_p}\bar{\phi}_p(y)\gamma^\alpha\phi_p(x)\phi_q(y)\,. \quad (143)$$

The interaction term on the right hand side consists of a direct and an exchange contribution. Their influence can be understood calculating the total change in energy which they produce. In first order we have

$$\Delta E_{ex}=-\tfrac{1}{2}e^2\int d^4xd^4yD_o(y-x)\tilde{\sum}\tilde{\sum_{pq}}\bar{\phi}_p(y)\gamma_\alpha\phi_p(x)\bar{\phi}_q(x)\gamma^\alpha\phi_q(y) \quad (144)$$

$$\Delta E_d=\tfrac{1}{2}e^2\int d^4xd^4yD_o(y-x)\tilde{\sum}\tilde{\sum_{pq}}\bar{\phi}_p(y)\gamma_\alpha\phi_p(y)\bar{\phi}_q(x)\gamma^\alpha\phi_q(x)\,. \quad (145)$$

The double tilde sum allows a decomposition into three terms

$$\tilde{\sum_p}\tilde{\sum_q}=\sum_p^{Uh}\sum_q^{Uh}+\Big(\sum_p^{HF}\sum_q^{Uh}+\sum_q^{HF}\sum_p^{Uh}\Big)+\sum_p^{HF}\sum_q^{HF}\,. \quad (146)$$

The first term leads to an energy shift independent of the electron occupation (vacuum correlation energy) and does not lead to observable energy differences between electronic states. The last expression contains the usual Hartree-Fock energy discribing the mutual interaction of bound electrons. The mixed double sum in (146) leads to electron self-energy in the exchange term and vacuum polarization in the direct term. (For further discussion see [37]).

Eq. (143) may be simplified since we are interested in time independent (or adiabatic) fields. Separating the time dependence by $\phi_q(x) = e^{-iE_q t}\,\phi_q(\vec{x})$ and taking the Fourier transform of the photon propagator (133)

$$\int_{-\infty}^{\infty} d(t_y - t_x)\, D_o(y-x)\, e^{-i(E_q - E_p)(t_y - t_x)} = \frac{e^{i|E_p - E_q||\vec{x}-\vec{y}|}}{4\pi|\vec{x}-\vec{y}|}, \quad (147)$$

we arrive at the following self-consistent equation

$$[(E - eV_{ex}) - (\vec{\alpha}\cdot\vec{p} - e\vec{\alpha}\cdot\vec{A}_{ex}) - \beta m]\phi_q(x) = \frac{e^2}{4\pi}\gamma_o\gamma_\alpha \int d^3y \frac{\tilde{\Sigma}\phi_p^+(y)\gamma^o\gamma^\alpha\phi_p(y)}{|\vec{x} - \vec{y}|}\phi_q(x$$

$$- \frac{e^2}{4\pi}\gamma_o\gamma_\alpha \int d^3y \tilde{\sum_p} \frac{e^{i|E_p - E_q||\vec{x}-\vec{y}|}}{|\vec{x}-\vec{y}|}\phi_p^+(y)\gamma^o\gamma^\alpha\phi_p(x)\phi_q(y).$$

(148)

This equation unifying the treatment of real and virtual electron-electron-interaction was obtained by Reinhard [46], Reinhard et al. [41] who also discuss its derivation from the Schwinger equation for the electron and photon Green function and the proper renormalization procedure.

Eq. (148) may be represented graphically if one defines the photon propagator $\sim\!\sim\!\sim$, the exact Fermion propagator $\equiv$ and the single particle electron propagator in external field ——— :

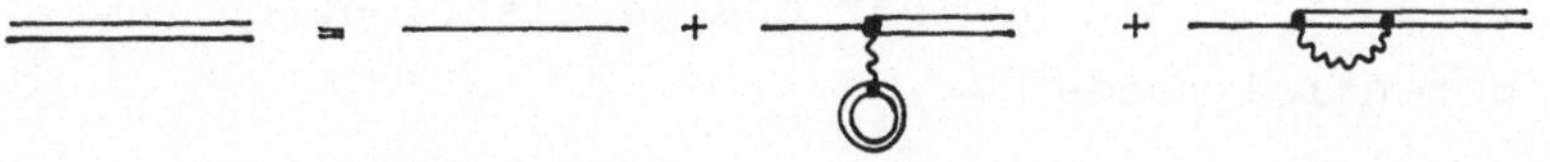 (149)

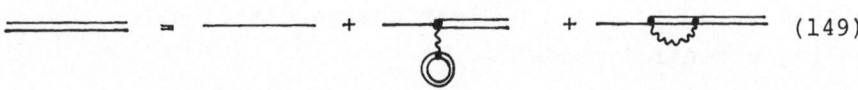

This relation follows from the exact Schwinger equations when the exact photon propagator is substituted by the free one. The exact vertex function is approximated in the same way (a similar procedure leads to the Hartree-Fock equations in the nonrelativistic case [47]).

The self-consistent equation (148) has not yet been fully exploited numerically. Its extension to two fermionic fields, namely electrons and muons and the consequences for muonic atoms are discussed by Rafelski et al.[42]. Here we only remark that the electron-electron interactions contained in the self-consistent terms change smoothly as a function of the strength of V_{ex}. The tilde sum is well defined in the undercritical case. Therefore it was justified to take eigenfunctions of the single particle Dirac Hamiltonian when qualitatively discussing the diving process in Sect. 4.

The remainder of this section will be devoted to the explicit results on vacuum polarization in high Z atoms recently obtained by several authors. The early result of Serber [45] and Uehling [44] applies to a pointlike nucleus. The vacuum polarization charge density and its corresponding potential produced by the nuclear Coulomb field was calculated to first order in $Z\alpha$. The resulting formula is

$$V_{Uh}(r) = -\frac{2\alpha Z\alpha}{3\pi r}\int_m^\infty d\tau\; e^{-2\tau r}\;(1+\frac{m^2}{2\tau^2})\,(\tau^2-m^2)^{1/2}\tau^{-2}. \quad (150)$$

In case of an extended nuclear charge distribution the Uehling potential reads

$$V_{Uh}(\vec{x}) = -\frac{2\alpha^2}{3\pi}\int d^3y \int_m^\infty d\tau\; e^{-2\tau|\vec{x}-\vec{y}|}\,(1+\frac{m^2}{2\tau^2})\,(\frac{(\tau^2-m^2)^{1/2}}{|\vec{x}-\vec{y}|}\rho(\vec{y}). \quad (151)$$

This result may be obtained from the first order vacuum polarization loop correction to the photon propagator. In the momentum representation using the Feynman gauge

$$D(q^2) = -\frac{1}{q^2+i\varepsilon}[1+\frac{\alpha}{3\pi}\int_0^1 dz\;(1-z)\;\ell n(1-q^2\frac{1-z}{m^2-i\varepsilon})] \;. \quad (152)$$

The change in interaction energy due to the second term leads to the Uehling potential.

It is easily seen from (150) that V_{Uh} vanishes exponentially for $r >> 1/m$. Therefore vacuum polarization has only an extremely small influence on wave functions which have an extension large compared with the electron Compton wavelength. For example, the energy shift due to vacuum polarization in the Lamb-shift of hydrogen is only -27 MHz (compared with +1079 MHz from the lowest order vertex correction). However, it increases strongly if the wave function becomes more localized. This is true for muonic atoms where it is the dominant correction and also for atomic states at very high nuclear charge.

The magnitude of the contribution of vacuum polarization in all cases remains small. This can be estimated from the relation of $V_{Uh}(r)$ to the inducing Coulomb potential $V(r)$. In the small distance limit of (147) we have

$$\frac{V_{Uh}(r)}{V(r)} = \frac{Z\alpha}{3\pi} \ln \frac{1}{rm} \; . \tag{153}$$

This quantity remains small for any reasonable value of r. In one of the highest atomic systems accessible for spectroscopy, $_{100}Fm$, vacuum polarization produces an energy shift of 155 eV out of 142 keV total binding energy for the $1s_{1/2}$-state [48]. At even higher Z the Uehling energy-shift was calculated by Pieper and Greiner [17]. It approximately doubles if Z is increased by 10 charge units and reaches E = -11.83 keV for the $1s_{1/2}$-state at Z = 171 (Ref. 21). Thus it slightly favours the diving process.

The rapid increase of the Uehling energy-shift might lead to the apprehension that higher order vacuum polarization becomes dominant. The expansion in $Z\alpha$ might break down. In fact at one time it was believed (based on erroneous arguments) that vacuum polarization increases infinitely as $Z \to Z_{cr}$ (Ref. 49).

Explicit calculation has proved that higher order contributions remain exceedingly small even for $Z\alpha > 1$. The primary work on the problem was done by Wichmann and Kroll [50]. These authors developed a method to calculate vacuum polarization to all orders $\alpha(Z\alpha)^n$ employing the exact (single particle) solutions of the Dirac equation in the external field. We will follow the presentation of Gyulassy [51,52].

The induced charge density may be expressed by an integral over the trace of the electron Green function. The density ρ_{VP} is the vacuum expectation value of the current operator (129). As already mentioned it can be wirtten

$$\rho_{VP} = ie \, \mathrm{Tr} \, [S_F \, (x,x')\gamma_o]_{x' \to x} \quad . \tag{154}$$

It is useful to consider the electron propagator in an intermediate representation transformed from time to energy variable. For time independent potentials S_F depends only on the difference t-t',

$$i \, S_F(\vec{x},\vec{x}'; \, t-t') = \frac{1}{2\pi i} \int_C dZ \, e^{-i(t-t')Z} \, G(\vec{x},\vec{x}';Z) . \tag{155}$$

The Green function G then satisfies

$$[\vec{\alpha}\cdot(\vec{p}-e\vec{A}(\vec{x}))-Z+eA_o(\vec{x})+\beta m] \, G(\vec{x},\vec{x}';Z) = \delta^3(\vec{x}-\vec{x}') \quad . \tag{156}$$

Then (150) can be written as

$$\rho_{VP} = \frac{e}{2\pi i} \int dZ \, \mathrm{Tr} \, G(\vec{x},\vec{x}';Z)\Big|_{\vec{x}' \to \vec{x}} \quad . \tag{157}$$

G exhibits branch cuts at $Z < -m$, $Z > m$ and simple poles at the various discrete eigenvalues. Otherwise it is a single valued analytic function of Z. As demonstrated in Fig. 11 the position of the contour C determines which states are occupied. At $Z > Z_{cr}$ the pole corresponding to the $1s_{1/2}$-state moves off the physical sheet of the Riemann surface for the Green function G. This again leads to the introduction of a charged vacuum since the contour C is not able to follow the $1s_{1/2}$-pole and has to remain inside the gap. To evaluate the trace entering (157) a spherical wave decomposition can be used [53]:

$$\mathrm{Tr}\, G(\vec{x},\vec{x}';Z)\Big|_{x'\to x} = \sum_{\kappa} \frac{2|\kappa|}{4\pi}\, \mathrm{Tr}\, G_{\kappa}(|x|,|x'|;Z)_{x'\to x}. \tag{158}$$

Furthermore an expansion in $Z\alpha$ may be employed for the Green function. Denoting the resolvent in the case of vanishing external potential by G_{κ}^{o} we have the Neumann series

$$G_{\kappa} = \sum_{n} (Z\alpha)^{n}\, G_{\kappa}^{n} = \sum_{n} (Z\alpha)^{n}\, G_{\kappa}^{o}(V_{o}G_{\kappa}^{o})^{n} \tag{159}$$

with

$$\mathrm{Tr}\, G_{\kappa}^{n}(r,r') = \int \prod_{i=1}^{n} (dr_i r_i^2 V_o(r_i))\, \mathrm{Tr}[G_{\kappa}^{o}(r,r_1)\ldots G_{\kappa}^{o}(r_n,r')]. \tag{160}$$

Deforming the contour C to the imaginary axis Gyulassy [53] finally obtains the n-th order vacuum polarization charge density for $\kappa = \pm|\kappa|$,

$$\rho_{|\kappa|}^{n}(r) = \frac{e|\kappa|}{(2\pi)^2} 2(Z\alpha)^{n} \int_{-\infty}^{\infty} dy\, \mathrm{Tr}\, G_{\kappa}^{n}(r,r';y)\Big|_{r'\to r}. \tag{161}$$

In accordance with Furry's theorem the expression is only valid for n even, otherwise $\rho_{|\kappa|}^{n}(r)$ vanishes. The Feynman

Diagrams corresponding to ρ_{VP} are

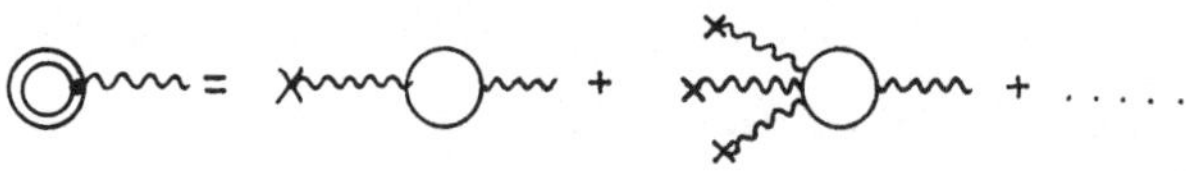

As it stands the expression for $\rho(r)$ is not meaningful since it is not uniquely defined. For the contributions in $(Z\alpha)$, $(Z\alpha)^3$ neither the limit $\vec{x}' \to \vec{x}$ nor the integral over Z exists. Therefore the results must be made unambiguous by regularization. One may employ the regulator method of Pauli and Villars introducing auxiliary masses which make the expressions finite and gauge invariant. At the end of the calculation the auxiliary masses are taken infinite. In the case of an extended nucleus this procedure is not necessary for $(Z\alpha)^n$, $n \geq 5$, since the decreasing nuclear charge form factor makes the integral convergent.

The explicit form of the Green function is constructed from the solutions of the Dirac equation with cut-off Coulomb potential discussed in Sect. 3, namely Whittaker functions in the exterior and spherical Bessel functions (for model I nuclei) in the interior region.

Gyulassy [52] investigated the energy shifts due to the term $\alpha(Z\alpha)^3$ alone and $\alpha(Z\alpha)^n$ for all $n \geq 3$ for the $1s_{1/2}$-level near the diving point. He neglected contributions from $|\kappa| \neq 1$. With a nuclear radius R = 10 fm the critical charge obeys $Z\alpha = 1.27459$. At $Z\alpha = 1.27445$ the $1s_{1/2}$-energy is just above the negative continuum, namely $E_{1s_{1/2}} = -.999$. Here the energy shift due to vacuum polarization is $\Delta E^3 = 0.570$ keV and $\Delta E^{3+} = 1.150$ keV. This demonstrates, that the Uehling potential leads by far to the strongest energy shift (ΔE^1) and higher orders do in no way qualitatively change the behaviour of the energy level.

The calculations of Gyulassy also verified the smooth transition of the K-shell charge distribution between undercritical and overcritical potentials.

An independent and somewhat different approach to vacuum polarization in strong realistic electric fields was taken by Rinker and Wilets [54,55]. They explicitly performed the summation in (140) using exact electron wave functions. Renormalization was achieved by subtracting the contributions linear in Z. Using a monopole approximation to simulate the U+U quasimolecule Rinker and Wilets [55] near the diving point found an energy shift of -3.98 keV consisting of $\Delta E^1 = -4.62$ keV from the Uehling potential $\Delta E_1^{3+} = +609$ eV, $\Delta E_2^{3+} =$ $= +34$ eV for $|\kappa| = 1,2$.

Extending the Wichmann-Kroll-results a third group treated the higher order vacuum polarization anyltically with the approximation of vanishing electron mass. The results of Brown et al. [56] on muonic atom energy levels are in agreement with those of the already mentioned works.

Thus we come to the conclusion that vacuum polarization, although rapidly increasing with the nuclear charge does not behave anomalously near Z_{cr} and does not significantly influence the diving process. (Cf. in this connection Migdal [57] who considers the Green function G in regions of locally homogeneous field strength i.e. linearly increasing potentials.)

The influence of electronic self-energy is known

less reliably. Application of a series expansion of Erickson [58] gives $\Delta E = +2.9$ keV (Ref. 21). Thus the self-energy correction is of comparable size to the vacuum polarization shift but of opposite sign. More reliable new results have been based on the work of Brown et al.[59]. The nonperturbative calculation [60] recently has been applied to superheavy systems. In this region self-energy increases abnormally. For a pure Coulomb potential it diverges at $Z\alpha = 1$. Taking into account the finite nuclear size Cheng and Johnson [61] find approximately $E = +12$ keV. Thus the energy-shift due to the self-energy is approximately cancelling that of vacuum polarization.

6. POSSIBLE INFLUENCE OF NONLINEAR FIELD EFFECTS

The physics of very strong electrostatic field permits the study of phenomena not encountered in ordinary atomic physics. In the following section we will briefly study a few conceivable "exotic" possibilities of modification of the usal QED.

At first we will revisit the limiting field electrodynamics best known from the theory discussed in the 1930's by Born [62] and Born and Infeld [63]. This theory started from a Lagrangian density

$$L = E_0^2 \left[\left(1 - \frac{E^2 - B^2}{E_0^2} - \frac{(\vec{E}\cdot\vec{B})^2}{E_0^4}\right)^{1/2} - 1 \right] \qquad (162)$$

formed in analogy to the relativistic mechanical Lagrangian which leads to a limiting velocity c.

Generalizing the model [64] for the electrostatic case introduced a class of Hamiltonian densities ($L = H - \vec{E}\cdot\vec{D}$)

$$H = \frac{E_o^2(n)}{2n}\left[\left(1 + \frac{D^2}{E_o^2(n)}\right)^n - 1\right] \tag{163}$$

which reduce to the theory of Infeld and Hoffmann [65] for $n = 0$, of Born and Infeld ($n = 1/2$) or to Maxwell's theory ($n = 1$). $\vec{D} = -\partial L/\partial\vec{E}$ is the electric displacement satisfying $\vec{\nabla}\cdot\vec{D}(\vec{r}) = \rho(\vec{r})$. While the displacement $\vec{D}$ superimposes linearly also in limiting field electrodynamics this is no longer true for the field strength $\vec{E}$. The interaction energy of a two particle system is then computed according to

$$V = \int d^3x\,\{H(\vec{D}_1 + \vec{D}_2) - H(\vec{D}_1) - H(\vec{D}_2)\}\ . \tag{164}$$

The nonlinear theories of electrodynamics have the agreeable property that they lead to finite values of the self-energy of a point charge (when $n < 3/4$). Originally they were proposed in the hope to explain the rest mass of the electron in the frame of classical electrodynamics. As long as $n < 1/2$ the model (163) leads to an upper limit of the electric field strength E determined by the parameter $E_o(n)$.

To find the effect of the modified electromagnetic interaction the Dirac equation was solved with (164) for high Z atoms using the Thomas-Fermi model. Insertion of the limiting field E_{max} derived from the electron rest mass leads to strong energy shifts [66].

The diving point is shifted up to higher Z (see Fig.12). However, Soff, Rafelski and Greiner found from a comparison with experimental data of high precision (0.005%) spectroscopic measurements on upper limit of $E_{max} \geq$ $\geq 1.7 \cdot 10^{20}$ V/cm. This value is 140 times larger than the Born result which is therefore completely ruled out.

The realistic value of E_{max} may shift the critical charge by at most 2 units. It is interesting to note that the limiting field theory is a model of maximal vacuum polarization. This is illustrated in Fig. 13 which shows the effective field as a function of the applied field. It led Soff et al.[67] to conclude that vacuum polarization cannot alter the diving process in any significant way.

In general, limiting field theories are not capable to prevent diving because the potential which is responsible for the binding energy of the electrons has no upper bound. A limiting potential theory could provide this. It may be based on the Lagrange function

$$L = \frac{1}{4} f^{k\ell} f_{k\ell} V(A_\sigma A^\sigma) - \psi\pi j^k \phi_k (A_i A^i) \qquad (165)$$

with $\quad f_{k\ell} = \partial_k A_\ell - \partial A_k \quad$ and $\quad \phi_k = \int_{A_k(\infty)}^{A_k} dA_k \sqrt{V(A_\sigma A^\sigma)}$.

Müller considered the parametrization

$$V(A_\sigma A^\sigma) = \left(1 \pm \frac{A_\sigma A^\sigma}{\phi_o^2}\right)^n \quad . \qquad (166)$$

Again, the decisive parameter (ϕ_o) has to be chosen so large, $\phi_o \geq 1000\ m_e$, as to eliminate any significant observable effects in high Z systems [21].

Another possible source of deviation from the normal, weak field QED is the influence of nonlinear terms in the spinor field, e.g.

$$\gamma^{\mu}(p_{\mu}-eA_{\mu})-m\psi + \lambda_1(\bar{\psi}\psi)\psi + \lambda_2(\gamma_{\mu}\gamma_5\bar{\psi}\gamma^{\mu}\gamma^5\psi)\psi = 0 \quad . \tag{167}$$

The third order terms can act as a kind of counter potential, most prominent in strongly bound systems where the electron density is highly localized. The values of λ_1, λ_2, however, have to be very small to avoid inconsistency with spectroscopic data.

After discussing nonlinearities in the electromagnetic and the spinor field one finally could think of a modification in the coupling. The usual electromagnetic coupling is introduced in the Dirac equation by minimal substitution $p_{\mu} \rightarrow p_{\mu} - eA_{\mu}$

$$(\vec{\alpha}\cdot\vec{p} + \beta m - (E-V))\psi = 0 \ . \tag{168}$$

The Coulomb potential is the fourth component of a 4-vector. Interactions of a different origin might couple with a scalar potential, namely,

$$(\vec{\alpha}\cdot\vec{p} + \beta(m+V) - E)\psi = 0 \ . \tag{169}$$

The Dirac equation with a scalar 1/r-potential has been solved by Soff et al.[68]. Its spectrum (Fig. 14) shows positive and negative energy branches, both electrons and positrons are attracted for $V < 0$. The levels asymptotically approach E = 0. Thus diving is avoided for a scalar coupling [36]. The argument following Eq. (23) had shown that in the case of an electromagnetic potential binding energy drops continuously with finite slope as a function of potential strength. For a scalar potential Eq. (24) reads

$$\frac{dE}{dZ} = \langle\psi|\beta U(r)|\psi\rangle = \int_0^\infty dr(u_1^2 - u_2^2)U(r) \quad . \tag{170}$$

If the "small" component of the wave function becomes comparable to u_1 in the potential well, the slope of E(Z) will decrease. Of course, Fig. 14 has nothing to do with observed atomic energies. But it might be possible that a weak scalar interaction is present together with the Coulomb potential. The magnitude of its coupling constant compared to the electromagnetic coupling is limited from experimental knowledge to about 10^{-8}.

Sundaresan and Watson [69] had propsed scalar coupling caused by a massive scalar boson of about 8 MeV to account for discrepancies in muonic atoms (which have been cleared meanwhile).

At any rate, neither of the proposed phenomenological alterations of QED has been verified by experiment and neither will be able to give rise to essential new contributions even in very strong external fields.

7. STATISTICAL DESCRIPTION OF THE CHARGED VACUUM

In the previous sections the creation of a charged vacuum via diving of strongly bound electronic levels into the negative energy continuum has been discussed. We mainly stressed the region of field strength which is hoped to be in the scope of ordinary atomic physics. Then only a single ($1s_{1/2}$, $Z_{cr} = 172$) or at most a few ($2p_{1/2}$, $2s_{1/2}$, $Z_{cr} = 185, 215$) levels are involved. It is of considerable principal and theoretical interest to study the opposite limit of a highly charged vacuum state.

Müller and Rafelski [70] have considered a relativistic Thomas-Fermi statistical model which should be appropriate to describe the real vacuum polarization in very strong external fields.

The charge density of electrons is related to the Fermi momentum k_F by

$$\rho = \frac{e}{3\pi} k_F^2 \ ; \tag{171}$$

k_F^2 has to be a positive quantity. Therefore a unit step function $\theta(E_F - V - m)$ is introduced in the relation between Fermi momentum and Fermi energy E_F,

$$k_F^2 = ((E_F - V)^2 - m^2)\theta(E_F - V - m) \ . \tag{172}$$

These equations lead to the following ground state expectation values of the charge density

$$\langle 0|\rho(x)|0\rangle = -\frac{e}{3\pi^2}\,[(E_F-V)^2 - m^2]^{3/2}\theta(E_F-V-m) \quad . \qquad (173)$$

As usual in the Thomas-Fermi model one can employ the Poisson equation

$$\Delta V(\vec{r}) = e\rho_T(\vec{r}) \qquad (174)$$

linking the electrostatic potential energy V and the total charge density

$$\rho_T = \rho_N + \rho \qquad (175)$$

to obtain a nonlinear second order differential equation. This reads

$$\Delta V = e\rho_N - \frac{e^2}{3\pi^2}\,[(E_F - V)^2 - m^2]^{3/2}\theta(E_F - V - m) \quad . \qquad (176)$$

To describe ordinary neutral atoms one has to take $E_F = m$. Neglecting V compared with m Eq. (176) is the usual Thomas-Fermi equation

$$\Delta V = e\rho_N - e^2 m\,\frac{2\sqrt{2}}{3\pi^2}\,V^{3/2} \quad . \qquad (177)$$

In the present context our interest lies in the charge distribution of the autoionized states in the negative energy continuum. Therefore the Fermi energy is chosen $E_F = -m$. Then Eq. (176) takes the form

$$\Delta V = e\rho_N - \frac{e^2}{3\pi^2}\,(2mV + V^2)^{3/2}\theta(-V-2m) \quad . \qquad (178)$$

Müller and Rafelski [70] solved this equation prescribing an external charge distribution $\rho_N(\vec{r})$ of an abnormally large nucleus represented by a homogeneously charged sphere. They took $R_o = r_o A^{1/3}$ for the radius with $r_o =$ $= 1.2$ fm corresponding to the density of ordinary nuclear matter and assumed symmetry $N = Z$, i.e. $A = 2Z$.

The differential Eq. (178) has to be complemented with suitable boundary conditions. Since the vacuum charge will be locally confined (the autoionized positrons have escaped to infinity) the potential asymptotically has to behave like

$$V(r) \xrightarrow[r\to\infty]{} -\gamma \frac{\alpha}{r} \ . \qquad (179)$$

Here γ is the total charge of the system (nucleus + vacuum), $Z - \gamma$ is the screening charge. The second boundary condition which allows the determination of γ is

$$\left.\frac{dV}{dr}\right|_{r=0} = 0 \ . \qquad (180)$$

Fig. 15 shows the numerically calculated eigenvalues γ as a function of "nuclear" charge Z. The screened charge γ starts on the curve $\gamma = Z$ but when diving sets in at Z-values of a few hundred it increases much more slowly than linear. This means that the vacuum charge (curve $Z-\gamma$) leads to nearly total screening of the pure nuclear Coulomb potential.

The statistical model is only justified if the vacuum charge is large and comparable in size with the external charge Z. It is interesting (and reassuring), however, that the exact single particle results, namely the values of Z_{cr} for the first (three) diving states denoted by crosses in Fig. 15, agree quite well with the Thomas-Fermi result.

The effect of the vacuum screening on the electrostatic potential is demonstrated in Fig. 16. It turns out that V obtains a limiting value if Z becomes very large. The magnitude of V_{lim} can be calculated from the condition $\rho_T(0) = 0$ or

$$V_{lim} = -m-(m^2+ \frac{3\pi^2\rho_0}{e})^{1/2} \sim - \frac{(q\pi)^3}{2\gamma_0} = - 300 \frac{\mathrm{MeVfm}}{r_0} . \qquad (181)$$

Contrary to the earlier conjectures V_{lim} depends on the size and density of the prescribed nuclear charge distribution. In ordinary nuclear matter ($r_0 = 1.2$ fm) $V_{lim} \approx -250$ MeV. The reason of the potential limit is easily understood. With increasing Z the single particle binding energy increases only linearly while the electron-electron correlation energy grows with Z^2.

The charge distribution of ρ_T essentially obtains the structure of a dipole-layer. Except in the vicinity of the nuclear surface the induced charge just cancels the nuclear charge (Fig. 16). The assumption of symmetric nuclear matter is thereby justified.

The screening of the nuclear Coulomb field leads to a limit of the repulsive energy H/A which has to be paid when a nucleon is added to a large nucleus. Constructing a Hamiltonian from Eq. (178), Müller and Rafelski [70] obtained the value $H/A = 3/8|V_{lim}| = 112$ MeV fm/r_o.

This result if of interest in connection with speculations on abnormal nuclear states. In the Lee-Wick model [71], see also [72], the energy functional $E[\rho]$ for nuclear matter obtains a deep second minimum with 130...500 MeV binding energy per absorbed nucleon. The above result indicates, that abnormal nuclei of the Lee-Wick-type could not be stopped from growing by the Coulomb-repulsion.

FIGURE CAPTIONS

Fig. 1: Lowest bound states of the Dirac-equation for nuclei with charge Z. While the Sommerfeld-eigenenergies (dashed lines) for $\kappa = -1$ end at $Z = 137$ the solutions with extended Coulomb potential (full lines) can be traced down to the negative energy continuum which is reached at critical charge Z_{cr}. The states entering the continuum obtain a spreading width as indicated by the bars (magnified by a factor of 10). If the state was previously unoccupied two positrons will be emitted spontaneously.

Fig. 2: Illustration of the autoionization mechanism in nuclear physics. A 2p-2h-state can decay into an energetically degenerate 1p-1h continuum state.

Fig. 3: Spreading of the bound state (solid lines) over the negative energy continuum. The system is quantized in a finite box. For $Z > Z_{cr}$ the continuum contains one more state.

Fig. 4: The $1s_{1/2}$ bound state resonance as exhibited in the scattering phase shift δ. The embedded state is located at $E_o = -926.4$ keV and has a width $\Gamma = 4.8$ keV for the nuclear charge $Z = 184$.

Fig. 5: Diving energies of the $1s_{1/2}$ and $2p_{1/2}$ state in an overcritical extended Coulomb potential. The dashed curves indicate the linear diving model while the full curves have been extracted from the exact phase shift analysis.

Fig. 6: The finite extension R_o of the nuclear charge determines the critical charge Z_{cr}. As $R_o \to 0$ (point nucleus) Z_{cr} approaches $1/\alpha = 137$.

Fig. 7: Density $u_1^2 + u_2^2$ of continuum wave functions for an overcritical extended nuclear Coulomb potential, $Z = 180$. Note the difference between the curves which belong to energies at ($E = -1.7455\ mc^2$) and off ($E = -1.5\ mc^2$) the resonance. The figure illustrates the tunneling between bound state and continuum region of the potential.

Fig. 8: Electron charge distribution for various overcritical states compared with the K-shell distribution for the still undercritical charge $Z = 172$. The transition between these cases is smooth. Also shown is the virtual vacuum polarization density $\Delta\rho$ multiplied by a factor 100.

Fig. 9: Position of the maximum of the $1s_{1/2}$ wave function showing a rapid shrinking of the spatial extension for increasing Z.

Fig.10: Single particle spectrum of the Dirac equation for an adiabatic potential and position of the Fermi surface at E_F. The Hartree-Fock-, tilde- and Uehling sum are defined.

Fig.11: Choice of the integration path specifying the Fourier transform of the Feynman propagator. The negative energy continuum and the states 1, 2, and 4 are occupied.

Fig.12: Energy levels for superheavy atoms obtained from the limiting field electrodynamics of Born and Infeld (solid lines) compared to the result of linear theory. Although Z_{cr} is increased, diving will be not avoided.

Fig.13: Sketch of the effective field E_{eff} in dependence of the applied field E. Limiting field theory is a model of maximum vacuum polarization.

Fig.14: Solutions of the Dirac equation with a purely scalar 1/r potential as a function of the scalar coupling constant $g = \alpha'$.

Fig.15: Unscreened charge γ and charge of the vacuum $Z-\gamma$ as a function of the bare nuclear charge Z. For very high Z the screening becomes nearly total.

Fig.16: Solutions of the relativistic statistical potential equation (178) for nuclear charge numbers $Z = 600, 1000, 2000, 5000, 10^4, 10^5, 10^6$.

a) The self-consistent potential;
b) the corresponding charge distribution of the vacuum;
c) the total charge densities, scaled with γ. Note the large screening of the nuclear charge.

Lecture II

in collaboration with:
W. Betz, G. Heiligenthal, V. Oberacker,
H. Peitz, J. Reinhardt, G. Soff

Experimental Tests: QED Effects in Heavy Ion Collisions

8. INTERMEDIATE QUASIMOLECULES IN HEAVY ION SCATTERING

In the first lecture QED modifications of atomic physics in the presence of highly charged nuclei have been discussed. The effect of main interest urges for a Z of at least $Z_{cr,1s} = 172$. Nature does not provide sufficiently large nuclei, if one discards the possible existence of abnormal, condensed states. Even stable "ordinary" superheavy nuclei (which cannot be produced artificially by fusion) in the region of the predicted island of stability around Z = 164 would not create strong enough fields.

A way out of this dilemma is posed by the idea to use the field of two colliding heavy ions. For high enough bombarding energies the nuclei approach closely, so that the surrounding electron shells experience their combined Coulomb potential. Thus, for a short period of time electronic quasimolecules with varying internuclear distance R(t) are formed. In the limit of vanishing R, corresponding to the point of closest approach in collisions at the Coulomb barrier, even the electronic structure of a quasiatom with combined central charge $Z_1 + Z_2$ can be reached. Quasimolecular states as a basis set for expanding the time-dependent electronic wave function in scattering processes have been applied for a long time (see for example the monograph of Mott and Massey [73] and its earlier editions). Heavy ion physics provides a means of directly testing the formation of quasi-molecules. Measurements of the spectra of emitted X-rays, positrons and electrons allow to draw conclusions on the shape of the molecular correlation diagrams (i.e. the dependence of binding energy on separation) and especially to investigate details of the diving process. These investigations will be impeded somewhat by non-adiabatic and smearing-out effects as a consequence of the finite collision time which is dictated by the Coulomb repulsion of the nuclei.

The idea of superheavy electronic quasimolecules has been proposed in interuniversity GSI-seminars during 1969 (see e.g. the references in the GSI-seminar-report by P. Mokler [74]) and published by Rafelski et al. [66]. A discussion of the adiabacity of the electronic motion was given in [75]. For the system of U-U the collision time, defined as the time during which the nuclei advance from, and recede to, a separation of 100 fm is about 1.8×10^{-19} sec. Roughly assuming half the velocity of light for the K-shell-electron the relation of collision- and orbiting-time is of the order

$$\frac{\tau_{coll}}{\tau_{orb}} \approx 20 \ . \tag{182}$$

Therefore the electron configuration should be able to adjust to the nuclear motion and a molecular description of the scattering process is meaningful. The topic of excitation and rearrangement processes occurring in atomic collisions merits a thorough treatment of its own.

We refer especially to the reviews [76-78]. K-shell excitation in heavy systems is discussed by Taulbjerg and Meyerhof in a forthcoming article. Here we restrict our discussion to a few points which are in our line of interest to test atomic physics for very strong fields.

A general description of the collision of complex ions would have to start from a many-body Hamiltonian for the system of nuclei and electrons and to solve the quantum mechanical scattering problem. For an extensive discussion of the formalism cf. Smith et al. [79], who stress the analogy with the strong coupling model facilitates the solution, since one can apply the Born-Oppenheimer approximation and separate nuclear and electronic motion. A further simplification is achieved by using the semiclassical approximation. Here the nuclear motion is assumed to follow a definite trajectory, corresponding to a localized wave packet in the wave function of relative nuclear position. This treatment if justified if the Sommerfeld-parameter (which can be viewed as the relation between distance of closest approach and de Broglie wavelength)

$$\eta = \frac{Z_1 Z_2 e^2}{\hbar v}$$

is large compared to unity. In the collision of heavy ions this condition is fulfilled very well (for instance $\eta > 500$ for U-U at the Coulomb barrier). It is justified to use a Rutherford hyperbola for the nuclear trajectory. Even for light systems a classical path may be assumed, but an influence of the electronic state on the motion may be important.

With the indicated approximations the problem reduces to the time dependent Schrödinger equation

$$i\frac{\partial}{\partial t}|\psi(\vec{r}_i,\vec{R}(t))\rangle = H(\vec{r}_i,\vec{R}(t))|\psi(\vec{r}_i,\vec{R}(t))\rangle \ . \tag{183}$$

The nuclear separation enters this equation only via the time-dependent parameter $\vec{R}(t)$, not as a dynamical variable. In principle one has to use anti-symmetrized electron wave functions (Slater-determinants) and a Hamiltonian including the electron-electron interaction

$$H(\vec{r}_i,\vec{R})=\sum_i(\vec{\alpha}_i\cdot\vec{p}_i+\beta_i m)+\frac{1}{2}\sum_{i\neq j}\frac{e^2}{|\vec{r}_i-\vec{r}_j|}-\sum_{i,n}\frac{Z_n e^2}{|\vec{r}_i-\vec{\rho}_n|} \ . \tag{184}$$

$\vec{r}_i$ are the coordinates of the electrons, $\vec{\rho}_1,\vec{\rho}_2$ those of the nuclei with charge Z_1e, Z_2e. (This Hamiltonian is appropriate for the relativistic problem since the kinetic energy is described by the Dirac operator $\vec{\alpha}\cdot\vec{p}+\beta m$. It does, however, not contain retardation effects and magnetic energies since it is restricted to the static instantaneous Coulomb interaction.) As a first step when treating the innermost shells of heavy ions the electron-electron-interaction may be neglected.

In view of the slow motion of the nuclei it is natural to use the eigenfunctions of the stationary one electron two centre Dirac (TDC) problem as a basis for the solution of the time-dependent problem

$$[\vec{\alpha}\cdot\vec{p} + \beta m + V_1(\vec{r},\vec{\rho}_1) + V_2(\vec{r},\vec{\rho}_2) - E_\alpha(R)]|\phi_\alpha(R)\rangle = 0. \tag{185}$$

The basis functions $|\phi_\alpha>$ form a complete orthonormal set which is parametrically time-dependent. Taking into account the variation of the energy eigenvalue the wave function will be expanded like

$$|\psi(t)> = \sum_\alpha a_\alpha(t)\, e^{-i\int_0^t dt' E_\alpha(t')} |\phi_\alpha(R(t))> . \quad (186)$$

By projection the time-dependent Schrödinger equation leads to a coupled system of differential equations for the coefficients

$$\dot{a}_\alpha = - \sum_{\beta \neq \alpha} a_\beta\, e^{i\int_0^t dt'(E_\alpha(t')-E_\beta(t'))} <\phi_\alpha|\frac{\partial}{\partial t}|\phi_\beta>. \quad (187)$$

Note that the summation over β has to include continuum-states. The operator $\partial/\partial t$ acts on the states $|\phi_\alpha>$ only via their dependence on $R(t)$. Since the $|\phi_\alpha>$ will be molecular states oriented along the internuclear axis one has to divide the differentiation in a radial and an angular part:

$$\frac{\partial}{\partial t} = \dot{R}\frac{\partial}{\partial R} + \dot{\theta}_{rot}\frac{\partial}{\partial \theta_{rot}} = v_R \frac{\partial}{\partial R} - i\vec{\omega}\cdot\vec{J} , \quad (188)$$

where V_R is the radial and $\vec{\omega}$ the angular velocity. Thus in (187) transitions are caused by a radial as well as a rotational coupling. The latter originates from the noninertial character of the used coordinate system. It is well-known under the name Coriolis coupling also in nuclear physics. Both types of transitions, which display different radial dependence and different selection rules, are important for electron excitation and ionization.

The coupled channel problem could be solved numerically to obtain a complete description of the scattering process (in terms of the independent electron approximation). There is, however, a complication since no simple asymptotic boundary conditions ($t\to\pm\infty$) can be specified. This is due to the fact that the base functions in (186) are eigensolutions of $H(R(t))$ in the asymptotic region. The $|\phi_\alpha\rangle$ represent stationary molecules while the real ions are moving. Thus the operator $\partial/\partial t$ has to effect a Galilean transformation and its matrix elements do not vanish as $R\to\infty$. The first to note this problem were Bates and McCarroll [80] who introduced travelling orbitals my multiplying their (atomic) basis functions with a plane wave factor

$$e^{i(P_i\vec{v}.\vec{r} - \frac{1}{2}P_i^2v^2t)} \tag{189}$$

where $P_i = Z_i/(Z_1 + Z_2)$. In this way the spurious transitions at large distances are eliminated. Translational factors have been commonly adopted in atomic collision theory (their use has recently been discussed also for nuclear physics, Pauli and Wilets [81]). There remains a certain arbitrariness because factors of the form (189) give the electron a boost of the nuclear velocity. This is unrealistic in the interaction region where the electron correlates with both centres. By the introduction of an ad hoc constructed and optimized weight function $f(\vec{r},\vec{R})$ several authors (see for example Thorson and Levy [82]) tried to remedy this deficiency.

9. THE TWO CENTRE DIRAC EQUATION (TCD). MO RADIATION

It is clear from the short discussion of the last section, that any quantitative description of adiabatic atomic collisions must be based on the quasimolecular, i.e. two centre-wave functions. The solution of this kind of problem is essential for quantum chemistry; therefore it had been treated very early in the history of quantum mechanics.

The nonrelativistic one-electron Hamiltonian

$$H_{NR} = \frac{p^2}{2m} + V_1(r_1) + V_2(r_2) \tag{190}$$

has been solved by Teller [83], Hylleraas [84] after first approximate results of Heitler and London [85]. It is essential that the two centre Schrödinger equation has two constants of motion. Trivially, one of them is the projection L_z of angular momentum on the symmetry axis. A second operator commuting with the Hamiltonian has been constructed by Erikson and Hill [86]. In prolate spherical coordinates

$$\xi = \frac{r_1 + r_2}{R}, \quad \eta = \frac{r_1 - r_2}{R}, \quad \phi$$

the Schrödinger equation separates producing two coupled eigenvalue equations which can be solved for the energy.

The TCD equation is more difficult to handle since there exists no orthogonal coordinate system in which it separates. Correspondingly, no second constant of motion besides the projection of the total angular momentum exists. The solution, then, can only be obtained by numerical methods. Except for a calculation of relativistic corrections in perturbation theory [87], this problem has been attacked only recently. Müller et al. [88] treated Eq. (185) in prolate spheriodal coordinates and chose an expansion in analogy with the nonrelativistic case. The basis functions were

$$\psi^{m}_{n\ell s}(\xi,\eta) = e^{-x/2}\, L_n^{m+\varepsilon_s}(x)\, P_\ell^{m+\varepsilon_s}(\eta)\, \chi_s \quad . \tag{191}$$

Here $\varepsilon_s = \mp 1/2$ for s odd (even) and x is a scaled coordinate $x = (\xi - 1)/a$. The variable a serves to satisfy the asymptotic behaviour of the wave function

$$\psi \to e^{-\sqrt{m^2 - E^2}\, r}$$

and therefore is energy dependent. The matrix elements of the TCD Hamiltonian with the (nonorthogonal) basis (191) may be obtained analytically. The eigenvalue problem then was diagonalized numerically (for further details and results see Müller and Greiner [89]).

In an actual calculation the basis, of course, must be truncated and it is difficult to judge the accuracy. In particular, the question of convergence is more critical than in the usual nonrelativistic case because the Dirac Hamiltonian is not bounded from below and states of the negative continuum may be admixed.

Recently Rafelski and Müller [90] have solved the TCD equation by numerical integration. They have discretized one of the coordinates by a multipole expansion of the wave function

$$\psi_\mu(\vec{r}) = \sum_\kappa \psi_{\mu\kappa}(\vec{r}) = \sum_\kappa \begin{bmatrix} g_\kappa(r) \ \chi_\kappa^\mu \\ i f_\kappa(r) \ \chi_{-\kappa}^\mu \end{bmatrix} \quad . \tag{192}$$

Using a multipole expansion of the two centre potential

$$V(\vec{r},R) = \sum_\ell V_\ell(r,R) \ P_\ell(\cos\theta) \tag{193}$$

the radial equations become

$$\frac{d}{dr} f_\kappa(r) = \frac{\kappa-1}{r} f_\kappa(r) - (E-m) g_\kappa(r) + \sum_{s\ell} g_s(r) V_\ell(r,R) A_{\kappa\ell s}$$

$$\frac{d}{dr} g_\kappa(r) = (E+m) f_\kappa(r) - \frac{\kappa+1}{r} g_\kappa(r) - \sum_{s\ell} f_s(r) V_\ell(r,R) A_{-\kappa\ell-s} \, . \tag{194}$$

The coefficients $A_{\kappa\ell s} = \langle\chi_\kappa^\mu|P_\ell|\chi_s^\mu\rangle$ are determined by angular momentum algebra. After the sum in Eq. (194) had been truncated at a sufficiently large angular momentum j_{max} the $2(2j_{max}+1)$ coupled differential equations were integrated. The energy eigenvalue E is determined by an iteration to produce vanishing of the wave function at a large distance.

This highly accurate numerical integration method is also convenient since it easily allows the use of modified potentials (193) which e.g. may contain electron screening effects. The influence of finite nuclear radii too is easily calculated (cf. in this connection Marinov and Popov [91], who used perturbation theory).

Fig. 17 shows a numerically calculated diagram of the energy levels as a function of distance R for the relatively light system Ni-Ni. This type of "correlation diagram" for the two centre Schrödinger equation has been extensively discussed since the 1920's. The molecular states are classified by the good quantum number j_z with eigenvalues $|m_j| = 1/2, 3/2, 5/2, \ldots$ which are named by $\sigma, \pi, \delta, \ldots$. One often assigns the quantum numbers of the united atomic state ($R \to 0$) to the molecular wave function to which it correlates ($1s_{1/2}\sigma$, $2p_{3/2}\sigma, \ldots$). For identical nuclei the molecule has a further constant of the motion. Since the parity operator $\beta\vec{P}$ commutes with the Hamiltonian and with j_z one can distinguish even ("gerade") and odd ("ungerade") molecular states.

The principal behaviour of the molecular states can be understood by general symmetry arguments [92]. Of special interest are those points where the two levels approach each other closely. If they belong to different symmetry they are allowed to cross. States of equal symmetry, however, may not cross due to the Neumann-Wigner rule (which is not applicable in the nonrelativistic case [93]).This rule is truly obeyed by the numerically evaluated molecular levels. As a typical example we examine the $2p_{1/2}\sigma$ and $3d_{3/2}\sigma$ states near 8000 fm which are repelled and do not cross (see Fig. 17).

This interesting behaviour is only of limited practical consequence. A closer analysis of the character of the wave functions actually reveals, that they cross. In any collision with finite velocity the strong dynamical coupling will cause the electrons not to follow the adiabatic correlation diagram but a "diabatic" molecular orbital. An explicit example of this transition was treated for the asymmetric system NeO (Ref. 94).

Fig. 18 shows the level diagram of the heavy asymmetric system I - Au with $Z_1 + Z_2 = 132$. The states possess no good parity, therefore many pseudocrossings occur.

The level diagram of U-U, which is of greatest importance to us, is presented in Fig.19. Comparison with the Ni-Ni case shows drastic differences due to relativistic correlations. Contrary to the light quasimolecules in U-U all inner-shell levels strongly gain energy when the internuclear distance sinks below a few hundred fm. This is particularly noticeable for the state which normally ascends in energy to a weakly bound L-shell of the united atom. Relativistic binding is also responsible for a level crossing of the $3s_{1/2}\sigma$ and $2p_{3/2}\sigma$ and a pseudocrossing between $2p_{3/2}\sigma$ and $3p_{1/2}\sigma$. It is quite clear already from these results that any simple kind of scaling of nonrelativistic calculations (e.g. H^+-H) will not work for the supercritical systems.

A result vital for the discussion of positron production in Sect. 11 is the magnitude of the critical distance R_{cr}, where the $1s_{1/2}\sigma$-level joins the lower continuum. Here the autoionization process of positrons sets in just as in our previous gedankenexperiment where Z had been increased beyond Z_{cr}. Numerical solution of the TCD equation leads to $R_{cr} = 37$ fm for pointlike nuclei and $R_{cr} = 35$ fm for extended nuclei. Fig. 20 gives a magnified picture of the $1s_{1/2}\sigma$ and $2p_{1/2}\sigma$ states of three supercritical molecules U-U, U-Cf and

Cf-Cf. For the latter two systems the $1s_{1/2}\sigma$-level dives at 48 fm and at 61 fm, respectively. The $2p_{1/2}\sigma$ level remains undercritical (for distances below the nuclear Coulomb barrier) at U-U and reaches -m at R_{cr} = 16 fm for U-Cf and 25 fm at Cf-Cf.

These critical distances are at variance with the values obtained by Popov et al. [95,96] who have employed several approximation methods to determine the critical distance. The TCD problem has been solved analytically in the limit of low supercriticality $\delta = (Z_1+Z_2-Z_{cr})/Z_{cr} \to 0$. Popov further devised a method to obtain the two centre solution at E = -m from a variational principle. He started from basis functions embodying a singularity of the form (for $1s_{1/2}\sigma$)

$$\psi \sim (\xi^2 - \eta^2)^{-\sigma} \qquad \text{with} \qquad \sigma = 1 - \sqrt{1-(\tfrac{1}{2}Z\alpha)^2} \tag{195}$$

which emerges in the relativistic two centre problem. Variation was preformed on a set of one, two or three trial functions. The convergence of this procedure is not quite clear. The latest results are Z_{cr} = 51 fm for U-U and 77.7 fm for Cf-Cf (point nuclei). Since Popov contends that these numbers form a lower limit to the true R_{cr}, there is a clear disagreement with our result. It seems to us, however, that the good agreement between two independent methods of numerical solution of the TCD equation stands in favour for the above proposed numbers.

The discribed relativistic MO's are important for many applications in heavy ion collisions, e.g. excitation and ionization of electrons, quasimolecular X-rays, deviations from Rutherford scattering due to effects of electronic molecular binding [97] and, finally, autoionization of positrons. For many purposes the one-electron approximation is quite satisfactory. The electron-electron interaction could be incorporated by a relativistic extension of the two centre Hartree-Fock or Thomas-Fermi method. Larkins [98] published the first quasimolecular nonrelativistic Hartree-Fock correlation diagram. These calculations are very time-consuming on the computer. They are perhaps "too accurate" since they assume total relaxation of the electron shell at every distance R, leading to a discontinuous appearance of the level diagrams.

The quantum statistical approach for molecules recently has been revived by Eichler and Wille [99]. Gross and Dreizler [100] obtained a single particle Thomas-Fermi potential with the object of constructing two centre orbitals which include electron screening.

Especially when dealing with the outer shell region, however, one has to bear in mind that the adiabacity condition is not fulfilled in heavy ion collisions. Since the mean number of electron-electron collisions [101] is too small, it is disputable to what extent an average potential does build up. If the quasi-molecular states do exist, it should be possible to observe radiative transitions among them. Such transitions have indeed been seen and complete spectra of a heavy ion collision have been obtained during the last few years. Besides gaining valuable insight in the dynamics of atomic excitation and deexcitation processes, the ultimate aim of these studies is to push atomic physics forward into a new region and to investigate the electronic states in superheavy systems. Ideally it might even be possible to follow the change of binding of the 1sσ level with increasing nuclear charge and to identify the diving process by looking at the produced X-rays. It is, however, quite difficult to extract detailed information from the measured spectra due to the finite collision time (Heisenberg groadening). Furthermore background processes may seriously interfere with the MO effect.

From a theoretical point of view photon emission is easily incorporated in the formalism sketched in Sect. 8. The electromagnetic potential is introduced in the electronic Hamiltonian via minimal coupling $\vec{p} \to \vec{p} - e/c\,\vec{A}$. Expanding the time dependent wave function in a way analogous to Eq. (186) one has now also to include photon channels. Since the radiative coupling is a weak perturbation it is sufficient to include one-photon states.

$$|\psi(t)\rangle = \sum_\alpha a_\alpha(t)\, e^{-i\int^t dt'\, E_\alpha(t')} |\phi_\alpha\rangle |0\rangle + \sum_{\alpha,\vec{k}\mu} C_{\alpha,\vec{k}\mu}(t)\, e^{-i\int^t dt' (E_\alpha(t')+\omega)} |\phi_\alpha\rangle \hat{a}_{\vec{k}\mu} |0\rangle . \quad (196)$$

$C_{\alpha,\vec{k}\mu}$ denotes the amplitude for finding the electron configuration $|\phi_\alpha\rangle$ together with an emitted photon with wave vector $\vec{k}$ and polarization μ. By projection one obtains coupled-channel equations

$$\dot{a}_\beta = - \sum_\alpha a_\alpha(t) \langle\phi_\beta|\frac{\partial}{\partial t}|\phi_\alpha\rangle\, e^{i\int^t dt'(E_\beta - E_\alpha)} - i \sum_\alpha \langle\phi_\beta|H_{rad}|\phi_\alpha\rangle\, e^{i\int^t dt'(E_\beta - E_\alpha - \omega)} C_{\alpha,\vec{k}\mu} \quad (197)$$

$$\dot{C}_{\beta,\vec{k}\mu} = -i\sum_\alpha \langle\phi_\beta|H_{rad}|\phi_\alpha\rangle\, e^{i\int^t dt'(E_\beta - E_\alpha + \omega)} a_\alpha(t) -$$

$$- \sum_\alpha \langle\phi_\beta|\frac{\partial}{\partial t}|\phi_\alpha\rangle\, e^{i\int^t dt'(E_\beta - E_\alpha)} C_{\alpha,\vec{k}\mu} , \qquad (198)$$

with the interaction

$$H_{rad} = \vec{j}\cdot\vec{A} = -e\vec{\alpha}\cdot\vec{A} . \qquad (199)$$

The differential cross section for photon emission simply is the scattering cross section multiplied by an incoherent sum of emission probabilities corresponding to different final states and averaged over photon polarization

$$\frac{d\sigma}{d\Omega}_{ion} = \left(\frac{d\sigma}{d\Omega}_{ion}\right)_{scatt} \sum_{\beta,\mu} |C_{\beta,\vec{k}\mu}(\infty)|^2 k^2\, dk . \qquad (200)$$

Eq. (198) teaches that the emission spectrum is determined by a Fourier transform with variable frequency. To be more specific, let us neglect the small second term on the right hand side of (198) and employ the dipole approximation. The vector potential reduces to $\vec{A} = \vec{\varepsilon}_\mu$ and the photon amplitudes becomes

$$C_{\beta,\vec{k}\mu}(\omega) = i\int_{-\infty}^{\infty} dt \sum_\alpha \left(\frac{2\pi}{\omega}\right)^{1/2} \vec{\varepsilon}_\mu \cdot \langle\phi_\beta|R(\theta_N)\vec{j}|\phi_\alpha\rangle e^{i\int dt'(\omega_{\beta\alpha}+\omega)} a_\alpha(t) . \qquad (201)$$

With the identification

$$<\phi_\beta|\vec{j}|\phi_\alpha> = -ie<\phi_\beta|[H_o,\vec{r}]|\phi_\alpha> = -ie\omega_{\beta\alpha}<\phi_\beta|\vec{r}|\phi_\alpha> \qquad (202)$$

the amplitude C is seen to be a Fourier transform of several time dependent factors: the dipole strength $\omega_{\beta\alpha}\cdot<\phi_\beta|\vec{r}|\phi_\alpha>$, the rotation matrix $R(\theta_N(t))$ mediating between laboratory and rotating coordinate system and the amplitude $a_\alpha(t)$ describing the time evolvement of the electron configuration.

Even apart from these varying factors the spectrum is influenced by the phase integral $\int^t (\omega_{\beta\alpha}(t')-\omega)dt'$ which contains the energy difference $\omega_{\alpha\beta}(t) = E_\alpha(t) - E_\beta(t')$ changing with internuclear separation. The effect of a time-dependent transition frequency has been of interest for a long time in connection with collisional line-broadening [102]. It is obvious that in heavy ion collisions it will lead to a broad continuum of noncharacteristic X-rays located between the corresponding lines of the separated and united atomic system (Fig. 21).

An exact treatment of the Fourier integral leads to cross sections uniformly decreasing with ω. The high energy tail (beyond the united atom limit) has the approximate shape [103]

$$\frac{d\sigma}{d\omega} \sim \omega e^{-\frac{\omega}{2\Gamma}} \quad . \qquad (203)$$

The width Γ can be simply related to the parameters of the colliding system. Interestingly the width is proportional to the square root of the projectile velocity contrary to simple $\Delta E.\Delta t$ Heisenberg-broadening arguments which indicate a linear dependence. H.D. Betz et al. [104] have obtained a simple analytical formula for the MO tail which quantitatively describes the shape and magnitude of a large number of systems.

The first observation of a molecular X-ray spectrum was due Saris et al. [105]. Mokler et al.[106] shortly afterwards observed M-shell radiation in the superheavy system I - Au (Z = 132). Meyerhof et al.[107] pioneered the investigation of K-radiation by studying Br-Br collisions. He was followed by Greenberg et al. [108] with studies on the Ni-Ni-system in particular and by Wölfli et al.[109] with experiments on Al, Ca, Fe, Ni etc. Studies of the K-radiation in superheavy systems were first published by the group of Kaun et al. [110]. A convincing proof for the quasimolecular nature of the observed continua was given by Meyerhof, Saylor and Anholt [111] and - following them - by Folkmann et al. [112]. Using the Doppler shift they determined the velocity of the source of X-ray emission. While the characteristic lines are shifted in accordance with the respective ion velocities the MO continuum source moves with the common centre of mass velocity.

Fortunately, there is a simple way to draw more information from the experiments. Measurement of the angular distribution of noncharacteristic radiation provides further details of the collision process. The first of these measurements were made by Greenberg, Davis and Vincent [108] and for M-shell transitions by Kraft, Mokler and Stein [113]. Greenberg's results and those obtained later by a large number of other groups all exhibited a huge bump of positive anisotropy which peaks in the vicinity of the united atom transition energy (see Fig. 22). At present there does not exist an a priori theory to explain this behaviour. However, already Müller and Greiner [114] have noted that the anisotropy strongly depends on the specific admixture of dipole transitions from states with different angular momentum quantum numbers. This alignment is held responsible for the observed angular distribution of MO X-rays (Ref.[115]). To obtain good agreement with experiment one of the dynamical couplings of Eq. (196) has to be included: The strong rotational coupling between $2p\sigma$ and $2p\pi$ at close distances. This reflects the fact that the electron cannot follow the swift rotation of the internuclear axis at small distances. Parametrizing this "slippage" effect and assuming a certain alignment W.Betz et al. [116] were able to explain the experimental data of Greenberg (Ni-Ni) very well. A recent calculation [117] using perturbation theory and - restricted to slow collisions and light nonrelativistic systems verified those results. Assumption of an empty $2p\pi$ suchell produces the characteristic anisotropy bump. For the system $H^{+}H$ a calculation [118] leads to a similar distribution. In addition to the alignment it is necessary to treat correctly the various interfering paths which lead to the same final configuration.

The angular distribution may leave a door open for the quasimolecular spectroscopy of superheavy two centre orbitals [119]. In a very thorough systematic study, Wölfli [120] has investigated the position of the MO X-ray asymmetry-bump (defined by the point of maximal positive slope) in dependence of the total nuclear charge $Z = Z_1 + Z_2$. For all included systems (with charge between 26 and 93) an accurate scaling with Z^2

is observed so that the bump position always lies slightly above the united atom energy (see Fig. 23). However, experimental problems largely increase when entering the superheavy region due to increasing background and small cross sections. It remains to be seen how far the investigations can be pushed.

10. DECAY OF THE NEUTRAL VACUUM IN HEAVY ION COLLISIONS

The decisive experimental test for the diving of strongly bound electronic states in heavy ion collisions will be the observation of emitted positrons. Apart from disturbances by background processes (Sect. 13), a complete dynamical calculation of quasi-molecular positron formation mechanism in e.g. U-U collisions at present is not yet available. All previous investigations assumed a certain probability L_o (which was held constant) for the presence of a K-hole. Except for systems of two naked nuclei or a naked projectile and an atomic target, or for the possibility of nuclear conversion during the collision, unfortunately no K-hole is originally available. It has to be created dynamically via Coulomb excitation. Postponing this problem we at first assume a constant value of $L_o = 0.01$.

In a very slow and truly adiabatic collision positrons could only be produced via the discussed autoionization process. Theory then becomes particularly simple since a static prescription can be applied [91, 121,122]. According to [121], the probability per time interval to emit a positron of energy E at a given distance R is governed by the width of the quasistationary level

$$p(R,E) = \Gamma(R)\,|a(E,R)|^2 = \frac{\frac{1}{2\pi}\Gamma(R)^2}{[E-(E_o+\Delta E(R))]^2+\frac{1}{4}\Gamma(R)^2} \tag{204}$$

in accordance with the results of Sect. 4. The differential cross section for positron production is obtained from the emission probability by a time integration along the hyperbolic trajectory. We have

$$\frac{d\sigma}{dE\, d\Omega_{ion}} = \left(\frac{d\sigma}{d\Omega_{ion}}\right)_{scatt} L_o\, W(E,\theta) \tag{205}$$

with

$$W(E,\theta) = \int_{-t_{cr}}^{t_{cr}} dt\, p\,[R(t,\theta),E]\ . \tag{206}$$

This result may be improved by considering the filling up of holes during the collision which leads to a small reduction and smearing out of the spectrum. A simple estimation of the width Γ compared with collision time reveals, that only a small fraction of holes will decay.

Using the monopole approximation for $\Gamma(R)$ and $\Delta E(R)$ Peitz et al. [121] performed the quasi-static calculation and obtained a total positron cross section of $2.8\cdot 10^{-5}$b for U-U and 10^{-3}b for Cf-Cf. The positron energy distribution exhibits a very sharp peak at the energy which belongs to the point of deepest diving where the radial velocity vanishes. A behaviour of this kind would be very desirable in order to identify the effect experimentally. Popov and coworkers obtained comparable quasi-static results employing two-centre solutions valid in the limit of low supercriticality. Also in their model $|a(E)|^2$ in Eq. (204) is replaced by a delta function.

Further investigations have turned out that the quasi-static results will be grossly modified by dynamical effects. To illustrate this point Fig. 24 schematically depicts several types of transitions encountered during the collision process. a and b denote the Coulomb excitation or ionization which produce a hole in the $1s\sigma$ level. This hole may be set free spontaneously (c) during the diving without investment of energy. Due to the scattering dynamics energy can be drawn from the nuclear motion leading to nonadiabatic filling of the hole even at distances larger than R_{cr}. This effect may be called induced transitions and will enhance the cross sections. From the shortness of collision time which is about $2\cdot 10^{-21}$ sec and the corresponding large energy uncertainty the importance of induced transitions is evident. Their influence may be seen in analogy with the dynamical broadening of MO X-ray spectra (except for the energy threshold for positron creation).

As earlier discussed the description of positron production starts from the time dependent Schrödinger equation (183). In the expansion (186) of the wave function positron states have to be included. Special care has to be taken of the region of diving $R < R_{cr}$, $-t_{cr} < t < t_{cr}$. To follow the time development of the diving state we change the representation at $R = R_{cr}$ and expand the overcritical wave function in the stationary basis $|\phi_{\alpha_{cr}}\rangle \; |\phi_\alpha(R_{cr})\rangle$. Time dependence then is contained in the R-dependent Hamiltonian H(R) rather than in matrix elements of the operator $\partial/\partial t$. Coupling to the positron levels is quite weak, so that first order perturbation theory gives good results. Restricting our discussion to the interaction of 1s-level and negative continuum the positron amplitude is determined by the integral

$$C(E) = \int_{-\infty}^{\infty} dt \, e^{i\int_{-\infty}^{t} dt'(E-E_{1s}(t')) - \frac{1}{2}\int_{-\infty}^{t} dt'\Gamma(t')} M_E(t) \qquad (207)$$

with the matrix element

$$M_E(t) = \begin{cases} -\langle\psi_E(t)|\frac{\partial}{\partial t}|\phi(t)\rangle & (|t|>t_{cr}) \\ iV_E(t) & (|t|\leq t_{cr}) \end{cases} \; . \qquad (208)$$

We use the definitions

$$E(R) = \langle\phi_{cr}|V(R)|\phi_{cr}\rangle \qquad V_E = \langle\psi_E^{cr}|V(R)|\phi_{cr}\rangle \qquad (209)$$

where $V(R) = H(R) - H(R_{cr})$ is the overcritical potential. The decay width Γ in the exponential may be calculated from an averaged coupling matrix element $\Gamma(t) = 2\pi|V(t)|^2$ (Ref.79). Having calculated the amplitude C(E) the positron cross section follows immediately from (203) with the probability $W(E,\theta)dE = |C(E)|^2 dE$ for producing a positron of (total) energy between E and E + dE.

In the work of Smith et al. [123] the width of the diving level has been extracted numerically from the one-centre resonance. Fig. 25 shows the scaled width $\gamma(E)$ defined by

$$\gamma(E) = 2\pi |V_E|^2 (\frac{m}{\varepsilon^2(R)}) .$$

This quantity is nearly charge independent and may be conveniently parameterized. Note the influence of electrostatic repulsion at low kinetic energy E_p.

The resulting energy distribution of positrons obtained by Smith et al. [123] for the system U-U is exhibited in Fig. 26. Here the probability W has been divided into two parts. The integral (207) was calculated separately with the limits $-\infty < t < -t_{cr}$, $t_{cr} < t < \infty$ (pre- and after-diving, W_{PA}) and $-t_{cr} < t < t_{cr}$ (during diving W_D). Coherent addition of these contributions yielded the total probability W_T. The curves demonstrate that, after all, most of the positrons are emitted in the overcritical region, especially at higher positron energies. The variation of ion energy or scattering angle has no essential influence on the shape of the spectra. Their decrease at high energies is mainly governed by the Fourier transform of the changing two-centre potential, i.e. by the characteristic frequencies of the nuclear motion. The availability of high Fourier frequencies is the reason for strong induced transitions and even brings about direct pair creation between upper and lower continuum (process f in Fig. 24), see Sect. 12.

Integration over impact parameter leads to a further broadening of the energy distribution. The cross section

$d\sigma/dE_p$ is given in Fig. 27a at several values of the distance of closest approach (note the logarithmic scale) while 27b shows the total cross section for positron production. $\sigma(E_p)$ has a kind of threshold at an ion energy (300 MeV) which corresponds to a distance of closest approach of the order R_{cr}. Fig. 28 demonstrates how the positron production increases with the amount of overcritical charge brought into the collision.

As indicated by the described results dynamical autoionization of positrons in heavy ion collisions leads to rather broad and not very characteristic spectra. Hence the experimental study of this process will demand careful systematic measurement of the dependence on nuclear charge, ion energy and positron energy. This is particularly important since strong background processes will be present.

Also the theory has to be refined. Up to now a fixed K-vacancy probability, $L_o = 0.01$ was assumed. However, the K-vacancy is produced during the collision by direct ionization through the collision dynamics. Until recently, no calculation of this process was available which could predict either the magnitude or the time-dependence of the ionization amplitude along the collision trajectory. As we will discuss in the next section new calculations of the ionization and excitation of inner shell electron in very heavy collisional systems indicate that the K-hole probability during the collision indeed reaches several percent, rapidly decreasing with impact parameter. Although the above calculations with constant L_o have to be modified; the result gives the assurance that the induced decay of the vacuum leads to an observable positron production rate.

11. "DYNAMICAL THEORY OF IONIZATION". SUPERSTRONG MAGNETIC FIELDS

The solution of the time-dependent Schrödinger equation (183) for a single electron state can be expanded in terms of the wave functions obtained from the stationary two-centre Dirac equation

$$\psi(\vec{r},t) = \sum_n a_n(t)\, \phi_n(\vec{r},R(t))\, e^{-i\int E_n dt} . \tag{210}$$

Upon substitution of this ansatz one is left with an infinite system of coupled channel equations [79]

$$\dot{a}_n(t) = - \sum_{m\neq n} a_m(t) \exp(i\int^t dt'(E_n - E_m)) \cdot$$

$$\cdot\langle\phi_n|\dot{R}\frac{\partial}{\partial R} - i\Omega(R)\cdot\hat{j}|\phi_m\rangle . \tag{211}$$

Note that the summation over m includes the continuum states.

The first part in the interaction matrix element is commonly called radial, the second one rotational or Coriolis coupling. The two operators exhibit different angular momentum selection rules: the radial coupling acts between states with the same μ-quantum number, whereas the Coriolis term couples states with angular momentum projection μ differing by $\pm$ 1.

In a superheavy quasimolecule like U+U one must be careful in applying Eq. (211), because for distances $R \leq R_{cr}$ (with $R_{cr} \sim 35$ fm in this example), the 1sσ

level is degenerate with the negative energy continuum. The resultant overcritical continuum states are in no way adiabatic due to a sharp distortion when being traversed by the original bound state [1].

The set of coupled equations (211) is equivalent to the full Schrödinger equation and able to describe all aspects of the collision dynamics. In practical calculations [124], however, one faces two limitations: (1) The restriction to a finite subset of Hilbert space. Whether this is justified can be tested from the convergence behaviour by including more states in the calculation, which we have done for quasi-molecular bound states and continuum states separately. (2) If continuum states are included, Eqs. (211) can only be solved with reasonable effort up to first order, i.e. taking $a_m(t) = \text{const.}$. Thereby unitarity is violated to the order of $P_{cont} = dE|a_E|^2$, the total excitation probability into the continuum. Our calculations yield $P_{cont} \leq 0.05$ for the 1sσ state.

Thus we have computed the ionization amplitude into the continuum states $|E s\sigma\rangle$ only in the first order:

$$a_E(t) = -\int_{-\infty}^{t} dt' \langle E s\sigma|\dot{R}\frac{\partial}{\partial R}|1s\sigma\rangle \exp(i\int^{t'} dt''(E-E_{1s\sigma})). \qquad (212)$$

It should be stressed that this is <u>not</u> a kind of "never-come-back" approximation because it allows for oscillations in the excitation amplitude a(t) which indicates a "jumping" of the vacancy between the 1sσ and the n-state until it approaches the final value $a(t = \infty)$ after the collision. If more than one electron is present, the expansion (210) should be over configurations rather than wave functions, in order to account for the Pauli principle. Since the operators in Eq. (211) are one-body operators, for the determinants as basis states the matrix elements reduce to the same form as in Eq. (211).

Our first calculations were performed in the U+U quasi-molecule [125]. Because it is a symmetric system, the wave functions have a good parity and the coupled equations (211) decouple into two sets for the even and odd parity states, respectively. We shall concentrate on the set of even parity states in the dynamical calculations.

We are most interested in the low-lying molecular states, so that the chosen adiabatic basis can be expected to give a reasonable description of the collision physics. Starting from the 1sσ level, the rotational part couples to the 3dπ state as lowest state and - most important - the corresponding matrix element vanishes for $R \to 0$. This is easily seen, since in the united atom limit the wave functions acquire good angular momentum and $\dot{R}\frac{\partial}{\partial R}$ is a monopole operator. On the other hand, the radial part couples to all nsσ, ndσ, etc. states, with large matrix elements $\langle ns\sigma|\frac{\partial}{\partial R}|1s\sigma\rangle$ at close distances. We have therefore limited our basis to the 1sσ, 2sσ, 3sσ, 4sσ bound states and the Esσ continuum states. The calculation of the radial matrix elements can be obtained by use of

$$(E_m-E_n)\langle\phi_n|\frac{\partial}{\partial R}|\phi_m\rangle = \langle\phi_n|[\frac{\partial}{\partial R},H]|\phi_m\rangle =$$

$$= \langle\phi_n|\frac{\partial V_{TC}}{\partial R}|\phi_m\rangle \tag{213}$$

from the two-centre potential V_{Tc}, provided the two wave functions ϕ_n and ϕ_m are orthogonal. The matrix elements between bound states as obtained from Eq.(213) are shown in Figs. 29 and 30 (dashed lines). Between neighbouring states, e.g. 1sσ-2sσ, 2sσ-3sσ, 3sσ-4sσ, etc., they are almost equal. Hence the matrix elements with the 1sσ state and those with the 2sσ state are shown separately. To a good approximation we find that the matrix elements between the nsσ and the (n+k)sσ state scale as $1/(k+1)^2$. This rapid fall-off the transition strength makes all bound states above the 4sσ level unimportant for 1sσ ionization.

In order to obtain constant occupation amplitudes $a_n(t)$ for $t\to\infty$, it is necessary that the coupling matrix elements vanish asymptotically. The dashed curves in Fig. 29 exhibit almost constant tails which give rise to unphysical oscillations in the amplitudes at large separations. This behaviour is known [126] to be due to the failure of the stationary molecular states to be asymptotic solutions to the scattering equation with two atoms moving apart with velocity $\vec{v}$. Instead one has to multiply each separated atoms wave function by a "translation" factor $\exp(\pm\, i/2\, m_e \vec{v}\cdot\vec{r})$ depending in sign on which side it is centred. For symmetric systems one has first to form combinations of an even and the corresponding odd parity state to describe a single separated atoms wave function and then again to take the (anti-or) symmetric combination of the boosted states:

$$\tilde{\phi}_n^{(\pm)} = \frac{1}{2}(\phi_n^{(+)}+\phi_n^{(-)})e^{\frac{i}{2}m\vec{v}\cdot\vec{r}} \pm \frac{1}{2}(\phi_n^{(+)}-\phi_n^{(-)})e^{-\frac{i}{2}m\vec{v}\cdot\vec{r}} =$$

$$= \frac{\phi_n^{(+)} \cos(\frac{1}{2}m\vec{v}\cdot\vec{r}) + \phi_n^{(-)} \sin(\frac{1}{2}m\vec{v}\cdot\vec{r})}{\phi_n^{(+)} \sin(\frac{1}{2}m\vec{v}\cdot\vec{r}) + \phi_n^{(-)} \cos(\frac{1}{2}m\vec{v}\cdot\vec{r})} \quad . \tag{214}$$

For our U-U collision at $v/c \sim 0.1$ we find that the sin/cos factors oscillate once (through 2π) in 50000 fm which makes only a small influence on the $1s\sigma$ level which is localized within less than 1000 fm. It is therefore reasonable to include the translation factors only up to first order in the projectile velocity v. The matrix elements of the so corrected wave functions (214) are [124]:

$$\langle\tilde{\phi}_n^{(+)}|\frac{\partial}{\partial R}|\tilde{\phi}_m^{(+)}\rangle = \langle\phi_n^{(+)}|\frac{\partial}{\partial R}|\phi_m^{(+)}\rangle -$$

$$- \frac{m}{2}(E_m^{(+)} - E_n^{(-)})\langle\phi_n^{(-)}|z|\phi_m^{(+)}\rangle \quad . \tag{215}$$

If $\phi_n^{(+)} = |1s\sigma\rangle$ then $\phi_n^{(-)} = |2p_{1/2}\sigma\rangle$ etc., z is the intrinsic coordinate along the molecular axis. The matrix elements (215) are shown in Fig. 29 by full lines. They vanish rapidly beyond R = 2000 fm, indicating the point where the two- K-shells just begin to influence each other.

In the truly molecular region ($R \lesssim 1000$ fm) the use of the full velocity v in the translating phase factor does not seem to be justified since the electron belongs to neither nucleus. Various authors have tried to remedy this by introduction of a weight function $f(R,\vec{r})$ in front of v. The U-U system is very favourable in this respect, because of small R the effect of the phase factor tends to become a very small correction to the matrix element. In consequence, our choice has been $f(R) \equiv 1$.

The most noteworthy feature of the matrix elements is, however, the steep increase even at very small R. For comparison we have scaled a nonrelativistic matrix element to U+U (dashed-dotted line in Fig. 29). The difference is due to the fact that for $Z_1 + Z_2 > 137$ the $ns\sigma$ and $np_{1/2}\sigma$ wave functions are extremely sensitive on R, especially for small separations (this is the lack of a "runway" in the correlation diagram). The smooth behaviour of the matrix elements suggested calculating them also in the monopole approximation, substituting a blown-up nucleus of radius $\frac{1}{2}R$ for the two nuclei. The results are in agreement within 2 percent with the exact curves in the range 20 fm $\leq R \leq$ 400 fm, so that they cannot be shown separately in the same figure. This makes it possible to obtain the matrix elements with continuum states without actually computing relativistic two-centre continuum wave functions [124]. However, it should be kept in mind that, if the final state is calculated in the monopole approximation, so must be the initial (bound) state. Fig. 31 shows some $\partial/\partial R$ matrix elements from the $1s\sigma$ state of U+U to electron scattering states. They are seen to fall off nicely to larger separations R, so that the Fourier integral (212) converges rapidly.

Obviously, it is impossible to obtain the asymptotic translation factors in the monopole approximation. We shall therefore present a method to ensure convergence of the Fourier integral (Eq.(212)) even if the matrix elements approach constant values for $R\to\infty$. To this end we integrate between large times (−T) and T and let afterwards $T\to\infty$. Integrating by parts we have:

$$a_E(T) = -\int_{-T}^{T} dt \langle f|\frac{\partial}{\partial t}|i\rangle\, e^{i\int_0^t dt'(E_f-E_i)} =$$

$$= -\frac{\langle f|\frac{\partial}{\partial t}|i\rangle}{i(E_f-E_i)}\, e^{i\int_0^t dt'(E_f-E_i)}\Bigg|_{-T}^{T} +$$

$$+\int_{-T}^{T} dt\, \frac{\partial}{\partial t}\left[\frac{\langle f|\frac{\partial}{\partial t}|i\rangle}{i(E_f-E_i)}\right] e^{i\int_0^t dt'(E_f-E_i)} \quad . \tag{216}$$

If the matrix element $\langle f|\frac{\partial}{\partial t}|i\rangle$ vanishes for $T\to\infty$, the first term disappears in the limit; if it approaches a constant value, $C_{fi}(\infty)$, then the boundary term corresponds to an asymptotic, unphysical coupling between states $|i\rangle$ and $|f\rangle$:

$$a_E^{(b)}(T) \to i\,\frac{C_{fi}(\infty)}{E_f(\infty)-E_i(\infty)} \cos\left(\int_0^T dt\,(E_f - E_i)\right) \, . \tag{217}$$

This contribution is explicitly eliminated and the physical result for the transition amplitude is:

$$a_E^{(phys)} = -i\int_{-\infty}^{\infty} dt\, \dot{R}\frac{\partial}{\partial R}\left[\frac{\langle f|\dot{R}\frac{\partial}{\partial R}|i\rangle}{(E_f - E_i)}\right] \exp\left(i\int_0^t dt'(E_f-E_i)\right). \tag{218}$$

The integrand now vanishes faster than 1/R, redering the integral absolutely convergent.

Eqs. (211), (212) respectively (218) have been solved numerically for the U+U system and, in a few cases, for the Pb+Pb quasi-molecule. For the nuclear trajectory R(t) we have invariably prescribed Rutherford hyperbolae through the parametric representation

$$R = a(\varepsilon \cosh\xi + 1), t = \left[\frac{\mu}{2E_{CM}}\right]^{1/2} a(\varepsilon \sinh\xi + \xi)$$

with

$$\varepsilon = \left(1 + \frac{b^2}{a^2}\right)^{1/2}, \quad a = \frac{Z_1 Z_2 e^2}{2E_{CM}}, \quad \mu = \frac{M_1 M_2}{M_1 + M_2} .$$

We first discuss the calculations including only bound states. The four states 1sσ, 2sσ, 3sσ, 4sσ were taken as bound state channels. The 16 possible configurations are denoted for the states in the above order, by a '1' for occupied states and a '0' if vacant. E.g. if two electrons are in the 1sσ and 2sσ states we write: (1,1,0,0), etc. . Only one spin projection is treated, since we neglect any coupling between $j_z = \pm 1/2$ states, such as magnetic fields.

Let us first look at the amplitudes a(t). Fig.32 shows the probabilities $|a|^2$ developing from the one-electron configuration (1,0,0,0) at zero impact parameter. As expected, the curves show oscillations with a maximum twice as high as the final excitation probability $P(b) = |a(\infty)|^2$. The steep rise of $|a(t)|^2$ right beyond the distance of closest approach can be understood like that: During the approach of the two nuclei, a major portion of the 1sσ electronic density flows radially inward following the nuclear motion. (This is characteristic for system with $(Z_1 + Z_2)\alpha > 1!$). When this motion is suddenly reversed the electron is left behind and redistributes over other states. A more realistic configuration to start with is (1,1,1,0). The binding energy of a 4sσ electron in U is 1.4 keV, corresponding to an incident energy of 2.8 MeV/nucl. From the general argument that electrons which are slower than the projectile nucleus are stripped away in the target, it can be inferred that the 4sσ state will be vacant to a considerable degree. The amplitudes developing during the collision are plotted in Fig. 33. The crucial curve is the probability for exciting a K-vacancy. It reaches 4 percent after the collision, but it is much smaller in the region R < 35 fm, where the 1sσ state has dived into the negative energy continuum.

Next, we include 1sσ ionization into the continuum. The summed probability $\int dE|a_E(t)|^2$ is shown in Fig. 34 as dashed line in comparison to the 1sσ ionization, if the 2sσ, 3sσ or 4sσ state is vacant before the collision. After the collision, continuum ionization is comparable to excitation into the 4sσ state (if vacant), but in the diving region it is the dominant process (because the 2sσ state will be practically filled and cannot contribute). The projectile energy and impact parameter dependence of the four processes leading to 1sσ vacancy production is shown in Figs. 35 and 36. $P_n(b)$ denotes the final 1sσ ionization probability, if the nsσ state was vacant in the beginning of the collision, $P_E(b)$ the ionization into the continuum. With decreasing incident ion energy the excitation probabilities fall off very steeply (Fig.35), until at tandem energies (E_{lab} < 200 MeV) they are much less than 10^{-6}. The same is true for increasing impact parameters: Typically the probability shrinks to one half at b = 20 fm (Fig. 36). This will facilitate coincidence experiments, because all the cross section goes into large angle collisions anyway.

The excitation of vacancies in higher states cannot be calculated reliably without inclusion of more bound state channels (higher nsσ, ndσ, ndπ). Also, the negative parity states must be investigated. This extension of the coupled channel code is in progress and will be published soon. Thus, we can only report results on the ionization into the continuum for the $2p_{1/2}\sigma$, 2sσ and 3sσ states in U+U (Fig. 37a) and Pb+Pb (Fig. 37b) in perturbation theory. Integrated over all impact parameters the 1sσ ionization cross section in U+U is $\sigma(1s\sigma) \sim 2$ barns, the $2p_{1/2}\sigma$ ionization gives $\sigma(2p\sigma) \sim 20$ barns. On the basis of this, we can estimate the total K-vacancy production cross section in a 1600 MeV U+U collision from direct ionization to be $\sigma_k^{dir} \sim 22$ barns. (It must be remembered that the cross section is $4\pi \cdot \int bdb\, P(b)$ because two spin states contribute.) For Pb+Pb at 1380 MeV the corresponding numbers are obtained from Fig. 37a: $\sigma(1s\sigma) \sim 3.5$ barns, $\sigma(2p\sigma) \sim 60$ barns. Obviously, the K-vacancy production is dominated by the ionization of the molecular $2p_{1/2}\sigma$ level. In order to determine experimentally the 1sσ ionization rate one has to take refuge to asymmetric systems in which the Meyerhof-sharing between the 1sσ and 2sσ levels has dropped below, say, 5 percent.

In a heavy ion collision the two nuclei are not only the source of a (time-dependent) Coulomb potential, but they also create a strong (transverse) magnetic field [90]. Since the adiabatic basis functions include the effect of the Coulomb potential only, it is convenient to work in the Coulomb (transverse) gauge:

$$\nabla^2 V_c = -4\pi\rho \tag{219a}$$

$$(\nabla^2 - \frac{\partial^2}{\partial t^2})\vec{A}_c = -4\pi\vec{j} + \frac{\partial}{\partial t}(\nabla V_c) \equiv -4\pi\vec{j} \quad . \tag{219b}$$

The potentials V_c and $\vec{A}_c$ could also be obtained from the retarded potentials V_L and $\vec{A}_L$ in Lorentz-gauge (the solid so-called Lienard-Wiechert potentials) by a gauge transformation requiring div $\vec{A} = 0$. Eq. (219a) gives simply the instantaneous Coulomb potential, but (219b) is not easily solved. In a first approximation we have neglected the retardation effect and the correction to the current. Then

$$e\vec{A}_c = -\frac{Z_1 e^2 \vec{V}_1}{|\vec{r}-\vec{R}/2|} - \frac{Z_2 e^2 \vec{V}_2}{|\vec{r} + \frac{\vec{R}}{2}|} \quad , \tag{220}$$

where $\vec{v}_i$ are the heavy ion velocities. We shall restrict the further discussion to symmetric collisions: $Z_1 = Z_2 = Z$.

The heavy ion velocity can be divided into a radial part v_ρ and an angular one v_ϕ (the ρ-direction is along the internuclear axis)

$$V_\rho = \frac{V_\infty}{2}\left(1 - \frac{2Z^2e^2}{RE_{kin}} - \frac{b^2}{R^2}\right)^{1/2} \tag{221a}$$

$$v_\phi = \frac{bv_\infty}{2R} \quad . \tag{221b}$$

Here E_{kin} and v_∞ are the (lab system) projectile kinetic energy and velocity before the collision, and b is the impact parameter. To gain an estimate of the magnetic fields involved, we take only the angular contribution

$$B(0) = \frac{2Zev_\phi}{(R/2)^2} = \frac{4Zev_\infty b}{R^3} \sim 3.3 \times 10^{16} \text{ gauss} \tag{222}$$

for a U+U collision at E_{lab} = 1600 MeV and R = 20 fm, b = 10 fm.

The magnetic interaction Hamiltonian of the relativistic electrons bound to the molecular system is given by (α_ρ, α_ϕ are Dirac matrices)

$$H_{mag} = H_\rho + H_\phi = -Ze^2 (v_\rho \alpha_\rho + v_\phi \alpha_\phi)[\,|\vec{r} - \tfrac{\vec{R}}{2}|^{-1} - |\vec{r} + \tfrac{\vec{R}}{2}|^{-1}] \, . \tag{223}$$

In the basis of Eq. (158) the magnetic states of opposite spin are degenerate. Therefore, the calculation of the magnetic splitting due to H_{mag} proceeds via degenerate state perturbation theory. In the simplest case of spin $-\frac{1}{2}$ states ($1s\sigma$ and $2_{1/2}\sigma$) the energy change of the spin-up and spin-down states is

$$E_{\uparrow,\downarrow} - E_0 = \pm [\,|\langle\uparrow|H_\rho|\uparrow\rangle|^2 + |\langle\downarrow|H_\phi|\uparrow\rangle|^2]^{1/2} \quad . \tag{224}$$

It is evident from the definition of H_{mag} (Eq. (223)) that the dynamical parameters E_{kin} and b of a particular collision enter only via v_ϕ and v_ρ. Let us therefore introduce the matrix elements

$$M_\phi = \langle\downarrow| -Ze^2\alpha_\phi (|\vec{r}-\vec{R}/2|^{-1} - |\vec{r}+\vec{R}/2|^{-1})|\uparrow\rangle \tag{225a}$$

$$M_\rho = \langle\uparrow| -Ze^2\alpha_\rho (|\vec{r}-\vec{R}/2|^{-1} - |\vec{r}+\vec{R}/2|^{-1}|\uparrow\rangle \tag{225b}$$

which depend only on Z and the nuclear separation R. They are shown in Figs. 39a,b as a function of R for the systems U+U (1sσ and $2p_{1/2}\sigma$), Pb+Pb ($1s_{1/2}\sigma$) and Xe+Xe ($1s_{1/2}\sigma$). In the region 20 fm < R < 200 fm the velocity components v_ϕ and v_ρ are comparable (except for the unlikely case of backward collisions). Therefore the rotational magnetic coupling is dominant, M_ϕ being 30 times larger than M_ρ. We find that the matrix element scales as the fourth power of Z. The relativistic corrections enhance the dependence on Z due to the partial collapse of the wave function to the centre which brings the electrons to the region of strong magnetic field strength.

We have calculated the magnitude of the splitting ΔE of the opposite spin states following from Eq. (224) using the collision parameters: b = 13 fm and E_{kin}/N = 9 MeV/nucl. The relative splittings $\Delta E/E_B$ are shown in Fig. 39 as a function of R for the above mentioned scattering systems. We find that the relative magnetic splitting of the $2p_{1/2}\sigma$ level in U+U is very large and exceeds that of the 1sσ state because the binding energy is considerably smaller, while the matrix elements are of the same size. These results indicate that any experiment to test the behaviour of electrons in the strong field should possibly be carried out in a system as heavy as Pb+Pb. As the system becomes heavier an investigation of the splitting of the $2p_{1/2}$-level seems to be advantageous, since it is much easier to have this state ionized in the collision.

The energy difference between the two magnetic substates will lead to a difference in the ionization probability for the two states. This will cause the polarization degree of the subsequent K_α radiation to be sensitive to the ionization mechanism for the molecular states. If the ionization occurs in the region of strong magnetic splitting ($R \lesssim 100$ fm), then the ratio between the two photon polarization states after the collision will be $\sigma_\uparrow/\sigma_\downarrow \approx e^{+\Delta E/\Gamma} \sim (1+ \frac{\Delta E}{\Gamma}) > 1$. Γ is the dynamic collision broadening of the molecular $1s\sigma$ state, with an estimate $\Delta E/\Gamma \sim 0.2$. On the other hand, if ionization occurs in the periphery of the collision then no measurable polarization can be built up. This qualitative argument [90] is fully borne out in our calculation of the direct ionization probability P(b) for the two magnetically split $1s\sigma$ states in a U+U collision (Fig. 40). The average polarization is 20% at E_{lab} = 1600 MeV. If it could be measured, one would (1) explore magnetic fields of more than 10^{14} gauss, (2) provide an experimental test of the prediction that ionization takes place at small distances. Of course, a coincidence experiment is required to fix the direction of the molecular magnetic field.

12. SHAKE-OFF OF THE VACUUM POLARIZATION

Having discussed the decay of the vacuum (Chap.10) and the dynamical treatment of Coulomb excitation, let us now explain the induced direct transitions between the positive and negative energy continua. This will lead to another new effect of QED, namely a collective type of electron-positron creation, due to the coherent action of the strong, extended time-dependent electric field.

We again work in the framework of the quasiadiabatic molecular model and start from a complete set of solutions of the stationary two-centre Dirac equation

$$E_i(R)\phi_i(\vec{x},R) = H(\vec{x},R)\phi_i(\vec{x},R) \quad . \tag{226}$$

The functions ϕ_i depend on the internuclear distance R as a continuous parameter. The time-dependent collision is described in the semiclassical approximation by introducing a definite time-dependence R = R(t). The part of the time evolution operator not pre-diagonalized by Eq.(226) is then expressed as (neglecting magnetic fields)

$$i\dot{R}\,\frac{\partial}{\partial R} = i\,\frac{\partial}{\partial t} - H(R(t)) \quad . \tag{227}$$

The initial electron configuration is specified by a set F of occupied states, where F includes all states of the negative energy continuum (definition of the vacuum) and all occupied bound states [41]. The state vector of this configuration will be denoted by $|F\rangle$. During the collision, electron-positron configurations are excited. Since Eq. (227) is a single particle operator, we can restrict the basis in first order to one electron-one positron excitations. The time-dependent state vector of the system can then be written as

$$|\psi(t)\rangle = c_o(t)|F\rangle + \sum_{j\varepsilon F}\sum_{i\notin F} c_{ij}(t) b_i^+ d_j^+ |F\rangle , \tag{228}$$

b_i^+, d_j^+ being creation operators for particles and holes, respectively. The normalization condition is

$$|c_o|^2 + \sum_{j\varepsilon F}\sum_{i\notin F} |c_{ij}|^2 = 1 . \tag{229}$$

In the first order perturbation theory, the amplitudes c_{ij} are given by

$$c_{ij}(T) = -\int_{-\infty}^{T} dT \langle\phi_i(R)|\dot{R}\frac{\partial}{\partial R}|\phi_j(R)\rangle \cdot$$

$$\cdot\, e^{i\int_{-\infty}^{t} d\tau (E_i(R(\tau)) - E_j(R(\tau)))} . \tag{230}$$

In order to compute the only observable $|c_{ij}(T\to\infty)|^2$, we introduce the time-dependent density matrix

$$\rho_i(\vec{x},t,R(t);\vec{x}',t',R(t') \equiv R') =$$

$$= \phi_i^+(\vec{x},R)\phi_i(\vec{x}',R')\exp[i\int_{t'}^{t} d\tau E_i(R(\tau))] . \tag{231}$$

One verifies by straightforward calculation

$$|c_{ij}(\infty)|^2 = -\int_{-\infty}^{\infty}dt \int_{-\infty}^{\infty}dt' \iint d^3x d^3x' [\dot{R}\frac{\partial}{\partial R}\rho_i(\vec{x},t,R;\vec{x}',t',R')]^+ \cdot$$

$$\cdot [\dot{R}'\frac{\partial}{\partial R'}\rho_j(\vec{x},t,R;\vec{x}',t',R')] . \tag{232}$$

The total number of excited particle-hole pairs N_{ph} after the collision is obtained by summing over i and j. We introduce the density matrices ρ_+ of occupied (positron) and ρ_- of vacant (electron) states by

$$\rho_o(\vec{x},t,R;\vec{x}',t',R') = \sum_{j\in F} \rho_j \; ; \qquad \rho_- = \sum_{i\notin F} \rho_i \; , \tag{233}$$

the charge symmetrized density matrices $\tilde{\rho} = \frac{1}{2}(\rho_+ - \rho_-)$, and $\rho = \frac{1}{2}(\rho_+ + \rho_-)$. The integrals in Eq. (232) can be interpreted as trace in the coordinate representation, and we can write in matrix formulation

$$N_{ph} = \sum_{j\in F} \sum_{i\notin F} |c_{ij}(\infty)|^2 = -\mathrm{Tr}[(\dot{R}\frac{\partial}{\partial R}\rho_-(R,R'))^+ \dot{R}'\frac{\partial}{\partial R'}\rho_+(R,R')] =$$

$$= + \mathrm{Tr}[(\dot{R}\frac{\partial}{\partial R}\tilde{\rho}(R,R'))^+ \dot{R}'\frac{\partial}{\partial R'}\tilde{\rho}(R,R')]$$

$$- \mathrm{Tr}[(\dot{R}\frac{\partial}{\partial R}\rho(R,R')^+ \dot{R}'\frac{\partial}{\partial R'}\rho(R,R')] \; . \tag{234}$$

The last form of Eq. (234) clearly exhibits the nature of the process: The shake-off of vacuum polarization [127] and of bound electrons. Namely, $\tilde{\rho}$ can be decomposed into the vacuum polarization charge density matrix and a contribution from the bound electrons

$$\tilde{\rho} = \frac{1}{2}[\sum_{E_i<-m} \rho_i - \sum_{E_i>-m} \rho_i] + \sum_{\substack{i\in F \\ E_i>-m}} \rho_i \; . \tag{235}$$

The term involving ρ is independent of the initial electron configurations and serves to regularize the divergent expression involving $\tilde{\rho}$. The meaning of Eq. (234) is illustrated in Fig. 41. The usual two-photon pair production diagram is shown in part (a), the process described by the theory outlined in this paper is depicted in part (b). Because $(Z_1 + Z_2)\alpha > 1$ in the collision of very heavy nuclei, infinitely many interactions with the combined electric field of both nuclei must be taken into account. In this sense, diagram (b) stands for an infinite series of Feynman diagrams like Fig.41a. By coherent summation of these diagrams the virtual photon field acquires a collective nature. For $Z_1 + Z_2 << \alpha^{-1}$ this collectivity is small, meaning that the pair production is well described by diagram (Fig.41a). However, for $(Z_1 + Z_2)\alpha > 1$ the collective effect becomes dominant. This is expressed in the Z-dependence of the total cross section: The lowest order diagram (Fig.41a) increases as Z^4, whereas in the superheavy region at constant ion velocities the dependence is approximately Z^n, with $n \approx 18$. This immediately sheds a light on the average number of photon interactions in the production process.

To evaluate Eq. (230) numerically, it is useful to integrate by parts, yielding

$$|c_{ij}(\infty)| = 2 \int_0^\infty dt \frac{\cos(E_i - E_j)t}{(E_i - E_j)^2} [\ddot{R} + \dot{R}^2 \frac{\partial}{\partial R}] \langle \phi_i | \frac{\partial H}{\partial R} | \phi_j \rangle . \quad (236)$$

It is sufficient to keep only $\kappa = \pm 1$ states in Eq.(236) because matrix elements with higher angular momentum states are smaller by at least one order of magnitude. This is analogous to vacuum polarization in heavy atoms [51,55]. For these states the matrix elements peak strongly for small values of R. They can be calculated in the monopole approximation without introducing a large error (cf. Sect. 12).

We checked the accuracy of every step carefully. However, the calculation of the final cross section

$$\sigma_{e^+e^-} = 2\pi \int_0^\infty b\, db \int_{-\infty}^{-m} dE \int_m^\infty dE' |c_{E'E}(b)|^2 \tag{237}$$

may produce an uncertainty of ± 50% due to the multiple integrations involved.

(a)	Z	σ[μb]	(b)	R_{min}[fm]	σ[μb]
	146	5.7		25	3.0
	158	33.4		20	19.4
	164	76.4		16	76.4
	168	135		13	193.4

(a) The total pair production cross section $\sigma_{e^+e^-}$ for various superheavy collision system $Z = Z_1 + Z_2$. The distance of closest approach (16 fm) is kept constant, so that the increase in ion energy contributes somewhat to the sharp rise with Z.

(b) The cross section for Pb+Pb collisions for various ion energies, characterized by the distance of closestapproach.

The table shows the total pair production cross section for symmetric collisions with a total nuclear charge $Z = Z_1 + Z_2$, at ion energies so that the distance of closest approach is always 16 fm. $\sigma_{e^+e^-}$ is seen to increase rapidly with Z. The energy spectrum of the created positrons and electrons is shown in Fig. 42a for a Pb+Pb collision (Z = 164). The main difference between the electron and the positron distribution, viz. the behaviour at small energies, is explained by the Coulomb repulsion of the positron states. The positron distribution peaks at ca. 400 keV kinetic energy with $d\sigma/dE_p \sim 0.1$ µb/keV. Also more high energy positrons than electrons are produced.

The impact parameter dependence of pair creation is plotted in Fig.42b up to b = 40 fm. bP(b) peaks at 3 fm and falls off exponentially for larger impact parameters. Thus the shake-off of the vacuum polarization cloud should be experimentally observable in the collision of very heavy ions. Its study in undercritical systems ($Z \leq 170$) is advisable so that it is not intermixed with the decay of the overcritical vacuum. Pb-Pb collisions should be favoured, because positrons stemming from internal conversion of Coulomb excited nuclei should be minimal in this case. The shake-off of the vacuum polarization cloud, which is a collective type of e^+e^- creation, is a new process of quantum electrodynamics. It is characteristic for the region of strong binding $(Z_1+Z_2)\alpha > 1$ where the electron and positron wave functions and therefore also the vacuum polarization charge density strongly depend on the internuclear separation and thus are very sensitive to the collision dynamics.

13. BACKGROUND: BREMSSTRAHLUNG AND NUCLEAR CONVERSION

Several processes are known to produce X-rays and positrons in the following of heavy ion collisions. Some of them will be able to severly disturb the observation of quasi-molecular phenomena and thus impede the investigation of QED of strong fields. To conclude the second lecture we will report on two particularly interesting and important background effects, nuclear bremsstrahlung and conversion of Coulomb-excited nuclear states.

Nucleus-nucleus bremsstrahlung leads to rather small radiative cross sections. Its energy dependence, however, often if weaker than that of electronic processes like MO radiation, Radiative Electron Capture, Radiative Ionization or Electron bremsstrahlung. Therefore there exist combinations of projectile and target for which the high-energy tails of the measured spectra are dominated by nuclear bremsstrahlung.

Intensity and angular distribution of this radiation may be obtained from a straightforward classical calculation, provided that the Sommerfeld parameter η is large. The resulting radiative cross sections for coincidence and singles experiments are presented by Reinhard, Soff and Greiner [128]. Employing a multipole expansion up to second order the cross section integrated over impact parameter has the general form

$$\frac{d\sigma}{dE_x d\Omega_x} = \frac{1}{E_x} \frac{\alpha^3}{4\pi} \left(\frac{c}{v}\right)^2 Z_1^2 Z_2^2 \frac{1}{M^2} \nu \{ f_1^2 (g_1^{(0)} + g_1^{(2)} P_2(\cos\theta))$$

$$-f_1 f_2 A_R (\frac{v}{c}) (g_{12}^{(1)} P_1(\cos\theta) + g_{12}^{(3)} P_3(\cos\theta)) +$$

$$+f_2^2 A_R^2 (\frac{v}{c})^2 (g_2^{(0)} + g_2^{(2)} P_2(\cos\theta) + g_2^{(4)} P_4(\cos\theta))\} \quad . \tag{238}$$

Here M is the nucleon mass, A_R is the reduced nucleon number, $\nu = \omega/\omega_o$ where ω_o is a reciprocial collision time, and $f_\lambda = (Z_1/A_1^\lambda + (-)^\lambda Z_2/A_2^\lambda)$ denotes the electric multipole moment of order λ. The $b(\nu)$ are uniformly decreasing functions of photon frequency ω.

In general the multipole series expansion (238) converges rapidly. However, the first contribution ($\lambda=1$) is multiplied by the dipole factor $(Z_1/A_1 - Z_2/A_2)$ which is small for all heavy systems or even vanishes for equal projectile and target nuclei. Therefore the quadrupole radiation plays an important role. Even if the dipole radiation is much larger than the quadrupole contribution the inteference term, which originates from their coherent superposition, influences the angular distribution of X-rays. For example, in the slightly asymmetric system ^{58}Ni-^{60}Ni a forward-backward anisotropy of about 5 is introduced (at E_{lab} = 60 MeV, E_x = 30 keV). Recently Trautvetter, Greenberg and Vincent [129] observed this predicted constructive and destructive interference effect for several asymmetric collision systems like ^{13}O-^{58}Ni. The results of this experiment are shown in Fig. 43. This proves that nuclear bremsstrahlung can contribute to the MO X-ray spectra (especially to the anisotropy) and should be subtracted according to ref. [238]. Fig. 44 demonstrates the general dependence of $d\sigma/dE_x$ on nuclear charge number for symmetric (Fig. 44a) and representative asymmetric (Fig. 44b) systems. The cross section was evaluated at the K_α united atom transition energy, therefore it drops with increasing Z, despite the factor Z^4.

In fast collisions of highly charged nuclei a certain fraction of X-rays with energy above $2mc^2$ is produced. Fourier frequencies larger than the e^+e^- threshold are well contained in the nuclear motion. Using the bremsstrahlung spectrum and the well known conversion coefficients $\beta_{E\lambda}(\omega)$ for electron positron creation the total pair production cross section via bremsstrahlung can be calculated according to

$$\sigma^{e^+e^-} = \sum_\lambda \int_{2m}^{\infty} \frac{d\sigma^\lambda_{Brems}}{d\omega} \beta_{E\lambda}(\omega)\, d\omega \; . \tag{239}$$

Using the conversion coefficients in Born approximation, Reinhardt, Soff and Greiner [128] obtained a cross section of $3.8 \cdot 10^{-8}$b for ^{132}Xe-^{238}U and $5.8 \cdot 10^{-8}$b for ^{238}U near the Coulomb barrier. This background process can therefore be neglected.

The described calculation is an approximate treatment of the pair production mechanism in the first of the graphs.

For sake of completeness and clarity we present with the fourth graph also the decay of the vacuum, which, of course, should only by understood schematically since a proper Feynman type description of the vacuum rearrangement is inadequate. The contribution of the second graph often discussed in high energy physics

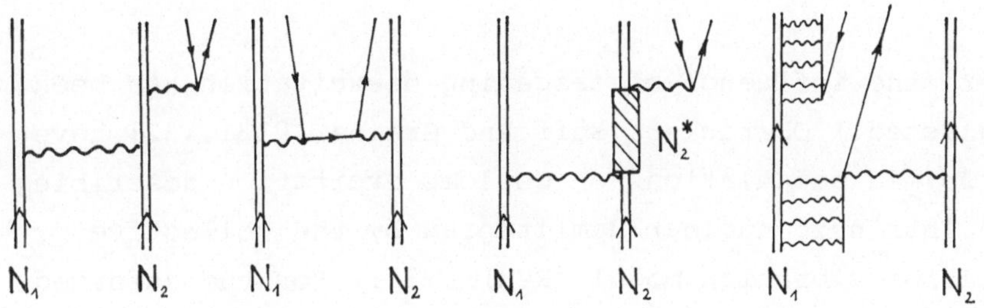

[130] is expected to be small in nonrelativistic collisions. Substantially more important is the background due to nuclear Coulomb excitation. In close collisions of heavy nuclei, collective degrees of freedom are excited with a probability of nearly one. Within about 10^{-3} sec the populated higher states decay under the emission of a photon. This photon can be converted to emit an atomic electron or create an electron-positron pair.

The differential pair formation cross section is given by the product of the scattering cross section and the positron emission probability of both nuclei:

$$\frac{d\sigma^{e^+e^-}_{cb}}{dE_p d\Omega_{ion}} = \frac{d\sigma_{scatt}}{d\Omega_{ion}}(\theta_{ion}) \sum_{1,2} \frac{dW^{e^+e^-}_{cb}}{dE_p}(E_p,\theta_{ion}) \ . \qquad (240)$$

The probability dW/dE_p is a function of the positron kinetic energy E_p and the scattering angle θ_{ion}. It depends on the Coulomb excitation probability $P^{(b)}_{i_\gamma}$ of the initial nuclear level, the branching ratio p^{γ}_{if} for a photon transition into the final state, and the corresponding differential conversion coefficient $d\beta/dE_p$.

$$\frac{dW^{e^+e^-}_{cb}}{dE_p}(E_p,\theta_{ion}) = \sum_{i,f} P^{cb}_i(\theta_{ion}) P^{\gamma}_{if} \frac{d\beta_{if}}{dE_p}(E_p) \ . \qquad (241)$$

(Here the influence of cascading deexcitation has been neglected.) Oberacker, Soff and Greiner [131,132] have performed calculations on Coulomb excitation describing the intrinsic nuclear Hamiltonian by the collective rotation-vibration model (RVM) [133]. For the deformed even-even nucleus ^{238}U two calculations were done. In Method 1 all states of the RVM below the fission barrier were taken into account with the restriction to magnetic substates M = 0 which is correct for $\theta_{ion} = 180^o$. Method 2 included all magnetic substates but was restricted to the ground state band and the first β- and γ-vibrational bands. The excitation probabilities have a magnitude of the order 10^{-3}. They are obtained from the solution of the time-dependent Schrödinger equation with a coupling potential including Coulomb- and nuclear forces [134]. Due to destructive interference, excitation near the barrier may be weaker than at impact energies well below.

Fig. 45 shows the resulting positron cross sections for ^{238}U-^{238}U in dependence of the scattering angle (since the nuclei are not distinguishable $d\sigma/d\Omega$ had to be symmetrized). The dotted and dashed lines belong to model 1 or 2 respectively. These curves must be compared with the solid line which is the cross section for dynamical autoionization of positrons (L_o = 0.01). Obviously the ratio between desired and background positrons is most favourable at forward angles. Here, however, the nuclei do not, if at all, probe the diving region very deeply. For backward ion angles the cross sections only differ by a factor 2. The total positron cross sections are $\sigma_{vac}^{e^+e^-} = 5.0 \cdot 10^{-4}$b, $\sigma_{cb}^{e^+e^-} = 1.25 \cdot 10^{-4}$b (method 1) or $2.28 \cdot 10^{-4}$b (method 2).

The incorporation of dynamical K-hole formation should increase the first of these numbers somewhat. On the other hand, the cross section at low scattering angles will be depressed similar to the background.

A clear distinction between the positron spectra from both processes is the behaviour at high kinetic energies of the positron shown in Fig. 46. In the considered model positrons are produced from transitions between β- and γ- and the ground state band with a maximum energy difference of 1.8 MeV. Thus, the positron spectrum is cut off above $E_p \sim 800$ MeV. Inclusion of the rotation-vibration interaction leads to band mixing and allows for higher transition energies but the resulting cross section will be very small.

The magnitude of the nuclear excitation background precludes the possibility of quasimolecular X-ray spectroscopy in the region of very high Z since the nuclear lines are dominant by orders of magnitude. The only exception is the very stiff vibrator nucleus ^{208}Pb with a lowest lying 2^+ state at 4.086 MeV. On the other hand, the decay of the neutral vacuum can be separated from nuclear conversion pair production provided that the nuclear excitation is systematically studied for various colliding systems both in theory and experiment. Pending on the further results of a fully dynamical treatment, the first process probably can be also distinguished from the shake-off of the vacuum polarization (direct pair production). Experiments to study positron production in the collision of very heavy ions like Pb-Pb, U-U are presently performed at GSI (Darmstadt) by Backe, Kankeleit, Handschug, Makajama, Richter, Weik, Willwater (TH Darmstadt), Bohemeyer and Greenberg (GSI, Darmstadt), Berdermann, Kienle, Stettmeier (TH München).

REFERENCES

1. B. Müller, J.Rafelski, W. Greiner, Z.Physik 257, (1972a) 62.
2. B. Müller, J. Rafelski, and W. Greiner, Z. Physik 257 (1972b) 183.
3. P.A.M. Dirac, Proc. Roy. Soc. A 117 (1928) 610.
4. P.A.M. Dirac, Proc. Roy. Soc. A 126 (1930) 360.
5. J.D. Bjorken, and S.D. Drell, Relativistic Quantum Mechanics (1964) and Relativistic Quantum Fields (1965) (New York: McGraw Hill).
6. J.M. Jauch and F. Rohrlich, The Theory of Photons and Electrons, 1976, 2nd ed. (New York: Springer).
7. M.E. Rose, Relativistic Electron Theory (1961) (New York: Wiley).
8. A.I. Akhiezer, and N.B. Berestetzkii, Quantum Electrodynamics (1965) (New York: Interscience).
9. V.S. Popov, Sov. J. Nucl. Phys. 12 (1971) 235.
10. Ya. Zeldovich, and V.S. Popov, Sov.Phys.Usp. 14 (1972) 673.

11. R.E. Langer, Phys. Rev. 51 (1937) 669.

12. K.M. Case, Phys. Rev. 80 (1950) 797.

13. S.P. Alliluev, Sov. Phys. JETP 34 (1972) 8.

14. I. Pomeranchuk, and J. Smorodinsky, J.Phys.USSR 9 (1945) 97.

15. F.G. Werner, and J.A. Wheeler, Phys. Rev. 109 (1958) 126.

16. V.V. Voronkov, and N.N. Kolesnikov, Sov. Phys. JETP 12 (1961) 136.

17. W. Pieper, and W. Greiner, Z. Physik 218 (1969) 327.

18. S.S. Gershtein, and Ya.B. Zeldovich, Sov. Phys. JETP 30 (1969a) 358 and Lett. Nuovo Cim. 1 (1969) 835.

19. B. Fricke and W. Greiner, Phys. Lett. 30B (1969).

20. B. Fricke, and G. Soff, Dirac-Fock-Slater Calculations of the Elements Fermium (Z=100) to Z=173, GSI-Bericht T1-74, (1974), Darmstadt, also Atomic and Nuclear Data Tables, in print.

21. G. Soff, B. Müller and J. Rafelski, Z. Naturforschung 29a (1974) 1267.

22. U. Fano, Phys. Rev. 124 (1961) 1866.

23. B. Müller, H. Peitz, J. Rafelski and W. Greiner, Phys. Rev. Lett. 28 (1972) 1235.

24. B. Müller, J. Rafelski and W. Greiner, Nuovo Cim. 18A (1973) 551.

25. G. Bastard, and P. Nozières, Phys. Rev. B13 (1976) 2560.

26. V.S. Popov, JETP Lett. 11 (1970a) 162.
V.S. Popov, Sov.J.Nucl.Phys. 12 (1970b) 235.

27. V.S. Popov, Sov.J. Nucl. Phys. 14 (1971) 257 and Sov. Phys. JETP 32 526, 33, 665.

28. W. Gordon, Z. Phys. 48 (1928) 11.

29. G.C. Darwin, Proc. Roy. Soc. A118 (1928) 654.

30. O. Klein, Z. Physik 53 (1929) 157.

31. E. Sauter, Z. Physik 73 (1931) 547.

32. K. Nikolsky, Z. Physik 62 (1929) 677.

33. W. Heisenberg, and H. Euler, Z. Physik 98 (1936) 714.

34. F. Hund, Materie als Feld (1954) (Berlin: Springer).

35. F. Beck, H. Steinwedel, G. Süßmann, Z. Physik 171 (1963) 189.

36. H.G. Dosch, J.H.D., Jensen, V.F. Müller, Physica Norvegica 5 (1971) 2.

37. J.Rafelski, B. Müller and W. Greiner, Nucl. Phys. B68 (1974) 585.

38. L. Fulcher, A. Klein, Phys. Rev. D8 (1973) 2455.

39. L. Fulcher, A. Klein, Ann. of Physics (NY) 84 (1974) 335.
W.M. Furry, Phys. Rev. 81 (1951) 115.

40. A. Klein and J. Rafelski, AIP Conf. Proc. 23 (1975a) 356.

41. P.G. Reinhard, W. Greiner and H. Arenhövel, Nucl. Phys. A166 (1971) 173.

42. J. Rafelski, B. Müller, G. Soff and W. Greiner, Ann. of Physics (NY) 88 (1974) 419.

43. J. Schwinger, Phys. Rev. 91 (1953) 713.

44. E.A. Uehling, Phys. Rev. 48 (1935) 55.

45. R. Serber, Phys. Rev. 48 (1935) 49.

46. P.G. Reinhard, Lett. Nuovo Cim. 3 (1970) 313.

47. A.B. Layzer, Phys. Rev. 129 (1963) 897.

48. B. Fricke, J.B. Desclaux, and J.T. Waber, Phys. Rev. Lett. 28 (1972) 711.

49. N. Panchapakesan, Phys. Lett. 35B (1971) 522.

50. E.H. Wichmann and N.M. Kroll, Phys. Rev. 96 (1954) 232 and E.H. Wichmann and N.M. Kroll, Phys. Rev. 101 (1956) 843.

51. M. Gyulassy, Phys. Rev. Lett. 32 (1974a) 1393.
M. Gyulassy, Phys. Rev. Lett. 33 (1974b) 921.

52. M. Gyulassy, Nucl. Phys. A244, (1975) 497.

53. P.J. Mohr, Ann. of Physics (NY) 88 (1974) 26 and 52.

54. G.A. Rinker and L. Wilets, Phys. Rev. Lett. 31 (1973) 1559.

55. G.A. Rinker and L. Wilets, Phys. Rev. A12 (1975) 748.

56. L.S. Brown, R.N. Cahn and L.D. McLerran, Phys. Rev. Lett. 32 (1974) 562.
L.S. Brown, R.N. Cahn and L.D. McLerran, Phys. Rev. D12 (1975) 581, 596, 609.

57. A.B. Migdal, Nucl. Phys. B52 (1973) 483.

58. G.W. Erickson, Phys. Rev. Lett. 27 (1971) 780.

59. G.E. Brown, J.S. Langer and G.W. Schaefer, Proc. Roy. Soc. A251 (1959) 92.

60. A.M. Desiderio and W.R. Johnson, Phys. Rev. A3 (1971) 1267.

61. K.T. Cheng, W.R. Johnson, Phys. Rev. A14 (1976) 1943.

62. M. Born, Ann. Inst. Henri Poincare 7 (1937) 155.

63. M. Born, L. Infeld, Proc. Roy. Soc. A144 (1934) 429.

64. J. Rafelski, L.P. Fulcher, and W.Greiner, Nuovo Cim. B7 (1972) 137.

65. L. Infeld, and B. Hoffmann, Phys. Rev. 51 (1937) 765.

66. J. Rafelski, L.P. Fulcher, and W. Greiner, Phys.Rev. Lett. 27 (1971) 958.

67. G. Soff, J. Rafelski and W. Greiner, Phys. Rev. A7 (1973) 903.

68. G. Soff, B. Müller, J. Rafelski, and W. Greiner, Z. Naturforschung 28a (1973) 1389.

69. M.K. Sundaresan, and P.J.S. Watson, Phys. Rev. Lett. 29 (1972) 15 and 1122.

70. B. Müller and J. Rafelski, Phys. Rev. Lett. 34 (1975) 349.

71. T.D. Lee and G.C. Wick, Phys. Rev. D9 (1974) 2291.

72. T.D. Lee, Rev. Mod. Phys. 47 (1975) 267.

73. N.F. Mott, H.S.W. Massey, The Theory of Atomic Collisions 3rd Ed. Oxford Clarendon 1965.

74. P. Mokler, GSI Bericht 72-7, Darmstadt, Germany.

75. J. Rafelski, B. Müller and W. Greiner, Lett. Nuovo Cimento 4 (1972) 469.

76. D.H. Madison and E. Merzbacher in: Atomic Inner Shell Processes, vol. 1, ed. B. Crasemann (1975) (New York: Academic Press) p.l.

77. W. Lichten in: Atomic Physics 4 ed. G. zu Putlitz (1975) (New York: Plenum Press) p. 249.

78. W. Scheid, and W. Greiner, Heavy Ion Atomic Physics in: Heavy Ion Physics, ed. R. Bock (1976/7) (Amsterdam: North Holland) in print;
J. Reinhard and W. Greiner, Quantum Electrodynamics of Strong Fields in: Progress on Reports in Physics (1977).

79. K. Smith, B. Müller and W. Greiner, J. Phys. B8 (1975) 75.

80. D.R. Bates, R. McCarroll, Proc. Roy. Soc. A245 (1958) 175.

81. H.C. Pauli, and L. Wiltes, Z. Physik A277 (1976) 83.

82. W.R. Thorson and Levy, Phys. Rev. 181 (1969) 252.

83. E. Teller, Z. Physik 61 (1930) 458.

84. F.A. Hylleraas, Z. Physik 71 (1931) 739.

85. W. Heitler, and F. London, Z. Physik 44 (1927) 455.

86. H.A. Erikson, E.L. Hill, Phys. Rev. 75 (1949) 29.

87. S.K. Luke, G. Hunter, R.P. McEachran, and M. Cohen, J. Chem. Phys. 50 (1969) 1644.

88. B. Müller, J. Rafelski, and W. Greiner, Phys. Lett. 47B (1973) 5.

89. B. Müller and W. Greiner, Z. Natuforschung 31a (1976) 1.

90. J. Rafelski and B. Müller, Phys. Rev. Lett. 36 (1976) 517.

91. M.S. Marinov and V.S. Popov, Sov. J. Nucl. Phys. 20, (1975a) 641.
M.S. Marinov and V.S. Popov, Sov. Phys. JETP 40 (1975b) 621.
M.S. Marinov and V.S. Popov, Sov. Phys. JETP 41 (1975c) 205.
M.S. Marinov and V.S. Popov, J.Phys. A8 (1975d) 1575.

92. M. Barat, W. Lichten, Phys. Rev. A6 (1972) 211.

93. K. Helfrich, and H. Hartmann, Theor.Chim.Acta 24 (1972) 271.

94. K. Taulbjerg, and J.S. Briggs, J.Phys. B8, 1895.

95. M.S. Marinov, V.S. Popov, and V.L. Stolin, JETP Lett. 19 (1974) 49.

96. V.S. Popov, JETP Lett. 18 (1973a) 29.
V.S. Popov, Sov.J. Nucl. Phys. 17 (1973b) 322.

97. W. Schäfer and G. Soff, Nuovo Cim. 31B (1976) 250.

98. F.P. Larkins, J. Phys. B5 (1972) 571.

99. J. Eichler and U. Wille, Phys. Rev. A11 (1975) 1973.

100. E Gross, and R.M. Dreizler, Phys. Lett. 57A (1976) 131.

101. J.F. Hofmann, H. Stöcker, W. Scheid, W. Greiner, V. Ceausescu, and E. Badralexe, Z. Phys. A 280 (1977) 131.

102. V. Weisskopf, Phys. Z. 34 (1933) 1.

103. B. Müller, Proc. of the IXth ICPEAC, Seattle, Eds. J. Risley, and R. Geballe (1975).

104. H.D. Betz, F. Bell, H. Panke, W. Stehling, E.Spindler, M. Kleber, Phys. Rev. Lett. 34 (1975) 1256.

105. F.W. Saris, W.F. von der Weg, H. Tawara and W.A. Laubert Phys. Rev. Lett. 28 (1972) 717.

106. P.H. Mokler, H.J. Stein and P. Armbruster, Phys.Rev. Lett. 29 (1972) 827.

107. W.E. Meyerhof, T.K. Saylor, S.M. Lazarus, W.A. Little, B.B. Triplett and L.F. Chase, Phys.Rev.Lett.30 (1973) 1279.

108. J.S.Greenberg, C.K.Davis and P. Vincent, Phys. Rev. Lett. 33 (1974) 473.

109. W. Wölfli, Ch. Stoller, G. Bonani, M. Suter, M.Stöckli, Phys. Rev. Lett. 35 (1975) 656.

110. W. Frank, P. Gippner, K.-H. Kaun, H. Sodan and Yu.P. Tretyakov, Phys. Lett. 58B (1975) 41.

111. W.E. Meyerhof, T.K. Saylor, R. Anholt, Phys. Rev. A12 (1975) 2641.

112. F. Folkman, P. Armbruster, S. Hagmann, G. Kraft, P.H. Mokler, H.J. Stein, Z. Physik A276 (1976) 15.

113. G. Kraft, P.H. Mokler, H.J. Stein, Phys. Rev. Lett. 33 (1974) 476.

114. B. Müller and W. Greiner, Phys. Rev. Lett. 33 (1974) 469.

115. R.K. Smith and W. Greiner, Spontaneous and Induced Radiation from Intermediate Molecules (1975), unpublished.

116. W. Betz, G. Heiligenthal, J. Reinhard, R.K. Smith and W. Greiner, Proc.IXth ICPEAC, (1976), Seattle, Eds.

117. M. Gros, P.T. Greenland, W. Greiner, Z. Phys. A280 (1977) 31.

118. J.S. Briggs, Continuum Emission in low-velocity ion-atom collisions, preprint (1976b).

119. B. Müller, R.K. Smith and W. Greiner, Phys. Lett. 53B (1975) 401.

120. W. Wölfli, Proceedings of the XIth In. School on Nuclear Physics, Predal, Romania (1976).

121. H. Peitz, B. Müller, J. Rafelski, W. Greiner, Lett. Nuovo Cim. 8 (1973) 37.

122. V.S. Popov, Sov. J. Nucl. Phys. 19 (1974a) 81.
V.S. Popov, Sov. Phys. JETP 38 (1974b) 18.

123. K. Smith, H. Peitz, B. Müller and W. Greiner, Phys. Rev. Lett. 32 (1974) 554.

124. W. Betz, B. Müller. G. Soff and W. Greiner, Phys. Rev. Lett. 37 (1976) 1046.

125. W. Betz, Diploma Thesis, Frankfurt am Main 1976.

126. D.R. Bates, R. McCarroll, Proc. Roy. Soc. A245 (1958) 175.

127. G. Soff, J. Reinhardt, B. Müller, W. Greiner, Phys. Rev. Lett. 38 (1977) 592.

128. J. Reinhardt, G. Soff and W. Greiner, Z. Physik A 276 (1976) 285.

129. H.P. Trautvetter, J.S. Greenberg, P. Vincent, Phys. Rev. Lett. 37 (1976) 202.

130. V.M. Budnev, F. Ginzburg, G.V. Meledin, V.G. Serbo, Phys. Reports 15C (1975) 181.

131. V. Oberacker, G. Soff and W. Greiner, Phys. Rev. Lett. 36 (1976a) 1o24.

132. V. Oberacker, G. Soff and W. Greiner, Nucl. Phys. A259 (1976b) 324.

133. J.M. Eisenberg, W. Greiner, Nucl. Theory vol. I (1970) (Amsterdam: North Holland).

134. V. Oberacker, H. Holm, W. Scheid, Phys. Rev. C10 (1974) 1917.

FIGURE CAPTIONS

Fig. 17: Relativistic correlation diagram for the system Ni-Ni showing the dependence of binding energy on internuclear distance for several quasimolecular states.

Fig. 18: Correlation diagram for the superheavy asymmetric system I-Au.

Fig. 19: Correlation diagram for the overcritical system U-U. At about R_{cr} = 35 fm for extended nuclei and 37 fm for pointlike (dashed line) nuclei the $1s\sigma$-state reaches $E = -m$.

Fig. 20: Energy of the most deeply bound states $1s\sigma$, $2p_{1/2}\sigma$ at small internuclear distances. Dashed lines belong to extended nuclei. The dependence of the critical distance R_{cr} on total nuclear charge is demonstrated.

Fig. 21: Quasimolecular K X-ray spectrum for Ni-Ni collisions measured by Greenberg. The photon energies lie above the characteristic K-lines of the Ni atom.

Fig. 22: X-ray energy dependence of the MO emission anisotropy for several collision systems measured by Wölfli. Note the broad bumps above the respective united atom energy.

Fig. 23: Systematic dependence of the position of the anisotropy bump of MO K X-rays on the total nuclear charge. Note the good agreement with the united atom transition energies. The figure was obtained by Wölfli.

Fig. 24: Dynamical processes connected with positron production in overcritical heavy ion collisions. a,b: electron excitation and ionization; c: spontaneous autoionization of positrons; d: induced decay of the vacuum; e: direct pair creation.

Fig. 25: Dependence of the scaled width γ on the diving energy E_p obtained from the one-centre phase shift analysis. $\gamma(E_p)$ clearly is independent of the amount of overcritical charge.

Fig. 26: Positron emission probability $W(E_p,\theta)$ for U-U-collisions at 812 MeV c.m. energy. The total probability W_T is divided into contributions from the trajectory regions during diving ($R < R_{cr}$), W_D, and before or after diving ($R > R_{cr}$), W_{PA}.

Fig. 27: a) Differential cross section for positron emission in U-U collisions $d\sigma/dE_p$ for several impact energies given by the distances of closest approach R_o.
b) Total cross section as a function of c.m. ion energy ($L_o = 0.01$).

Fig. 28: Positron emission cross section in dependence of the amount of overcritical charge (L_o=0.01).
a) Energy spectrum $d\sigma/dE_p$ at R_o = 15 fm distance of closest approach.
b) Total cross section versus distance of closest approach.

Fig. 29: Radial coupling matrix elements of the $1s\sigma$ to the $2s\sigma$, $3s\sigma$, $4s\sigma$, levels in U+U as a function of internuclear separation.The dashed lines are without the translation factor, the solid lines are with it. The dash-dotted line shows a non-relativistic 1sσ-2sσ matrix element scaled to U+U.

Fig. 30: Radial coupling matrix elements of the 2sσ to the 3sσ and 4sσ levels in U+U as a function of internuclear separation. The dashed lines are without the translation factors, the solid lines are with it.

Fig. 31: The radial coupling elements of the 1sσ levels to the continuum states (energies in electron masses). The continuum states have been calculated in monopole approximation.

Fig. 32: Coupled channel approach (1sσ, 2sσ, 3sσ, 4sσ). At $t = -\infty$ only the 1sσ state is occupied (1 electron problem). The curves 2,3,4 represent the excitation probabilities of the three possible configurations. The top curve, labeled TOTAL, shows the total K-vacancy production during the collision.

Fig. 33: Coupled channel approach (1sσ, 2sσ, 3sσ, 4sσ). At the $t = -\infty$ the 3 lowest states (1sσ-3sσ) are occupied (3 electron problem). The shown three curves show the change in the total K-, L- and M-vacancy production respectively.

Fig. 34: Coupled channel approach (1sσ, 2sσ, 3sσ, 4sσ). The total K-vacancy production has been plotted for the 1 electron problem (1,0,0,0), the 2 electron problem (1,1,0,0) and the 3 electron problem (1,1,1,0). The initally occupied states are 1sσ, 1sσ and 2sσ, 1sσ to 3sσ respectively. The dashed curve represents the excitation of the 1sσ electron into the continuum as calculated in perturbation theory.

Fig. 35: Coupled channel approach ($1s\sigma$, $2s\sigma$, $3s\sigma$, $4s\sigma$). The final K-vacancy probability of the 1-, 2- and 3 electron problem in dependence of the laboratory energy E_{lab} is shown. Again, the dashed line represents the excitation probability of the $1s\sigma$ electron into the continuum.

Fig. 36: Coupled channel approach ($1s\sigma$, $2s\sigma$, $3s\sigma$, $4s\sigma$). The final K-vacancy probability of the 1-, 2- and 3 electron problem in dependence of the impact parameter b is shown. Again, the dashed line represents the excitation probability of the $1s\sigma$ electron into the continuum.

Fig. 37: Impact parameter dependence of the Coulomb interaction of the $1s\sigma$, $2p_{1/2}\sigma$ and $2s\sigma$ electrons in Pb+Pb at E_{CM} = 192 MeV and of the $1s\sigma$, $2p_{1/2}\sigma$, $2s\sigma$ and $3s\sigma$ electrons in U+U at E_{CM} = 800 MeV into the continuum. The continuum states have been calculated in the monopole approximation.

Fig. 38: The magnetic-coupling matrix elements. (a) The rotational M_ϕ and (b) the radial magnetic coupling M_ρ as a function of internuclear distance R, for Xe+Xe, Pb+Pb, U+U $1s\sigma$, and U+U $2p_{1/2}\sigma$ electronic quasimolecular states.

Fig. 39: The relative magnetic splitting $(E_\downarrow - E_\uparrow)E_B$ for the four discussed quasimolecular states. Collision parameters are E_{kin}/N = 9 MeV/nucleon and b = 13 fm.

Fig. 40: Direct ionization probability P(b) for the two magnetically split $1s\sigma$ states in the supercritical U+U quasimolecule.

Fig. 41: (a) The two-photon Feynman diagram describing pair production. There is a second diagram with the electron and positron line interchanged.
(b) The multi-photon diagram representing the collective nature of pair production out of quasi-molecular states.

Fig. 42: (a) The positron (solid line) and electron (dashed line) spectrum of pair production in a 1210 MeV (lab) lead-lead collision.
(b) The impact parameter dependence bP(b) of pair-production in the same collision.

Fig. 43: High energy tail of the X-ray spectra produced in several heavy ion collision systems as measured by Greenberg et al. [129]. Magnitude and angular distribution agree well with the calculated curves for nuclear bremsstrahlung (solid lines).

Fig. 44: a) Bremsstrahlung cross section $d\sigma/dE_x$ in μb/keV for various symmetric systems in dependence of the charge number Z at 1, 2 and 5 MeV/nucleon projectile energy. The cross sections are calculated at the K X-ray energies of the united atom limit.
b) The same plot for asymmetric colliding systems. The dashed lines indicate quadrupole radiation while the solid lines give averaged values for the total (dipole plus quadrupole) cross sections.

Fig. 45: The differential pair-formation cross section (c.m.) with respect to ion angle for the symmetric system ^{238}U-^{238}U. The solid lines indicate the positron emission cross section from the spontaneous and induced decay of the vacuum

(L_o = 0.01 assumed). The dashed lines give the background from Coulomb and nuclear excitation calculated with method 1 (lower curve) and method 2 (upper curve) described in the text.

Fig. 46: Pair creation cross sections as a function of the positron kinetic energy E_p. Notation as in Fig. 33.

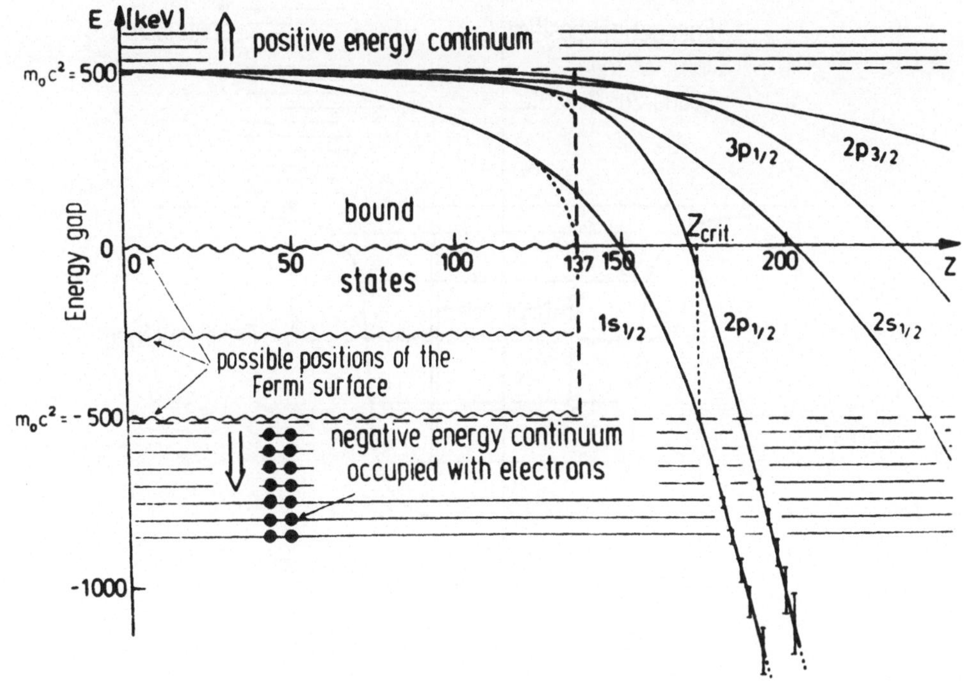
E [keV]
positive energy continuum
m_0c^2 = 500
3p1/2
2p3/2
bound
Energy gap
0
Zcrit.
0
50
100
137
150
200
Z
states
1s1/2
2p1/2
2s1/2
possible positions of the
Fermi surface
m_0c^2 = -500
negative energy continuum
occupied with electrons
-1000

Fig.1

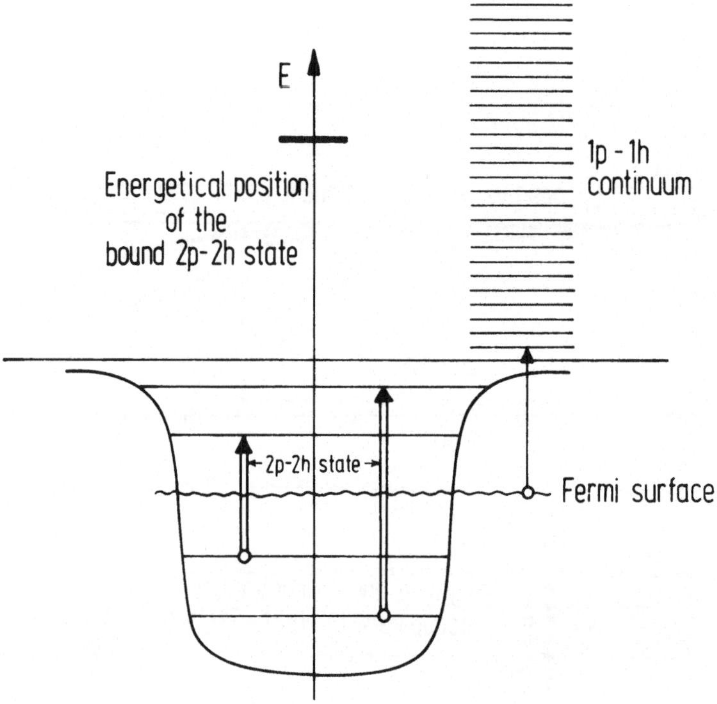
E
Energetical position
of the
bound 2p-2h state
1p - 1h
continuum
2p-2h state
Fermi surface

Fig. 2

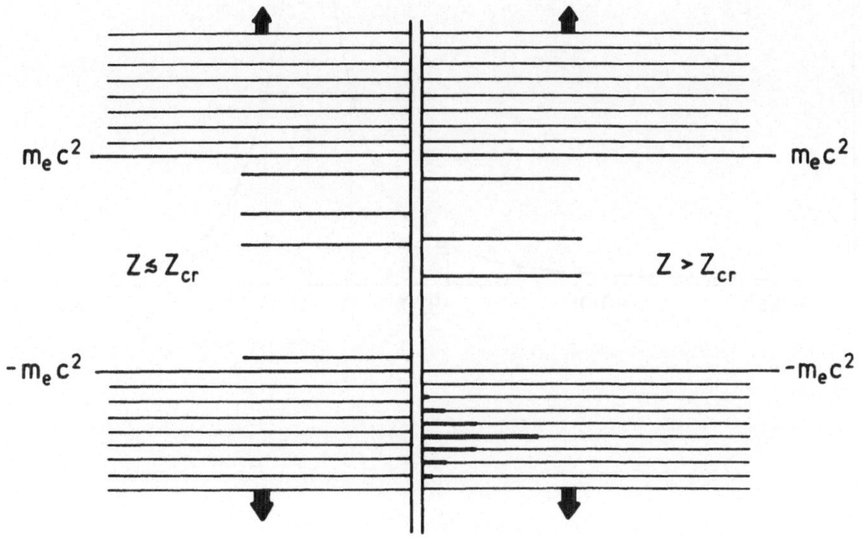

Fig. 3

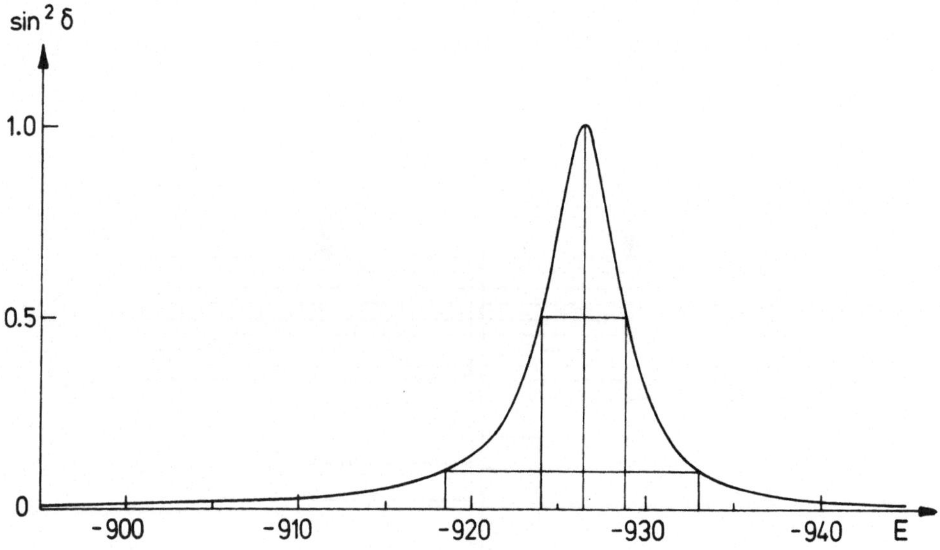

Fig. 4

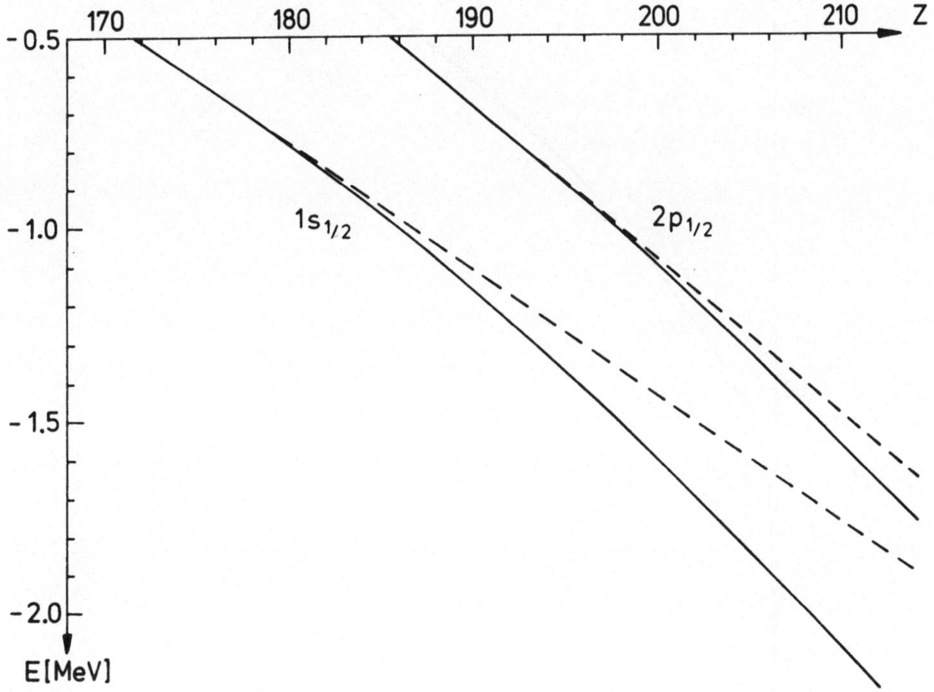

Fig. 5

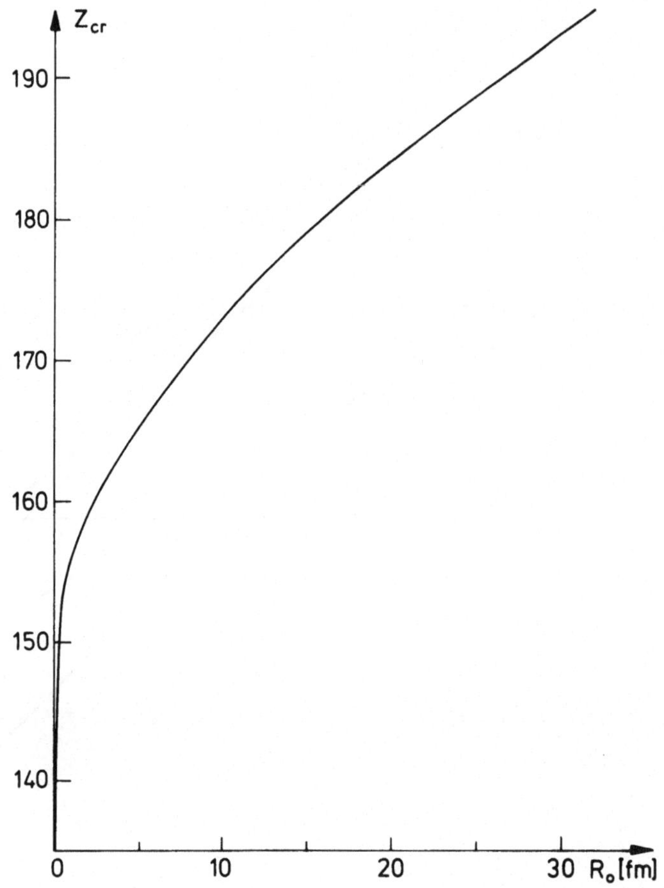

Fig. 6

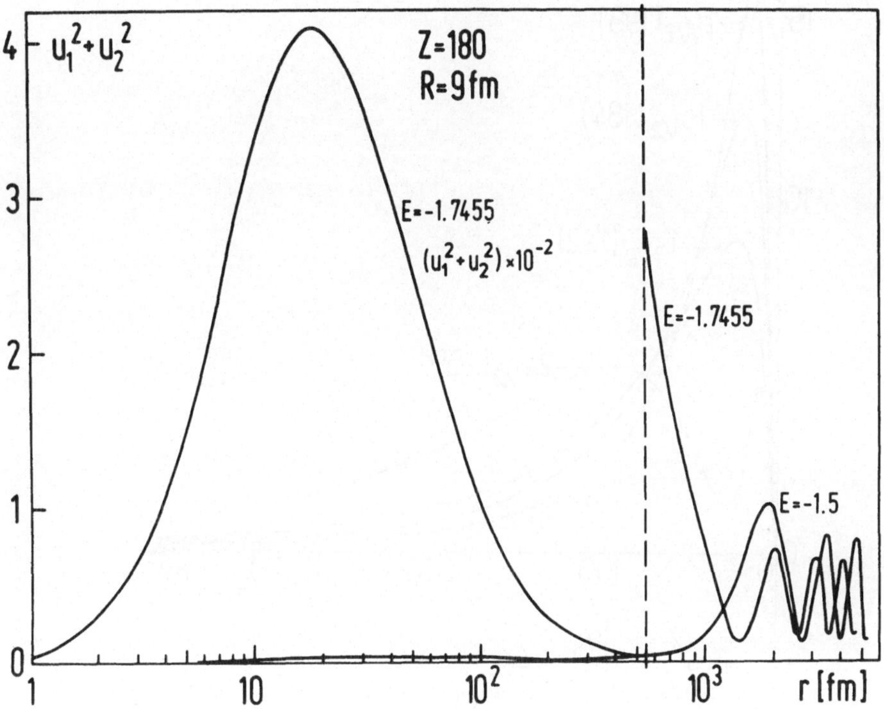

$u_1^2+u_2^2$
Z=180
R=9fm
E=-1.7455
$(u_1^2+u_2^2)\times10^{-2}$
E=-1.7455
E=-1.5
0
1
2
3
4
1
10
10^2
10^3
r [fm]

Fig. 7

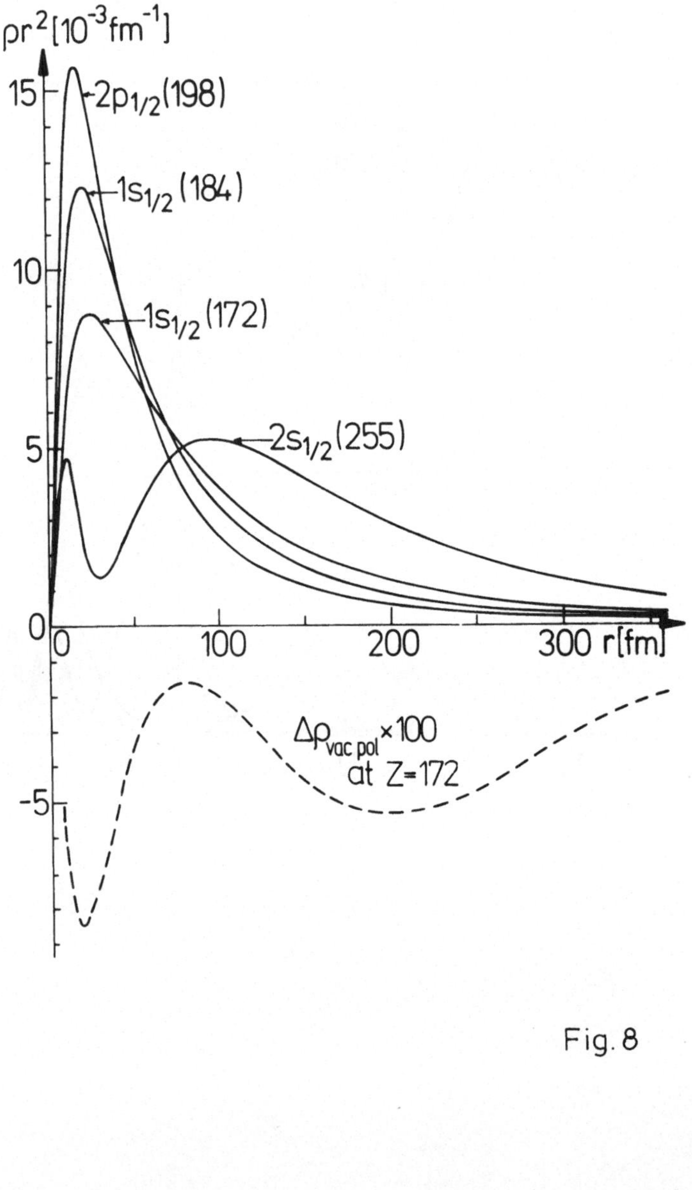

ρr² [10⁻³ fm⁻¹]
15
10
5
0
-5
2p1/2(198)
1s1/2(184)
1s1/2(172)
2s1/2(255)
0
100
200
300 r[fm]
Δρvac pol ×100
at Z=172

Fig. 8

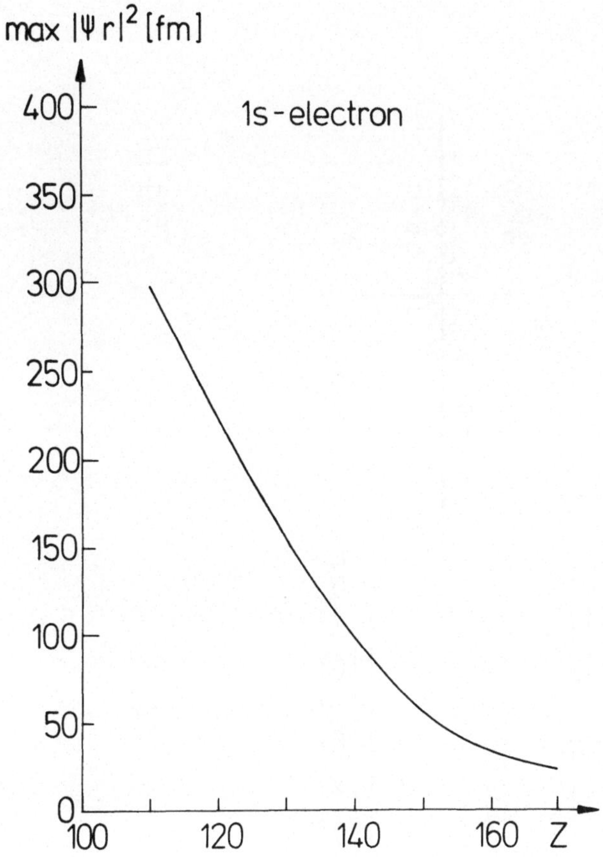
max |Ψ r|^2 [fm]
1s - electron
400
350
300
250
200
150
100
50
0
100
120
140
160
Z

Fig. 9

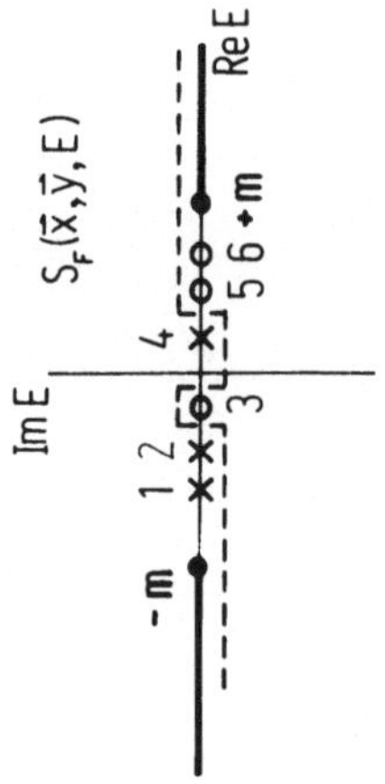
Im E
S_F(x⃗, y⃗, E)
-m
1 2
3
4
5 6
+m
Re E

Fig.11

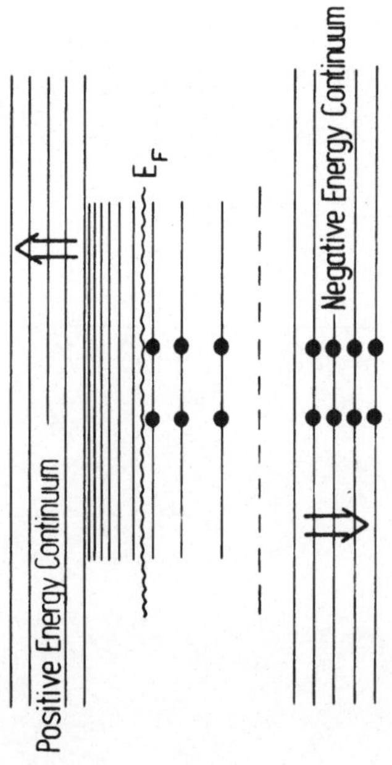
Positive Energy Continuum
E_F
Negative Energy Continuum

Fig.10

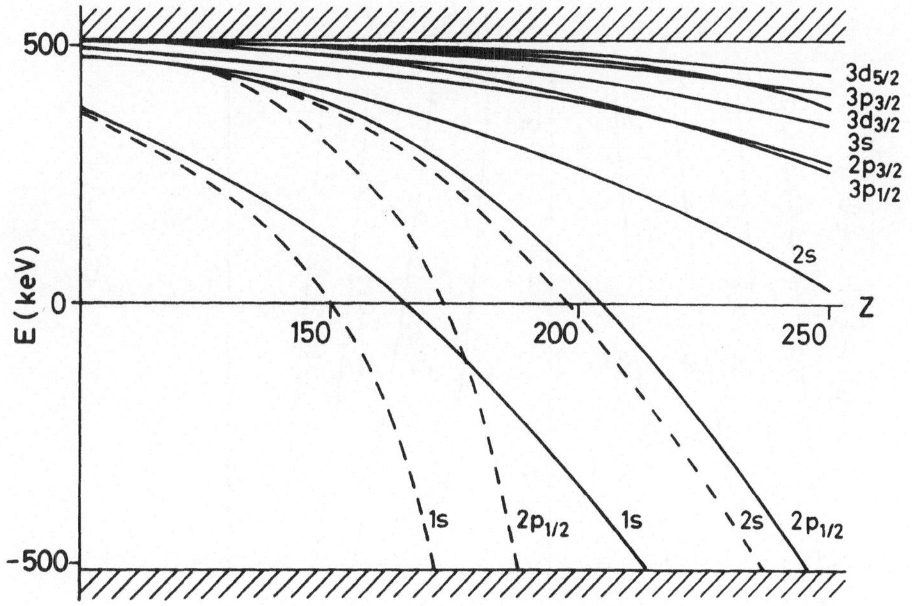
500
0
-500
E (keV)
150
200
250
Z
3d5/2
3p3/2
3d3/2
3s
2p3/2
3p1/2
2s
1s
2p1/2
1s
2s
2p1/2

Fig.12

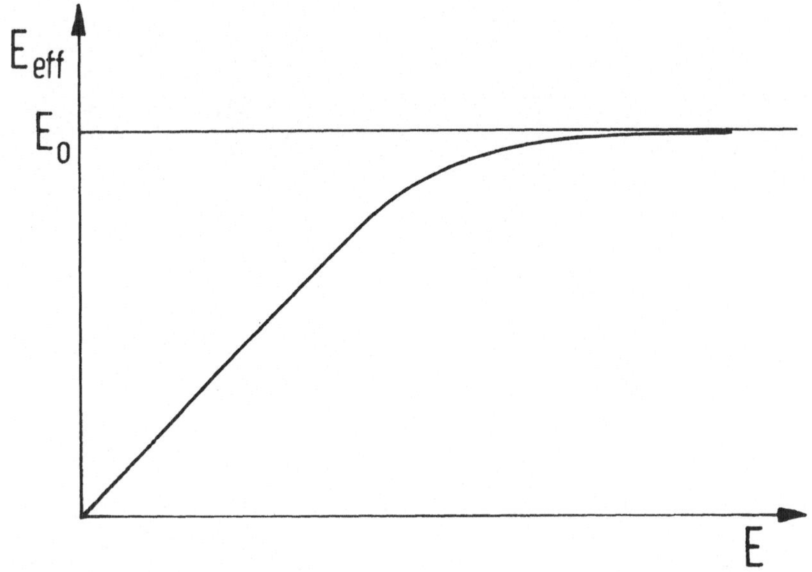
Eeff
E0
E

Fig.13

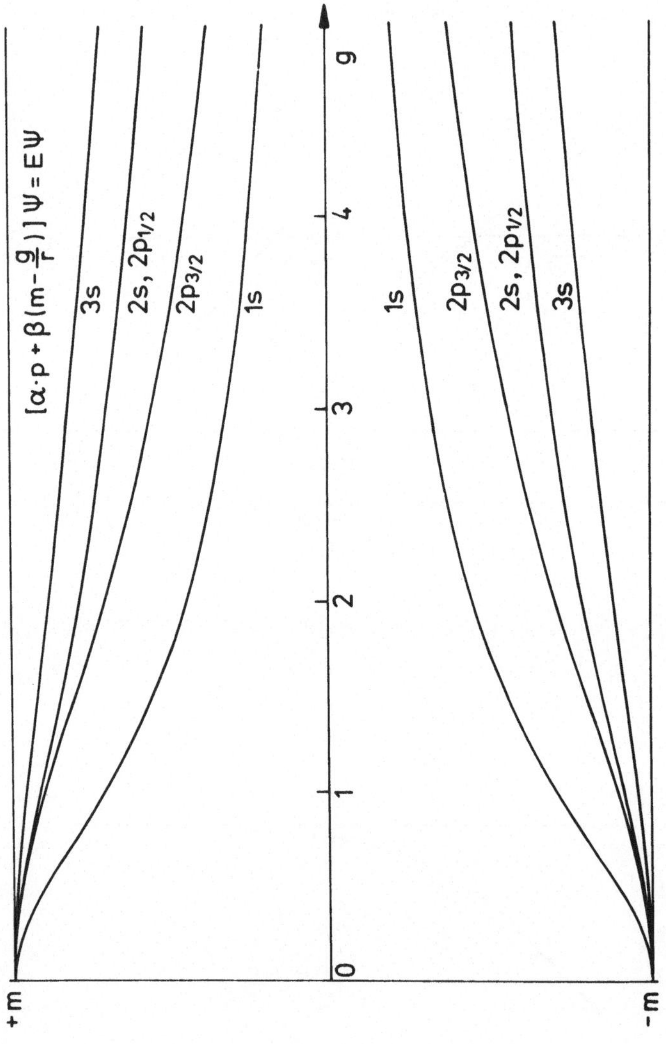
$[\alpha \cdot p + \beta(m - \frac{g}{r})]\Psi = E\Psi$
3s
2s, 2p1/2
2p3/2
1s
1s
2p3/2
2s, 2p1/2
3s
+m
-m
0
1
2
3
4
g

Fig.14

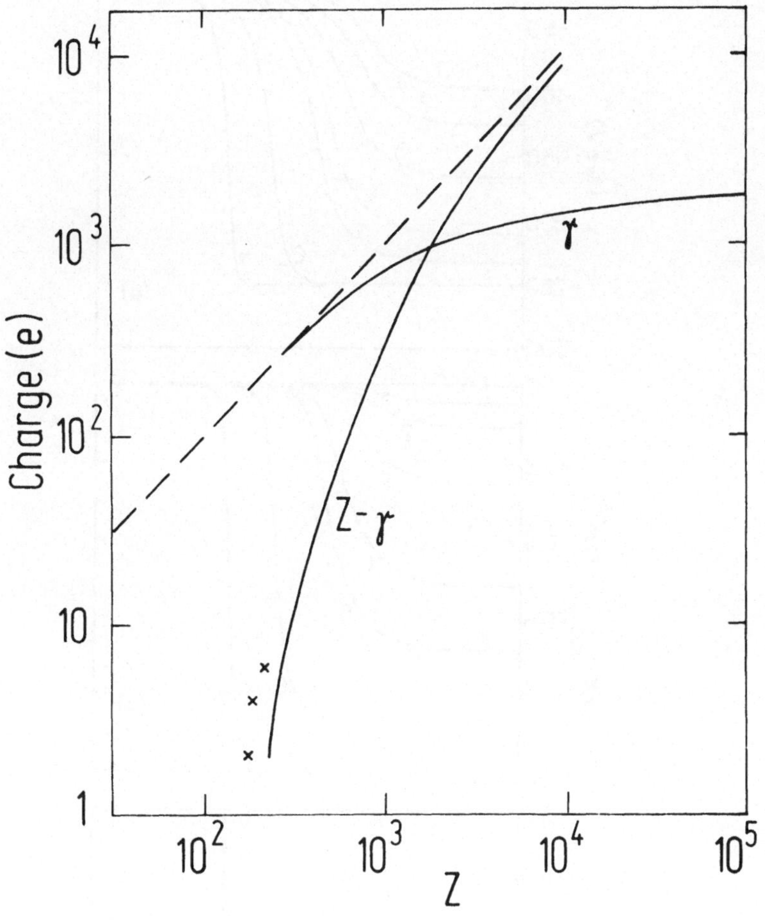

Fig. 15

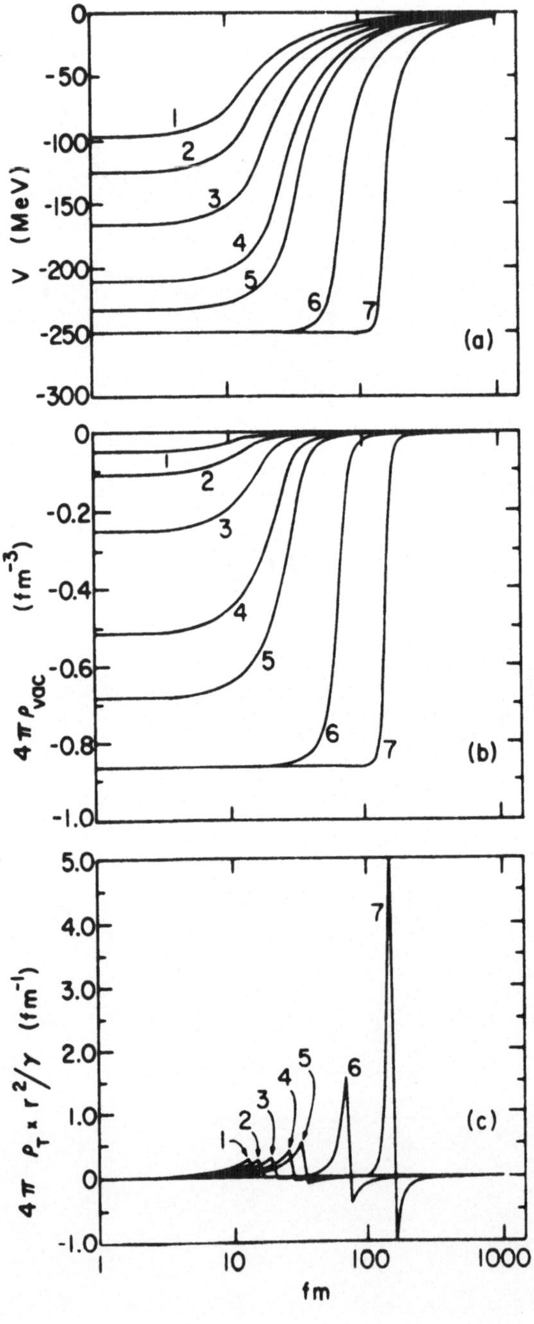

0
-50
-100
-150
-200
-250
-300
V (MeV)
1
2
3
4
5
6
7
(a)
0
-0.2
-0.4
-0.6
-0.8
-1.0
4π ρvac (fm-3)
(b)
5.0
4.0
3.0
2.0
1.0
0
-1.0
4π ρT x r2/γ (fm-1)
(c)
1
10
100
1000
fm

Fig.16

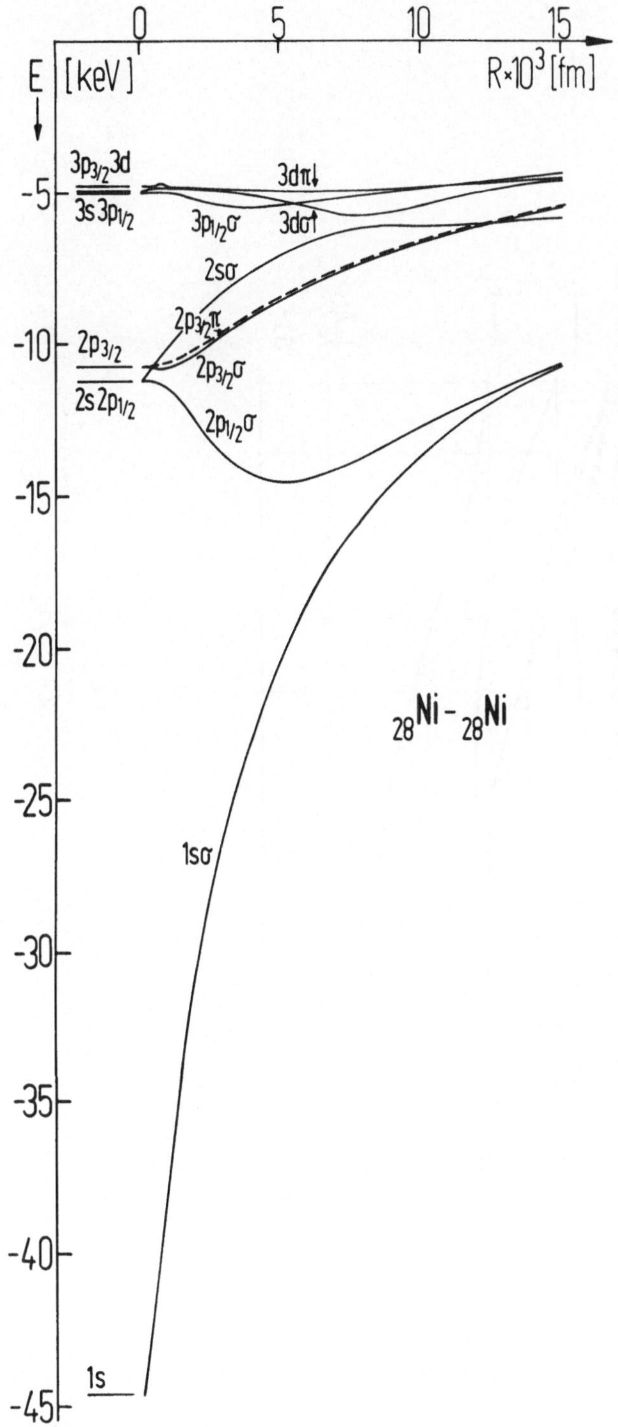

0
5
10
15
E [keV]
R×10³ [fm]
3p3/2 3d
3s 3p1/2
3dπ↓
3p1/2σ
3dσ↑
2sσ
2p3/2π
2p3/2
2s 2p1/2
2p3/2σ
2p1/2σ
-5
-10
-15
-20
-25
-30
-35
-40
-45
28Ni - 28Ni
1sσ
1s

Fig.17

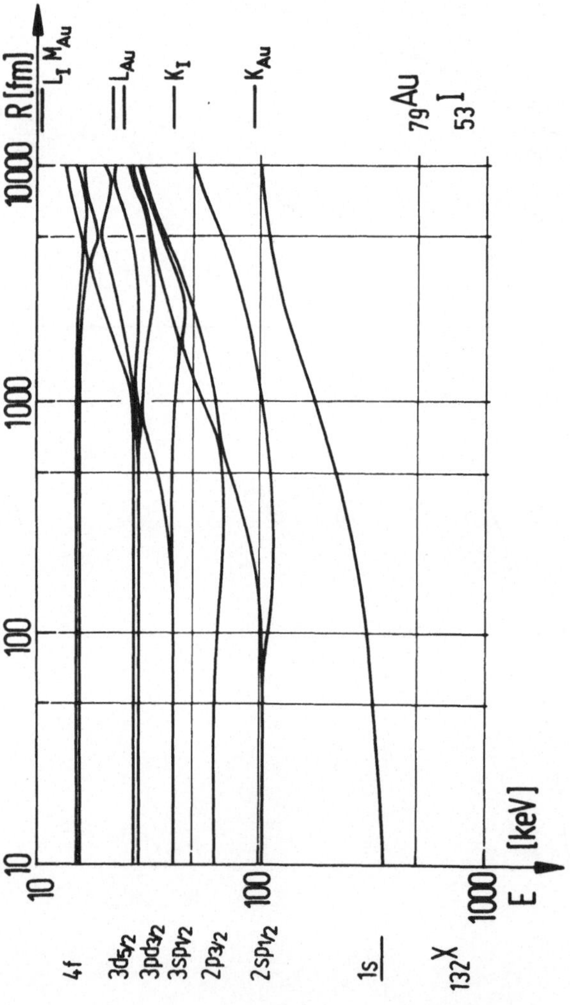
R [fm]
E [keV]
10
100
1000
10000
4f
1s
2sp1/2
2p3/2
3sp1/2
3pd3/2
3d5/2
132X
79Au
53I
K_Au
K_I
L_Au
L_I M_Au

Fig.18

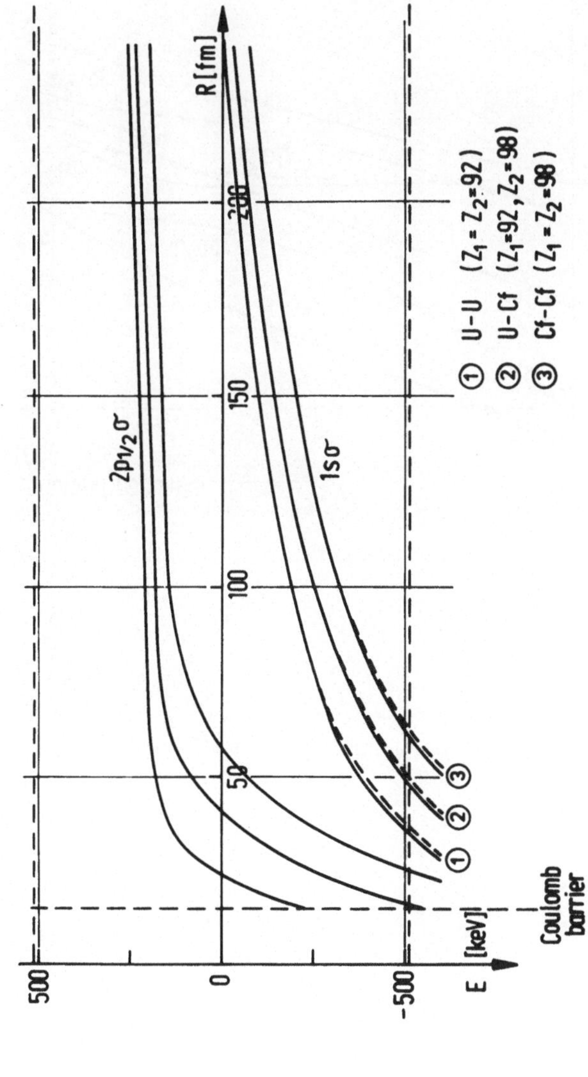

500
0
-500
E
[keV]
R [fm]
50
100
150
200
2p1/2σ
1sσ
Coulomb barrier
① ② ③
① U-U (Z1 = Z2 = 92)
② U-Cf (Z1 = 92, Z2 = 98)
③ Cf-Cf (Z1 = Z2 = 98)

Fig. 19

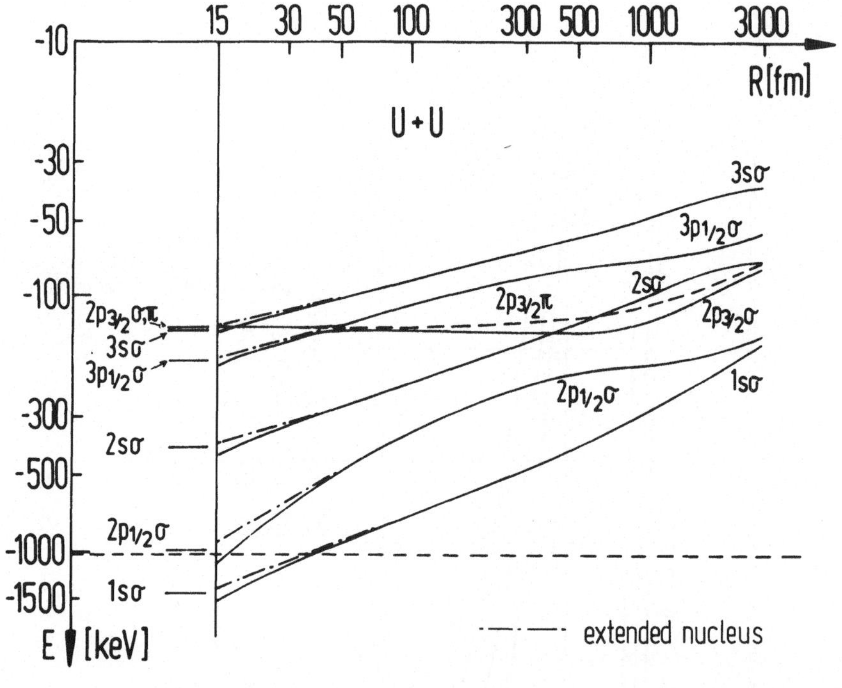
15
30
50
100
300
500
1000
3000
R[fm]
U+U
-10
-30
-50
-100
-300
-500
-1000
-1500
E [keV]
3sσ
3p1/2σ
2sσ
2p3/2π
2p3/2σ
2p1/2σ
1sσ
2p3/2σ,π
3sσ
3p1/2σ
2sσ
2p1/2σ
1sσ
extended nucleus

Fig. 20

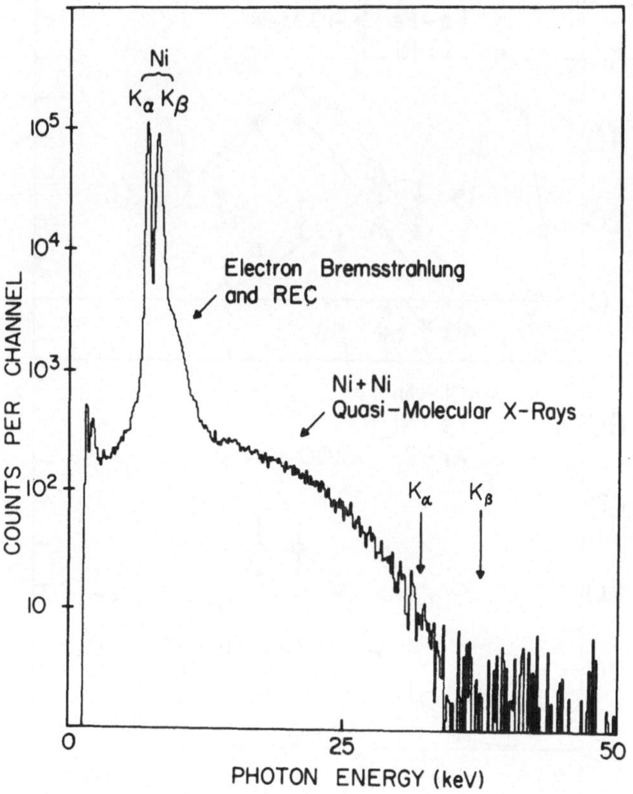
Ni
Kα Kβ
Electron Bremsstrahlung and REC
Ni + Ni
Quasi-Molecular X-Rays
Kα
Kβ
COUNTS PER CHANNEL
10⁵
10⁴
10³
10²
10
0
25
50
PHOTON ENERGY (keV)

Fig. 21

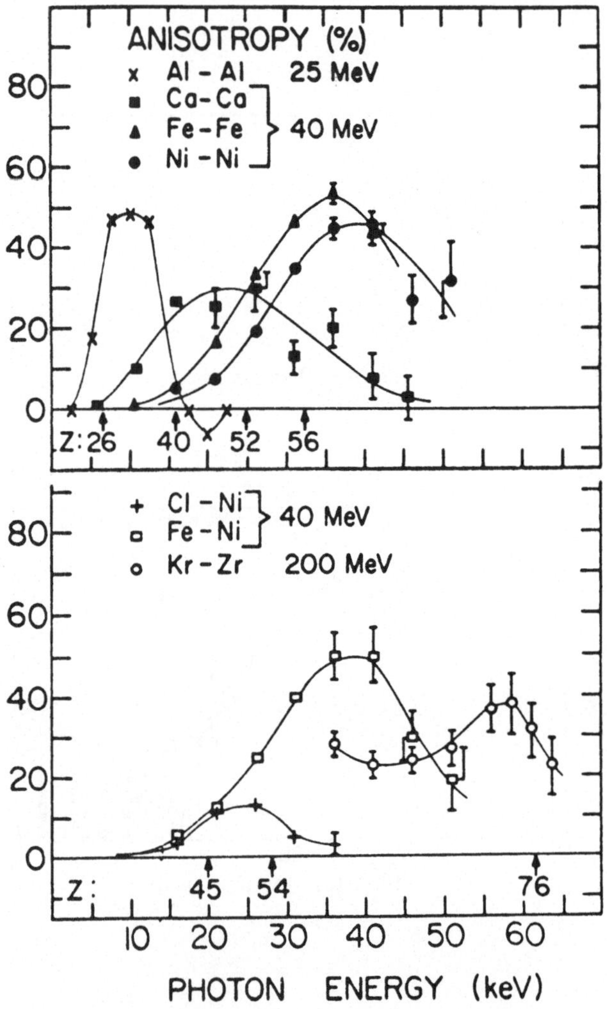

ANISOTROPY (%)
Al - Al 25 MeV
Ca-Ca
Fe-Fe
Ni-Ni
40 MeV
80
60
40
20
0
Z: 26 40 52 56
Cl - Ni
Fe - Ni
40 MeV
Kr - Zr 200 MeV
Z: 45 54 76
10 20 30 40 50 60
PHOTON ENERGY (keV)

Fig. 22

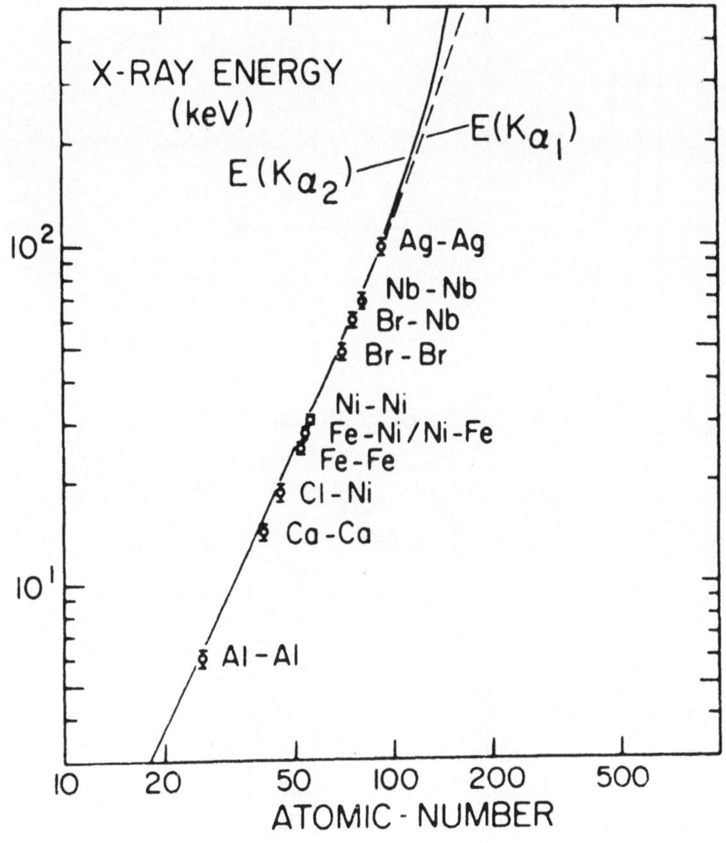
X-RAY ENERGY
(keV)
E(Kα1)
E(Kα2)
10²
10¹
Ag-Ag
Nb-Nb
Br-Nb
Br-Br
Ni-Ni
Fe-Ni/Ni-Fe
Fe-Fe
Cl-Ni
Ca-Ca
Al-Al
10
20
50
100
200
500
ATOMIC-NUMBER

Fig. 23

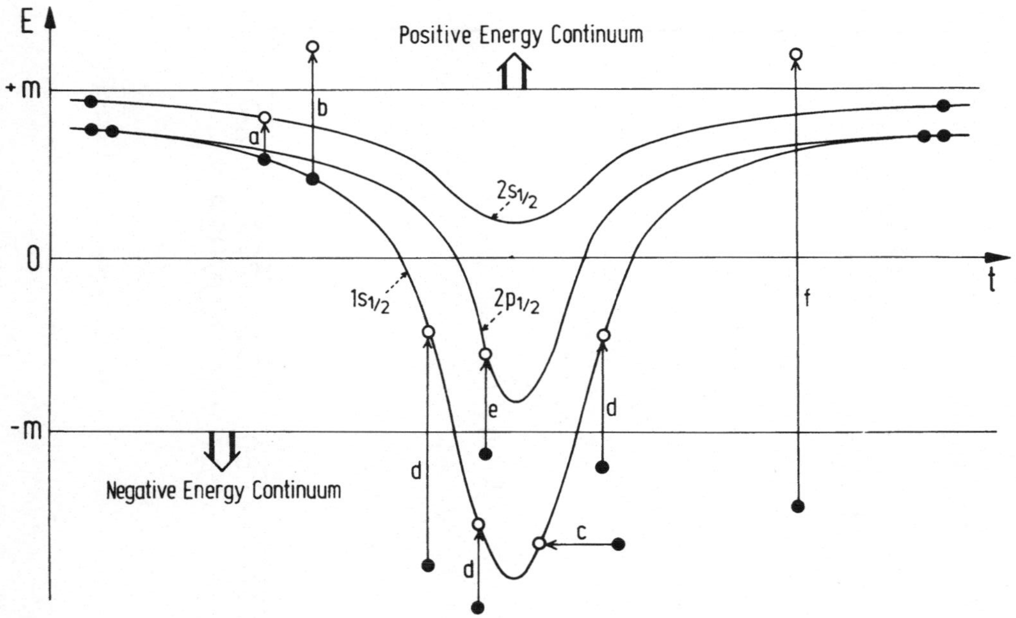

E
+m
0
-m
t
Positive Energy Continuum
Negative Energy Continuum
2s1/2
2p1/2
1s1/2
a
b
c
d
d
d
e
f

Fig. 24

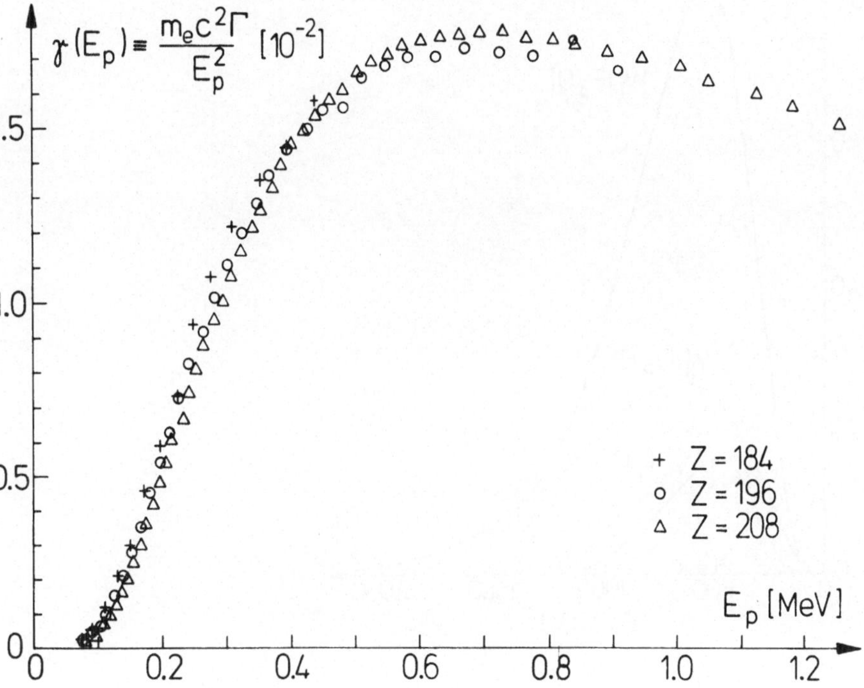

γ(E_p) ≡ m_e c²Γ / E_p² [10⁻²]
1.5
1.0
0.5
0
0
0.2
0.4
0.6
0.8
1.0
1.2
E_p [MeV]
+ Z = 184
o Z = 196
Δ Z = 208

Fig. 25

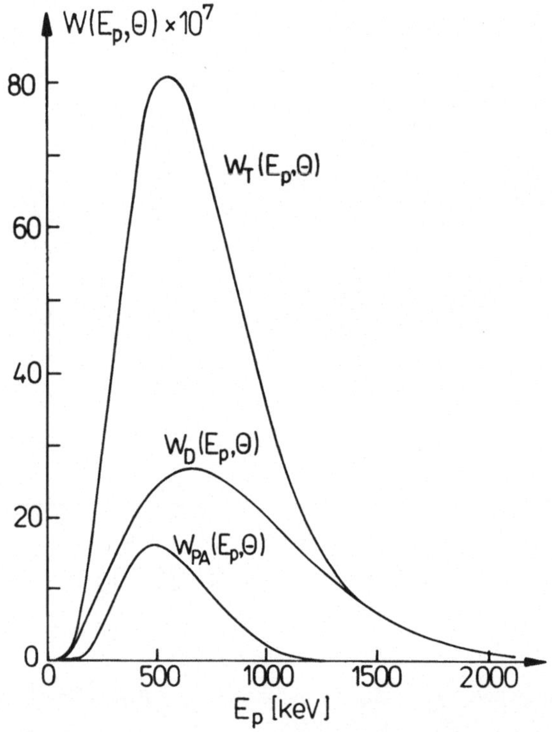

W(E_p,Θ) ×10^7
80
60
40
20
0
W_T(E_p,Θ)
W_D(E_p,Θ)
W_PA(E_p,Θ)
0
500
1000
1500
2000
E_p [keV]

Fig. 26

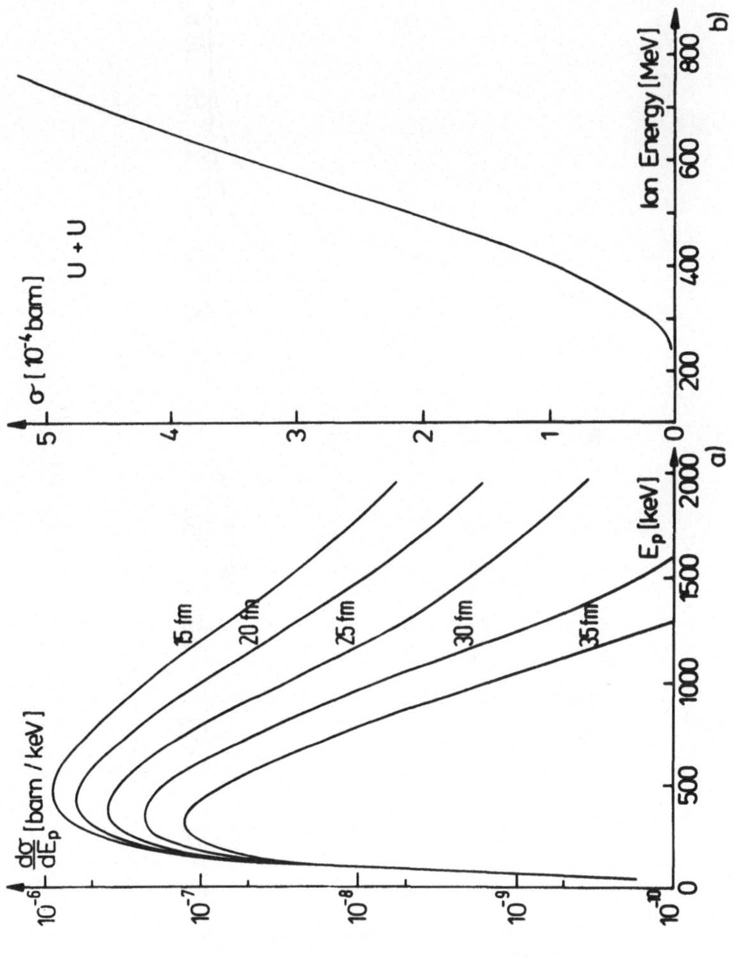
dσ/dE_p [barn / keV]
15 fm
20 fm
25 fm
30 fm
35 fm
E_p [keV]
a)
σ [10^-4 barn]
U + U
Ion Energy [MeV]
b)

Fig. 27

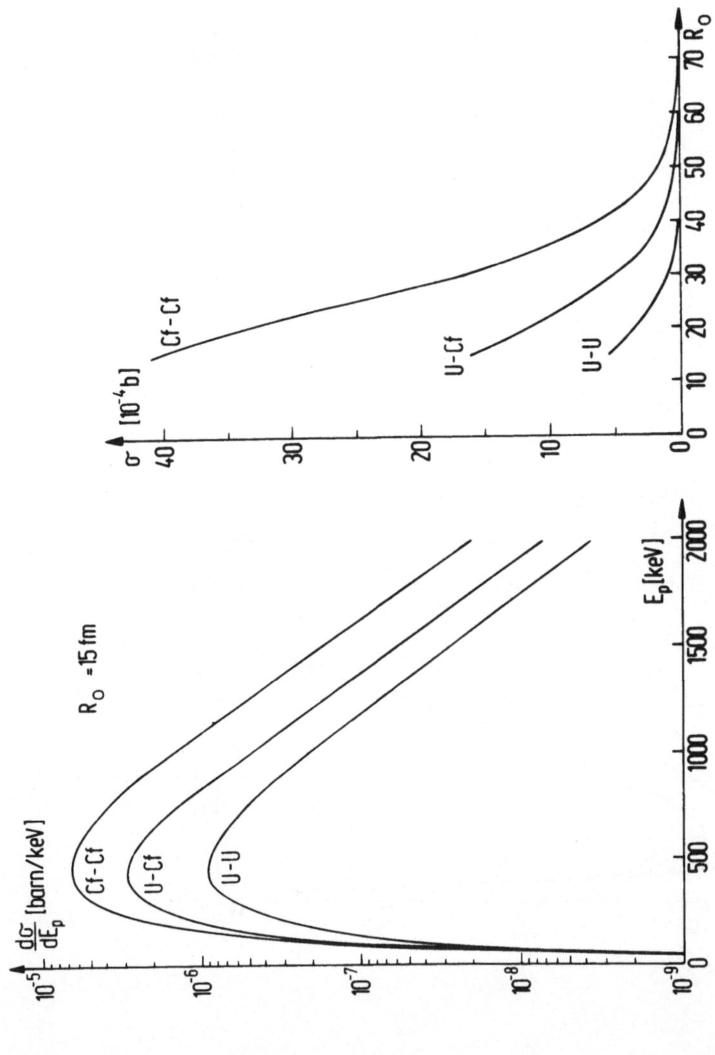

dσ/dE_p [barn/keV]
10^-5
10^-6
10^-7
10^-8
10^-9
R_0 = 15 fm
Cf-Cf
U-Cf
U-U
0
500
1000
1500
2000
E_p [keV]
σ [10^-4 b]
40
30
20
10
0
Cf-Cf
U-Cf
U-U
0
10
20
30
40
50
60
70
R_0

Fig. 28

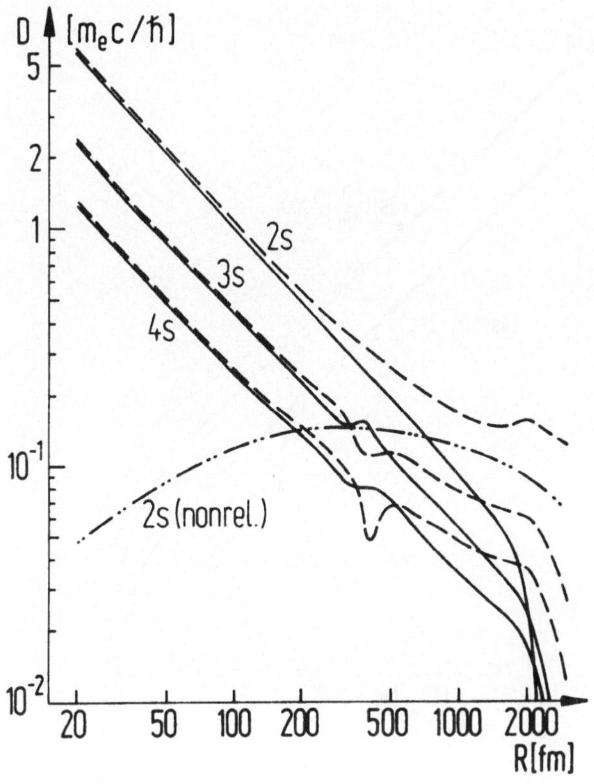
D [$m_e c/\hbar$]
5
2
1
10^{-1}
10^{-2}
2s
3s
4s
2s (nonrel.)
20
50
100
200
500
1000
2000
R[fm]

Fig. 29

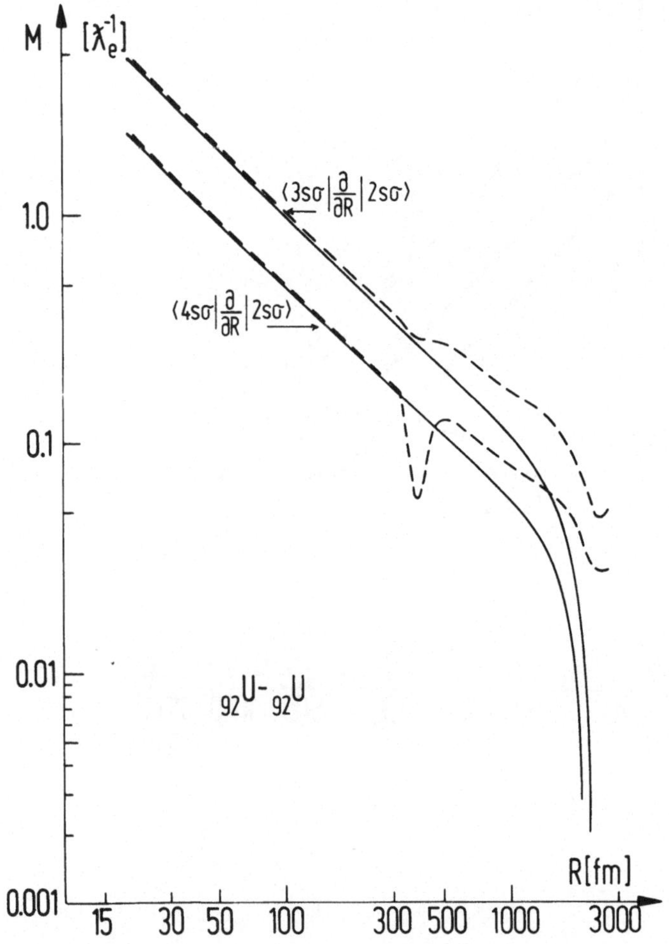
M
$[\lambda_e^{-1}]$
1.0
0.1
0.01
0.001
$\langle 3s\sigma|\frac{\partial}{\partial R}|2s\sigma\rangle$
$\langle 4s\sigma|\frac{\partial}{\partial R}|2s\sigma\rangle$
${}_{92}U-{}_{92}U$
R[fm]
15
30
50
100
300
500
1000
3000

Fig. 30

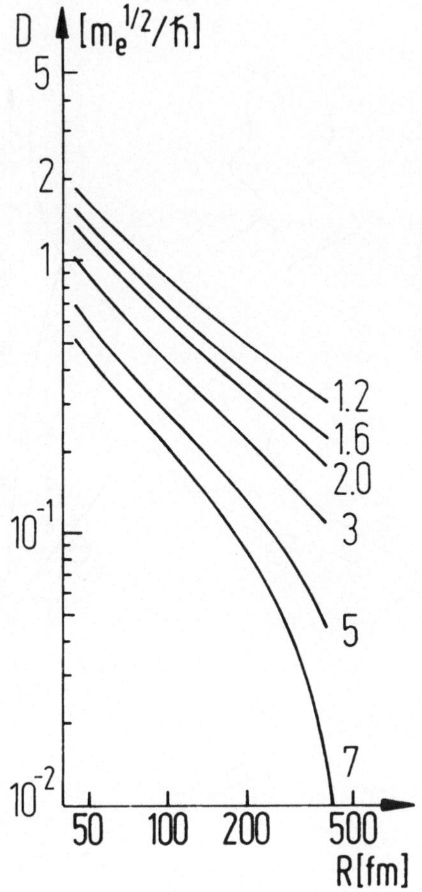

Fig. 31

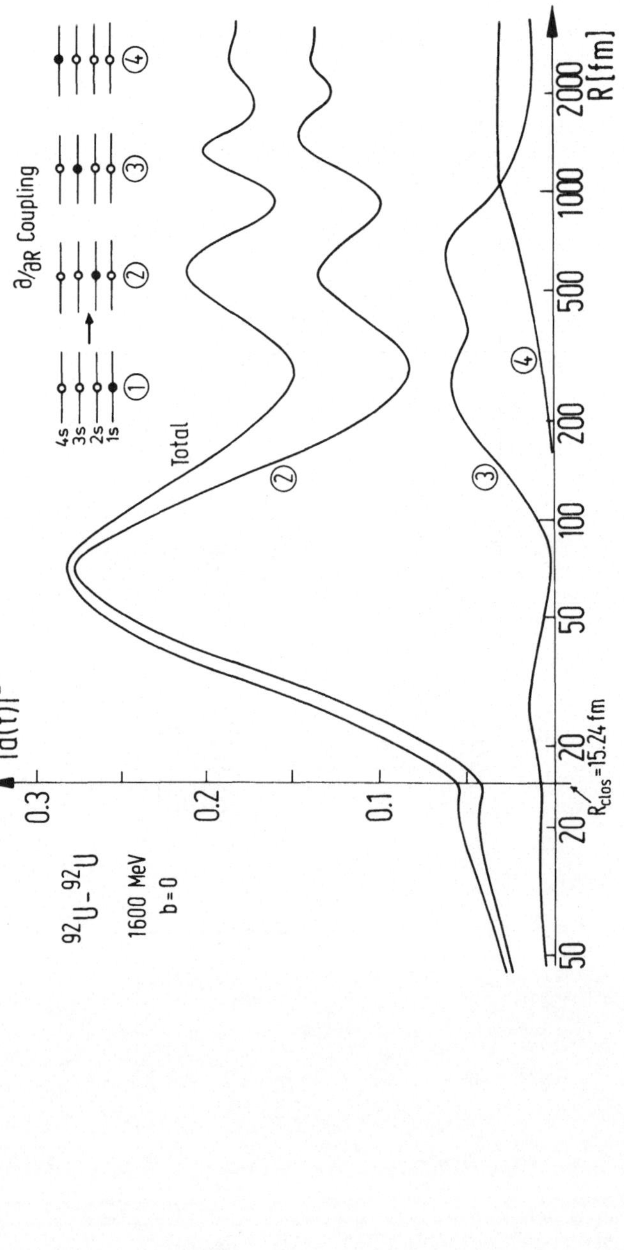

|a(t)|²
0.3
0.2
0.1
92U-92U
1600 MeV
b=0
∂/∂R Coupling
4s
3s
2s
1s
① ② ③ ④
Total
②
③
④
R_clos=15.24 fm
50
20
20
50
100
200
500
1000
2000
R [fm]

Fig. 32

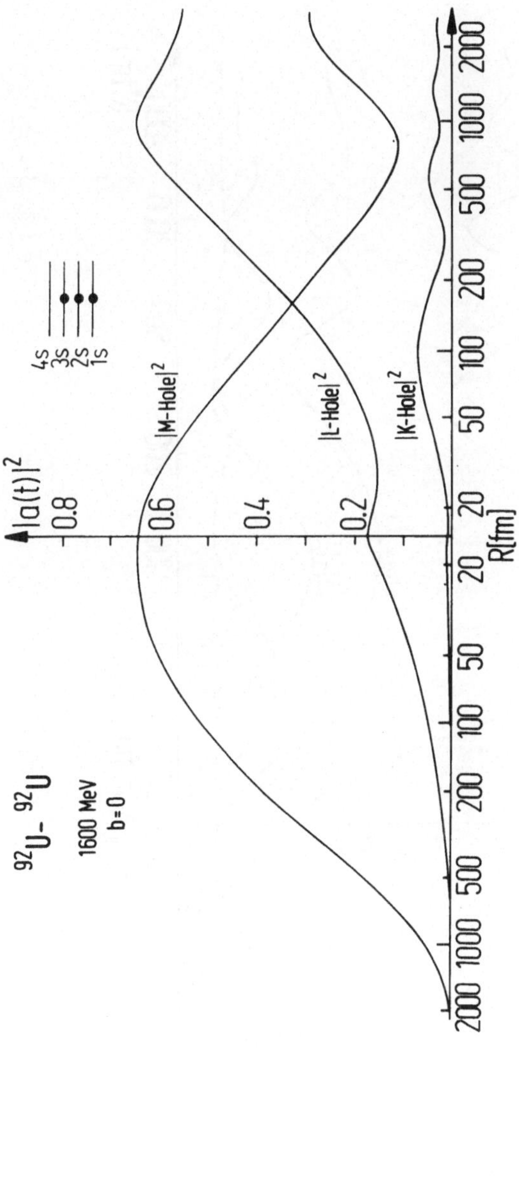

$^{92}U - ^{92}U$
1600 MeV
b=0
$|a(t)|^2$
0.8
0.6
0.4
0.2
$|M\text{-Hole}|^2$
$|L\text{-Hole}|^2$
$|K\text{-Hole}|^2$
4s
3s
2s
1s
2000 1000 500 200 100 50 20 20 50 100 200 500 1000 2000
R[fm]

Fig. 33

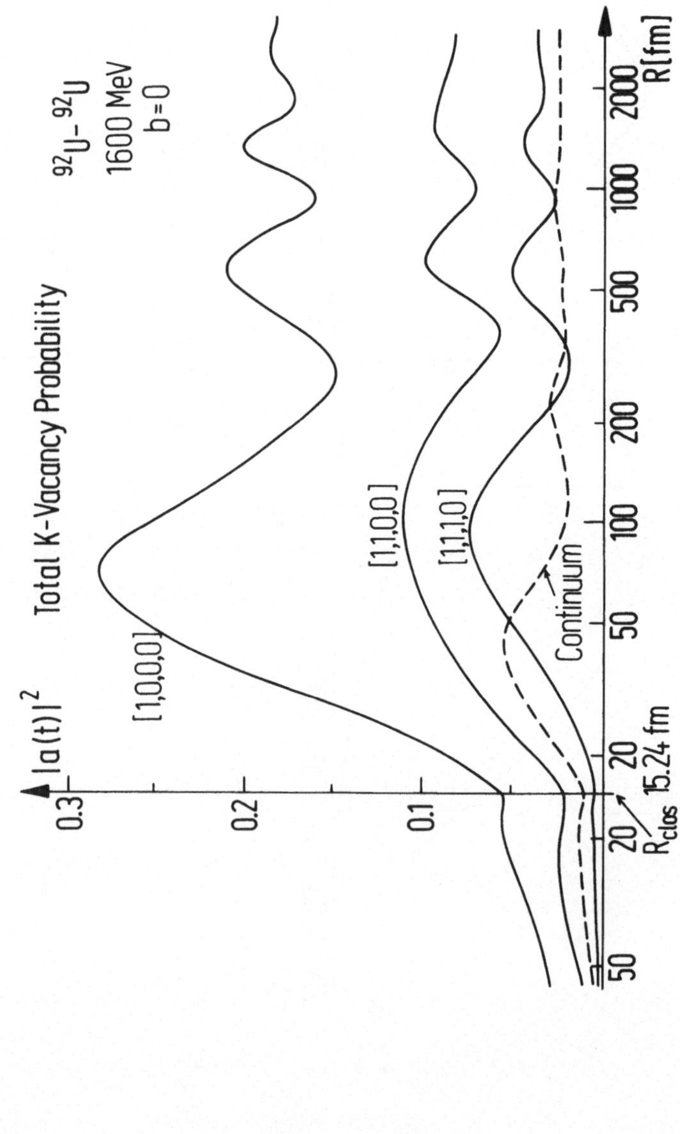

Total K-Vacancy Probability
$|a(t)|^2$
$^{92}U-^{92}U$
1600 MeV
b=0
[1,0,0,0]
[1,1,0,0]
[1,1,1,0]
Continuum
0.3
0.2
0.1
50
20
20
50
100
200
500
1000
2000
R[fm]
R_{clos} 15.24 fm

Fig. 34

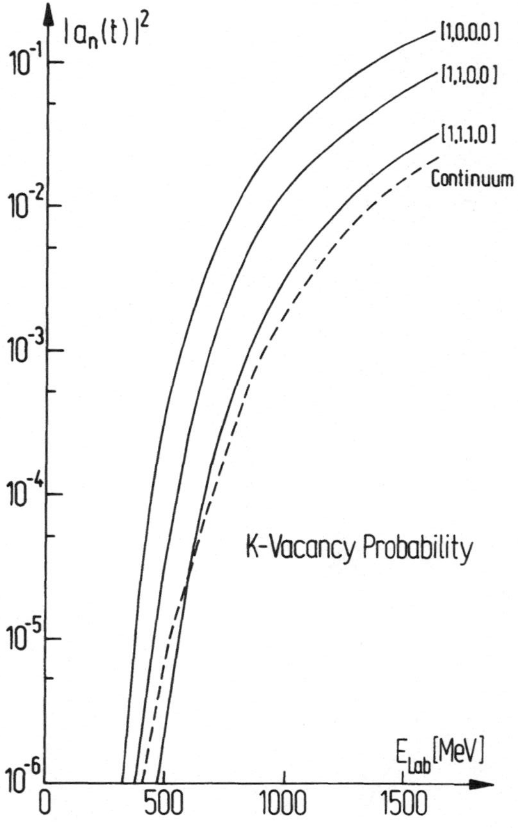
$|a_n(t)|^2$
K-Vacancy Probability
[1,0,0,0]
[1,1,0,0]
[1,1,1,0]
Continuum
E_{Lab}[MeV]
10^{-1}
10^{-2}
10^{-3}
10^{-4}
10^{-5}
10^{-6}
0
500
1000
1500

Fig. 35

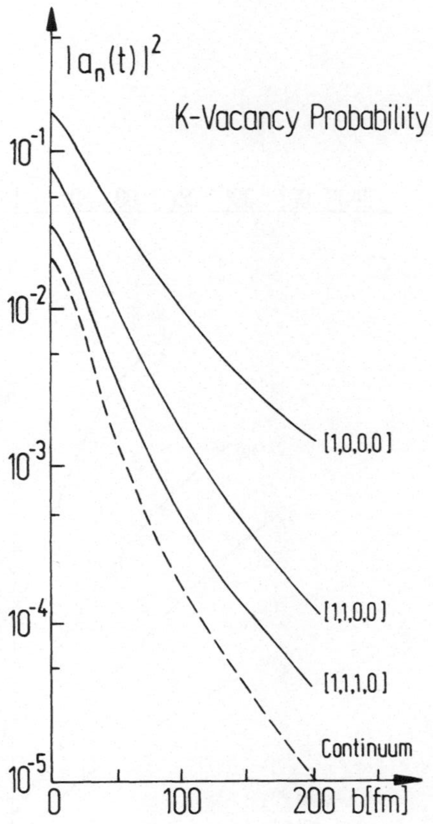
$|a_n(t)|^2$
K-Vacancy Probability
[1,0,0,0]
[1,1,0,0]
[1,1,1,0]
Continuum
b[fm]
10^{-1}
10^{-2}
10^{-3}
10^{-4}
10^{-5}
0
100
200

Fig. 36

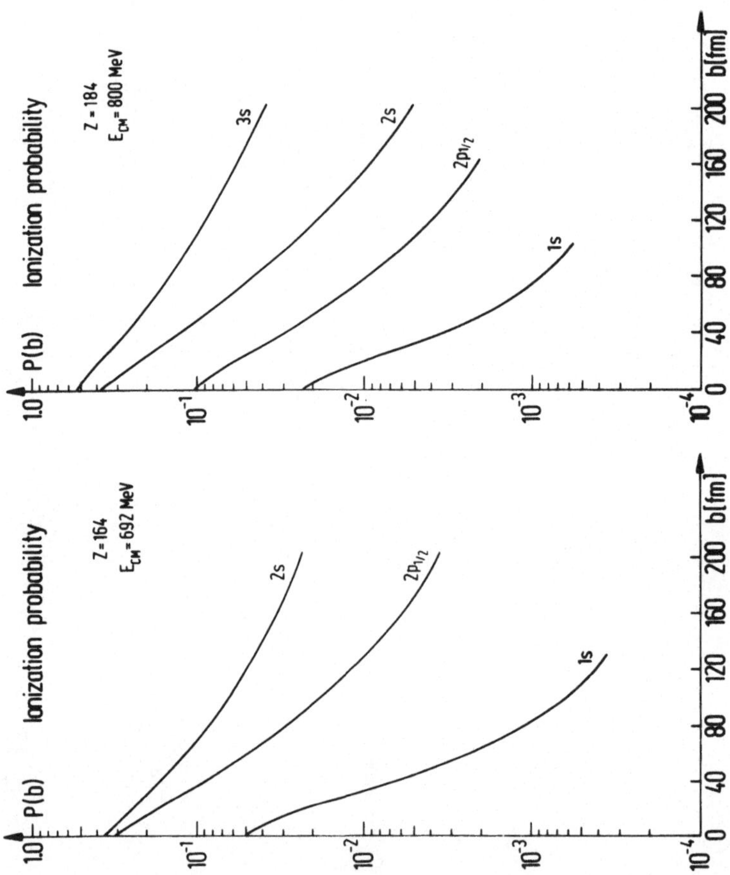

P(b)
Ionization probability
Z = 184
E_CM = 800 MeV
3s
2s
2p1/2
1s
1.0
10^-1
10^-2
10^-3
10^-4
0
40
80
120
160
200
b[fm]
P(b)
Ionization probability
Z = 164
E_CM = 692 MeV
2s
2p1/2
1s
1.0
10^-1
10^-2
10^-3
10^-4
0
40
80
120
160
200
b[fm]

Fig. 37

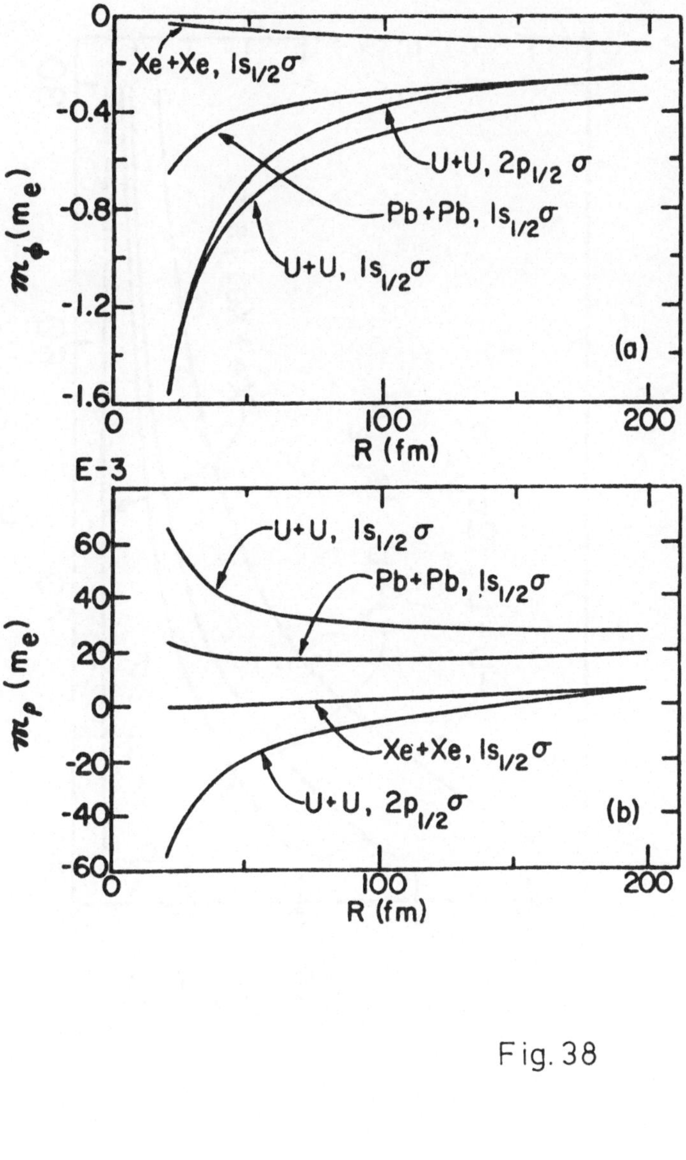

Xe+Xe, $1s_{1/2}\sigma$
U+U, $2p_{1/2}\sigma$
Pb+Pb, $1s_{1/2}\sigma$
U+U, $1s_{1/2}\sigma$
(a)
$m_\phi(m_e)$
R (fm)
E-3
U+U, $1s_{1/2}\sigma$
Pb+Pb, $1s_{1/2}\sigma$
Xe+Xe, $1s_{1/2}\sigma$
U+U, $2p_{1/2}\sigma$
(b)
$m_\rho\ (m_e)$
R (fm)

Fig. 38

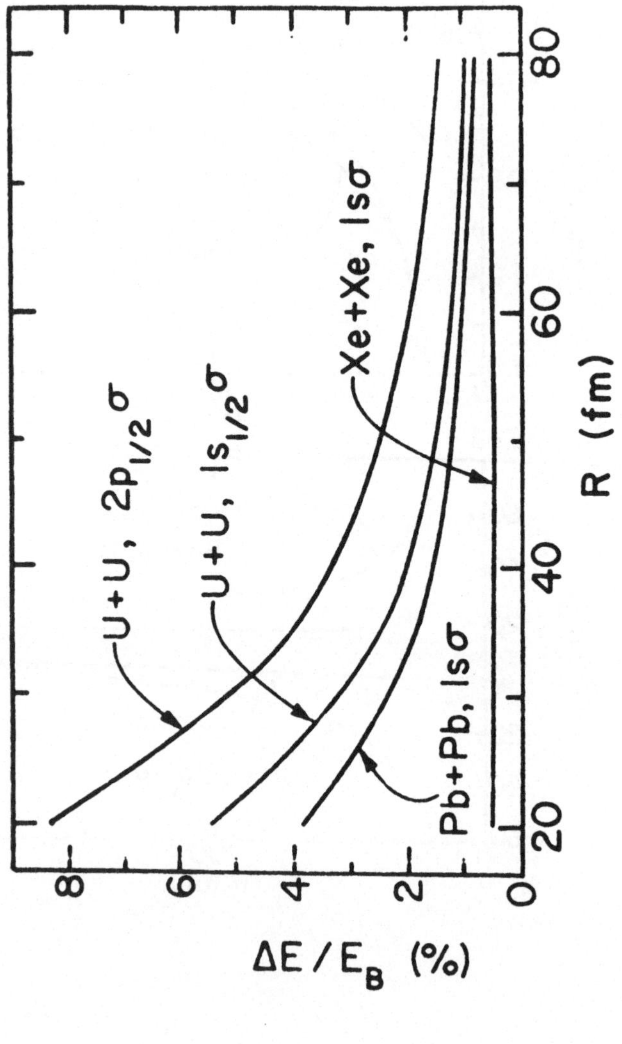

Fig. 39

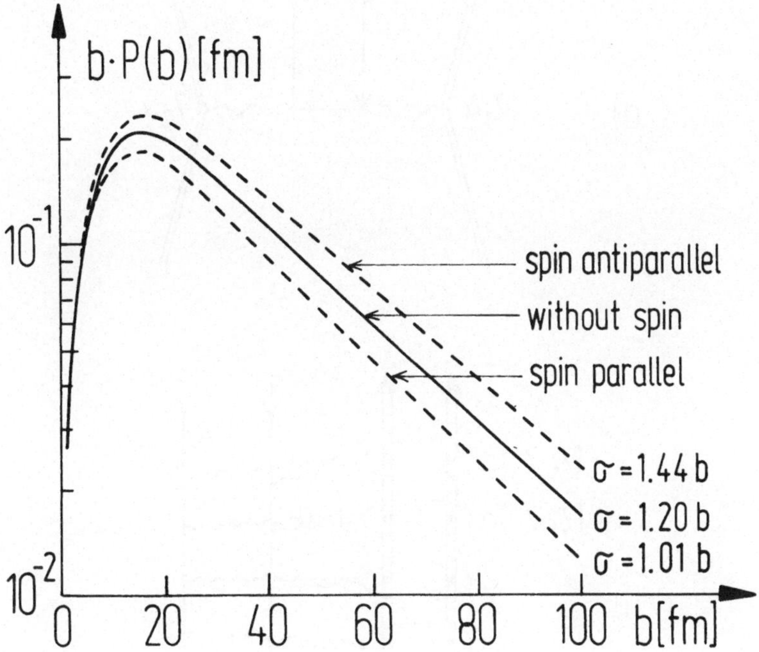
b·P(b) [fm]
10^{-1}
10^{-2}
spin antiparallel
without spin
spin parallel
σ = 1.44 b
σ = 1.20 b
σ = 1.01 b
0
20
40
60
80
100
b[fm]

Fig.40

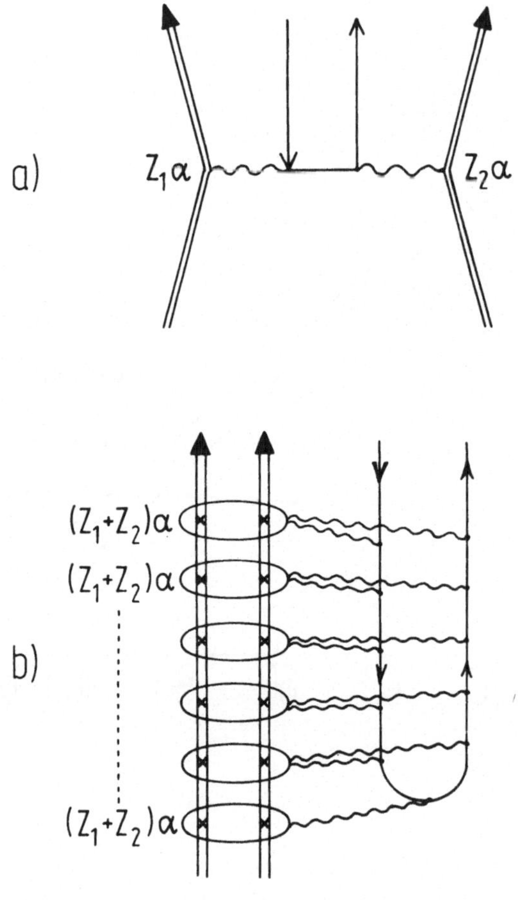

Fig. 41

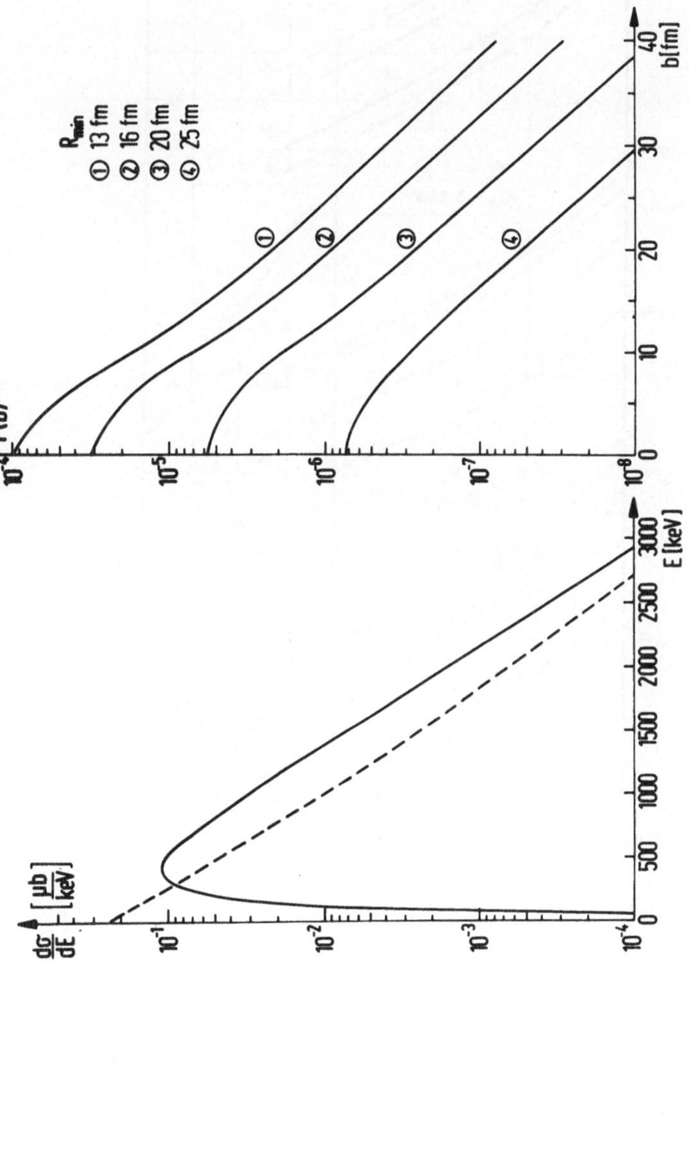

Fig. 42

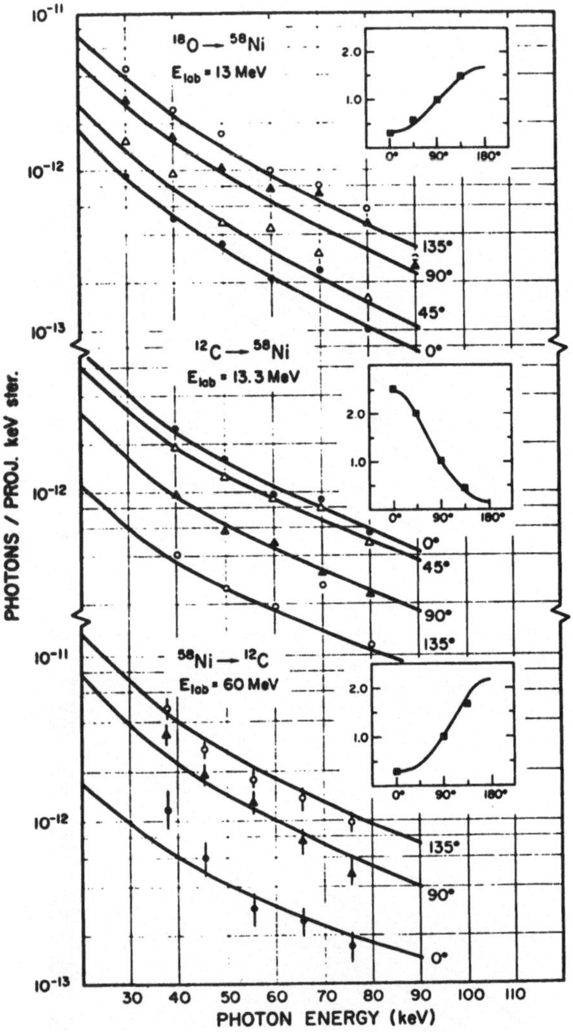

$^{18}O \rightarrow {}^{58}Ni$
E_{lab} = 13 MeV
$^{12}C \rightarrow {}^{58}Ni$
E_{lab} = 13.3 MeV
$^{58}Ni \rightarrow {}^{12}C$
E_{lab} = 60 MeV
PHOTONS / PROJ. keV ster.
PHOTON ENERGY (keV)
135°
90°
45°
0°

Fig. 43

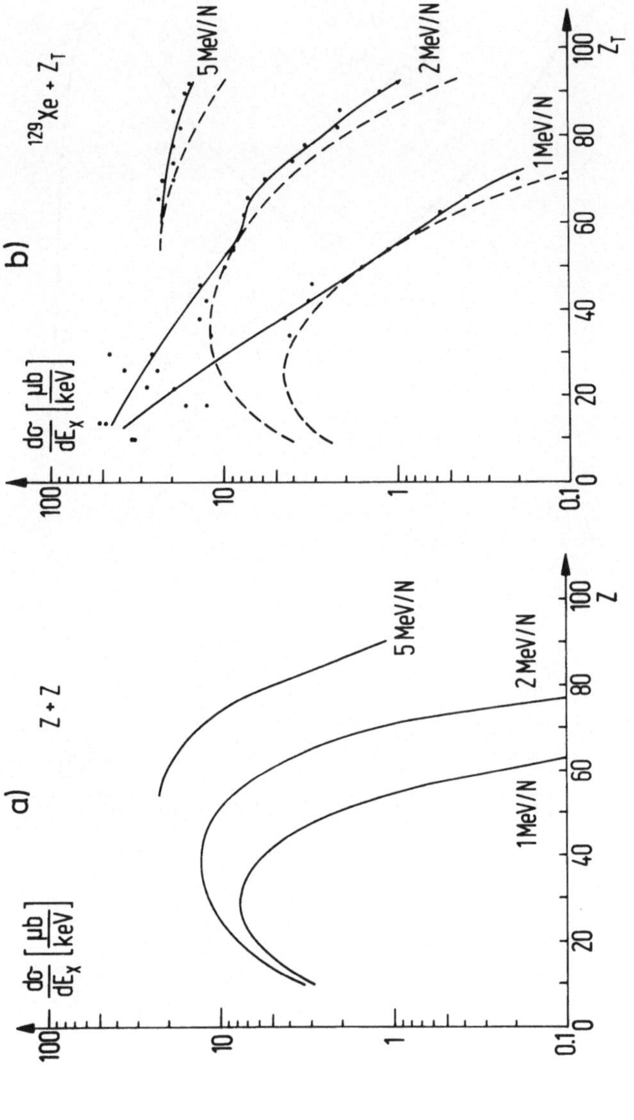
a)
Z + Z
$\frac{d\sigma}{dE_X}\left[\frac{\mu b}{keV}\right]$
100
10
1
0.1
0
20
40
60
80
100
Z
5 MeV/N
2 MeV/N
1 MeV/N
b)
$^{129}Xe + Z_T$
$\frac{d\sigma}{dE_X}\left[\frac{\mu b}{keV}\right]$
100
10
1
0.1
0
20
40
60
80
100
Z_T
5 MeV/N
2 MeV/N
1 MeV/N

Fig. 44

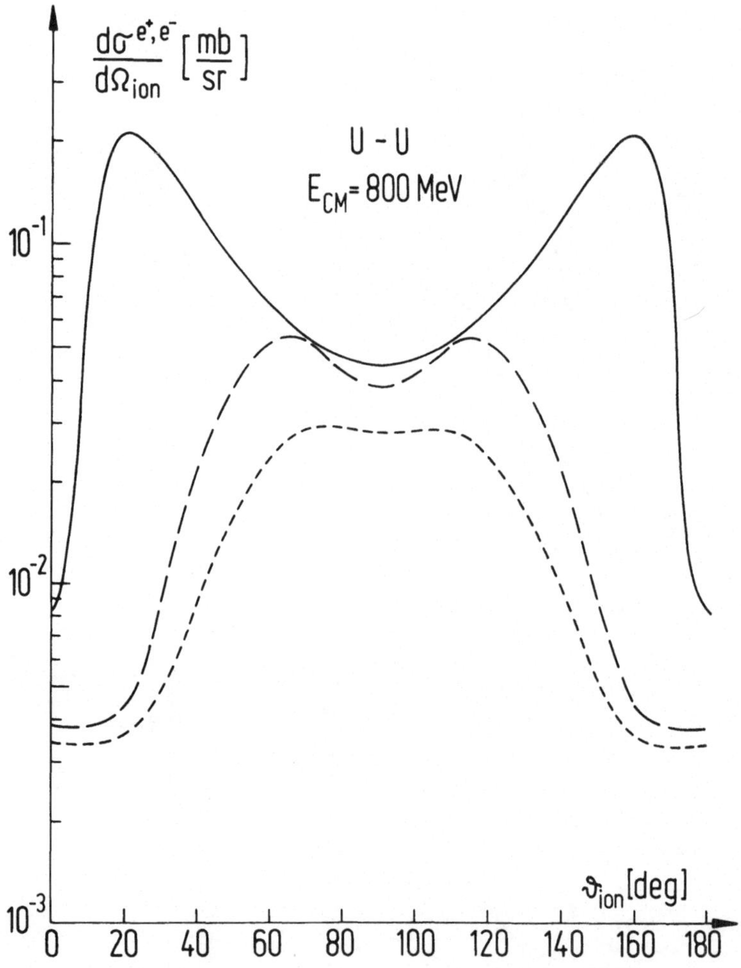
$\frac{d\sigma^{e^+,e^-}}{d\Omega_{ion}}\left[\frac{mb}{sr}\right]$
U - U
E_{CM} = 800 MeV
10^{-1}
10^{-2}
10^{-3}
ϑ_{ion}[deg]
0
20
40
60
80
100
120
140
160
180

Fig. 45

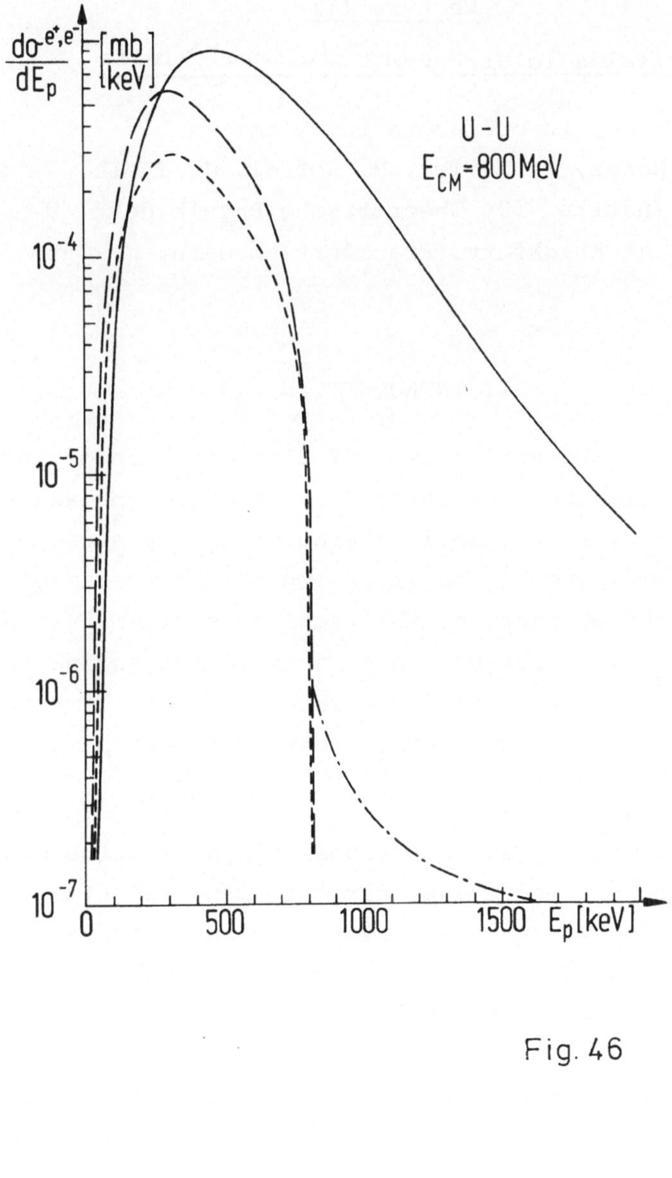
$\frac{d\sigma^{e^+,e^-}}{dE_p}$ [mb/keV]
U - U
E_{CM}=800 MeV
10^{-4}
10^{-5}
10^{-6}
10^{-7}
0
500
1000
1500
E_p[keV]

Fig. 46

Lecture III

Particles in Strong Gravitational Fields

in collaboration with:
U. Heinz, A. Müller, M. Soffel, J. Theis
Institut für Theoretische Physik der
Universität Frankfurt, Frankfurt am Main, W.-Germany

I. INTRODUCTION

One of the curiosities of Einstein's theory of gravitation is the appearance of event horizons and singularities. A horizon is characterized by a region in space-time, from which it is impossible to escape to infinity. The boundary of this region is given by a wave front (null hypersurface) which just cannot escape to infinity. This event horizon generates a strong limitation on the causal relationships between different parts of space-time.

Singularities are positions in a space-time manifold at which the normal picture of space and time breaks down, caused e.g. by an infinite Riemannian curvature. It has been shown [1] in a number of theorems that under very general conditions singularities and horizons appear in Einstein's theory. This circumstance will possibly speak against Einstein's Lagrangian among competitive theories. On the other hand the X-ray source X1 in Cygnus is because of the experimental data possibly a black hole [2]. Furthermore a number of experiments have proved many of Einstein's predictions just in the last few years up to an accuracy of about 1 %.

So one is led to the question, whether the inclusion of matter fields and/or the dynamical treatment of the gravitational collapse by use of Einstein's Lagrangian leads to manifolds free from pathologies. To investigate this, we shall couple matter as a first quantized field to the (classical) gravitational field. This is an approximation, since the gravitational field is not quantized. It is expected, however, that this approximation is good enough to study global gravitational effects as e.g. the macroscopic change of singularities like the classical Schwarzschild radius, under the influence of other fields e.g. electron-positron-, pion-, nucleon-antinucleon fields. Therefore, our program can be characterized as the attempt to investigate the interaction of a geometry with a quantized matter field. The phenomenon guiding us in our research is the meanwhile well known example of supercritical electric fields (Z = 137-problem) which can be generated by superheavy nuclei and molecules [3]. It has been shown for this case, that the normal neutral electron-positron-vacuum breaks down for overcritical charges leading to the new stable charged vacuum. The latter contains finite, quantized amounts of charge shielding to some extent the original central charge. The charge of the vacuum is generated by the creation of e^+e^- pairs for which the electrons are bound. This is due to the charge asymmetry of the central Coulomb interaction. The question we would like to answer is whether similar effects happen in superstrong gravitational fields, which are - in contrast to the electromagnetic case - charge symmetric as long as there is no charge of the central mass.

Particle creation in non-static geometries (an expanding or collapsing universe for example [4]) and in strong gravitational fields appearing in the vicinity of a black hole, has already been discussed. Among these works, Hawking's theory [5] of a "thermal emission", which studies the influence of the time-dependent metric during a gravitational collapse upon a scalar field, is probably best known. Hawking's result, that a collapsing star radiates with a certain temperature, has been confirmed by a number of other authors [6,7,8]. It was realized that the same result can be obtained by imposing suitable boundary conditions for the Green's function on the event horizon of a black hole [9-12]. However, only the radiation at large distances has been studied by most authors and not the local back reaction onto the metric. W. Gerlach claims that a collapsing star never falls through the event horizon because of its perpetual loss of energy [8], until a massless singularity is left containing all quantum numbers. This does not seem to be a very reasonable outcome.

Ruffini et al. [13] found the possibility of pair creation in the surrounding of a Kerr-Newman black hole, based on a static background geometry, in contrast to Hawking's theory. Pair creation occurs here as an example of "Klein's paradox", in analogy to the previously mentioned situation of superstrong fields in Quantum Electrodynamics (QED) that have been discussed by Müller, Rafelski and Greiner [3] during the last eight years.

For the Klein-Gordon equation in a Kerr-Newman geometry Ruffini found a continuum of states with resonances in the energy region from $+mc^2$ to $-mc^2$. Based on classical arguments, those resonances were interpreted as quasibound states, which decay with finite probability towards the singularity, the probability of decay being given by the width of the resonance. This is again in great analogy to the results of overcritical Coulomb fields where bound states have entered the negative energy continuum of the Dirac equation yielding quasibound resonance configurations. The width of those is a measure for the decay of the vacuum state.

Not clear, however, has been the connection between this continuum with the discrete energy levels of a scalar or bispinor field, which one would expect in the case of a weak gravitational field (without an event horizon). Therefore in the present work, we study the Dirac and Klein-Gordon equations in the field of an extended gravitating source (liquid drop) and investigate the structure of the energy spectrum for the source radius tending towards the Schwarzschild radius [14].

Furthermore we present in the solutions of Dirac's equation in a Schwarzschild and Reissner-Nordstrøm background and discuss the appearance of resonances in a "pseudo"-continuum, like in the Klein-Gordon case. One of the interesting results will be the appearance of a limiting charge to mass density (specific charge) for black holes [14].

In the final sections we discuss the problem of particle creation in the time-dependent metric of a collapsing star. We present the metric for such a collapse of incoherent matter in both co-moving and asymptotic coordinates. We then apply the quasi-adiabatic picture to describe particle-hole pair excitations and establish an analogy between the time- and a temperature bath.

Finally we estimate the "thermal" energy absorbed by the pair excitations and study its effect onto the time-dependence of the collapse. It turns out that the collapse motion must be modified after a short time in the 10^{-3} sec range.

II. THE DIRAC (KLEIN-GORDON) EQUATION IN A SPHERICALLY SYMMETRIC GRAVITATIONAL FIELD

We consider Dirac's equation in a Schwarzschild (Reissner-Nordstrøm) field or more general, in a spherically symmetric graviational field (see also [15,16,17]):

$$ds^2 = e^{\nu}\, dt^2 - e^{\lambda}\, dr^2 - r^2\, (d\theta^2 + \sin^2\theta\, d\phi^2) \qquad (1)$$

$$\nu = \nu(r,t)\,,\quad \lambda = \lambda(r,t)$$

where e.g.:

$$e^{-\lambda} = e^{\nu} = 1 - \frac{2M_G}{r} \qquad \text{(Schwarzschild)} \qquad (2)$$

or

$$e^{-\lambda} = e^{\nu} = 1 - \frac{2M_G}{r} + \frac{Q_G^2}{r^2} \equiv \Omega^2 \qquad \text{(Reissner-Nordstrøm)} \qquad (3)$$

In the last equations we have introduced mass and charge of the black hole, M_G and Q_G, in terms of "geometrized" units, which are connected to mass M and charge Q in ordinary units by $M_G = \frac{GM}{c^2}$, $Q_G^2 = \frac{GQ^2}{c^4}$. In general, we shall characterize "geometrized" units by a subscript G and set $\hbar = c = 1$. The action principle of the Dirac field reads:

$$\delta \int d^4x \sqrt{-g} \, (\bar{\psi} \gamma^k \psi_{:k} + im \bar{\psi}\psi) = 0 \quad . \tag{4}$$

The colon symbolizes the covariant derivative of a bispinor, and $\bar{\psi}$ denotes the conjugate of ψ, i.e.

$$\bar{\psi} = i \, \psi^{\dagger} \, \tilde{\gamma}^{o} \quad . \tag{5}$$

Eq. (4) leads to the covariant Dirac equation

$$\gamma^k \left(\frac{\partial}{\partial x^k} - \Gamma_k + ie A_k\right)\psi + im \psi = 0 \tag{6}$$

where A_k are the components of the electromagnetic field. The four Γ_k matrices are defined by

$$\gamma_{i \, k} - \left\{{\ell \atop ik}\right\} \gamma_\ell - \Gamma_k \gamma_i + \gamma_i \Gamma_k = 0 \tag{7}$$

up to an imaginary multiple of the unit matrix. From this follows that Γ_k can be chosen as

$$\Gamma_k = - \frac{1}{4} \gamma^j \left(\gamma_{j,k} - \gamma_\ell \left\{{\ell \atop jk}\right\}\right) \quad . \tag{8}$$

Noether's theorem applied to eq. (4) gives the current density j^k and the matter tensor T_{ij}:

$$j^k = e \, \bar{\psi} \, \gamma^k \, \psi \tag{9}$$

$$T_{ij} = \frac{i}{4}[\bar{\psi}(\gamma_i\psi_{ij} + \gamma_j\psi_{ii}) - (\bar{\psi}_{ii}\gamma_j + \bar{\psi}_{ij}\gamma_i)\psi]. \tag{10}$$

The generalized Dirac-matrices γ^i are defined by

$$\gamma^i\,\gamma^j + \gamma^j\,\gamma^i = 2\,g^{ij} \tag{11}$$

up to unitary transformations in Clifford space. We take

$$\begin{aligned}
\gamma_o &= e^{\nu/2}\,\tilde{\gamma}_o & \gamma^o &= e^{-\nu/2}\,\tilde{\gamma}_o \\
\gamma_1 &= e^{\lambda/2}\,\tilde{\gamma}_1 & \gamma^1 &= -\,e^{-\lambda/2}\,\tilde{\gamma}_1 \\
\gamma_2 &= r\,\tilde{\gamma}_2 & \gamma^2 &= -\,r^{-1}\,\tilde{\gamma}_2 \\
\gamma_3 &= r\sin\theta\,\tilde{\gamma}_3 & \gamma^3 &= -\,r^{-1}\sin\theta\,\tilde{\gamma}_3
\end{aligned} \tag{12}$$

where the $\tilde{\gamma}_i$ are the usual Dirac-matrices in Minkowski-space. Thus the charge density is

$$j^o = e\,e^{-\nu/2}\,\psi^\dagger\psi \tag{13}$$

and the energy density becomes

$$T^o_o = \frac{i}{2}\,e^{-\nu/2}\,(\psi^\dagger\,\frac{\partial\psi}{\partial t} - \frac{\partial\psi^\dagger}{\partial t}\,\psi)\quad . \tag{14}$$

In the spherically symmetric field (1) the four-component Dirac equation can be reduced to a two-component form in the standard way (see e.g. [19]).

We write:

$$\psi = e^{-\lambda/4} \frac{1}{r} \begin{bmatrix} \phi_1(r,t) & \chi^{\mu}_{\kappa} \\ \phi_2(r,t) & \chi^{\mu}_{-\kappa} \end{bmatrix} , \quad \phi = \begin{bmatrix} \phi_1 \\ \phi_2 \end{bmatrix} \tag{15}$$

with the spinor spherical harmonics

$$\chi^{\mu}_{\kappa}(\theta,\phi) = [Y_{\ell m} \otimes \chi^{m'}_{1/2}]^{\mu}_{j} \qquad (|\kappa| = j + \tfrac{1}{2}) \tag{16}$$

which satisfy the eigenvalue equation

$$K\, \chi^{\mu}_{\kappa} \equiv \tilde{\gamma}_o(\vec{\sigma}\cdot\vec{L} + 1)\chi^{\mu}_{\kappa} = -\, \kappa\, \chi^{\mu}_{\kappa} \quad . \tag{17}$$

Inserting this into eq. (6) we are left with

$$i\frac{\partial\phi}{\partial t} = [-i\alpha_r e^{\frac{\nu-\lambda}{4}} \frac{\partial}{\partial r} e^{\frac{\nu-\lambda}{4}} + i\beta\alpha_r e^{\nu/2} \frac{\kappa}{r} + \beta e^{\nu/2} m + V]\phi \equiv (H+V)\phi \tag{18}$$

with

$$\beta = \tilde{\gamma}_o = \begin{pmatrix} 1 & 0 \\ 0 & -1 \end{pmatrix} \qquad \alpha_r = \sigma_2 = \begin{pmatrix} 0 & -i \\ i & 0 \end{pmatrix}, \qquad V = eA_o \; ; \tag{19}$$

H is explicitly Hermitian.

$$\int dr\, \phi^{\dagger}_1 (H\phi_2) = \int dr\, (H\phi_1)^{\dagger}\, \phi_2 \quad . \tag{20}$$

The radial density is $\rho = \phi^{\dagger}\phi$ and the normalization condition becomes

$$\int dr\, \phi^{\dagger}\phi = 1 \tag{21}$$

$$\langle T^o_o \rangle = 4\pi \int r^2 dr\, T^o_o = \omega \int dr\, e^{-\frac{\lambda+\nu}{2}} \phi^\dagger \phi . \tag{22}$$

In the case of a static gravitational field, the wave equation (18) can be reduced further by solving for stationary states

$$\phi = e^{\frac{\lambda-\nu}{4}} \begin{bmatrix} f(r) \\ g(r) \end{bmatrix} e^{-i\omega t} . \tag{23}$$

The radial equations are:

$$\frac{d}{dr} \begin{bmatrix} f \\ g \end{bmatrix} = e^{\lambda/2} \begin{bmatrix} \frac{\kappa}{r} & m - e^{-\nu/2}(\omega - V) \\ m + e^{-\nu/2}(\omega - V) & -\frac{\kappa}{r} \end{bmatrix} \begin{bmatrix} f \\ g \end{bmatrix} \tag{24}$$

V(r) is the Coulomb potential of the collapsed stellar object (e.g. black hole, neutron star) given by

$$V(r) = \frac{eQ}{r} . \tag{25}$$

The Klein-Gordon equation in a static spherical metric can be derived in an analogous way from the covariant form

$$(-g)^{-1/2} \frac{\partial}{\partial x^k} \left(g^{ik} \sqrt{-g} \frac{\partial}{\partial x^i} \psi \right) + m^2 \psi = 0 . \tag{26}$$

Writing

$$\psi = \phi(r)\, Y_{\ell m}(\theta,\phi)\, e^{-i\omega t} \tag{27}$$

one obtains the equation

$$\frac{d}{dr}\left(r^2 e^{\frac{\nu-\lambda}{2}} \frac{\partial\phi}{\partial r}\right) - e^{\frac{\nu+\lambda}{2}} [\ell(\ell+1)+m^2r^2-\omega^2 e^{-\nu} r^2]\,\phi = 0 \;. \quad (28)$$

III. THE DIRAC (KLEIN-GORDON) EQUATION IN THE FIELD OF AN EXTENDED GRAVITATING LIQUID DROP

Since classically a periodic motion of particles in the field of a black hole is only possible outside the event horizon, one normally restricts the manifold M to the region outside the coordinate singularities of (2) or (3). Then one is able to interpret the theory within the frame of an asymptotic observer.

In order to understand better the results of the field equations in the restricted manifold, we now turn to the case without an event horizon which limits the solution of Dirac's (Klein-Gordon's) equation and consider the Dirac (Klein-Gordon) equation in the field of an extended gravitating source (liquid drop) of radius r_o. For the metric we take [20]

$$ds^2 = \begin{cases} -(1-\frac{r^2}{\hat{R}^2})^{-1}dr^2 - r^2(d\theta^2+\sin^2\theta d\phi^2) \\ +[\frac{3}{2}(1-\frac{r_o^2}{\hat{R}^2})^{1/2} - \frac{1}{2}(1-\frac{r^2}{\hat{R}^2})^{1/2}]dt^2 \end{cases} r \leq r_o$$

$$\qquad\qquad (29)$$

$$\begin{cases} -(1-\frac{2M_G}{r})^{-1}dr^2-r^2(d\theta^2 + \sin^2\theta\, d\phi^2) \\ +(1-\frac{2M_G}{r})\, dt^2 \end{cases} r > r_o$$

with $$\hat{R}^2 = \frac{r_o^3}{2M_G} \quad .$$

The radius r_o of the drop should be greater than its Schwarzschild radius. Indeed, the model breaks down already for $r_o = 9/8\ r_s$, because of an infinite pressure in the matter tensor. In order to study the process $r_o \to r_s$, we have analytically continued the metric to the case $r_s < r_o \leq 9/8\ r_s$,

$$e^{\nu} = a\ e^{br^2} \qquad r \leq r_o$$

$$b = r_o^{-2}\ (\frac{r_o}{M_G} - 2)^{-1} \ , \qquad a = (1 - \frac{2M_G}{r_o})\ e^{-br_o^2} \ . \tag{30}$$

For $r_o \leq 9/8\ r_s$ this metric does not account for a reasonable physical situation but should only serve us for a definition of an energy eigenvalue of a field equation in this region. The radial Dirac equation (24) has been numerically integrated and the energy eigenvalues have been obtained. The results are shown in Fig. 1. It is noticed, that the energy eigenvalues of all bound states (with finite number of modes in their radial functions) tend to zero as $r_o \to r_s$ and a quasi-continuum arises. We remark that the energy eigenvalues contain the average red shift in form of a factor $\langle\psi|e^{\nu/2}|\psi\rangle$ indicating the obtained results [22]. For comparison we calculated the energy states for the Klein-Gordon equation as we expect the spin-orbit coupling effects to be negligible. The results are nearly identical [23].

In Fig. 1 we have defined continuum states as those which are asymptotically free, i.e. $|E| \geq m$, as

E makes the energy as seen from an asymptotic observer. Now, one might argue, that the physically important quantities are the local ones as for example the "local" energy ($E < e^{-\nu/2}>$). This seems not to be the case. The reason for this is, that the notion of a particle in quantum field theory must be viewed from a global (and not a local) standpoint due to the uncertainty relation.

IV. THE SOLUTION OF THE DIRAC EQUATION IN A REISSNER-NORDSTRØM GEOMETRY. LIMITING BLACK HOLE CHARGE

In the case of the extended source (incompressible fluid model) we obtain the quasicontinuum and vanishing energy eigenvalues of all bound states as $r_o \to r_s$. So in the case of a coordinate singularity we expect to obtain a continuum with infinitely many modes in the radial functions for energies between zero and m. This is shown here for the Dirac equation. If the black hole is charged, we require that $Q_G^2 \leq M_G^2$ in order to have an event horizon. For a black hole of solar mass $M_\theta \approx 2 \times 10^{33}$ g, this means: $Q \lessapprox 1.7 \times 10^{20}$ coulomb.

The coordinate singularities lie at

$$r_\pm = M_G \pm (M_G^2 - Q_G^2)^{1/2} \quad . \tag{31}$$

We then restrict our manifold by introducing the r^*-coordinate (see equation (3))

$$\frac{dr^*}{dr} = \Omega^{-2} \quad . \tag{32}$$

This introduction of the r^*-coordinate means that the outer coordinate singularity (event horizon) r_+ is projected to $(-\infty)$. The wave functions in the interior of a black hole, i.e. for $r \leq r_+$, are therefore not defined. By this procedure one avoids the introduction of boundary conditions for the wave functions at $r = r_+$.

The radial Dirac equation (24) in r^*-coordinates are simply

$$\frac{dg(r^*)}{dr^*} = - \frac{\kappa}{r} \Omega g(r^*) + [(\omega-V) + \Omega m] f(r^*)$$

$$\frac{df(r^*)}{dr^*} = \frac{\kappa}{r} \Omega f(r^*) - [(\omega-V) - \Omega m] g(r^*) \qquad (33)$$

which have as solution a plane wave for $r^* \to -\infty$, namely

$$g(r^*) \simeq A_o \sin((\omega-\Lambda) r^* - \delta(\omega))$$

$$f(r^*) \simeq A_o \cos((\omega-\Lambda) r^* - \delta(\omega)) \ . \qquad (34)$$

Here $\Lambda \equiv \frac{eQ}{r_+}$ is the Coulomb potential of the black hole at its outer coordinate singularity r_+. Equation (33) has been solved numerically. By means of (32) it has been integrated in r^*-coordinates from large positive to negative values and we looked for resonances similar to those encountered by Deruelle and Ruffini [13] in the case of the Klein-Gordon equation. For convenience the resonance parameter P_r which is the ratio of the first $(r^* \to +\infty)$ amplitude to the resonance amplitude $(r^* \to -\infty)$

of the large component in eq. (33) has been introduced. The inverse of P_r is a measure for the probability that the resonance state decays towards the physical singularity, so resonances can be found by maximum values of P_r. P_r as a function of ω can also be used to determine the width of a resonance. Typical wave functions of resonating and non-resonating type are depicted in Figs.2. The results are in good agreement with those given in ref. [13] and the energies of the resonance as well as the radial density distributions of the electron can be understood with the help of the effective potential. The effective potential was first derived from the classical Hamiltonian Jacobi formalism by Christodoulou and Ruffini [24]. For spin 1/2 particles it must be rederived by transforming the Dirac-equation (33) into a second order differential equation and applying the WKB approximation. The result is the effective potential, which for the Reissner-Nordstrøm field reads:

$$V^{\pm}_{eff}(r) = \frac{eQ}{r} \pm \Omega(m^2 + \frac{\kappa^2}{r^2})^{1/2} \quad . \qquad (35)$$

For the model of the neutral extended gravitating source it is

$$V^{\pm}_{eff}(r) = \pm\, e^{\nu/2}\,(m^2 + \frac{\kappa^2}{r^2})^{1/2} \quad . \qquad (36)$$

One finds that pair creation in the case of a charged black hole is due to the decay of the vacuum in analogy to the problems of strong fields in QED. This is demonstrated in Figs. 3a-c.

Even in the case of a neutral black hole the gap between the positive and negative energy continuum (particle and antiparticle states) is narrowed by the attractive gravitational interaction and vanishes for $r = r_s$. Due to the charge conjugation invariance of the Schwarzschild field the picture (3a) is symmetric with respect to $\omega = 0$.

With the introduction of a charged centre the symmetry between the electron and positron states is broken. Speaking in the language of hole-theory the occupied negative continuum is raised by a negatively charged center and vice versa. If the Fermi energy is above $m_e c^2$ spontaneous electron emission (for negatively charged central objects), see Fig. (3b), or spontaneous positron emission (for positively charged centers) takes place. This manifests the phase transition to a charged electron-positron vacuum. In other words pair creation occurs if $\Lambda = \frac{eQ}{r_+} > m_e c^2$ which leads us to a limiting stable charge of a black hole, namely

$$\frac{Z_{lim}\, e^2}{r_+} = m_e c^2 \quad . \tag{37}$$

In contrast to the same phenomenon in superheavy atoms, the right side of (37) is only $m_e c^2$ and not $2m_e c^2$. Effectively only one particle must be created because the antiparticle is absorbed by the black hole. Since Z_{lim} is much smaller than the limit $G^{-1/2}M_G$, we can write with $r_+ = \frac{2GM}{c^2}$:

$$Z_{lim} = \frac{2\, GM\, m_e}{e^2} \quad . \tag{38}$$

For black holes of one solar mass this means $Z_{lim} \gtrsim 10^{18}$ or a limiting charge of $Z_{lim}e \sim 0.16$ Coulomb. This corresponds to a limiting charge to mass ratio per atom in a black hole of

$$\frac{Z_{lim}\,e}{M} = \frac{2Gm_e}{e} = \frac{2Gm_e^2}{e^2}\left(\frac{e}{m_e}\right) = 0.48 \times 10^{-42}\left(\frac{e}{m_e}\right)$$

$$= 8.4 \times 10^{-35}\ \mathrm{C/g}$$

reflecting the double ratio of the gravitational ($\gamma = \frac{G\,m_e^2}{\hbar c}$) to the electromagnetic ($\alpha = \frac{e^2}{\hbar c}$) coupling constant. The same result was also obtained by Gibbons [24], who discussed it in the WKB approximation.

V. GRAVITATIONAL COLLAPSE

In order to connect our results for the extended stellar object (densely spaced, deeply bound states) with those for the singular metric with event horizon (continuum with high angular momentum resonances), we have to find a smooth transition between the two situations. This cannot be achieved with static space times, since a star with $r_o < 9/8\ r_s$ is known to be unstable, but only within a time dependent collapse of the star. Physically, this is a meaningful assumption, because black holes with a mass of stellar size, - if they exist at all - are conceivably formed by collapse following a supernova [25].

The gravitational field describing a collapse of a mass distribution has to be determined by Einstein's equation

$$R_{ik} - \frac{1}{2} g_{ik} R = - 8\pi G T_{ik} . \qquad (39)$$

In the following we discuss the case of incoherent matter for which T^i_k is simply:

$$T^i_k = h u^i u_k \qquad (40)$$

where h is the proper energy density and u^i is the four-velocity. A comoving frame $x^i = \{\tau, \rho, \theta, \phi\}$ has been found to be particularly well adapted to the present problem [25]. The line element of space-time is of the form:

$$ds^2 = d\tau^2 - a(\rho,\tau) d\rho^2 - r^2(\rho,\tau) d\Omega^2$$

$$d\Omega^2 = d\theta^2 + \sin^2\theta \, d\phi^2 . \qquad (41)$$

In such a comoving Gaussian coordinate system the components of the four-velocity are

$$u^i = u_i = \{1,0,0,0\} . \qquad (42)$$

Thus from (39), (40) and (42) Einstein's equation leads to:

$$R^o_1 = R^1_1 - \frac{1}{2} R = 0$$

$$R^o_o - \frac{1}{2} R = - 8\pi G h . \qquad (43)$$

For a given initial distribution of matter it is sufficient to calculate the functions $a(\rho,\tau)$, $r(\rho,\tau)$ and $h(\rho,\tau)$ up to coordinate transformations.

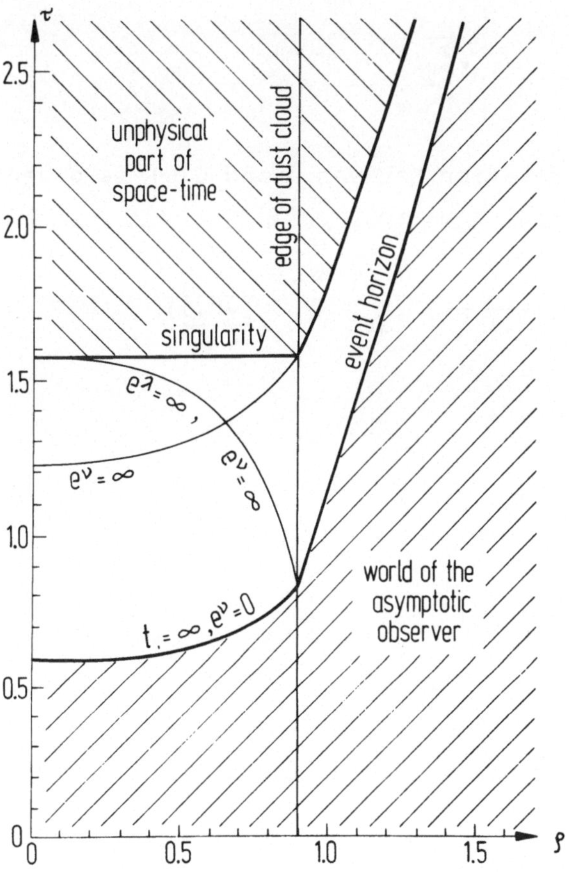

Fig. 4

Thus we obtain the complete solution of our problem [25]:

$$r = \rho \cdot C(\mu) \; ; \qquad \frac{dC}{d\mu} = - \sqrt{\frac{1-C(\mu)}{C(\mu)}}$$

$$\mu = \mu(\rho,\tau) = \tau \sqrt{\frac{\psi}{\rho^2}} \tag{44}$$

where the relation between μ and $C(\mu)$ is given by the transcendental equation

$$\mu = \sqrt{C(\mu)(1-C(\mu))} + \text{arc tan} \sqrt{\frac{1-C(\mu)}{C(\mu)}} \tag{45}$$

with

$$\psi(\rho) = \frac{3}{\rho} \int_0^\rho \bar{\rho}^2 \lambda(\bar{\rho}) \, d\bar{\rho}$$

$$\lambda(\rho) = \frac{8\pi G \; h(\rho,0)}{3} \tag{46}$$

where $h(\rho,0)$ is the initial distribution of rest energy at $\tau = 0$. This energy density at any time is

$$h(\rho,\tau) = \frac{\rho^2 h(\rho,0)}{r^2 r'} \; . \tag{47}$$

Here the prime denotes the derivative with respect to ρ. Now our line element is of the form:

$$ds^2 = d\tau^2 - \frac{(r')^2}{1-\psi(\rho)} d\rho^2 - r^2(\rho,\tau)\, d\Omega^2 . \tag{48}$$

Assuming that $\lambda(\rho)$ is a never increasing function of ρ, then in any physically meaningful case $\psi(\rho)$ must be in the interval

$$0 \leq \psi(\rho) < 1 \quad . \tag{49}$$

For a gravitational collapse starting at $\tau = 0$ the solution (48) is regular everywhere in the part of space-time for which $\mu(\rho,\tau)$ lies in the interval

(see also eq.(45))
$$0 \leq \mu < \frac{\pi}{2} \quad . \tag{50}$$

For $\mu \to \frac{\pi}{2}$ we have
$$\begin{cases} C(\mu) \to 0 \\ \dfrac{dC}{d\mu} \to -\infty \\ r \to 0 \end{cases} \tag{51}$$

thus for any fixed value of ρ in the interval $0 \leq \rho < \infty$ the metric becomes singular at a finite time τ_s. We obtain τ_s from

$$\mu(\rho,\tau_s) = \frac{\pi}{2} \longrightarrow \tau_s = \frac{\pi\rho}{2\sqrt{\psi(\rho)}} \quad . \tag{52}$$

By the pair of coordinates (ρ,τ_s) a timelike curve in space-time is marked, because it is impossible to extend lines beyond this hypersurface.

This is the reason why we call the limit $\mu \to \frac{\pi}{2}$ <u>an invariant singularity</u>, which occurs for every value of ρ because we use comoving coordinates, thus a fixed value of ρ "reaches" the singularity at $\tau = \tau_s$ [26],[27].

We are now looking for a system of asymptotic coordinates (i.e. coordinates suitable for an asymptotic observer) in which the line element is of the form:

$$ds^2 = e^{\nu}\, dt^2 - e^{\lambda}\, dr^2 - r^2\, d\Omega^2$$

$$d\Omega^2 = d\theta^2 + \sin^2\theta\, d\phi^2 \quad . \tag{53}$$

We want to find a connection between (48) and (52) by finding a transformation

$$t = W(\rho, \tau) \tag{54}$$

which brings (48) into the desired form.

The transformation $W(\rho,\tau)$ is characterized by the following equations:

$$(1 - \psi)\, W' - \dot{r}\, r'\, \dot{W} = 0$$

$$e^{\lambda} = \frac{1}{1 - \frac{\rho\psi}{r}}$$

$$e^{\nu} = \frac{1-\psi}{\dot{W}^2 (1 - \frac{\rho\psi}{r})} \quad . \tag{55a-c}$$

The dot denotes the derivative with respect to τ.

We now assume the collapsing "dust cloud" to be confined within a radius ρ_b.

In the empty space outside the matter distribution (i.e. $\rho > \rho_b$) we have

$$\lambda(\rho) = 0 \longrightarrow \psi(\rho) = \frac{2M_G}{\rho} \quad . \tag{56}$$

From (55a) we obtain

$$\dot{W} = \frac{\sqrt{1-\frac{2M_G}{\rho}}}{1-\frac{2M_G}{r}}$$

$$W' = \frac{\dot{r}r'}{\sqrt{1-\frac{2M_G}{\rho}}\left(1-\frac{2M_G}{r}\right)} \tag{57}$$

and by integration

$$W(\rho,\tau) = \tau\sqrt{1-\frac{2M_G}{\rho}} + 2\sqrt{2M_G(\rho-2M_G)}\ \text{arc tan}\ \sqrt{\frac{\rho-r}{r}}$$

$$+ 4M_G\ \ln \frac{\sqrt{r(\rho-2M_G)} + \sqrt{2M_G(\rho-r)}}{\sqrt{\rho\,|r-2M_G|}} \quad . \tag{58}$$

With (56) and (57a) equations (55b,c) - according to Birkhoff's theorem - reduce to the well known Schwarzschild expressions:

$$e^{\lambda} = \frac{1}{1-\frac{2M_G}{r}}$$

$$e^{\nu} = 1 - \frac{2M_G}{r} \quad . \tag{59}$$

In the inner region (i.e. $\rho \leq \rho_b$) of the dust cloud we assume $\lambda(\rho)$ to be a constant λ_o for all $\rho \leq \rho_c < \rho_b$, so we get with equation (46)

$$\psi(\rho) = \lambda_o \ \rho^2 \quad . \tag{60}$$

If we match a smooth transition from the constant value λ_o for $\rho \leq \rho_c < \rho_b$ to the value $\lambda(\rho) = 0$ for $\rho > \rho_b$ in the interval $\rho_c \leq \rho \leq \rho_b$, the components of the metric tensor in (48) will be continuous. At this point it should be remarked that in the limit $\rho_c \to \rho_b$, where $\lambda(\rho)$ is a step function at $\rho = \rho_b$ and therefore

$$\lambda_o \ \rho_b^2 = \frac{2M_G}{\rho_b} \quad , \tag{61}$$

the quantities $\frac{\partial r}{\partial \rho}$, $a(\rho,\tau)$ etc. are discontinuous, according to the discontinuity of $h(\rho,\tau)$ at $\rho = \rho_b$. In the following calculation we will work with the limit $\rho_c = \rho_b - 0^+$, so that $\lambda(\rho)$ is a step function. In the inner region ($\rho < \rho_b$) of the dust cloud we now can solve the partial differential equation (55a) by the method of characteristic equations.

By this way we find the required function $\dot{W}(\rho,\tau)^2$:

$$\dot{W}^2(\rho,\tau) = \sqrt{1-\lambda_o\rho^2}\ \frac{\{\sqrt{1-\lambda_o\rho_b^2} - \sqrt{1-\lambda_o\rho^2}\,(1-C(\mu))\}^3}{C(\mu)\{\sqrt{1-\lambda_o\rho_b^2}^{\,3} - \sqrt{1-\lambda_o\rho^2}\,(1-C(\mu))\}^2} \, . \tag{62}$$

With (60) and (62) we find

$$e^{\lambda} = \frac{1}{1 - \frac{\lambda_o \rho^3}{r}}$$

$$e^{\nu} = \frac{1 - \lambda_o \rho^2}{\dot{W}^2 \left(1 - \frac{\lambda_o \rho^3}{r}\right)} \quad . \tag{63}$$

We recognize that for $r \to 2M_G$ the metric in the outer region becomes singular, because in this limit e^{λ} goes to infinity while e^{ν} vanishes.

For an asymptotic observer (at rest) this situation is only established at $t = \infty$, and this is clearly to be seen by

$$\lim_{r \to 2M_G} \dot{W} \equiv \lim \frac{dt}{d\tau} = + \infty \ . \tag{64}$$

We will call this the <u>formation of an event horizon,</u> and it is characterized by $t = \infty$.

We want to express ν and λ in asymptotic coordinates by inverting the relationship (44): $r = \rho C(\mu)$.

In the limit $r \to 2M_G$ the first order expansion in e^{ν} and e^{λ} yields

$$\lim_{r \to 2M_G} e^{\nu(r,t)} \sim e^{-t/2M_G} \tag{65}$$

$$\lim_{r \to 2M_G} e^{\lambda(r,t)} \sim e^{t/2M_G} \quad . \tag{66}$$

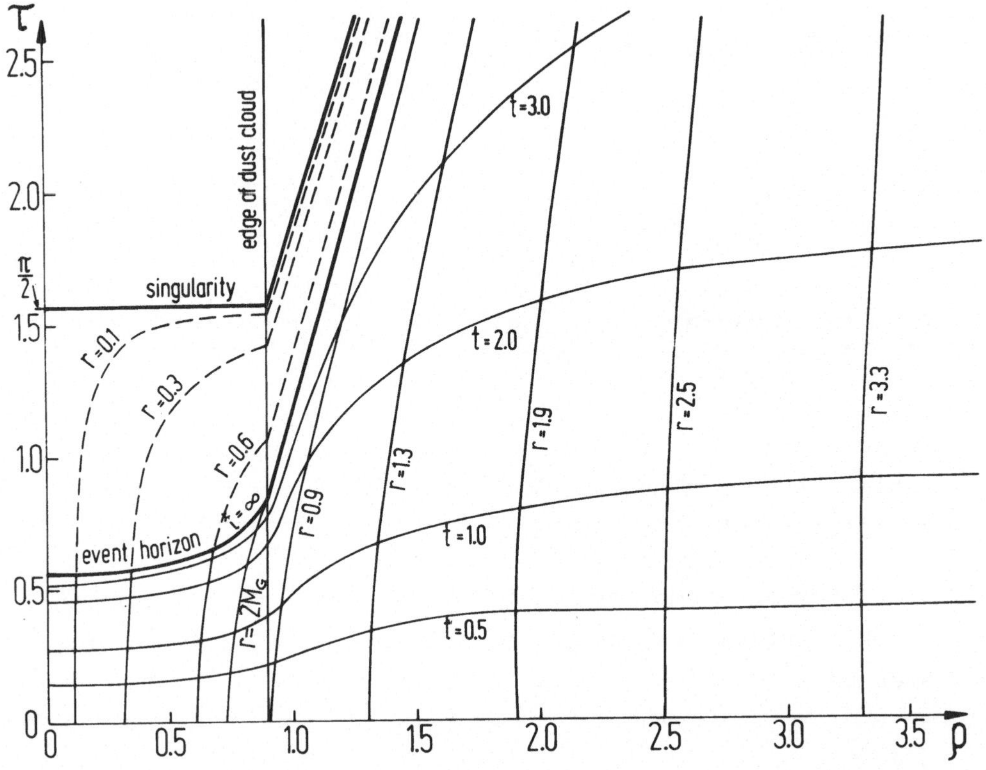

Fig. 5

In figure 5 the numerical results for $e^{\nu(r,t)}$ and $e^{\lambda(r,t)}$ are shown, where the exact dependence of $\rho(r,t)$ was used.

VI. PARTICLE PRODUCTION IN TIME-DEPENDENT GRAVITATIONAL FIELDS

It is a general rule in quantum mechanics that time-dependent potentials $V(t)$ will lead to particle pair creation as soon as the Fourier-transform $V_F(\omega)$ contains frequencies $\omega > 2m$. We can therefore expect that a time-varying metric will also produce particles. Furthermore, since gravitation couples to all known forms of energy, all particles will be produced as long as the necessary Fourier frequencies are available. As a first step, we formulate the pair production in the same way as we have done in heavy ion collisions. We start from the adiabatic solutions of the instantaneous stationary Dirac equation

$$\omega_n(t)\phi_n = H(t)\ \phi_n \tag{67}$$

with the Hamiltonian from eq. (18):

$$H(t) = -i\alpha_r\ e^{\frac{\nu-\lambda}{4}} \frac{\partial}{\partial r} e^{\frac{\nu-\lambda}{4}} + i\beta\alpha_r\ e^{\nu/2} \frac{\kappa}{r} + \beta e^{\nu/2}\ m\ . \tag{68}$$

The hermiticity of H, eq. (20), ensures that the adiabatic eigenstates $\phi_n(r,t)$ form an orthogonal and complete basis for every t. Hence the true solution of eq. (18) can be written as

$$\phi(r,t) = \sum_n a_n(t)\ \phi_n(r,t)\ \exp\left(-i \int dt\ \omega_n(t)\right) \tag{69}$$

with occupation amplitudes $a_n(t)$ for the stationary states. Inserting (69) into the time dependent Dirac equation (18) leads to a set of coupled equations:

$$\dot{a}_n(t) = - \sum_{m\neq n} a_m(t)\ [\int dr \phi_n^\dagger \frac{\partial}{\partial t}\phi_m] \exp(i\int dt(\omega_n-\omega_m))\ . \tag{70}$$

All the physics is contained in the matrix elements ($n\neq m$):

$$\langle n|\frac{\partial}{\partial t}|m\rangle = \int dr \phi_n^\dagger \frac{\partial\phi_m}{\partial t} = (\omega_m - \omega_n)^{-1} \langle n|\frac{\partial H}{\partial t}|m\rangle\ . \tag{71}$$

Now $\partial H/\partial t$ can be written in the following way:

$$\frac{\partial H}{\partial t} = -i[\frac{\dot{\nu}-\dot{\lambda}}{4}, e^{\frac{\nu-\lambda}{4}} \alpha_r \frac{\partial}{\partial r} e^{\frac{\nu-\lambda}{4}}]_+ + \frac{\dot{\nu}}{2} e^{\nu/2}(i\beta\alpha_r \frac{\kappa}{r} + \beta m)$$

$$= [\frac{\dot{\nu}-\dot{\lambda}}{4}, H]_+ + \frac{\dot{\lambda}}{2} e^{\nu/2}\beta \; (i\alpha_r \frac{\kappa}{r} + m) \; . \qquad (72)$$

For states ϕ_n with dominant contribution of the kinetic energy to the total energy ω_n, or for particles with small rest mass m, the second term in eq. (72) is negligible. For these states,

$$<n|\frac{\partial}{\partial t}|m> \sim \frac{\omega_n + \omega_m}{\omega_n - \omega_m} < n|\frac{\dot{\lambda}-\dot{\nu}}{4}|m> \quad . \qquad (73)$$

Without explicitly evaluating the matrix elements, it is clear that the particle creation will depend on the value of $\frac{\dot{\lambda}-\dot{\nu}}{4}$.

For the free collapse described in the preceding section it turns out that inside the collapsing star

$$\frac{\dot{\lambda}-\dot{\nu}}{4} \xrightarrow{t\to\infty} \frac{1}{4GM} \quad . \qquad (74)$$

In general at the surface of the collapsing star one will have $\lambda = -\nu$ and therefore $\frac{\dot{\lambda}-\dot{\nu}}{4} = -\frac{\dot{\nu}}{2}$.

If many particle-antiparticle pairs are created, the approach based on perturbation theory is not viable. Instead it seems desirable to develop a formalism which deals directly with macroscopic quantities, i.e. a hydrodynamical form of the Dirac equation. In order to describe pair production, a two-fluid model is needed. We are presently investigating the feasibility of such an approach.

In the meantime, we propose the following qualitative discussion: Hawking [5] has shown that in free collapse the particle-production mechanism due to the time variation of the metric is equivalent to a temperature bath with temperature $kT = (8\pi\ GM)^{-1}$. It seems natural to generalize this result in view of eqs. (73-74) to

$$kT = \frac{\dot{\lambda}-\dot{\nu}}{8\pi} \sim -\frac{\dot{\nu}}{4\pi} \quad . \tag{75}$$

Here T is the temperature measured by an observer at infinity. This is also in agreement with the result of Ruffini [28], that a charged particle freely falling into a Schwarzschild black hole emits radiation with an average energy $\sim M_G^{-1}$. The spectrum has all characteristics of an evaporation spectrum of a body with the Hawking temperature.

We now calculate the thermal energy for a certain particle degree of freedom due to the temperature (75). The WKB approximation for the solutions of eq. (67) predicts that

$$\int_{r_1}^{r_2} dr\, e^{\lambda/2} [e^{-\nu}\, \omega_n^2 - m^2 - \frac{\ell(\ell+1)}{r^2}]^{1/2} = (n + \tfrac{1}{2})\pi \tag{76}$$

with r_1 and r_2 being the zeroes of the integrand (the

effective momentum). For a particle with negligible rest mass m one finds that the energy density of states in the volume $V = \int r^2\, dr\, d\Omega$ becomes

$$dn = \frac{1}{2\pi^2}\, e^{\frac{\lambda-3\nu}{2}}\, \omega^2\, d\omega \cdot V \quad . \tag{77}$$

At the same time it is obvious from eqs. (21,22) that

$$T^o_o \sim \omega\, e^{-\frac{\nu+\lambda}{2}} \cdot V^{-1} \quad . \tag{78}$$

For a given temperature T the occupation probability of a state with energy ω is

$$W(\omega) = [e^{\omega/kT} \mp 1]^{-1} \tag{79}$$

for bosons and fermions, respectively. The total energy density absorbed in thermal excitation of particles is therefore

$$(T^o_o)_{therm} \sim s \int_0^\infty dn(\omega)\; T^o_o(\omega) W(\omega) = s\, \frac{15\pm1}{16}\, \frac{(\pi k T\, e^{-\nu/2})^4}{30\, \pi^2} \quad , \tag{80}$$

s being the number of degrees of freedom (spin, etc.). The effective temperature at a certain point close to the collapsing star or inside clearly is

$$T_{eff} = e^{-\nu/2}\, T \tag{81}$$

which is in accord with the usual transformation property of the temperature [29].

Inserting eq. (75) we end up with

$$(T^o_o)_{therm} \sim \frac{s}{480\pi^2} \left(\frac{d}{dt} e^{-\nu/2}\right)^4 \quad . \tag{82}$$

With a radius larger than the Schwarzschild radius, the total asymptotically visible thermal energy (of the vacuum, not of the particles that constitute the collapsing star) is

$$M_{therm} = \int r^2 dr\, d\Omega\, (T^o_o)_{therm} \geq \frac{4\pi}{3} (2GM)^3 (T^o_o)_{therm}$$

$$\sim s\, (45\pi)^{-1} (MG)^3 \left(\frac{d}{dt} e^{-\nu/2}\right)^4 \quad . \tag{83}$$

Now we have two choices. Either we assume the free collapse of the preceding section and the Hawking temperature. Then

$$M_{therm} \geq \frac{s\, e^{t/MG}}{45.256\pi\, MG} \quad . \tag{84}$$

The thermal mass must be smaller than the total available mass M, therefore we find that the collapse must stop or slow down before the time

$$t^* \sim 2\, MG\, \ln\, (190\, MG^{1/2}) \quad . \tag{85}$$

Taking the suns's mass $M = M_\theta$, the critical time scale becomes

$$t^*_\theta \sim 10^{-5} \text{ sec.} \cdot \ln (1.73 \times 10^{40}) \sim 10^{-3} \text{ sec.} \qquad (86)$$

We can therefore argue that an outside observer will see strong modifications in the collapse (e.g. slowing down, oscillations, etc.) on time-scales in the millisecond region. The time depends essentially linearly on the mass of the star.

On the other hand we can start from eq. (83), without an assumption on $\nu(t)$, by simply requiring that $M_{therm} \leq M$:

$$\frac{d}{dt} (e^{-\nu/2}) \leq [\frac{45\ \pi}{M^2\ G^3}]^{1/4} \equiv \eta\ M^{-1/2} \quad . \qquad (87)$$

This gives a stringent inequality for the metric:

$$e^{-\nu/2} \leq \eta\ M^{-1/2}\ t\ , \qquad (88)$$

which allows to calculate the total proper time of the collapse

$$\tau = \int d\tau = \int e^{\nu/2}\ dt \geq \frac{M^{1/2}}{\eta} \ln t \to \infty \quad . \qquad (89)$$

This proves that the collapse is slowed down sufficiently so that no event horizon will form even within a finite <u>proper</u> time of a co-moving observer!

A rigorous theory would couple $(T^o_o)_{therm}$ back into Einstein's equation (39), and study the modifications induced in the collaps. What can be expected is a new structure in the millisecond range which might give rise to experimentally detectable signals (e.g. pulses, etc.). Also, one has to establish the justification for the temperature bath model. This probably must be done by showing an equivalence between time-dependent and temperature Green's functions.

Whatever the outcome of such a more accurate theory will be, it seems to be justified to expect a complete destruction of the zero-particle vacuum in gravitational collapse. If the formation of an event horizon is circumvented, as we have made plausible, it would mean both a revolution of black hole physics and a salvation of Einstein's gravitation theory from causal structures that are not compatible with our ideas about the necessary reproducibility of physical experiments. And it would also mean that the singularities, proven to be unavoidable in the classical theory by Hawking and Penrose [28], will not harm us.

VII. NONLINEAR EFFECTIVE LAGRANGIANS

It was first shown by Heisenberg and Euler [30] that the deviations from the classical electromagnetic theory due to polarization effects in the quantum theory could, as a first approximation, be incorporated into an effective Lagrangian:

$$L_{EM} = \frac{E^2-B^2}{2} + \frac{2\alpha^2}{45\ m^4}[(E^2-B^2)^2 + 7(E\cdot B)^2] + O(\alpha^3) \ . \quad (90)$$

This expression reflects the fact that a linear classical equation acquires a nonlinear nature in the second quantized theory.

In the same spirit one could hope to describe the nonlinear effects of (virtual and real) particle production in strong gravitational fields (as long as the gradient remains limited) by an effective Lagrangian which contains nonlinear terms. In analogy to the electromagnetic model of Born and Infeld [31] we have chosen

$$(-g)^{-1/2} L_{eff} = \frac{R_o}{n} [(1-R/R_o)^{-n} - 1] \equiv f(R) . \qquad (91)$$

For $R/R_o \to 0$ this Lagrangian goes over into the usual Einsteinian one. R_o plays the role of an upper limit to the scalar curvature R. Nonlinear extensions of the theory of general relativity have been investigated by a number of authors [32-35], but so far only quadratic terms were considered which cannot be sufficient in very strong fields.

The field equations following from (91) are:

$$H_{ik} = -8\pi\, G\, T_{ik} \qquad (92)$$

with the divergenceless tensor:

$$H_{ik} = \frac{\partial^3 f}{\partial R^3} [R_{,i}\ R_{,k} - g_{ik}\ R_{,s}\ R_{,r}\ g^{rs}] + \frac{\partial^2 f}{\partial R^2}[R_{;k;i} - \delta_{ik} g^{rs}$$

$$R_{;r;s}] + \frac{\partial f}{\partial R} R_{ik} - \frac{1}{2}\, f(R)\, g_{ik} \quad . \qquad (93)$$

These equations differ from Einstein equations by the fact that they contain the fourth derivatives of the metric tensor g_{ik}, thus exhibiting certain nonlocal features in comparison with the usual theory. In the empty space situation ($H_{ik} = 0$) it means that additional boundary conditions must be imposed in order to obtain a unique solution. We have chosen the requirement [36] that the velocity of light, as seen by an asymptotic observer, should everywhere remain finite and below the vacuum value $\underline{1}$.

In Fig. 8 we present the metric functions e^{λ} and e^{ν} for a point mass GM = 1, n = -1/2 and for various values of the nonlinearity parameter $a = (\frac{2}{3}R_o)^{-1/2}$. In contrast to the Schwarzschild solution, no event horizon ($e^{\nu} = e^{\lambda} = 0$) is formed. At the origin both e^{ν} and e^{λ} tend to zero but with a finite ratio, so that the velocity of light remains finite. The asymptotic ($r \to \infty$) deviation from the Schwarzschild solution is proportional to exp(-r/a). Thus, if one takes $a \sim 10$ km (the radius of a large black hole), then no measurable influence could be seen in the classical tests of general relativity, which are all performed at $r/a \gg \underline{1}$.

The situation is somewhat different if solutions to an extended star are sought. The deviations from the Einsteinian theory start at distances comparable to $\Delta r = a$ above the stellar surface. Since they show up primarily in R(r) and not so much in the metric itself it is not quite clear, what precisely the limit on R_o is that can be set from present experiments. Also, it has not been clarified how the solutions for a collapsing star would go over into the vacuum solutions shown in Fig. 8. This work is still in progress. As a first result we present in Fig. 9 the metric of a star composed of incompressible fluid matter with mass GM = 1.6, for n = -1/2 and a = 1. We realize that the discontinuity of R at $r = r_o$ is smeared out in the nonlinear theory and that $R \neq 0$ already outside the matter distribution. This deviation at $r \sim r_o$ from the Schwarzschild solution could give one a possibility to decide about the magnitude of R_o experimentally (e.g. by measurements with the gradiometer described in ref. [37]).

VIII. OUTLOOK

We believe that the theory of the gravitational interaction is unsatisfactory in its present form because of the appearance of singularities in the space-time manifold and of event horizons which forbid a physics of reproducible results in a certain region of space-time. Up to now many efforts have been made to overcome this crisis of todays gravitational physics, and we think there is justified hope that very soon one will be able to treat the interaction of gravitation with any other quantum fields consistently and thus avoid the present pathologies in the theory. Whether this can be achieved on the basis of Einstein's Lagrangian for a gravitational theory or if a different approach (such as nonlinear effective Lagrangians, manifolds with torsion etc.) will be necessary, is not clear yet. We also hope that such a theory can provide a more transparent picture of the particle aspect of nature when the observer is in an arbitrary state of motion or in a gravitational field.

REFERENCES

1. C.W. Misner, K.S. Thorne, J.A. Wheeler, "Gravitation", W.H. Freeman and Co., San Francisco 1973.

2. See De Witt, De Witt, "Black Holes", Les Houches (1972), Gordon and Breach.

3. We refer to the preceding lectures I and II and the references cited therein.

4. L. Parker, Phys. Rev. 183 (1969) 1057 and D3 (1971) 346, Ya.B. Zel'dovich, A.A. Starobinsky, JETP 34 (1972) 1159, Zh. E.T.F. 61 (1971) 2161.

5. S. Hawking, Nature 248 (1974) 30.
S. Hawking, Comm.Math. Phys. 43 (1975) 199-200.

6. L. Parker, Probability Distribution of Particles Created By a Black Hole, University of Wisconsin preprint UWM - 4867 - 75 - 6 (1975).

7. W.G. Unruh, Phys. Rev. D14 (1976) 870.

8. U.M. Gerlach, Phys. Rev. D14 (1976) 3290.

9. S.A. Fulling, Phys. Rev. D7 (1972) 2850.

10. P. Candelas, D.J. Raine, J. Math. Phys. 17 (1976) 2101.

11. T. Damour, R. Ruffini, Phys. Rev. D14 (1976) 332.

12. J.S. Dowker, Z. Phys. A10 (1977) 115.

13. N. Deruelle, R. Ruffini, Phys. Lett. 52B, 4 (1974) 437-41.
T. Damour, N. Deruelle, R. Ruffini, Lett. al Nuovo Cimento 15 (1976) 227.

14. M. Soffel, B. Müller, W. Greiner, Particles in an Stationary Spherically Symmetric Gravitational Field, J. Phys. A10 (1977) in press.

15. D. Brill, J.A. Wheeler, Rev. Mod. Phys. 29 (1957) 465.

16. D.G. Boulware, Phys. Rev. D12 (1975) 350.

17. S. Persides, Comm. Math. Phys. 48 (1976) 165.

18. E. Schmutzer, Symmetrien und Erhaltungssätze der Physik, (Vieweg, Berlin 1972).

19. M.E. Rose, Relativistic Electron Theory (Wiley, New York 1961).

20. R. Adler, M. Bazin, M. Schiffer, "Introduction to General Relativity", (MacGraw Hill, 1965).

21. A. Papapetrou, Ann. Phys. 17 (1956) 214.

22. J. Rafelski, B. Müller, W. Greiner, "Bose Condensates and Gravitational Collapse", unpublished article, Inst. für Theoret. Physik der Universität Frankfurt am Main 1976.

23. D. Christodoulou, R. Ruffini, Phys. Rev. D4 (1971) 3552.

24. G.W. Gibbons, Comm. Math. Phys. 44 (1975) 245.

25. C. Möller, Vgl. Dansk. Wid. Selskab, Mat.-Fys. Medd. 39 (1975) 7.

26. R. Penrose, Phys. Rev. Lett. 14 (1965) 57.
S. Hawking, G. Ellis, Astrophys. J. 152 (1968) 25.

27. G.E. Lemaitre, Ann. Soc. Scient, Bruxelles, A53 (1933) 51.

28. R. Ruffini, Phys. Lett. 41B (1972) 3.

29. L. Landau, E.M. Lifshitz, Statistische Physik, § 27, Berlin 1966.

30. W. Heisenberg, R. Euler, Z. Physik 98 (1936) 714.

31. M. Born, L. Infeld, Proc. Roy. Soc. A144 (1934) 425.
J. Rafelski, L.P. Fulcher, and W. Greiner, Phys. Rev. Lett. 27 (1971) 958.
J. Rafelski, L.P. Fulcher, and W. Greiner, Nuovo Cim. B7 (1972) 137.

32. C. Lanczos, Z. Physik 73 (1932) 147.

33. C. Gregory, Phys. Rev. 72 (1947) 71.

34. E. Pechlaner, R. Sexl, Comm. Math. Phys. 2 (1966) 165.

35. F.C. Michel, Ann. Phys. 76 (1973) 281.

36. A. Müller, Diploma thesis, Frankfurt 1977.

37. See ref. 1, p. 401.

FIGURE CAPTIONS

Fig. 1: The energy eigenvalues of the Dirac equation in the field of an extended gravitating source. The energies are shown as a function of the source radius r_o (in units of the Schwarzschild radius r_s).

Fig.2a: The wave function for the first (solid lines) and the second (dashed lines) resonance of the Dirac equation in the Schwarzschild metric. Parameters: $M_G = 1$, $\kappa = 4$.

b: The large component $g(r^*)$ of a typical wave function off resonance. The extremum at $r^* \approx 6$ occurs at the innermost of the two classical turning points, i.e. $E = V_{eff}(r)$. At this point df/dr^* and dg/dr^* can vanish simultaneously to produce the extraordinary bump. In resonance wave functions the bound and the continuum parts join smoothly in phase.

Fig. 3a: The effective potential for the Schwarzschild field ($M_G = 2$, $\kappa = 50$).

b: The effective potential for the Reissner-Nordstrøm field ($Q_G = 1000$, $M_G = 2$, $\kappa = 4$). Pair creation occurs for $|\Lambda| \geq m_e c^2$. The center is negatively charged and the decay of the vacuum happens via electron emission.

c: If the center is positively charged ($Q_G = -1000$), positrons are emitted for $|\Lambda| \geq m_c c^2$. The horizontal lines indicate the states of the upper continuum.

Fig. 4: Shows the space-time for a grav. collapse (with $\lambda_o = 1$, $R_b = 0.9$, $2M_G = 0.729$) in comoving coordinates. The invariant singularity S_e is given by (52) which for our choice of λ_o means $\tau_s = \frac{\pi}{2}$ for the inner region. S_e marks the boundary of the unphysical part of space-time. The event horizon, given by $t = \infty$ is clearly to be seen. Thus the dashed region beyond $t = \infty$ indicates the measurable world of an asymptotic observer.

Fig. 5: Same as Fig. 1, but now lines corresponding to const. values of asymptotic coord. r and t are depicted.

Fig. 6: The asymptotic time coord. is plotted against values of comoving coord.. The event horizon H_e is given by the projection of $t = \infty$ into the (ρ, τ)-plane.

Fig. 7: The metrical components e^{ν} and e^{λ} are plotted in asymptotic coordinates for different values of t. e^{λ} remains finite for $t \to \infty$. $r_b(t)$ markes the free falling edge of the dust cloud and converges against the event horizon ($r = 2M_G$) as $t \to \infty$.

Fig. 8: The metric components $g_{oo} = e^{\nu}$ and $-g_{rr} = e^{\lambda}$ as well as the curvature scalar R are shown for various values of the parameter $a = (\frac{2}{3}R_o)^{-1/2}$. The Schwarzschild solution is exhibited for comparison.

Fig. 9: The metric components $g_{oo} = e^{\nu}$ and $-g_{rr} = e^{\lambda}$ as well as the curvature scalar R and the pressure p for a star of mass GM = 1.6 and radius $r_o = 1.875 \cdot r_s = 1.875 \times (2GM)$. The Einstein solution is exhibited for comparison.

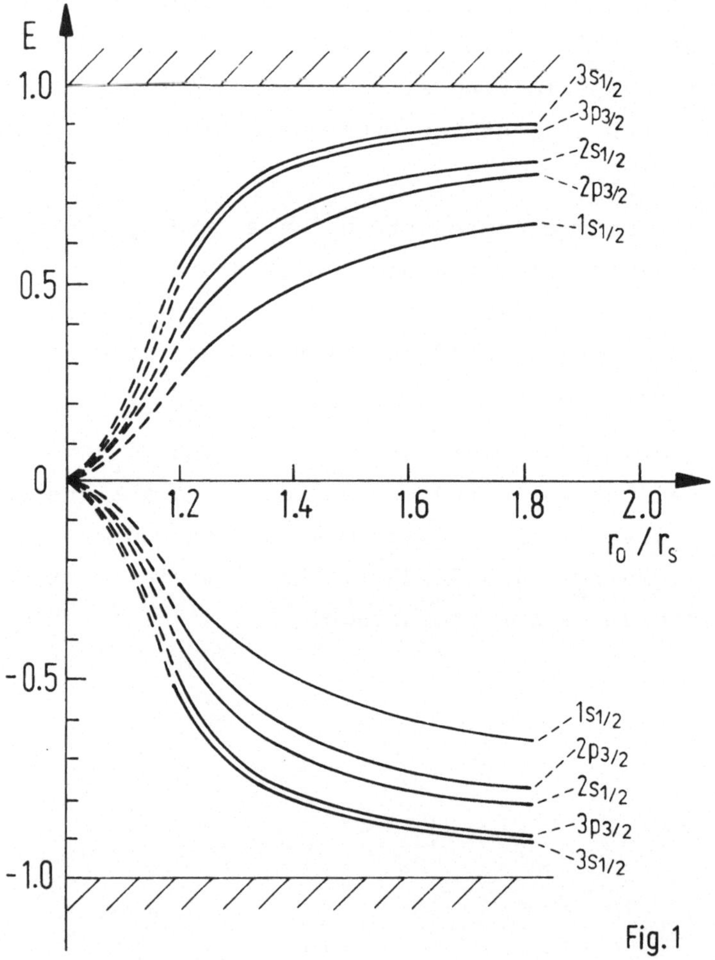
E
1.0
0.5
0
-0.5
-1.0
1.2
1.4
1.6
1.8
2.0
r0 / rs
3s1/2
3p3/2
2s1/2
2p3/2
1s1/2
1s1/2
2p3/2
2s1/2
3p3/2
3s1/2

Fig. 1

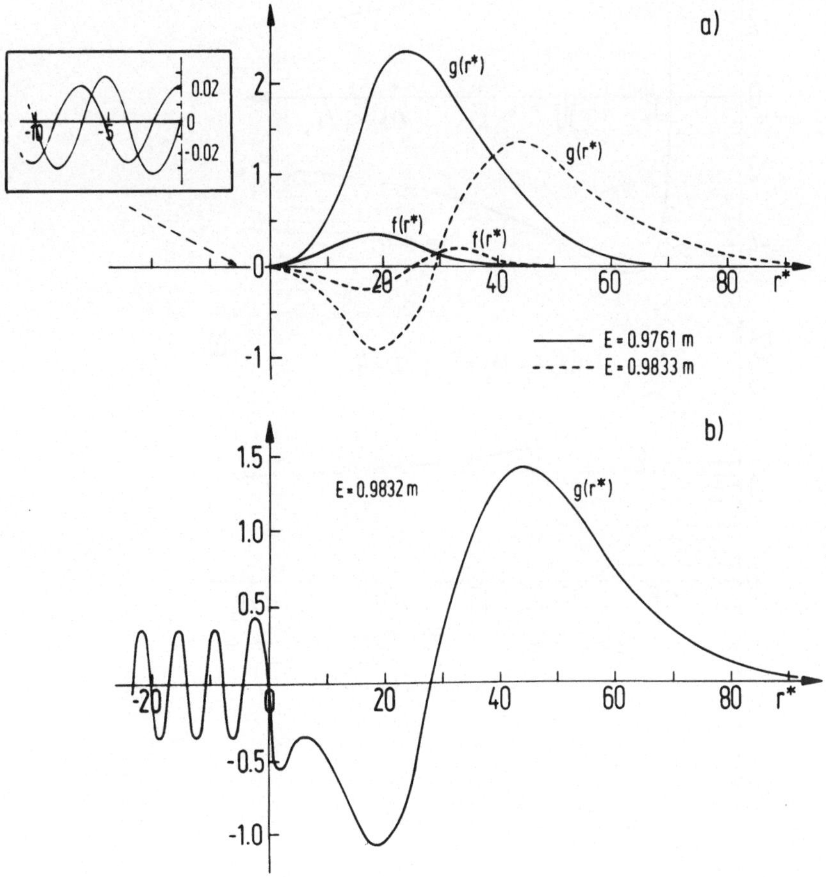
a)
g(r*)
g(r*)
f(r*)
f(r*)
0.02
0
-0.02
-10
-5
2
1
0
-1
20
40
60
80
r*
E = 0.9761 m
E = 0.9833 m
b)
1.5
1.0
0.5
-0.5
-1.0
E = 0.9832 m
g(r*)
-20
0
20
40
60
80
r*

Fig. 2

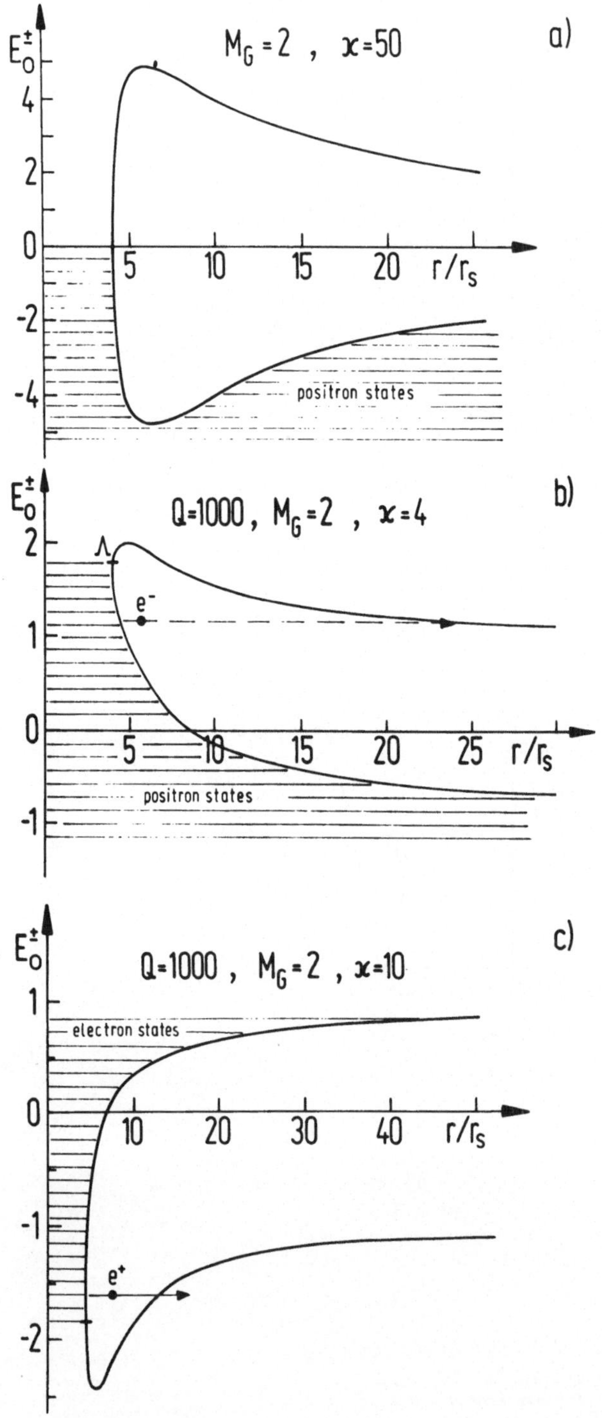
a)
$M_G = 2$, $\varkappa = 50$
$E_0^{\pm}$
4
2
0
-2
-4
5
10
15
20
r/r_s
positron states
b)
$Q = 1000$, $M_G = 2$, $\varkappa = 4$
$E_0^{\pm}$
2
1
0
-1
Λ
e^-
5
10
15
20
25
r/r_s
positron states
c)
$Q = 1000$, $M_G = 2$, $\varkappa = 10$
$E_0^{\pm}$
1
0
-1
-2
electron states
10
20
30
40
r/r_s
e^+

Fig. 3

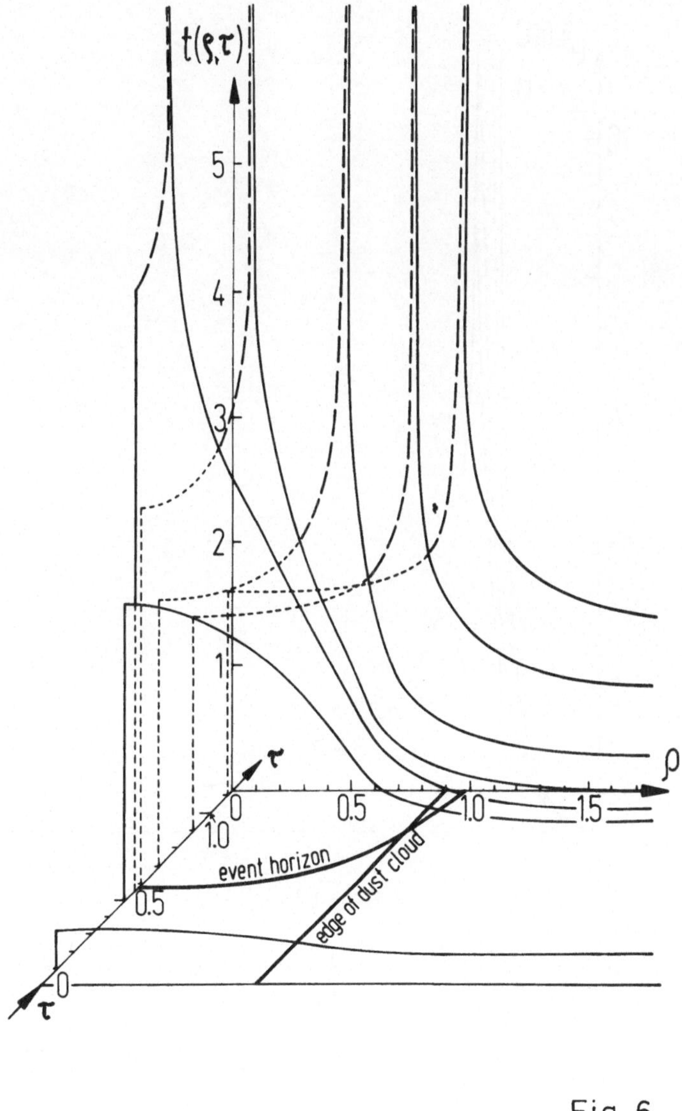
t(ρ,τ)
5
4
3
2
1
τ
ρ
0
0.5
1.0
1.5
1.0
event horizon
0.5
edge of dust cloud
0
τ

Fig. 6

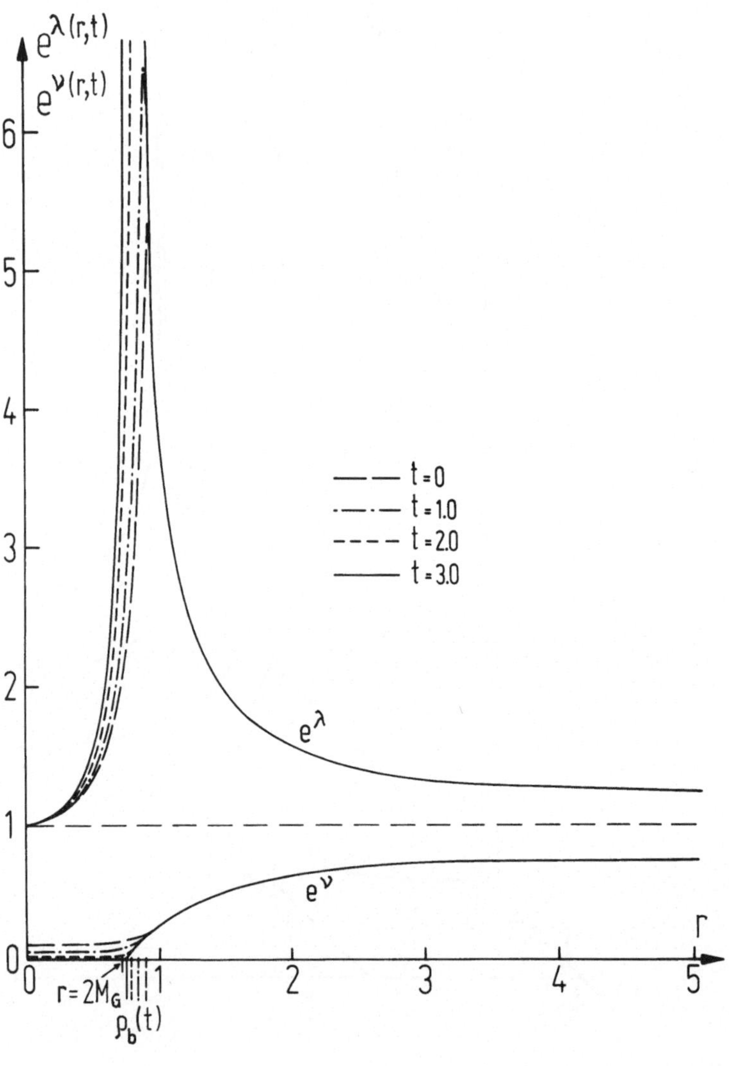

$e^{\lambda(r,t)}$
$e^{\nu(r,t)}$
6
5
4
3
2
1
0
t = 0
t = 1.0
t = 2.0
t = 3.0
e^{λ}
e^{ν}
r
0
1
2
3
4
5
$r = 2M_G$
$\rho_b(t)$

Fig.7

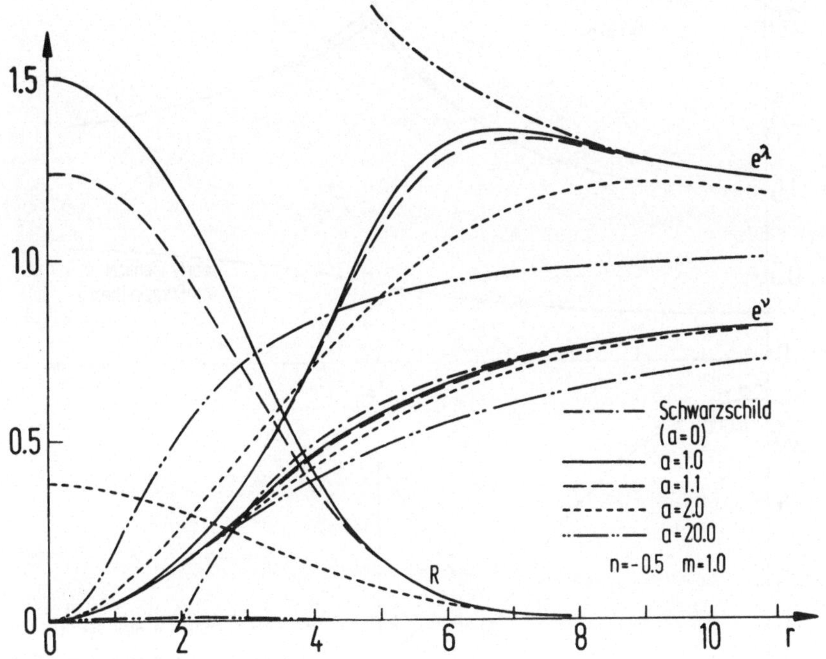
1.5
1.0
0.5
0
0
2
4
6
8
10
r
e^λ
e^ν
R
Schwarzschild
(a=0)
a=1.0
a=1.1
a=2.0
a=20.0
n=-0.5 m=1.0

Fig. 8

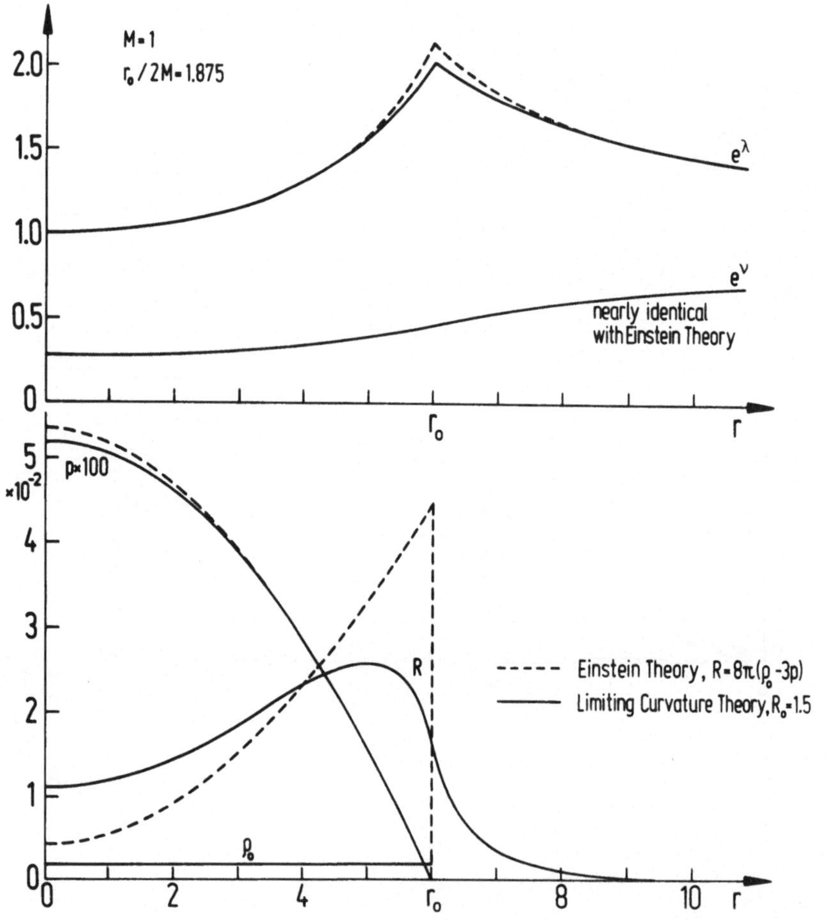
M=1
r₀/2M=1.875
2.0
1.5
1.0
0.5
0
e^λ
e^ν
nearly identical with Einstein Theory
r₀
r
5
4
3
2
1
0
×10⁻²
p×100
R
ρ₀
0
2
4
r₀
8
10
r
Einstein Theory, R=8π(ρ₀-3p)
Limiting Curvature Theory, R₀=1.5

Fig. 9

Acta Physica Austriaca, Suppl. XVIII, 385–428 (1977)

SOLITONS AND INSTANTONS IN QUANTUM FIELD THEORY+

by

J.L. GERVAIS
Laboratoire de Physique Théorique
de l'Ecole Normale Supérieure
Paris, France

INTRODUCTION

Those lectures are concerned with recent interesting developments of the last few years which took place as it was gradually realized that the existence of non-trivial classical solutions of field theories has rather striking implications at the quantum level. The common exciting feature of these implications is that they are beyond reach of the usual perturbative approach based on Feynman rules and thus represent a major break through in our understanding of quantum field theory.

+Lecture given at XVI.Internationale Universitätswochen für Kernphysik,Schladming,Austria,February 24-March 5, 1977.

The importance of classical solutions was revealed by a few pioneering works. Nielsen and Olesen [1] pointed out that in the Higgs model there exist classical solutions of the vertex type which behave very much like relativistic strings and thus resemble the hadron picture which emerges from dual models. Dashen, Hasslacher and Neveu [2] extended WKB methods to field theory and showed that localized classical solutions in Minkowski space correspond to the existence of new quantum states which could not be described by standard perturbation. Thus, field theories may have a richer structure than expected and in view of ref. [1], this raises the hope that the hadrons, with their complicated non-perturbative structure, can be fitted into the scheme of local quantum field theory. These new objects generally called solitons will be the subject of the first part of my lectures.

More recently, a different direction was advocated by Polyakov [3], [4] who pointed out the importance of classical solutions of finite action in the Euclidean space obtained after continuation to purely imaginary time. Such solutions, called instantons, only exist in theories with degenerate vacua where standard perturbation exhibits spontaneous symmetry breaking. They are the signal of tunneling between these different vacua so that symmetry is restored and long-range correlation may be destroyed.

As a result, Goldstone bosons may be avoided in such a way that the so-called η problem can be solved. The long-range correlation may disappear if the instantons are able to screen the long-range forces. In this case quarks become confined in the sense that the criterion of Wilson is satisfied. Here again the earlier picture based on standard perturbation turn out to be misleading and the new picture which emerges is much closer to the physical reality. This will be the subject of the second part of my lectures.

The common feature of both approaches is to be based on semi-classical approximation, i.e. small $\hbar$. If γ is a typical coupling constant and ϕ is a typical field one always find that $\hbar \propto \gamma$ and that the classical solution ϕ_{cl} is of order $\gamma^{-1/2}$. This is why the results one gets are non-perturbative in γ.

Standard perturbation is in general established by letting $\phi = \phi_o + \bar{\phi}$ where ϕ_o is a constant which is a minimum of the potential; ϕ_o is zero or order $\gamma^{-1/2}$ and $\bar{\phi}$ is considered O(1) in the way Feynman perturbation is built up. The effect of non-trivial classical solution is studied by letting $\phi = \phi_{cl} + \tilde{\phi}$. If $\tilde{\phi}$ is considered O(1) this leads again to an expansion in γ where the dominant term is the contribution of ϕ_{cl}. The non-perturbative effect with respect to the usual perturbation expansion $\phi = \phi_o + \bar{\phi}$ is under control as it is entirely contained in ϕ_{cl}. Because $\hbar \propto \gamma$, the perturbation expansion in γ one obtains is simply an expansion in $\hbar$ i.e. an estimation of quantum corrections to classical results. The fact that ϕ_{cl} is a classical solution, that is a minimum of the action, ensures that no term linear in $\tilde{\phi}$ will appear. This is necessary since otherwise those terms would give tadpole terms of order $\gamma^{-1/2}$ into the perturbation expansion of $\tilde{\phi}$ which would lead to corrections of the same order as the classical term.

Those semi-classical methods are most easily developed using Feynman path integrals. In this case we will be simply making change of variables in the path integral and the general idea is to consider the contribution of the fluctuation around non-trivial minima of the action.

For solitons we want the soliton energy to be finite compared to vacuum energy. This then must also be true at the classical level. Thus the classical soliton solution should, at large space distance, tend to one of the possible vacua of the theory that is to a minimum of the potential. The one-soliton state in its rest frame corresponds to a static solution and will thus be a minimum of the energy. Unless one imposes some additional condition, a minimum of the energy can only be a classical vacuum, which is a constant field. To obtain non-trivial solution one can either look for a minimum with a non-zero value of some conserved charge (non-topological soliton [5]) or in the case of degenerate minima impose boundary conditions which cannot be continuously deformed into the vacuum state boundary condition. If this is possible there always exists a quantity called topological charge which is conserved irrespective of the equations of motion because it can only take discrete values classically and one is looking at a minimum of the energy with non-zero value of this topological charge. The instanton being a solution in Euclidean field theory is a minimum of the action with all kinetic terms positive. Hence an instanton solution in d dimensional space time is also a time-dependent soliton solution in d + 1 dimensions and the classification of possible solutions can be done simultaneously for both cases. It is based on homotopy theory and we will not repeat it here [6].

The last general point is that these classical solutions always involve arbitrary parameters so that the change of field $\phi = \phi_{cl} + \tilde{\phi}$ is not well defined. The solution to this problem is to consider these parameters as dynamical variables which therefore should be determined from ϕ itself [7], [8]. Those dynamical variables are called collective coordinates and have natural physical interpretation as soliton position and momenta or Euclidean space time regions where tunneling occurs. I will mainly concentrate on this aspect of the semi-classical method mostly reviewing works performed over the last two years in collaboration with A. Jevicki and B. Sakita. Complementary reviews have been written by R. Rajaraman [9], S. Coleman [6] and R. Jackiw [10]. For additional references see ref. [11].

I - SOLITONS

I-A. One soliton in two dimensions

We consider the Lagrangian

$$L = -\frac{1}{2}\partial_\mu\phi\partial_\mu\phi - V(\phi) \tag{1.1}$$

with the potential of the form

$$V(\phi) = \frac{1}{g^2} U(g\phi) \tag{1.2}$$

which has a classical solitary wave solution

$$\phi_c(x,t) = \phi_o\left(\frac{x-vt-x_o}{\sqrt{1-v^2}}\right), \quad -\frac{\partial^2}{\partial x^2}\phi_o(x) + \frac{\delta V}{\delta\phi_o(x)} = 0$$

with finite energy. At the classical level this solution can be interpreted as a particle with the mass $M_o = \int dx\ (\partial\phi_o/\partial x)^2$, since the energy and momentum operators (' means $\partial/\partial x$)

$$H = \int dx\ [\frac{1}{2}\dot{\phi}^2 + \frac{1}{2}\phi'^2 + V(\phi)]$$

$$P = \int dx\ \dot{\phi}\phi'$$

give

$$H(\phi_c) = \sqrt{p^2 + M_o^2}\ , \qquad p = -P(\phi_c)\ .$$

We will present a method for quantization of such classical solutions. In the case of weak coupling, when the soliton mass is large, we developed a systematic perturbation expansion for the one-soliton sector [7], [12]. With the corresponding Feynman rules, one can make perturbative calculations of transition matrix elements for the initial and final states containing one soliton and an arbitrary number of mesons associated to the field ϕ.

The transition amplitude between initial and final states described by the wave functionals $\psi_i[\phi]$ and $\psi_f[\phi]$ is given by the following path integral:

$$S_{fi} = \int D\Pi D\phi\, \psi_f^*[\phi(x,t_f)]\psi_i[\phi(x,t_i)]\exp\{i\int dxdt[\Pi\dot{\phi} - H(\Pi,\phi)]\} \tag{1.3}$$

If, in order to develop a perturbation expansion for the one-soliton sector, one simply expands around the classical solution ϕ_o, as in the case of spontaneous symmetry breaking, one finds divergences connected with the translation invariance of our theory. Namely, the propagator of this perturbation expansion would be the inverse of the following differential operator:

$$-\frac{d^2}{dt^2} + \Omega^2 \equiv -\frac{d^2}{dt^2} + \frac{d^2}{dx^2} + V''(\phi_o) \qquad (1.4)$$

where

$$V''(\phi_o) = \left.\frac{\delta^2 V}{\delta\phi^2}\right|_{\phi = \phi_o} .$$

Taking the space derivative of the field equation satisfied by ϕ_o one immediately sees that ϕ_o' is eigenstate of Ω^2 with eigenvalue zero. Thus the propagator is ill defined since the differential operator (1.4) has a zero eigenvalue.

To solve this difficulty and develop a consistent perturbation expansion for the one-soliton sector, we will first separate the center-of-mass motion, extracting the total momentum and the center-of-mass coordinate. We insert the following identities into the path integral expression for the S-matrix element:

$$\int \prod_t \{Dp(t)\,\delta(p(t) + P)\} = 1 \quad , \qquad P \equiv \int dx \Pi \phi'$$

$$\int \prod_t \{DX(t)\,\delta(Q[\phi(x+X), \Pi(x+X)])\,\frac{\partial Q}{\partial X}\} = 1 . \qquad (1.5)$$

The first identity which we call the constraint, serves to identify the variable p(t) with the total momentum of the system while the second is the gauge condition associated with the constraint. Q can be arbitrary. We notice that $\partial Q/\partial X$ is given by Poisson bracket:

$$\frac{\partial Q}{\partial X} = \{Q,P\} \equiv \int dx \left[\frac{\partial Q}{\partial \phi}\frac{\partial P}{\partial \Pi} - \frac{\partial Q}{\partial \Pi}\frac{\partial P}{\partial \phi}\right] . \tag{1.6}$$

Next, we make a change of variables

$$\phi(x,t) = \tilde{\phi}(x-X(t),t) \equiv \tilde{\phi}(\rho,t)$$

$$\Pi(x,t) = \tilde{\Pi}(x-X(t),t) \equiv \tilde{\Pi}(\rho,t)$$

$$\rho = x - X(t) \tag{1.7}$$

so that, using $\dot{\phi} = \dot{\tilde{\phi}} - \dot{X}\tilde{\phi}'$ and the constraint, we get

$$\int dx\{\Pi\dot{\phi}-H(\Pi,\phi)\} = p(t)\dot{X}(t) + \int d\rho\,(\tilde{\Pi}\dot{\tilde{\phi}}-H(\tilde{\Pi},\tilde{\phi}))$$

$$\psi_{i,f}[\phi] = \exp(ip_{i,f}X(t_{i,f}))\psi_{i,f}[\tilde{\phi}] . \tag{1.8}$$

From the first expression, one sees that X is the variable conjugate to p, i.e. the C.M. position and (1.7) is a change to the moving frame attached to this center of mass. Thus we have explicitly exhibited the total momentum and center of mass position associated with a given field configuration. If it corresponds to quantum fluctuation around the one-soliton classical solution, X and p will be automatically the position and momentum of this soliton.

As X appears only in the term $p\dot{X}$, we can immediately integrate over X and p which leads to

$$S_{fi} = \delta(p_f - p_i)\int D\tilde{\Pi} D\tilde{\phi}\psi_f^*[\tilde{\phi}]\psi_i[\tilde{\phi}]\delta(p+P)\delta(Q)$$

$$\times \{Q,P\} \exp\{i\int d\rho\, dt(\tilde{\Pi}\dot{\tilde{\phi}} - H(\tilde{\Pi},\tilde{\phi}))\} \tag{1.9}$$

where $p = p_i = p_f$. The stationary point of the action with constraints is given by the following variational equation

$$\delta\{\int dt[\int d\rho\,(\tilde{\Pi}\dot{\tilde{\phi}} - H) + \alpha(t)(p+P)]\} = 0$$

where α is a Lagrange multiplier. One obtains, for the lowest energy stationary point ($\dot{\phi}_c = 0$), exactly the soliton solution

$$\phi_c = \phi_o\left(\sqrt{1+\frac{p^2}{M_o^2}}(\rho - a)\right) \qquad \Pi_c = \frac{-p}{\sqrt{p^2+M_o^2}}\phi_c' \tag{1.10}$$

where ϕ_o is solution of

$$-\phi_o'' + \frac{\delta^2 V}{\delta\phi_o^2} = 0 \tag{1.11}$$

and the constant a is fixed by the Q condition. The corresponding classical energy is found to be

$$E_c = \sqrt{p^2 + M_o^2} \quad . \tag{1.12}$$

At this point, we observe that, due to the property (1.2) of our potential, ϕ_c is of the order of $1/g$; accordingly, M_o is of the order of $1/g^2$. We can develop the perturbation expansion in g around the classical solution

$$\tilde{\phi} = \phi_c(\rho) + \chi(\rho,t) \qquad \tilde{\Pi} = \Pi_c(\rho) + \Theta(\rho,t) \ . \tag{1.13}$$

Here χ and Θ are considered order zero and represent small quantum fluctuations around the classical solution. In the case when the initial and final states contain only one soliton, the shift (1.13) gives, in the first approximation, the relativistic form for the soliton energy and we may develop a perturbation theory for the soliton energy.

On the other hand, if the momentum is considered O(1), one can alternatively shift by the zero momentum classical solution $\phi_o(\rho)$, because our method only requires that the function used in the shift be a classical solution to leading order, so that, in the final expression, one does not get zeroth order terms linear in χ. Since this last case is much simpler, we shall only discuss the corresponding perturbation expansion briefly in what follows. (See ref.[12] for details.)

Next, one has to choose the gauge condition. Although an arbitrary choice leads to the consistent perturbation expansion free of infrared divergences, we prefer to choose a linear gauge condition

$$Q[\phi(x+X(t),t)] \equiv \int dx \ f(x)\phi(x+X(t),t)$$

$$\frac{\partial Q}{\partial X} = \int dx\ f(x)\ \phi'(x+X,t) \tag{1.14}$$

in order to eliminate the zero-energy mode in a simplest possible way. Here f is still an arbitrary function and, identifying it later with the zero-frequency eigenfunction, we will completely eliminate the zero-energy mode from our functional integral. Now, before making the shift $\tilde{\phi} = \phi_o + \chi$, we linearize the constraint which is quadratic in fields, by making the following change of variables

$$\tilde{\Pi}(\rho,t) = -f(\rho)\ \frac{p+\int \bar{\Pi}(\rho,t)[\tilde{\phi}'-f_c]d\rho}{\int f\tilde{\phi}'d\rho} + \bar{\Pi}(\rho,t)\ . \tag{1.15}$$

Then the constraint becomes:

$$\delta(p + \int \tilde{\Pi}\tilde{\phi}'d\rho) = \delta(c\int f(\rho)\bar{\Pi}(\rho,t)d\rho)\ .$$

Computing the Jacobian of this transformation, we get

$$\det\left(\frac{\delta\tilde{\Pi}}{\delta\bar{\Pi}}\right) = \prod_t \left(\int d\rho\ f\tilde{\phi}'\right)^{-1}\ ;$$

so, it exactly cancels out the $\partial Q/\partial X$ which is given by (1.14). Now the Hamiltonian becomes more complicated:

$$H \equiv \int d\rho H = \frac{(p+\int\bar{\Pi}(\tilde{\phi}'-f_c))^2}{2(\int f\phi'd\rho)^2} + \int d\rho\left[\frac{\bar{\Pi}^2}{2} + \frac{\tilde{\phi}'^2}{2} + V(\tilde{\phi})\right]. \tag{1.16}$$

We used the normalization $\int d\rho f(\rho)^2 = 1$. The transition amplitude is now of the form

$$S_{fi} = \delta(p_f - p_i) \int D\bar{\Pi} D\tilde{\phi} \psi_f^*[\tilde{\phi}] \psi_i[\tilde{\phi}]$$

$$\times \delta(\int f\tilde{\phi} d\rho)\, \delta(\int f\bar{\Pi} d\rho)\ \exp\{i\int dt[\int d\rho \bar{\Pi}\dot{\tilde{\phi}} - H]\} \tag{1.17}$$

and since both the gauge condition and the constraint are linear in fields, one can easily develop a perturbation expansion.

We now continue our discussion by making the shift

$$\tilde{\phi}(\rho,t) = \phi_o(\rho) + \chi(\rho,t) \qquad \bar{\Pi}(\rho,t) = \Theta(\rho,t) \tag{1.18}$$

with the choice of f,c,

$$f = \phi_o'/\sqrt{M_o} \equiv \psi_o \quad , \qquad c = \sqrt{M_o} \quad . \tag{1.19}$$

The final result reads

$$H = M_o + \frac{(p + \int \Theta\chi' d\rho)^2}{2M_o(1+\xi/M_o)^2} + \int d\rho [\frac{\Theta^2}{2} + \frac{\chi'^2}{2} + V - V(\phi_o) - \frac{\delta V}{\delta\phi}\Big|_{\phi_o} \chi] + \Delta V \tag{1.20}$$

with

$$\xi = \int d\rho\, \phi_o'(\rho) \chi'(\rho,t) \tag{1.21}$$

$$\Delta V = \frac{1}{8}[-3 \frac{(\psi_o', \psi_o')}{(\tilde{\phi}', \psi_o)^2} + 2 \frac{(\psi_o', \tilde{\phi}'')}{(\psi_o, \tilde{\phi}')^3} + \frac{(\psi_o', \phi')^2}{(\psi_o, \phi')^4} +$$

$$+ \sum_{n,m\neq o} \frac{|(\psi_n, \psi_m')|^2}{(\psi_o, \phi')}] \tag{1.22}$$

$$\Omega^2 \psi_n = \omega_n^2 \psi_n \quad \text{(see below)}$$

where we noted in general for two arbitrary functions h_1, h_2,

$$(h_1, h_2) = \int dx \, h_1(x) \, h_2(x) \; ;$$

$\Delta V(\chi)$ is an additional potential which starts contributing at the two-loop level. It is not obtained if one performs the changes of variables (1.7), (1.15), (1.18) into the action as we have done here.

We have presented the procedure in these three steps because we want to illustrate the general method on this particular case. Here, on the other hand, one can directly obtain the result by making the change of field [12]

$$\Pi = \Pi_o + \Theta(x-X,t)$$

$$\phi = \phi_o(x-X) + \chi(x-X,t)$$

$$\Pi_o \equiv - \frac{\phi_o'(x-X)}{M_o+\xi} \, (p + \int \Theta\chi' dx) \qquad (1.23)$$

with the condition

$$o = \int d\rho \, \Theta \phi_o' = \int d\rho \, \chi \phi_o' \; . \qquad (1.24)$$

The need of an additional term was first pointed out by Tomboulis [13] who performed the canonical transformation (1.23) in the operator formalism. In the functional method ΔV also arises if the change of variables (1.23) is done with enough care [14]. The essential point is that in a functional integral, the derivatives which appear in the exponential of the action are not derivatives in the usual sense. Consider for instance a transition probability in quantum mechanics

$$\int Dq(t)\ \exp\{i\int dt\ (\tfrac{1}{2}\dot{q}^2 - V(q))\}$$

the real meaning of this integral is that we should take time intervals ε and write ($q_\ell = q(\ell\varepsilon)$)

$$\lim_{\varepsilon\to 0} \int \prod_\ell dq_\ell\ \exp\{i\varepsilon \sum_\ell (\frac{q_{\ell+1}-q_\ell}{\varepsilon})^2 - V(q_\ell)\} \quad .$$

The q_ℓ's are independent variables. However the first term of the exponent has a finite limit only for $q_{\ell+1} - q_\ell = o(\sqrt{\varepsilon})$. Hence, if we perform a canonical transformation $Q = F(q)$, we must expand F up to second order derivative when we compute $Q_{\ell+1} - Q_\ell$ in terms of $q_{\ell+1} - q_\ell$. The additional term $\frac{1}{2}\, d^2F/dq_\ell^2 (q_{\ell+1}-q_\ell)^2$ being of order ε gives a correction to the potential. This is in contrast with the naive computation where one would just write

$$\dot{Q} = \frac{dF}{dq}\,\dot{q} \; .$$

The Feynman rules of perturbation expansion can now be obtained easily [12]. The propagator is determined from the quadratic part of the Hamiltonian by expanding in terms of eigenfunctions of the following differential equations

$$\Omega^2\psi_n \equiv \left(-\frac{d^2}{d\rho^2} + V''(\phi_o)\right)\psi_n = \omega_n^2\,\psi_n \quad . \qquad (1.25)$$

The zero-energy eigenfunction is given by ψ_o. We choose f to be precisely given by ψ_o in such a way that $\omega_o = 0$ mode disappears from the eigenfunction expansion of Θ and χ because of the δ condition in (1.17).

Since we use first-order formalism, this perturbation expansion involves three different propagators $\langle 0|T(\chi,\chi)|0\rangle$, $\langle 0|T(\chi,\Theta)|0\rangle$, $\langle 0|T(\Theta,\Theta)|0\rangle$. The Hamiltonian (1.20) contains products of χ and Θ at the same point and therefore there are ordering problems if we want to write H as an operator as was done by Tomboulis [13]. In the functional formalism this ordering problem appears also in practice because the perturbation expansion will contain the mixed propagator $\langle 0|T(\chi,\Theta)|0\rangle$ with zero time separation which is ambiguous. This ambiguity is seen if one looks more carefully into the meaning of (1.17) again because derivative terms in the action are to be handled with care. Indeed, if we take discrete time intervals ε, it is not clear how we should interpret the term $\int\Theta\dot{\chi}dt$ of (1.17). Different choices lead to different expressions for ΔV. The expression (1.22) corresponds to the so-called mid-point definition namely we choose field variables $\chi(x,t_{2\ell})$, $\Theta(x,t_{2\ell+1})$; $t_{\ell+1}-t_\ell = \varepsilon$ and write

$$\int\Theta\dot{\chi}dt \equiv \sum_\ell \int d_3x\ \Theta(x,t_{2\ell+1})[\chi(x,t_{2\ell+2})-\chi(x,t_{2\ell})] \ .$$

In operator formalism ΔV is the term associated with

Weyl's ordering for the expression (1.20) of H. In perturbation theory this means that the mixed propagator $\langle 0|T(\chi,\theta)|0\rangle$ for zero-time separation is taken to be zero, i.e. all closed loops of mixed propagators are to be dropped. This choice is different from the one of ref.[13], however, it is more suitable for deriving Feynman perturbation theory explicitly. See ref.[14] for details.

In this perturbation expansion Lorentz invariance is not manifest, but one can show that higher order corrections in coupling constant, sum up to restore Lorentz invariance at least at the tree level [12]. Since $M_o = o(g^{-2})$ while p is considered O(1) we get the non relativistic expansion for the energy. At the lowest order we obtain, for instance,

$$E = M_o + \frac{p^2}{2M_o} \simeq \sqrt{p^2 + M_o^2} .$$

The renormalization of the one-soliton sector can be carried out in a straightforward manner, by adding the mass renormalization counter term

$$H_{\delta m} \equiv - \frac{1}{2} \delta m^2 \int d\rho \, (\phi_o + \chi)^2$$

where δm^2 is the mass counter term computed in the zero-soliton sector.

In the example of ϕ^4 theory, the first two one-loop contributions to the soliton energy have been computed [12]; after renormalization one obtains the finite answer

$$E \simeq M_o+\Delta M + \frac{p^2}{2M_o} - \frac{p^2}{2M_o^2}\Delta M \simeq (p^2+(M_o+\Delta M)^2)^{1/2}$$

where ΔM is the first quantum correction to the soliton mass calculated originally by Dashen, Hasslacher and Neveu [2]. For the diagrammatic representation of the Feynman rules, and also for the specific computations the reader is referred to ref.[12].

With this systematic perturbation expansion, one can make perturbative calculations of other quantities in the one-soliton sector besides the soliton energy and mass corrections. As an example, we consider the $\hat{\phi}$ field matrix elements investigated by Goldstone and Jackiw [15]. Their assumption about the leading order of $\langle p',\{k'\}|\hat{\phi}|p,\{k\}\rangle$ can be easily seen to be true [16]. Let us compute the leading term in the simplest matrix element $\langle p'|\hat{\phi}(x,0)|p\rangle$. From the path integral expression for this matrix element, we obtain the corresponding operator form

$$\langle p'|\hat{\phi}(x,0)|p\rangle = \langle p'|\phi_o(x-\hat{X})|p\rangle + \langle p'|\hat{\chi}(x-\hat{X},0)|p\rangle . \qquad (1.26)$$

Inserting identity, we obtain the leading term to be

$$\langle p'|\phi_o(x-\hat{X})|p\rangle = \int dy\, e^{iy(p-p')}\phi_o(x-y) \qquad (1.27)$$

which is the result of ref.[15]. Higher order corrections can now be systematically computed, and similarly, one has a complete perturbation expansion for the arbitrary $\hat{\phi}$-field matrix element in the one-soliton sector [12].

The two-loop correction to soliton mass has been computed [17] in Sine-Gordon theory and found to agree with the conjecture of Dashen, Hasslacher and Neveu [18] that WKB is exact for the ratio between soliton mass and fundamental field mass.

Note that expression (1.17) a priori looks like a highly non-renormalizable Hamiltonian since it involves vertices with an arbitrary number of legs. It is remarkable that finite results are in fact obtained to any order by just using the same counter term as in the usual sector. At the two-loop level this already involves remarkable cancellation among highly divergent integrals.

Let us now turn to different possible choices for Q. First of all, we can choose Q in such a way that the zeroth mode disappears if we perform the shift (1.13) in order to avoid tadpole terms completely. We have to do that if the momentum is of order 1/g, i.e. if the velocity of the soliton is finite for small g. The function f and constant c are now, instead of (1.19)

$$f = \frac{\phi_c'}{\sqrt{p^2+M_o^2}} \qquad c = \sqrt{p^2 + M_o^2} \tag{1.28}$$

where ϕ_c is given by (1.10).

One obtains

$$H = \sqrt{p^2+M_o^2} + \frac{1}{2\sqrt{p^2+M_o^2}} \frac{(p+\int \Theta\chi' d\rho)^2}{(1+\xi/\sqrt{M_o^2+p^2})^2} +$$

$$+ \int d\rho \left[\frac{\Theta^2}{2} + \frac{\chi'^2}{2} + V - V(\phi_o) - \frac{\delta V}{\delta \phi_o} \chi\right] + \Delta V$$

$$\xi \equiv \int d\rho\, \phi_c'(\rho)\chi_c'(\rho,t) \ .$$

Now, to zeroth order we get $E = \sqrt{p^2 + M_o^2}$ directly. This perturbation theory is however more difficult to handle since the quadratic term reads

$$H^{(2)} = v\int d\rho \Theta\chi' + \frac{1}{\sqrt{p^2+M_o^2}} \frac{3}{2} v^2 (\phi_c', \chi')$$

$$+ \int d\rho \left[\frac{\Theta^2}{2} + \frac{\chi'^2}{2} + V''[\phi_c]\frac{\chi^2}{2}\right] ,$$

$$p = \frac{M_o v}{\sqrt{1-v^2}} \ . \qquad (1.29)$$

This quadratic form can be diagonalized by introducing Lagrange multipliers for the constraints and working in first order formalism. The basic tool is the introduction of boosted solutions $\psi_n(x\sqrt{1+p^2/M_o^2})$ instead of (1.28). The real problem is the determination of ΔV because (1.23) is now replaced by

$$\Pi = \Pi_o + \Theta$$

$$\phi = \phi_o\left((x-X)\sqrt{1 + \frac{p^2}{M_o^2}}\right) + \chi \ . \qquad (1.30)$$

Hence, it is no more a point canonical transformation since ϕ involves both X and p. At the quantum level the meaning of

$$\phi_o\left((x-\hat{X})\sqrt{1+\frac{\hat{p}^2}{M_o^2}}\right)$$

is rather unclear. However, since everything worked out fine with the choice (1.18), (1.23), one should be able to handle (1.28), (1.30) exactly as well. Note that this problem only arises at the two-loop level where ΔV starts to contribute.

More generally, one can choose f to be different from the zeroth mode one wants to eliminate. This mode then appears in the propagator but because the result does not depend on Q there will be Ward-like identities which will tell us that the zeroth mode contribution vanishes. Indeed Faddeev and Korepin [19] have pointed out that the singularity of the propagator due to the zeroth mode is rather mild. They remarked that if we look at the resolvent

$$R(\varepsilon) = (\hat{H}^{(2)} + i\varepsilon)$$

$$\hat{H}^{(2)} \equiv \frac{\partial^2}{\partial t^2} - \frac{\partial^2}{\partial x^2} + V''(\phi_o) \qquad (1.31)$$

one gets

$$R(x,t;x',t') = \frac{i}{2\sqrt{i\varepsilon}}\, e^{-i\sqrt{i\varepsilon}|t'-t|}\ \psi_o(x)\psi_o(x') + \tilde{R} \qquad (1.32)$$

where $\tilde{R}$ is well behaved for $\varepsilon \to o$ and corresponds to the contribution of the other modes besides the zeroth mode. Thus the singularity is not a pole in ε, it is rather of the type $1/\sqrt{\varepsilon}$ and in ref.[19] it was proposed to approximate

$$\frac{e^{-i|t-t'|\sqrt{i\varepsilon}}}{2\sqrt{i\varepsilon}} \simeq \frac{1}{2\sqrt{i\varepsilon}} - \frac{i}{2}|t-t'| + o(\sqrt{i\varepsilon}) \tag{1.33}$$

and to use as propagator

$$\frac{1}{2\sqrt{i\varepsilon}} \psi_o(x)\psi_o(x') - \frac{i}{2}|t-t'|\psi_o(x)\psi_o(x') + \lim_{\varepsilon\to o} \tilde{R} \quad .$$

Up to two-loop level for a soliton mass, it has been checked that the $1/\sqrt{\varepsilon}$ term cancels. This reflects the existence of the Ward identities which we mentioned earlier as it shows that the zeroth mode indeed drops out.

More recently, Jevicki [20] pointed out that this procedure is ambiguous as higher order term in (1.33) can give finite contribution when multiplied by several terms of order $1/\sqrt{\varepsilon}$. He proposed a different method which does not introduce X(t) as a bona fide quantum variable but only extracts a constant parameter a from ϕ by letting

$$1 = \int da\, \delta(\int dxdt\, \phi(x,+t)\, \phi_o(x-a))J \tag{1.34}$$

instead of (1.5).

Up to two-loop level for soliton mass the zeroth mode again cancels and the result agrees with what one gets from (1.20). This method however can only be used if the soliton momentum is conserved. It does not apply, for instance, to matrix elements of the $\hat{\phi}$ field as in (1.26).

1-B. Several solitons - General collective coordinate method

For the case of several solitons we obviously need to extract more collective coordinates. We thus discuss a method for doing this which, in fact, is very general [21] and can be applied to any problem in which collective coordinates are relevant.

Let $\phi(x)$ and $\Pi(x)$ be a canonical field and its conjugate momenta respectively. Let $H[\Pi,\phi]$ be the Hamiltonian of a system under consideration. We consider a group G of transformations generated by a set of n generators $P_a[\Pi,\phi]$ through Poisson brackets. We assume that the Lie algebra closes namely

$$\{P_a, P_b\} = C^d_{ab} P_d \quad . \tag{1.35}$$

For arbitrary group element specified by parameters X_α, the transform of $A[\Pi,\phi]$ is given by

$$\tau_{[X]}(A) \equiv A_{[X]} = \sum_{n=o}^{\infty} \frac{1}{n!}\{\{\ldots\{\{A,G_{\{X\}}\},G_{\{X\}}\}\ldots\},G_{\{X\}}\} \tag{1.36}$$

where

$$G_{\{X\}} = \sum_\alpha X_\alpha P_\alpha[\Pi,\phi] \quad . \tag{1.37}$$

Since G is a Lie group, we have

$$\tau_{[Z]} = \tau_{[Y]} \cdot \tau_{[X]}$$

where

$$Z^a = f^a \; (Y^1, \ldots Y^n; \; X^1, \ldots X^n) \; . \tag{1.38}$$

Let us define V^a_b and its inverse U^a_b by

$$V^a_b(X) = \left. \frac{\partial f^a(Y,X)}{\partial Y^b} \right|_{Y = o}$$

$$V^a_b \; U^b_c = \delta^a_c \; .$$

The structure constant C^d_{ab} is related to U and V by

$$C^d_{ab} = - \; V^e_a \; V^f_b \; \left(\frac{\partial U^d_f}{\partial X^e} - \frac{\partial U^d_e}{\partial X^f}\right) \; .$$

Considering an infinitesimal transformation, one obtains

$$\{\phi_{[X]}, \; P_a\} = \frac{\partial \phi_{[X]}}{\partial X^b} \; V^b_a(X) \quad . \tag{1.39}$$

Now that we have enough machinery, we introduce X as a dynamical variable through making the change of variable:

$$\tilde{\phi}(x,t) = \phi_{[X(t)]}(x,t) \; , \; \tilde{\Pi}(x,t) = \Pi_{[X(t)]}(x,t) . \tag{1.40}$$

This is the generalization of (1.7).

The choice of P_α is a priori arbitrary. We assume that it is such that the X_α are the relevant collective coordinates as we shall discuss below. Hence we denote by X_α the collective coordinates, while one may say that $\tilde{\phi}$ and $\tilde{\Pi}$ are fields in the body fixed coordinate system .

Next, we insert (1.40) into the Hamiltonian to obtain a new Hamiltonian as a function of $\tilde{\phi}, \tilde{\Pi}$ and X's:

$$\tilde{H}[\tilde{\Pi}, \tilde{\phi}, X] \equiv H[\tilde{\Pi}_{[-X]}, \tilde{\phi}_{[-X]}] \qquad (1.41)$$

where [-X] is the inverse transformation of (1.40).

Let us now consider a new system with Hamiltonian $\tilde{H}$, canonical variables $X, \tilde{\phi}$, and canonical momenta p and $\tilde{\Pi}$. We now show that this new system is equivalent to the old one if we impose the contraints [22]

$$p_\alpha + P_\beta[\tilde{\Pi}, \tilde{\phi}] U_\alpha^\beta \equiv F_\alpha[\tilde{\Pi}, \tilde{\phi}, p, X] = 0 \quad .$$

It is obvious that we have to impose n constraints since the new system has n more dynamical variables than the old one.

In the sense of Dirac, those constraints are first class since one can check that

$$\{F_\alpha, F_\beta\} = 0 \qquad \{\tilde{H}, F_\alpha\} = 0 \; .$$

The existence of first class constraints, precisely, reflects the gauge invariance of the new system under the canonical transformations generated by F_α, which is simply

$$\tilde{\phi} \to \tilde{\phi}_{[-Y]}, \tilde{\Pi} \to \tilde{\Pi}_{[-Y]}, [X] \to [X + Y]$$

where the sign [X+Y] means group multiplication of the elements and where Y is an arbitrary function of t. That this transformation leaves $\tilde{H}$ invariant is obvious since it does not change the fields ϕ and Π we started from.

The effective Hamiltonian has the form

$$H_{eff} + \sum_{\alpha} \lambda_{\alpha} F_{\alpha} \qquad (1.42)$$

where λ_{α} are Lagrange multipliers. λ_{α} is determined from the equation $\dot{X}_{\alpha} = \delta H_{eff}/\delta p_{\alpha}$ which gives $\lambda_{\alpha} = \dot{X}_{\alpha}$. Choosing X_{α} determines λ_{α} through this equation and thus fixes the gauge.

For $X_{\alpha} \equiv o$, we find back the old system since then $\tilde{\phi} = \phi$, $\tilde{\Pi} = \Pi$, $\lambda_{\alpha} = o$. Therefore, since the physical contents of the theory is gauge independent, the new description is equivalent to the old one.

The quantization of the new system can be done following Faddeev's method [23]. One adds n additional gauge fixing conditions $Q_{\alpha} = o$, and write the transition matrix element as

$$\int D\tilde{\phi}D\tilde{\Pi}DXDp\prod_{\alpha}\delta(F_{\alpha})\prod_{\beta}\delta(Q_{\beta})\det\{F_{\alpha},Q_{\beta}\}\exp\{i\int dt\ (p_{\alpha}\dot{X}_{\alpha}$$

$$+ \int dx\tilde{\Pi}\dot{\tilde{\phi}}-H)\} \qquad (1.43)$$

assuming that the Q's are such that det $\{F_\alpha, Q_\beta\} \neq o$.

However, in order to be self-contained and to show the generalization of our one-soliton method, we rederive the quantization procedure starting from the transition matrix element

$$\int D\phi D\Pi \ \psi_f^*[\phi]\psi_i[\phi] \quad \exp\{i\int dt[\int dx\Pi\dot{\phi}-H]\}$$

and introducing into the functional integral

$$\int J \prod_{\alpha,\beta} dX_\alpha dP_\beta \delta(p_\alpha+P_\alpha[\Pi_{[X]},\phi_{[X]}])\delta(Q_\beta[\Pi_{[X]},\phi_{[X]}]) = 1$$

$$J = \prod_t \det(\frac{\delta Q_\beta}{\delta X_\gamma}) = \prod_t \det\{P_\alpha, Q_\beta\} \quad . \tag{1.44}$$

In this proof, we consider only, to simplify, the case of abelian canonical group. We immediately obtain

$$\int D\phi D\Pi \prod_{\alpha,\beta} Dp_\alpha DX_\beta \delta(p_\alpha+P_\alpha)\delta(Q_\beta)\psi_f^*\psi_i \ \det\{P,Q\}$$

$$\times \quad \exp\{i\int dt[\int dx\Pi\dot{\phi}-H]\} \ . \tag{1.45}$$

If we first integrate over ϕ and Π, we can make the change of variable (1.40) for fixed X and one sees that the Jacobian is one since (1.40) is a canonical transformation [24]. Next, consider the term

$$\int_{t_i}^{t_f} dxdt\Pi\dot{\phi} = \int_{t_i}^{t_f} dxdt\tilde{\Pi}_{[-X]} \frac{d}{dt}(\tilde{\phi}_{[-X]}) = F_{[-X]} \quad .$$

By an infinitesimal change δX, one can verify that

$$\delta F = \int_{t_i}^{t_f} dxdt\, P_\alpha \delta \dot{X}_\alpha + \int dx \,(\text{surface terms})\Big|_{t_i}^{t_f} ;$$

dropping the surface term we obtain $F = \int dt\, P_\alpha \dot{X}_\alpha + F_o$ where $F_o = F_{[X\equiv o]} = \int dxdt \tilde{\Pi}\dot{\tilde{\phi}}$ so that finally

$$\int dxdt \Pi \dot{\phi} = \int dt\, P_\alpha \dot{X}_\alpha + \int \tilde{\Pi}\dot{\tilde{\phi}} dxdt .$$

P_α and X_α are thus conjugate variables. We finally obtain

$$\int D\tilde{\Pi} D\tilde{\phi} DX_\alpha Dp_\beta \delta(p_\beta + P_\beta[\tilde{\Pi},\tilde{\phi}])\delta(Q_\beta[\tilde{\Pi},\tilde{\phi}])\det\{P,Q\}$$

$$\tilde{\psi}^*_f[\tilde{\phi}]\tilde{\psi}_i[\tilde{\phi}] \exp\{i\int dt[P_\alpha \dot{X}_\alpha + \int dx(\tilde{\Pi}\dot{\tilde{\phi}} - \tilde{H}[\tilde{\phi},\tilde{\Pi},X])\} \qquad (1.46)$$

which agrees with (1.43).

In this expression, we have replaced ψ_i, ψ_f, by $\tilde{\psi}_i$, $\tilde{\psi}_f$, in order to take into account the surface terms, which we dropped in the computation together with the change or argument of $\psi_{i,f}$. It is likely [23], though no general proof exists, that $\tilde{\psi}_{i,f}$ is simply transformed from $\psi_{i,f}$ by

$$\tilde{\psi}_{i,f}[\phi] = (\exp\{i \sum_\alpha \hat{P}_\alpha X_\alpha(t_{i,f})\})[\phi] \qquad (1.47)$$

namely, is, as one expects, obtained from $\psi_{i,f}$, by the unitary transformation associated with the canonical transformation introduced by (1.40).

The procedure we have followed is exactly the

generalization of our discussion for one soliton. There, for small g the soliton has a large mass compared to the mass associated with the quanta of χ. Hence for small g the soliton position moves much more slowly than the other degrees of freedom. This is the standard criterion for introducing collective coordinates, it is the so-called adiabatic approximation. In general, if we assume that the X_α in (1.46) vary much more slowly than the other degrees of freedom, we can determine an effective potential by first solving the dynamics of the other degrees of freedom with fixed X_α, p_β. In functional formalism this is formally done by assuming that the $\psi_{i,f}$ are eigenstates of P_α with eigenvalues $p_{\alpha,i}$, $p_{\alpha,f}$ and computing for fixed X_α, p_β,

$$\exp\{-i\int dt\ H_{eff}(X,p)\} \equiv \int D\tilde{\Pi} D\tilde{\phi}\,\delta(p_\alpha + P_\alpha)\,\delta(Q_\beta)\ \det\{P,Q\}$$

$$\psi_f^*[\tilde{\phi}]\psi_i[\tilde{\phi}]\ \exp\{i\int dt[\int dx \tilde{\Pi}\dot{\tilde{\phi}} - \tilde{H}[\tilde{\Pi},\tilde{\phi},X]]\}\ . \qquad (1.48)$$

In the adiabatic approximation, the transition probability is given by

$$\int DX_\alpha Dp_\beta\, e^{-ip_{\alpha,f}X_\alpha(t_f)}\, e^{ip_{\beta,i}X_\alpha(t_i)} \exp\{i\int_{t_i}^{t_f} dt(p_\alpha \dot{X}_\alpha - H_{eff})\}\ .$$

If the P_α are not constants of motion (i.e. $\{P_\alpha, H\} \neq o$) $\tilde{H}$ depends explicitly on X_α and the dynamics of the collective coordinates is complicated. If on the other hand $\{P_\alpha, H\} = o$, we have $\tilde{H} = H[\tilde{\Pi},\tilde{\phi}]$, independent of X and the dynamics of X_α, p_β is trivial. The eigenstates are plane waves. In this case, it is better to do the other way round as we did for the one soliton case, namely, choosing again $\psi_{i,f}$ to be eigenstates of P_α, we first integrate over X_α and p_β, immediately obtaining

$$\prod_\alpha \delta(p_{\alpha,i}-p_{\alpha,f}) \int D\tilde{\Pi} D\tilde{\phi}\ \delta(p_{\alpha,i}+P_\alpha)\ \delta(Q_\beta)\ \det\{P,Q\}$$

$$\psi_f^* \psi_i \exp\{i\int dt[\int dx \tilde{\Pi}\dot{\tilde{\phi}}-H]\} \quad . \tag{1.49}$$

In order to apply the semi-classical method, we look for the minimum of the action, but now we have constraints. At the classical level, if P_α are constants of motion, they can be given arbitrary values by a suitable choice of boundary conditions. In addition, starting from a classical solution with given p_α, we can generate an infinite set of classical solutions with the same p_α by applying an arbitrary transformation of G with parameters Y_α.

Hence, we have general classical solutions of the form $\phi_{cl}(x,t;p_\alpha,Y_\beta)$ and we can satisfy the constraints by choosing $p_\alpha = p_{\alpha,i}$ and by fixing Y_β such that all the Q vanish. In order to obtain a constant perturbation we will let

$$\tilde{\phi} = \phi_{cl}^n(x,t;\ p_{\alpha,i},\ Y_\beta) + \chi\ , \tag{1.50}$$

where ϕ_{cl}^n, which is the classical ground state of the sector considered, is the lowest energy classical solution satisfying the constraints. It will, in general, involve no additional parameters besides Y and p so that, because of the constraints, (1.49) is a well defined change of fields and no zeroth mode problem will be encountered in the perturbation theory for χ.

In the more general case when P_α are not constants of motion, semi-classical methods can be applied to (1.48) and again, one will be looking at a minimum of the action with constraints. Now, since the constraints do not commute with the Hamiltonian, this problem is totally different from looking at minima of the action alone. In particular, one may find solutions which exist only when the constraints are imposed on the system.

For solitons, only the case of the constant of motion has been used to discuss the scattering of solitons in two dimensions. The collective coordinate approach to soliton scattering has been developed [25] on the example of binary collisions in sine-Gordon theory [26]. The general case has been investigated following the same method for sine Gordon theory [27] and non-linear Schrödinger equation [28].

These works are based on the existence of an infinite set of constants of motion [26] at the classical level. Those are associated with general canonical transformations which mix "positions" and "momenta". Therefore, it is not clear whether they can really be performed at the quantum level and we have no way to determine the additional term ΔV which would arise from a more careful treatment in the one-soliton case. So far, however, only the one-loop approximation has been studied in which we do not expect ΔV to contribute.

Finally, there is a special feature of the soliton quantization in four-dimensional space-time which introduces a new type of collective coordinates. These solutions exist only for Higgs models with a non-trivial unbroken gauge group. They involve massless vector bosons as a consequence. If we look at a soliton monopole solution which has an electric charge [29], it follows from Gauss theorem that the electric field at infinity decreases like r^{-2}. As a result the usual field equations do not follow from the variations of the standard action since one cannot drop the surface term one obtains by partial integration of the kinetic term. This is a very general problem which arises whenever one studies field configurations with non-zero total electric charge. The solution [30] is to introduce new degrees of freedom at infinity such that with a modified action the field equations can really be derived. For the dyon solution one of these degrees of freedom is the collective coordinate conjugate to the charge and charge quantization conditions come out naturally, see ref.[30] for details.

II - INSTANTONS

II-A. The example of quantum mechanics

We first illustrate the general ideas on the example of a symmetric potential with two minima as drawn below:

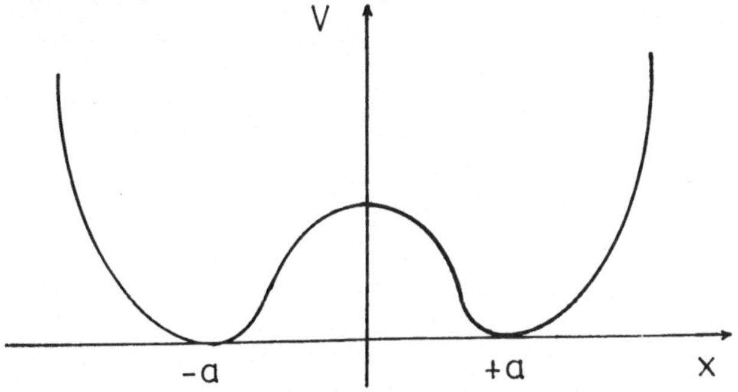

Fig. 1

Naively, one would quantize the theory by considering quantum fluctuations around one of the two minima of V. In this way one would conclude that there are two possible ground states for the theory. The usual reasoning would then say that since we must pick up one of the two in order to quantize the theory, we have a spontaneous symmetry breakdown of the invariance by $x \to -x$. Choosing for instance the right minimum one would let $x = a + \tilde{x}$ and develop perturbation by assuming $\tilde{x} \ll a$. Of course we all know that this is wrong since there is tunneling between states localized around $x=+a$ and states localized around $x = -a$. Hence it does not make sense to assume that $\tilde{x}$ is small since, if we wait long enough the important values of $\tilde{x}$ will be precisely $\tilde{x} \sim -2a$. The tunneling actually restores symmetry and the true ground state is in fact a symmetric combination of states localized around $x = +a$ and $x = -a$. If tunneling is a small effect, the two lowest eigenstates are just the symmetric and antisymmetric combinations of the two ground states we would have if there was no tunneling. The energy difference $\Delta E \propto e^{-c/\hbar}$ is a non-analytic function of $\hbar$ near $\hbar = o$ since classically it is impossible to go from $x = a$ to $x = -a$. It is well known that nonetheless ΔE can be estimated by semiclassical methods if one uses classical trajectories with imaginary time [31]. Starting from the classical equation with energy E

$$\frac{1}{2} \dot{x}^2 + V(x) = E$$

and letting $t = i\tau$ leads to

$$\frac{1}{2}\left(\frac{dx}{d\tau}\right)^2 - V(x) = -E .$$

Hence V and E have changed signs. The classically forbidden region becomes the classically allowed region. In particular there is a solution with E = o, $x_o(\tau)$ such that

$$\tau - b = \int_o^{x_o(\tau)} \frac{dy}{\sqrt{2V(y)}} \tag{2.1}$$

which is the instanton solution of this problem. It contains an arbitrary parameter b which is the "time" when classically we go from $x > o$ to $x < o$.

Standard WKB applied to this problem shows that if we introduce the classical action associated with $x_o(\tau)$

$$I = \int_-^+ d\tau \ [\frac{1}{2}(\frac{dx_o}{d\tau})^2 + V(x_o)]$$

we have

$$\Delta E \propto e^{-I/\hbar} \quad . \tag{2.2}$$

This result can also be obtained by using semi-classical methods in functional integral after continuing to pure imaginary time [31].

If we denote by $|\pm\rangle$ the two ground states we would have if there was no tunneling, eq. (2.2) shows that we cannot consider them as true ground states because $\langle +|H|-\rangle \propto e^{-I/\hbar} \neq o$. Until recently it was believed that no such phenomenon could occur in field theory because there is an infinite number of degrees of freedom. The feeling one had was that the transition probability between two ground states $|\alpha>, |\beta>$ would be essentially given by the product of the probabilities for each degree of freedom leading to $<\alpha|H|\beta> \propto \propto e^{-(\infty)/\hbar} = o$. We now know that this is not correct. If there is a classical solution for imaginary time with finite action S_u which goes from $|\alpha>$ to $|\beta>$ we have instead

$$<\alpha|H|\beta> \propto e^{-S_u/\hbar} \tag{2.3}$$

and we are in the same situation as in potential theory.

II-B. Tunneling in field theory

Vacuum to vacuum amplitudes are easily studied in functional methods by looking at the functional integral of the exponential of the action for very large time since this procedure automatically projects out all but the vacuum state matrix elements. This functional integral is continued to imaginary time by using standard arguments and the resulting integral is evaluated by expanding around the minima of the action which are now Euclidean classical solutions [3].

We discuss the example of pure Yang-Mills field theory with internal symmetry group SU(2).

$$L = -\frac{1}{4}(G^a_{\mu\nu})^2$$

$$G^a_{\mu\nu} = \partial_\mu A^a_\nu - \partial_\nu A^a_\mu + g\varepsilon_{abc} A^b_\mu A^c_\nu \quad . \tag{2.4}$$

Since we have to discuss solutions with imaginary time $x_4 = it$ we will distinguish Euclidean field theory quantities by putting a hat on them, for instance V_μ means V_o, $\vec{V}$ while $\hat{V}$ means $\vec{V}$, V_4.

Because the outcome will be of the form (2.3) we are only interested in classical solutions with finite Euclidean action. Hence for $\hat{x}^2 \equiv x_4^2 + \vec{x}^2 \to \infty$ all solutions should become pure gauges. Any field configuration of this type is classified by the topological charge (Pontryagin index)

$$\hat{Q} = \frac{g^2}{64\pi^2} \hat{\varepsilon}_{\mu\nu\rho\sigma} \int \hat{d}_4x \; \hat{G}^a_{\mu\nu} \hat{G}^a_{\rho\sigma} \quad . \tag{2.5}$$

As argued in ref. [4], since obviously

$$(\hat{G}^a_{\mu\nu} \pm \frac{1}{2} \hat{\varepsilon}_{\mu\nu\rho\sigma} \hat{G}_{\rho\sigma})^2 \geq 0$$

one has

$$\int \hat{d}_4x \; \hat{L} \geq \frac{8\pi^2}{g^2} \nu$$

where ν is the value of $\hat{Q}$. The equality sign is only reached when

$$\hat{G}^a_{\mu\nu} = \pm \frac{1}{2} \hat{\varepsilon}_{\mu\nu\rho\sigma} \hat{G}_{\rho\sigma} \quad . \tag{2.6}$$

Since it corresponds to the minimum of the action, any solution of (2.6) is also a solution of the Euclidean classical field equation. So far, all instanton solutions which have been exhibited [32] satisfy (2.6).

A particular solution is the one-instanton solution of ref. [4]

$$\hat{A}^p(\hat{x},\lambda) \equiv \hat{A}^{pa}\frac{\tau^a}{2} = \frac{i}{g}\,\omega^{-1}\partial_\mu\omega\,\frac{\hat{x}^2}{\hat{x}^2+\lambda^2} ,$$

$$\omega = \frac{x_4 - i\vec{x}\vec{\tau}}{\sqrt{\hat{x}^2}} ; \qquad (2.7)$$

λ is a scale parameter. It has $\nu = 1$.

The tunneling process is most easily seen in the gauge $\hat{A}_4^a = o$ where $\hat{Q}$ can be written as

$$\hat{Q} = \hat{q}(x_4 = +\infty) - \hat{q}(x_4 = -\infty) ,$$

$$\hat{q}(x_4) = \frac{g^2}{16\pi^2}\int d_3x\,[\hat{A}_i^a\partial_j A_k^a + \frac{9}{3}\,\varepsilon_{abc}A_i^a A_j^b A_k^c]\,\varepsilon_{ijk} . \qquad (2.8)$$

Using time independent gauge transformations one may choose $\hat{q}(-\infty)$ to be an integer. Then $\hat{q}(+\infty)$ will also be an integer. Since, for $x_4 = \pm\infty$ we must be at a classical ground state, we conclude that one has to consider a discrete set of vacua labeled by an integer $|n\rangle$. In functional integral we must integrate over field configurations with finite action which are also classified by the values of $\hat{Q}$. Selecting a particular value ν of the Pontryagin index we conclude that

$$\langle n+\nu | e^{-HT} | n \rangle \underset{T\to\infty}{\propto} \int (DA)_\nu \; e^{-\int \hat{d}_4 x \hat{L}} \times [\text{gauge fixing}] . \quad (2.9)$$

This is the argument developed in ref. [33] where the functional integral on the right-hand side was estimated by expanding around instanton classical solutions. The result is typically of the form (2.3) and shows that tunneling indeed occurs between the different ground states.

In this calculation, collective coordinates must be introduced for instanton sizes and positions. For instance the most general one-instanton solution is $A^p(\hat{x}-\hat{X},\lambda)$ where $\hat{X},\lambda$ are the instanton position and size and we must avoid the zeroth mode associated with small variations of $\hat{X}$ and λ. For the ground states however, this simply corresponds to a small gauge transformation so that the momenta conjugate to $\hat{X}$ and λ are always zero. Thus $\hat{X}$ and λ are not treated as true quantum mechanical operators defined for each time. They are rather numbers extracted from the field integrated over all space-time as in formula (1.34).

In the Minkowski field theory this tunneling process can be seen by introducing a collective co-ordinate [34] of the type discussed in (II-B) which is just q defined in the same way as in (2.8) but in Minkowski space. In the $A_o = o$ gauge, the conjugate to A_i^a is $E_i^a = G_{oi}^a$. It is easy to see that the variable P conjugate to q is given by

$$P = R/F$$

$$R = \frac{1}{2} \int d_3x \; E_i^a \, \varepsilon_{ijk} \, G_{jk}^a \qquad (2.10)$$

$$F = \frac{g^2}{16\pi^2} \int d_3x \, (G_{ij}^a)^2 \quad ; \qquad (2.11)$$

indeed one has

$$\{q,P\} = 1 \; .$$

Applying the method of sect. (I-B) one introduces $A_{[q]}$, $E_{[q]}$ defined as in (1.36). They are easily found to satisfy the differential equations

$$\frac{\partial A_i^a}{\partial q}[q] = \frac{1}{2} \{\frac{1}{F} \varepsilon_{ijk} \, G_{jk}^a\}_{[q]}$$

$$\frac{\partial E_i^a}{\partial q}[q] = \{\frac{1}{F} \varepsilon_{ijk} \, (D_j E_k)^a - \frac{g^2}{4\pi^2} \frac{P}{F} (D_k G_{ki})^a\}_{[q]}$$

$$(D_i)_{ac} \equiv \partial_i \delta_{ac} + g \, \varepsilon_{abc} \, A_i^b$$

with $E_{[o]} = E$, $A_{[o]} = A$. Let us look for a particular solution such that $E_i^a = \frac{\beta}{2} \varepsilon_{ijk} \, G_{jk}^a$. ($\beta$ is some constant). In that case, one can verify that $E_{i[q]}^a = \frac{\beta}{2} \varepsilon_{ijk} \, G_{jk[q]}^a$ and the above equation can be rewritten as

$$\frac{\partial}{\partial \tau} A_{i[q]}^a = \frac{1}{\beta} E_{i[q]}^a$$

$$\frac{\partial}{\partial\tau}\frac{E_i^a}{\beta}[q] = -(D_k\, G_{ki}^a)_{[q]}$$

where τ is such that

$$\frac{d\tau}{dq} = F\,[A_{[q]}, E_{[q]}] \quad .$$

These equations are the Euclidean classical field equations which are obtained in Minkowski space! This is why in ref.[34], tunneling is discussed in Minkowski space after shifting the field by a Euclidean classical solution.

When q is introduced as a collective coordinate one finds that the adiabatic approximation is valid for small λ which because of renormalization group, means small g. In this approximation the main features are given by the dynamics of q which has a periodic potential with minima at $q = n$. Each of these minima corresponds to the vacuum state $|n\rangle$ and the study of tunneling is essentially reduced to a quantum mechanical problem where standard WKB methods can be applied. In particular we have the equivalent of the band theory in a one-dimensional crystal so that a state of the band is given by

$$|\theta\rangle = \sum_n |n\rangle\, e^{in\theta}$$

with energy

$$E \propto E_0 \cos\theta .$$

We refer to ref.[34] for details. In this way, one directly obtains, in physical space-time the results derived in refs.[33] and [35] by analytic continuation to imaginary time.

In the Euclidean approach, one is led to the study of the equivalent of a partition function of statistical mechanics where g is replaced by the temperature [3], [36]. We illustrate the role of instantons on one-dimensional Ising model which corresponds to tunneling in quantum mechanics.

At T = o there are two ground states:

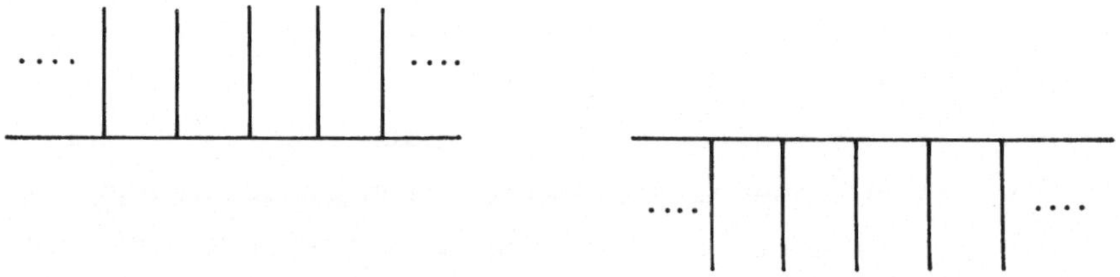

Let us pick up the one on the left. For small temperature we have "instanton" configurations:

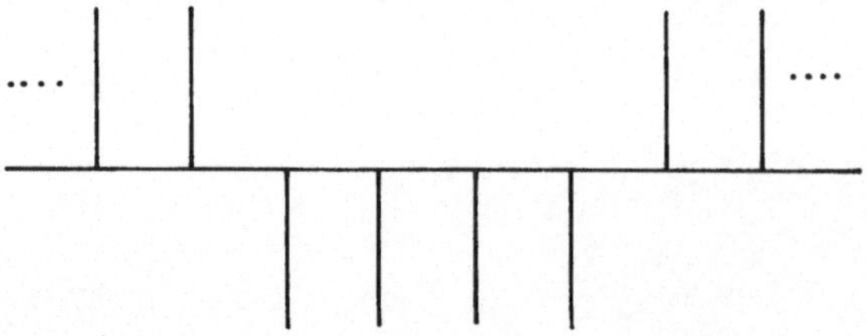

they have a higher energy than the ground state but there are many more such configurations since the flipping of spin can occur anywhere, any even number of times. In the partition function

$$Z = \sum_{\text{configurations}} e^{-E/T}$$

the instanton configurations in fact dominate the vacuum contribution. Indeed it is well known that the dominant contributions to Z are those with lowest free energy $F = E - TS$, $S \propto \ln$ (nb of states).

Thus instantons destroy the long-range order of the vacuum completely in this case (it is well known that there is no phase transition in one dimension). The ground state becomes symmetric between up and down in agreement with the well-known fact that in quantum mechanics the ground state is always symmetric as we recalled it in sect. (II-A).

In general, Goldstone bosons associated with spontaneous breakdown of a continuous symmetry reflect the existence of fluctuations around ground states with very long wavelength. By destroying long-range order in the vacuum state, instantons may therefore prevent Goldstone bosons from appearing. Indeed evidences have been given [37] that, in this way, the axial U(1) symmetry of quark model could be spontaneously broken without predicting a ninth axial boson with a small mass thus solving the so-called η problem.

Since they can spoil long-range order, instantons

may lead to quark confinement in the sense of Wilson, namely, the expectation value $<o|e^{-\oint d\ell_\mu A_\mu}|o>$ for a large closed loop mass decreases like the exponential of the area enclosed by the loop [3], [36].

Finally, by tunneling from the vacuum, pairs of fermions can be created leading to various possibilities of violation for baryon conservation laws [37].

The general problem of this method, when applied to pure Yang-Mills field theory, is that it is a small coupling approximation in a theory where the effective coupling constant depends on the scale considered. The final results involve an integration over instanton size which is out of control for large instantons where this effective coupling constant is not small [33], [37] so that the approximation presumably breaks down.

REFERENCES

1. H.B. Nielsen and P. Olesen, Nucl. Phys. B61 (1973) 45.
2. R. Dashen, B. Hasslacher and A. Neveu, Phys. Rev. D10, 4114;D10, 4130; D10, 4138 (1974).
3. A.M. Polyakov, Phys. Letters 59B (1975) 82.
4. A.A. Belavin, A.M. Polyakov, A.S. Schwartz, Y.S. Tyupkin, Phys. Letters 59B (1975) 85.
5. See e.g. T.D. Lee in ref. [11].
6. For a general discussion, see Coleman's 1975 Erice Lectures.

7. J.L. Gervais, B. Sakita, Phys. Rev. D11 (1975) 2943.

8. L.D. Faddeev, P.P. Kulish, V.E. Korepin, Pizma JETP 21 (1975) 302.

9. R. Rajaraman, Phys. Report 21C (1975) 227.

10. R.Jackiw, MIT preprint submitted to Rev. of Mod. Phys.

11. Extended Systems in Field Theory, edited by J.L. Gervais and A. Neveu, Phys. Report 23C (1976).

12. J.L. Gervais, A. Jevicki, B. Sakita, Phys. Rev. D12 (1975) 1038.

13. E. Tomboulis, Phys. Rev. D12 (1975) 1678.

14. J.L. Gervais, A. Jevicki, Nucl. Phys. B110 (1976) 93.

15. J. Goldstone, R. Jackiw, Phys. Rev. D11 (1975) 1486.

16. p,p',k,k' are soliton and meson momenta.

17. A. De Vega, Nucl. Phys. B115 (1976) 428.

18. R. Dashen, B. Hasslacher and A. Neveu, Phys. Rev. D11 (1975) 3424.

19. L.D. Faddeev, V.E. Korepin, CERN preprint.

20. A. Jevicki, I.A.S. preprint.

21. J.L. Gervais, A. Jevicki, B. Sakita in ref. 11 .

22. A. Hosoya and K. Kikkawa, Nucl. Phys. B101 (1975) 271.

23. L.D. Faddeev, Theor. and Math. Phys. 1 (1970) 1.

24. See e.g. A. Katz, Classical Mechanics, Quantum Mechanics, Field Theory (Academic Press, 1965).

25. J.L. Gervais, A. Jevicki, Nucl. Phys. B110 (1975) 113.

26. A.S. Scott, F.Y.F. Chu, D.W. McLaughlin, Proc. IEEE 61 (1973) 1443.

27. M.T. Jaekel, Nucl. Phys. B118 (1977) 506.

28. J. Honerkamp, M. Schlindwein, A. Wiesler, preprint.

29. B. Julia and A. Zee, Phys. Rev. D11 (1975) 2227.

30. J.L. Gervais, B. Sakita, S. Wadia, Phys. Letters 63B (1976) 55.

31. D. McLaughlin, J. Math. Phys. 13 (1972) 1099.

32. E. Witten, Harvard preprint; 't Hooft, unpublished; R. Jackiw and C. Rebbi, M.I.T. preprint.

33. C. Callan, R. Dashen, D. Gross, Phys. Letters 63B (1976) 334.

34. J.L. Gervais and B. Sakita, City College preprint and to be published.

35. R. Jackiw and C. Rebbi, Phys. Rev. Letters 37 (1976) 172.

36. A.M. Polyakov, Nordita preprint.

37. G. 't Hooft, Phys. Rev. Letters 37 (1976) 8 and Harvard preprint.

Acta Physica Austriaca, Suppl. XVIII, 429–444 (1977)

THE SCATTERING OF QUANTIZED SOLITONS IN THE NON-LINEAR SCHRÖDINGER THEORY+

by

M. SCHLINDWEIN
Fakultät für Physik
der Universität Freiburg

INTRODUCTION

Recently, the problem of interpreting classical particle-like solutions in the framework of quantum field theory has attracted much interest. There are two major point of views: the search for bound states and the bound state scattering. The former problem has successfully been attacked by means of the field theoretic WKB method [1] while the scattering problem has been treated by Gervais et al. [2] using the collective coordinate approach.

In this seminar I shall present the application of the general scattering formalism to the bound state scattering in the non-linear Schrödinger theory [3]. The latter model is of special interest as it can be

+Seminar given at XVI. Internationale Universitätswochen für Kernphysik, Schladming, Austria, February 24-March 5, 1977.

treated either as a second quantized field theory or as a quantum mechanical problem of N particles interacting via potentials of δ-function type. Therefore, by comparing the exact quantum mechanical results with the semi-classical treatment one may control the validity of the approximations.

The seminar is organized as follows: in the first section I briefly review the general scattering formalism of Gervais et al. [2]. For details the reader is referred to the lectures of Gervais contained in these proceedings. In the second part I describe how to calculate the classical phase shift for soliton-soliton scattering. The third section contains the treatment of the quantum correction up to the one-loop level and finally, in the last part, I compare the exact results with the semi-classical answers.

1. THE SCATTERING FORMALISM

The model to be treated is defined by the Hamiltonian:

$$H(\pi,\phi) = -i\pi'\phi' - i\,m_o\pi\phi + \frac{K}{2}(\pi\phi)^2 \qquad (1)$$

where π,ϕ are complex fields with $\pi = i\phi^*$. In order to describe the quantum scattering of solitons we use the transition amplitude in the functional language:

$$S_{fi} = \lim_{\substack{t_f\to+\infty \\ t_i\to-\infty}} \int[d\phi][d\pi]\Psi_f^*[\phi(x,t_f),\pi(x,t_f)]\Psi_i[\phi(x,t_i),\pi(x,t_i)]\cdot$$

$$\exp\{i \int_{t_i}^{t_f} dt \int_{-\infty}^{\infty} dx(\frac{1}{2}(\pi\dot{\phi}-\phi\dot{\pi}) - H(\pi,\phi))\}. \tag{2}$$

$\Psi_{i,f}$ are the wave functionals corresponding to the incoming (outgoing) two-soliton-states. It is our strategy to expand in (2) around the classical two-soliton-solution which is a stationary point of the functional integral. Since the two-soliton-solution has four degrees of freedom we want to introduce four collective coordinates and momenta. To this end we use four conserved quantities of the non-linear Schrödinger-equation (NLSE) which generate, by Poisson brackets, independent variations of the classical solution. Denoting by $\phi^{(2)}$ the classical two-soliton-solution and:

$$\phi_\alpha^{(2)} = \{\phi^{(2)}, P_\alpha\} = \frac{\delta P_\alpha}{\delta\pi}(\phi^{(2)}); \quad \alpha = 1\ldots4,$$

$$\{,\} \equiv \text{Poisson brackets}, \tag{3}$$

we choose the gauge conditions to be:

$$Q_\alpha[\phi] = \int_{-\infty}^{\infty} dx\, \phi_\alpha^{(2)}(x,t)\phi(x,t) \ ; \qquad \alpha = 1\ldots4 . \tag{4}$$

These conditions exclude the zero-modes $\phi_\alpha^{(2)}$ in the functional space we integrate over in (2).

The four conserved quantities P_α generate a four-parameter abelian group of transformations. Let $\phi_{[x]}$ be the transformed of ϕ with respect to the group element $\exp[i\, x_\alpha P_\alpha]$. Then we introduce the following identity into the transition amplitude (2):

$$1 = \int \Big(\prod_{\alpha=1}^{4} [dx_\alpha(t)][dp_\alpha(t)] \; \Big(\prod_t \prod_{\alpha=1}^{4} \delta(Q_\alpha[\phi_{[x(t)]}])$$

$$\delta(p_\alpha(t) + P_\alpha[\pi,\phi])J\Big) \tag{5}$$

where J is the Jacobian $\left|\frac{\partial(\ldots Q_\alpha \ldots)}{\partial(\ldots x_\beta \ldots)}\right|$. Next we perform a canonical transformation of the field variables:

$$\tilde{\phi} = \phi_{[x]} \; ; \qquad \tilde{\pi} = \pi_{[x]} \quad . \tag{6}$$

In general such a change of variables in the functional integral has to be handled with care [4] but at least in the one-loop approximation that we are dealing with, there are no complications.

We now assume that the wave functionals $\Psi_{i,f}$ describing the asymptotic states of widely separated solitons are eigenstates of the operators $\hat{P}_\alpha$, i.e.:

$$\Psi_{i,f}[\tilde{\pi}_{[-x]},\tilde{\phi}_{[-x]}] = \exp\{-i\, x_\alpha p_\alpha^{i,f}\}\; \Psi_{i,f}[\tilde{\pi},\tilde{\phi}] \quad . \tag{7}$$

Integrating out the variables $x_\alpha(t)$, $p_\alpha(t)$ we finally arrive at:

$$S_{fi} = \lim_{\substack{t_f\to+\infty \\ t_i\to-\infty}} (2\pi)^4 \Big(\prod_{\alpha=1}^{4} \delta(p_\alpha^f - p_\alpha^i)\Big) \int [d\tilde{\pi}][d\tilde{\phi}] \Psi_f^*[\tilde{\pi},\tilde{\phi}] \Psi_i[\tilde{\pi},\tilde{\phi}] \,\cdot$$

$$\cdot\, \exp\{iA[\tilde{\pi},\tilde{\phi}]\} \Big(\prod_t \prod_{\alpha=1}^{4} \delta(Q_\alpha[\tilde{\phi}]) \,\cdot$$

$$\cdot\, \delta(p_\alpha^i + P_\alpha[\tilde{\pi},\tilde{\phi}])J\Big) \tag{8}$$

$$A[\tilde{\pi},\tilde{\phi}] = \int_{t_i}^{t_f} dt \int_{-\infty}^{\infty} dx \; (\tfrac{1}{2}(\tilde{\pi}\dot{\tilde{\phi}} - \tilde{\phi}\dot{\tilde{\pi}}) - H(\tilde{\pi},\tilde{\phi})) \quad .$$

Next we shift the field variables:

$$\tilde{\phi} = \phi^{(2)} + \chi \; ; \qquad \tilde{\pi} = \pi^{(2)} + \omega \quad . \tag{9}$$

Restricting to lowest order in the fluctuations χ,ω we are led to:

$$\delta(Q_\alpha[\phi]) \longrightarrow \delta(\int_{-\infty}^{\infty} dx \; \phi_\alpha^{(2)} \chi)$$

$$\delta(p_\alpha^i + P_\alpha[\tilde{\pi},\tilde{\phi}]) \longrightarrow \delta(\int_{-\infty}^{\infty} dx \; \{\frac{\delta P_\alpha}{\delta\tilde{\phi}}\chi + \frac{\delta P_\alpha}{\delta\tilde{\pi}}\omega\})$$

$$A[\tilde{\pi},\tilde{\phi}] \longrightarrow A_{cl}[\pi^{(2)},\phi^{(2)}] + A_q[\chi,\omega] \tag{10}$$

$$A_q[\chi,\omega] = \tfrac{1}{2}\int_{t_i}^{t_f} dt \int_{-\infty} dx \{\omega\dot{\chi} - \chi\dot{\omega} + i\omega'\chi' + i\, m_o\omega\chi - K(\phi^{(2)})^2\omega^2 - K(\pi^{(2)})^2\chi^2 -$$

$$- 4K\, \pi^{(2)}\phi^{(2)}\omega\chi\} \quad .$$

In the spirit of the WKB-approximation we factorize the wave-functionals into classical and quantum parts:

$$\Psi_{i,f}[\tilde{\pi},\tilde{\phi}] = \Psi_{i,f}^{cl}[\pi^{(2)},\phi^{(2)}] \; \Psi_{i,f}^{q}[\omega,\chi] \quad . \tag{11}$$

Therefore our final transition amplitude reads:

$$S_{fi} = \lim_{\substack{t_f \to +\infty \\ t_i \to -\infty}} (2\pi)^4 (\prod_{\alpha=1}^{4} \delta(p_\alpha^f - p_\alpha^i)) \; \{\Psi_f^{cl*} [\pi^{(2)},\phi^{(2)}] \; .$$

$$\cdot\Psi_i^{cl}[\pi^{(2)},\phi^{(2)}]\ \exp\{i\, A_{cl}[\pi^{(2)},\phi^{(2)}]\}\ \cdot$$

$$\cdot\int[d\chi][d\omega]\ \Psi_f^{q*}[\omega,\chi]\ \exp\{i\, A_q[\chi,\omega]\}\ \cdot$$

$$\cdot(\prod_t \prod_{\alpha=1}^{4} \delta(\int_{-\infty}^{\infty} dx\phi_\alpha^{(2)}\chi)\delta(\int_{-\infty}^{\infty} dx(\frac{\delta P_\alpha}{\delta\tilde{\phi}}\chi + \frac{\delta P_\alpha}{\delta\tilde{\pi}}\omega))J)\Psi_i^q[\omega,\chi] \quad .(12)$$

Since the scattering process is purely elastic we have:

$$S_{fi} = (2\pi)^4 \ (\prod_{\alpha=1}^{4} \delta(p_\alpha^f - p_\alpha^i))\ \exp\{2i\delta\} \quad . \qquad (13)$$

Hence the phase shift δ may be separated into classical and quantum part according to:

$$e^{2i\delta_{cl}} = \lim_{\substack{t_f\to+\infty \\ t_i\to-\infty}} \Psi_f^{cl*}[\pi^{(2)},\phi^{(2)}]\,\Psi_i^{cl}[\pi^{(2)},\phi^{(2)}]\,e^{iA_{cl}[\pi^{(2)},\phi^{(2)}]} \qquad (14a)$$

$$e^{2i\delta_q} = \lim_{\substack{t_f\to+\infty \\ t_i\to-\infty}} \int[d\chi][d\omega]\ \Psi_f^{q*}[\omega,\chi]\ldots\ldots\Psi_i^q[\omega,\chi] \quad . \qquad (14b)$$

Let us postpone the treatment of the quantum correction and first determine the classical phase shift δ_{cl}.

2. THE CLASSICAL PHASE SHIFT FOR SOLITON-SOLITON SCATTERING

The classical phase shift δ_{cl} can in principle be

calculated from (14a). Since it seems, however, rather complicated to evaluate the action functional $A_{cl}[\pi^{(2)},\phi^{(2)}]$ explicitely, we prefer to derive δ_{cl} along the lines of ref. [5]. To this purpose we have to inspect the asymptotic behaviour of $\phi^{(2)}(x,t)$ for large time.

First we remind ourselves that the one-soliton-solution can be formulated in terms of canonical variables [6]:

$$\phi^{(1)}(x|\alpha,\beta,a,b) = \sqrt{\frac{K}{8}}\,\alpha\,\frac{\exp[-i(\beta-a\chi)]}{\cosh[\frac{K}{4}(\alpha\chi-b)]}$$

$$\alpha = \frac{8}{K}\eta \;; \qquad a = -2\xi \;; \qquad \{\beta,\alpha\} = 1$$

$$\dot{\beta} = 4(\xi^2-\eta^2) + m_o; \qquad \dot{b} = -\frac{32}{K}\xi\eta; \qquad \{b,a\} = 1 \tag{15}$$

with α,β representing the internal degrees of freedom and a, b describing the motion of the envelope. The momenta α,a are related to the mass M and the velocity v of the soliton by:

$$M = \frac{\alpha}{2} \;; \qquad v = 2a \quad . \tag{16}$$

We know from the semi-classical treatment of the one-soliton-sector [7] that the internal momentum α has to be quantized according to:

$$\alpha_N = N \;; \qquad N = 1,2,3.... \quad . \tag{17}$$

Therefore we will use in our subsequent considerations always quantized solitons in the sense of (17).

Let us now look at the two-soliton-solution. First we may construct the explicit solution by means of the inverse scattering transform [6]. Clearly the solution shows four momenta α_i, a_i; $i = 1,2$ describing the asymptotic motion of widely separated solitons. Furthermore, by inspecting the behaviour for large time, we find:

$$\lim_{t\to\pm\infty} \phi^{(2)}(x,t) = \sum_{i=1}^{2} \phi^{(1)}(x|\alpha_i^{\substack{out\\in}}, \beta_i^{\substack{out\\in}}, a_i^{\substack{out\\in}}, b_i^{\substack{out\\in}})$$

$$\alpha_i^{out} = \alpha_i^{in} \equiv \alpha_i \; ; \quad a_i^{out} = a_i^{in} \equiv a_i \; ; \quad a_2 > 0, \quad a_1 < 0$$

$$\beta_i^{out} = \beta_i^{in} + G_i \; ; \quad b_i^{out} = b_i^{in} + F_i \tag{18}$$

$$G_{1,2} = 2\left(\text{arctg}\left(\frac{K}{4}\,\frac{\alpha_1-\alpha_2}{a_2-a_1}\right) \mp \text{arctg}\left(\frac{K}{4}\,\frac{\alpha_1+\alpha_2}{a_2-a_1}\right)\right)$$

$$F_{1,2} = \pm\frac{4}{K}\,\ell n\,\frac{(a_1-a_2)^2 + (\frac{K}{4}(\alpha_1-\alpha_2))^2}{(a_1-a_2)^2 + (\frac{K}{4}(\alpha_1+\alpha_2))^2} \; .$$

The relation (18) can be viewed as a canonical transformation from in- to out-variables, i.e.:

$$\beta_i^{out} = \beta_i^{in} + \frac{\partial K}{\partial \alpha_i} \; ; \quad b_i^{out} = b_i^{in} + \frac{\partial K}{\partial a_i} \; . \tag{19}$$

The generator K is found by direct integration to be:

$$K(\alpha_i, a_i) = \frac{4}{K}(a_1-a_2)\,\ell n\,\frac{(a_1-a_2)^2 + (\frac{K}{4}(\alpha_1-\alpha_2))^2}{(a_1-a_2)^2 + (\frac{K}{4}(\alpha_1+\alpha_2))^2} + \tag{20}$$

$$+ 2(\alpha_1+\alpha_2)\ \mathrm{arctg}\ (\frac{K}{4}\,\frac{\alpha_1+\alpha_2}{a_1-a_2}) - 2(\alpha_1-\alpha_2)\ \mathrm{arctg}\ (\frac{K}{4}\,\frac{\alpha_1-\alpha_2}{a_1-a_2}).$$

Hence the classical phase shift reads [5]:

$$\delta_{cl}\ (\alpha_i, a_i) = -\frac{1}{2}\,K\ (\alpha_i,\ a_i\ )\ . \tag{21}$$

Notice that (21) coincides with the result of Dolan [8] for the special case $\alpha_1 = \alpha_2 = 1$. Let us now discuss the quantum correction.

3. THE QUANTUM CORRECTION

In calculating the quantum correction (14b) we first have to diagonalize the quadratic form $A_q[\chi, \omega]$ appearing in (10). This will conveniently be achieved by the concept of stability angles. In order to define stability angles we make the two-soliton-solution periodic by putting the solitons on a circle of length L. In the initial configuration the solitons are separated as widely as possible. Running then in opposite directions the solitons arrive simultaneously at their initial positions (in the CMS) after n_1 (resp. n_2) cycles. During these cycles there are $(n_1 + n_2)$ interactions between the solitons each of them giving rise to a time delay ΔT_{Ev}^{i} in the motion of the envelopes. Therefore, by imposing the conditions:

$$n_i\, L = v_i\ (T + (n_1 + n_2)\Delta T_{Ev}^{i}) \ ; \qquad T = t_f - t_i \tag{22}$$

the motion of the envelopes becomes periodic in time with period T. During a period the internal oscillations accumulate the following phases:

$$\phi_i = n_i |2\xi_i| L - 4(\xi_i^2 - n_i^2)(T + (n_1+n_2)\Delta T_{In}^{i}) - m_o T \tag{23}$$

where ΔT_{In}^{i} is the time delay of the internal motion due to one collision. In order to maintain periodicity of the two-soliton-solution we require:

$$\phi_i = 2\pi\, \ell_i \ ; \qquad \ell_i \quad \text{integer} \ . \tag{24}$$

In diagonalizing $A_q[\chi,\omega]$ we first transform to real fields u,v which simplifies our considerations:

$$\chi = \frac{1}{\sqrt{2}}\ (u + iv) \ ; \qquad \omega = \frac{1}{\sqrt{2}}\ (v + iu) \ . \tag{25}$$

Hence the quadratic form reads:

$$A_q\ [u,v] = \frac{1}{2} \int_{t_i}^{t_f} dt \int_{-\infty}^{\infty} dx\ (u,v) \begin{pmatrix} 0 & i \\ -i & 0 \end{pmatrix} K \begin{pmatrix} u \\ v \end{pmatrix} \ ;$$

$$K = i \begin{bmatrix} \partial_t + K|\phi^{(2)}|^2 \sin 2\theta^{(2)} & \partial_{xx} - m_o + K|\phi^{(2)}|^2 (2-\cos 2\theta^{(2)}) \\ -\partial_{xx} + m_o - K|\phi^{(2)}|^2 (2+\cos 2\theta^{(2)}) & \partial_t - K|\phi^{(2)}|^2 \sin 2\theta^{(2)} \end{bmatrix}$$

$$\phi^{(2)} = |\phi^{(2)}| \; e^{i\theta^{(2)}} \quad . \qquad (26)$$

Since K now contains only periodic coefficients, the solutions of:

$$K \binom{u}{v} = 0 \qquad (27)$$

can be classified by the corresponding stability angles. The actual computation of the stability angles is rather tedious, but elementary and may be found in ref. [3]. The result is:

$$K \binom{g}{f}_n^\pm = 0; \qquad \binom{g}{f}_n^\pm (x,t+T) = e^{-i\nu_n^\pm} \binom{g}{f}_n^\pm (x,t)$$

$$\nu_n^\pm = \pm\, \nu(k_n) \;; \qquad 2L\, k_n + \delta(k_n) = 2\pi n \;;\; n = 0, \pm 1, \pm 2 \ldots.$$

$$\nu(k) = 4\, k^2\, T + m_o\, T + n_1\, \delta_1(k) - n_2\, \delta_2(k) \qquad (28)$$

$$e^{i\delta_i(k)} = \left(\frac{1 - i\,\frac{n_i}{\xi_i - k}}{1 + i\,\frac{n_i}{\xi_i - k}} \right)^2 ; \quad i = 1,2 \;; \qquad \xi_1 > 0, \qquad \xi_2 < 0$$

$$\delta(k) = \delta_1(k) + \delta_2(k) \; .$$

From (28) we construct periodic eigenfunctions of K by:

$$\binom{G}{F}^{\pm}_{n} = e^{i \frac{\nu^{\pm}_{n}}{T} t} \binom{g}{f}^{\pm}_{n}$$

$$K \binom{G}{F}^{\pm}_{n} = - \frac{\nu^{\pm}_{n}}{T} \binom{G}{F}^{\pm}_{n} \quad . \tag{29}$$

Together with the zero-modes (3) the solution $(G,F)^{\pm}_{n}$ form a complete orthogonal base of periodic eigenfunctions of K with respect to the scalar product:

$$(V,W) = \int_{-\frac{L}{2}}^{\frac{L}{2}} dx \, V^{\dagger} \begin{pmatrix} 0 & i \\ -i & 0 \end{pmatrix} W \quad . \tag{30}$$

It turns out that, in order to fullfil the constraints in (10) we have to introduce new base vectors:

$$\binom{R}{S}^{\pm}_{n} = \binom{G}{F}^{\pm}_{n} + (\lambda^{\pm}_{n})_{\alpha} \begin{pmatrix} \mathrm{Re}\, \phi^{(2)}_{\alpha} \\ \mathrm{Im}\, \phi^{(2)}_{\alpha} \end{pmatrix} \tag{31}$$

where the Lagrange multipliers $(\lambda^{\pm}_{n})_{\alpha}$ have to be determined in such a way that $(R,S)^{\pm}_{n}$ respect the constraints. Finally we normalize the vectors (31) according to:

$$\int_{-\frac{L}{2}}^{\frac{L}{2}} dx \binom{R}{S}^{\varepsilon\dagger}_{n} \begin{pmatrix} 0 & i \\ -i & 0 \end{pmatrix} \binom{R}{S}^{\varepsilon'}_{n'} = \varepsilon \; \delta_{\varepsilon\varepsilon'} \; \delta_{nn'} \quad . \tag{32}$$

$$\varepsilon = \pm$$

The quadratic form (26) can now be diagonalized by expanding the fields (u,v) in the base (31):

$$\binom{u}{v}(x,t) = \sum_{n,\varepsilon} a^{\varepsilon}_{n}(t) \binom{R}{S}^{\varepsilon}_{n} (x,t) \quad . \tag{33}$$

Since the stability angles appear in pairs, i.e. $\nu^+ = -\nu^-$ we have $a_n^+(t) = (a_n^-(t))^*$, and the quadratic form reads:

$$A_q[a_n^+(t), a_n^{+*}(t)] = \int_{t_i}^{t_f} dt \sum_n \{\frac{1}{2i}(\dot{a}_n^{+*} a_n^+ - a_n^{+*} \dot{a}_n^+) - \frac{\nu_n^+}{T} a_n^{+*} a_n^+\} \quad . \tag{34}$$

Thus $A_q[a_n^+(t), a_n^{+*}(t)]$ becomes a sum of decoupled harmonic oscillators. It is therefore convenient to choose the quantum wave functionals $\Psi^q_{i,f}[\omega,\chi]$ such as to be the ground states of the harmonic oscillators (34), i.e.:

$$\Psi^q_{i,f}[\omega,\chi] = \exp\{-\frac{1}{2}\sum_n a_n^{+*}(t_{i,f})\, a_n^+(t_{i,f})\} \quad . \tag{35}$$

The quantum correction (14b) is now easy to evaluate, the result being:

$$e^{2i\delta_q} = e^{-\frac{i}{2}\sum_n \nu_n^+} \tag{36}$$

The procedure of summing up the stability angles is rather technical and may be found in ref. [3]. One has to count carefully the distinct modes and also to subtract the vacuum contributions. After all being done we find:

$$-\frac{1}{2}\sum_n \nu_n^+ = 2\pi k \quad ; \qquad k \quad \text{integer} \tag{37}$$

i.e. the quantum correction to the phase shift in the

one-loop approximation vanishes.

Fortunately the NLSE allows one to compare our approximate phase shift with the exact one which can be determined in a quantum mechanical N-body system. We now turn to this problem.

4. EXACT RESULTS VERSUS SEMI-CLASSICAL APPROXIMATION

It is well known that the NLSE has an associated exactly solvable quantum mechanical N-body problem with Hamiltonian:

$$H = - \sum_{i=1}^{N} \frac{\partial^2}{\partial x_i^2} - K \sum_{i<j}^{N} \delta(x_i - x_j) \quad . \tag{38}$$

The energy spectrum as well as arbitrary transition amplitudes of the system (38) can explicitely be calculated [9]. It turns out that there is only one N-particle bound state the binding energy being:

$$E_B = - \frac{K^2}{48} (N^3 - N) \quad . \tag{39}$$

Hence there is a one-to-one correspondence between a quantized soliton of the NLSE with internal quantum number N and the bound state (39). The scattering of quantized solitons with internal momenta N, M can therefore be treated as interaction between N- and M-particle bound states of the system (38). The corresponding transition amplitude has been calculated in ref. [3]. The process is purely elastic and the exact phase shift reads:

$$\delta_{ex}(E,N,M) = \text{arctg}\,\frac{K(M+N)}{4\sqrt{E_{MN}}} - \text{arctg}\,\frac{K(M-N)}{4\sqrt{E_{MN}}} -$$

$$- 2 \sum_{n=1}^{M-1} \text{arctg}\,\frac{K(M-N-2n)}{4\sqrt{E_{MN}}}$$

$$E_{MN} = \frac{M+N}{NM} E \quad ; \qquad M \geq N \tag{40}$$

where E is the kinetic energy in the CMS. Let us now compare (40) with the approximate result (21) in the weak coupling limit. We have:

$$\delta_{cl} = K\,\delta_{cl}^{(1)} + K^3\,\delta_{cl}^{(3)} + O(K^5)$$

$$\delta_{ex} = K\,\delta_{ex}^{(1)} + K^3\,\delta_{ex}^{(3)} + O(K^5)$$

$$\delta_{cl}^{(1)} = \delta_{ex}^{(1)} = \frac{MN}{2\sqrt{E_{MN}}} \tag{41}$$

$$\delta_{cl}^{(3)} = -\frac{1}{3}(16\,E_{MN})^{-3/2}\, 2\,MN\,(N^2 + M^2)$$

$$\delta_{ex}^{(3)} = -\frac{1}{3}(16\,E_{MN})^{-3/2}\,(2MN\,(N^2 + M^2) + 4\,MN)\ .$$

Thus both results agree in lowest order of K. Taking into account the explicit $\hbar$-dependence of the phase shifts one recovers the K-expansion to be effectively an expansion in powers of $\hbar$. As there are no K^2-contributions to δ_{ex} (i.e. no $\hbar^o$-contributions) we conclude that δ_q which is at least of order $\hbar^0$ i.e. K^2, should not contain terms of this order. We have shown by explicit calculation that δ_q vanishes at all. Non-zero corrections should arise at the two-loop level which may encourage further investigations along the lines of ref. [10].

REFERENCES

1. R.F. Dashen, B. Hasslacher and A. Neveu, Phys. Rev. D10, 4114, 4130, 4138 (1974), D11 (1975) 3424, D12 (1975) 2443.

2. J.L. Gervais and B. Sakita, Phys. Rev. D11 (1975) 2943. J.L. Gervais, A. Jevicki and B. Sakita, Phys. Rev. D12 (1975) 1038, Phys. Rep. 23C (1976) 281.

3. J. Honerkamp, M. Schlindwein and A. Wiesler, Freiburg Preprint, December 1976, to appear in Nucl. Phys. B.

4. J.L. Gervais and A. Jevicki, Nucl. Phys. B110 (1976) 113.

5. V.E. Korepin, P.P. Kulish and L.D. Fadeev, JETP Lett. 21 (1975) 138.

6. V.E. Zakharov and A.B. Shabat, JETP 34 (1972) 62. V.E. Zakharov and V. Manakov, teor. i. Mat. Fiz. 19 (1974) 332.
   D.J. Kaup, Journ. of Math. Phys. 16 (1975) 2036.

7. C.R. Nohl, Ann. of Phys. 96 (1976) 234.

8. L. Dolan, Phys. Rev. D13 (1975) 528.

9. J.B. Mc Guire, Journ. of Math. Phys. 5 (1964) 622. C.N. Yang, Phys. Rev. 168 (1968) 1920.

10. H.J. de Vega, Nucl. Phys. B115 (1976) 411.

Acta Physica Austriaca, Suppl. XVIII, 445–461 (1977)

BOSON-FERMION CORRESPONDENCE IN QUANTUM THEORY AND QUANTIZATION OF SPINOR FIELDS+

by

P. GARBACZEWSKI
Institute of Theoretical Physics
University of Wroclaw, Poland

1. INTRODUCTION

Recent developments on the connection between Thirring and Sine-Gordon systems in two space-time dimensions resulted in a couple of papers on the question of Fermion-Boson correspondence in quantum field theory (mysterious metamorphosis of Fermions into Bosons, as S. Coleman said), see e.g. [3]. The mentioned correspondence is not a particular feature of q.f.t.only. For example, under the name of the method of Boson expansions, it was employed to build a contemporary theory of spin waves in the low-temperature description of Heisenberg ferromagnet. In this last case, there was known for long time that the ideal magnon gas, in the weak ex-

+Seminar given at XVI.Internationale Universitätswochen für Kernphysik, Schladming, Austria, February 24-March 5, 1977.

citation limit, perfectly simulates the behaviour of the Heisenberg crystal itself, despite of the spin value assigned to the sites of the lattice.

A similar situation appears in the study of the weak excitation limit of the atomic nuclei, in the microscopic model, where the spectra of low lying excited states are similar to these of the weakly excited system of quadrupole Bosons. All that allows to expect that each quantum Boson in the weak excitation limit (not true for isolated systems, one needs any regulation mechanism establishing the needed excitation level), can exhibit Fermion properties, which then prevail the original Boson ones (Fermion-like behaviour). Here, the weak excitation (low temperature) limit of the Boson theory can be also considered as its strong coupling limit, provided the strong coupling potential (large distance phenomena in case of Heisenberg ferromagnet) prevents the Boson system from occupying more than a few, low lying, energy levels.

Quite conversely, if the higher excitations (weak coupling limit) are admitted, then starting from the Fermion system, we can expect that Boson properties will prevail the original Fermion ones (Boson-like behaviour of the Fermion).

Above conjectures have an unrestricted validity in the nonrelativistic quantum theory, or if the number of space-time dimensions is less than 4. In either case, the spin-statistics theorem must be taken into account.

2. ϕ_2^4 JUSTIFICATION: WHAT CAN BE DRAWN FROM THE BOSON SYSTEM

To support our thesis that, in a few cases at least, Bosons can be treated as more fundamental than Fermions, let us discuss the ϕ_2^4 example, following [1]. The ϕ_2^4 Hamiltonian is given by:

$$H = \int dx \ \{\frac{1}{2}\pi^2 + \frac{1}{2}(\nabla\phi)^2 + \lambda(\phi^2-f^2)^2\} \ . \tag{2.1}$$

This continuous model can be approximated by its <u>lattice version</u> (linear lattice, with the inverse spacing constant Λ and the number of 2N+1 sites). Due to the finite volume, the allowed momenta are

$$k = \frac{2\pi}{L} n, \ n = 0, \pm 1, \ldots, \pm N, \ L = \frac{2N+1}{\Lambda} \ ,$$

and

$$H = \frac{1}{\Lambda} \sum_s \{\frac{1}{2}\pi_s^2 + \frac{1}{2}(\nabla\phi_s)^2 + \lambda(\phi_s^2-f^2)^2\} \ , \tag{2.2}$$

where s enumerates lattice sites, and the gradient term should be still properly defined (in case of Bosons $\nabla\phi_s = \Lambda(\phi_{s+1} - \phi_s)$ can be introduced). In the rescaled form:

$$[\pi_s, \phi_t]_- = -i\Lambda\delta_{st} \rightarrow [p_s, x_t]_- = -i\delta_{st},$$

we have

$$H = \Lambda \sum_s \{(\frac{1}{2} p_s^2 + \frac{1}{2}(\nabla x_s)^2 + \lambda_o(x_s^2-f_o^2)^2\} \ . \tag{2.3}$$

There is useful to note here, that the gradient term carries an interaction between lattice sites:

$$H = \sum_s \{H^{(s)}_{self} + H^{(s)}_{int}\} \quad ,$$

while the single site term $H^{(s)}_{self}$ describes a Schrödinger problem of a particle in an anharmonic potential. Neglection of the gradient leaves us with the chain of noninteracting anharmonic solutions, for which a Fock construction exists, resulting in the single site basis

$$|\psi> = \prod_s |\psi_s> \ ,$$

$$<\psi_s|\psi_t> = \delta_{st}, \ |\psi_s> = \sum_{n_s=0}^{\infty} c_{n_s} \ |n_s> \ , \ |n_s> = \frac{1}{\sqrt{n_s!}}(a_s^*)^{n_s}|0_s>,$$

where $|0_s>$ is the s-th site vacuum.

Taking the expectation value $<\psi|H|\psi>$ of (2.3) in the single site trial state $|\psi>$, through minimization procedures one can calculate the ground state energy of the interacting system (2.3).

Let us now consider the lattice version of ϕ_2^4 system with the nearest neighbor coupling (periodic boundary conditions),

$$H = \Lambda\{\sum_s \frac{p_s^2}{2} + \frac{\mu^2+2}{2} x_s^2 + \lambda x_s^4 - x_s x_{s+1}\} \quad . \tag{2.4}$$

The single site terms describe anharmonic oscillators at each site, so that the single site basis can be introduced at once: $\otimes_s|n_s\rangle$, $0 \leq n_s \leq \infty$, and further the matrix form of the Hamiltonian (2.4):

$$H = \sum_s (E^s - X^s \otimes X^{s+1}) \; ; \tag{2.5}$$

$(H \equiv H/\Lambda)$, $E = \{E_n\}$ is a diagonal matrix with single site eigenvalues on the diagonal, $X = \{\langle n|x|m\rangle\}$ its elements do not vanish between even and odd parity states.

Truncation of the single site base: $0 \leq n_j \leq s-1$ to a finite number S of lowest energy levels corresponds to the approximation of the lattice system (2.4) by the coupled spin system (2s+1 = S, the <u>finite spin approximation</u> of (2.4) is achieved).

In special case of spin 1/2 approximation, the Hamiltonian matrix (2.5) reads:

$$H = \text{const} + \sum_s \{\frac{\varepsilon}{2}\sigma_s^z - \Delta(\sigma_s^+ + \sigma_s^-)(\sigma_{s+1}^+ + \sigma_{s+1}^-)\} \tag{2.6}$$

with

$$\varepsilon = (E_1 - E_o), \quad \Delta = |\langle 0|x|1\rangle|^2, \quad \sigma_{N+1} \equiv \sigma_{-N},$$

σ's are Pauli matrices. This is the case, when the vacuum and single excitation levels of the starting system (2.4) are mostly important (higher excitations appear with a negligible probability).

When Pauli matrices are involved, by the use of so-called Jordan-Wigner trick one can rewrite (2.6) in the equivalent form, where Fermi operators only appear (Fermion approximation of (2.4)):

$$H = LE_o + \varepsilon \sum_{s=-N}^{N} b_s^* b_s - \Delta \sum_{s=-N}^{N} (b_s^* - b_s)(b_{s+1}^* + b_{s+1}) +$$

$$+ \Delta (b_N^* - b_N)(b_{-N}^* + b_{-N}) \; (\exp(i\pi n) + 1) \qquad (2.7)$$

where

$$n = \sum_s n_s, \qquad n_s = b_s^* b_s \; .$$

In this place one can obviously state the question whether there exists any continuous Fermion theory, whose lattice approximation is (2.7).

Let us emphasize that in the above approximations of the starting Boson system (2.1) we did not bother what were exactly the mechanisms, whose influence could justify the choice of a concrete approximation. The question of interest was rather to identify the physical situations in which the starting Boson system like-transforms (in the approximate sense) into the finite spin or pure Fermion system.

If in addition to introduce into consideration the question of classical basis behind the quantum concepts (as e.g. the kind of correspondence principle realized via coherent state methods), then the diagram of current problems can be completed.

DIAGRAM OF PROBLEMS

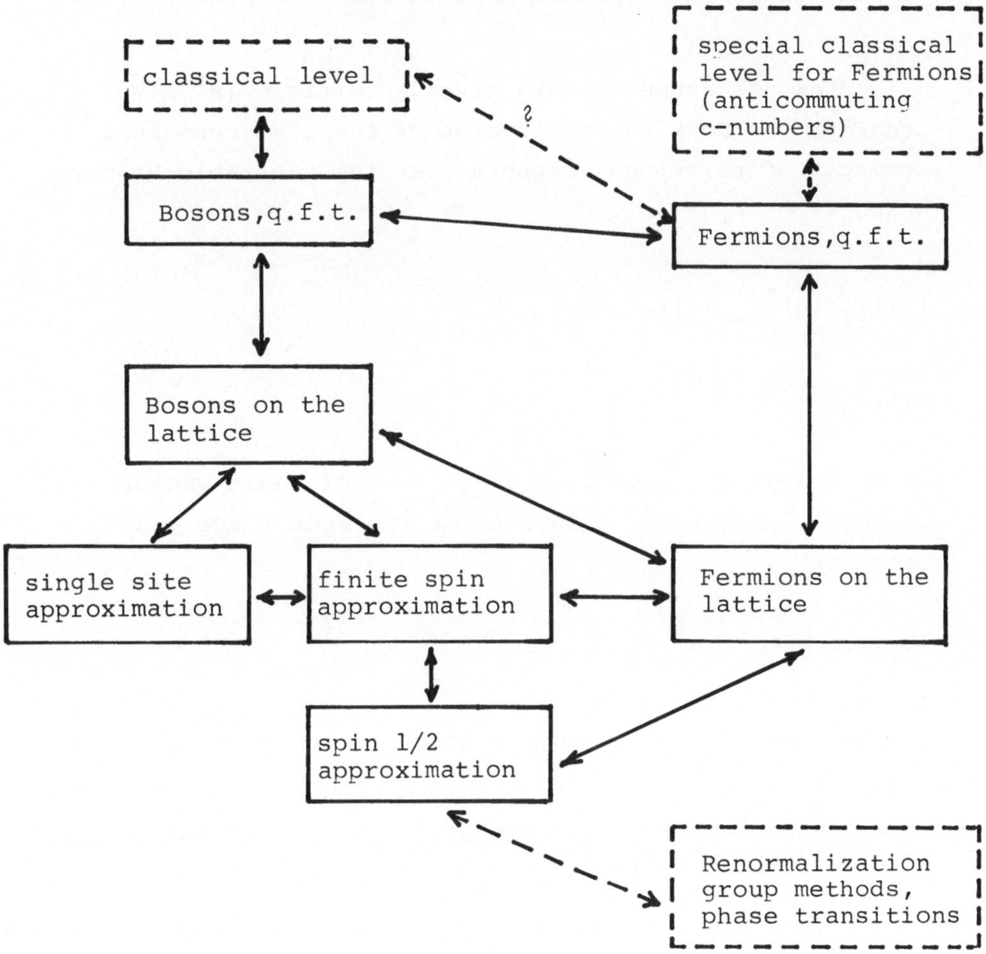

In the above, majority of steps can be realized by the use of Boson expansion methods, whose basic aim is to start from the given Boson system (q.f.t. or a corresponding classical level eventually), and generate further as many as possible from the indicated relations.

3. FERMION-BOSON CORRESPONDENCE IN THE FOCK CONSTRUCTION

Let us assume to have given the triple $\{a^*, a, \Omega_B\}_K$, generating a Fock representation of the CCR (canonical commutation relations) algebra over the separable Hilbert space $K \ni f,g$:

$$[a(f), a(g)^*]_- = (f,g)\, 1_B$$

$$a(f)\Omega_B = 0 \quad . \qquad (3.1)$$

If to choose a sequence $\{f_k\}_{k=1,2,\ldots}$ of basis vectors in K, then $a_k^* = a(f_k)^*$, allows to define Fock space basis vectors:

$$|k_1,\ldots,k_n>_B = a_{k_1}^* \ldots a_{k_n}^* \Omega_B \ ,$$

so that the Fock space vector is given in the form:

$$F_B \ni |F>_B = \sum_{n} \sum_{\{k\}} F^s_{k_1\ldots k_n} |k_1,\ldots,k_n>_B = \sum_n (F^s_n, |n>_B) . \qquad (3.2)$$

In the same way one can proceed in the Fermi case: $\{b^*, b, \Omega_F\}_K$ is given by

$$[b(f), b(g)^*]_+ = (f,g)\, 1_F$$

$$b(f)\Omega_F = 0 \ , \qquad (3.3)$$

so that $b_k^* = b(f_k)^*$ implies

$$|k_1,\ldots,k_n>_F = b_{k_1}^* \ldots b_{k_n}^* \Omega_F \ ,$$

and further

$$F_F \ni |F\rangle_B = \sum_n \sum_{\{k\}} F^a_{k_1 \ldots k_n} |k_1, \ldots, k_n\rangle_F = \sum_n (F^a_n, |n\rangle_F) \; . \tag{3.4}$$

In the above superscripts s,a denote symmetric and anti-symmetric tensors respectively, while F_B, F_F the Boson and Fermion Fock spaces respectively.

If to introduce now a discrete version $\varepsilon_{k_1 \ldots k_n}$ (Levi-Civitta tensor in n-dimensions) of the continuous Friedrichs-Klauder sign function $\sigma_n(k_1, \ldots, k_n)$, see e.g. [2], then one easily notices that in the Fock construction there is no essential difference between vectors of the form (3.2) and (3.4). For example one can consider $F^a_{k_1 \ldots k_n}$ in the form $F^s_{k_1 \ldots k_n} \varepsilon_{k_1 \ldots k_n}$ so that:

$$|F\rangle_F = \sum_n \sum_{\{k\}} F^s_{k_1 \ldots k_n} \{\varepsilon_{k_1 \ldots k_n} |k_1 \ldots k_n\rangle_F\} \; , \tag{3.5}$$

suggesting that $|F\rangle_F$ can be as well the element of F_F and F_B, provided suitable restrictions on representations of the CCR and CAR algebra are given.

This is exactly the case, when "schizons" (see Schroer's lectures [3]) are needed. There are the Boson and Fermion representations, whose vacuum and one-particle sectors coincide. In case of $K = L^2(R^n)$, one can even get a very simple example of Boson constructed Fermions:

$$k \in R^n, \; b(k) = \exp(-i\pi \int_k^\infty a^*(p) a(p) dp) \times a(k) \quad .$$

This representation can be always closed on F_F to a more sophisticated example, constructed in [5] (we prefer here a discrete language, but the transition to a continuous one is nearly immediate if in the place of $\varepsilon_{k_1 \dots k_n}$ to put $\sigma_n(k_1 \dots k_n)$ and in the place of summations with respect to k's to consider respective integrations):

$$b(f) = :\exp(-\sum_s a_s^* a_s) \cdot \sum_{n,m} \frac{1}{\sqrt{n!m!}} \sum_{\{r\}} \sum_t \cdot$$

$$\cdot \sqrt{n+1}\ \delta_{m,1+n}\ \varepsilon_{r_1 \dots r_n}\ \bar{f}_t\ \varepsilon_{tr_1 \dots r_n}\ a_{r_1}^* \dots a_{r_n}^* \cdot$$

$$\cdot \quad a_t a_{r_1} \dots a_{r_n} \cdot \tag{3.6}$$

One can easily check that

$$a(f)\Omega_B = b(f)\Omega_B$$

$$a(f)^*\Omega_B = b(f)^*\Omega_B$$

$$[b(f), b(g)^*]_+ = (f,g)\ 1_F \tag{3.7}$$

where

$$1_F = \sum_n \frac{1}{n!} \sum_{\{r\}} :\ a_{r_1}^* \dots a_{r_1}^*\ \ \varepsilon_{r_1 \dots r_n}^2\ a_{r_1} \dots . a_{r_n} \cdot$$

$$\cdot \exp(-\sum_s a_s^* a_s): \tag{3.8}$$

is a projection in F_B, so that

$$F_F = 1_F\, F_B \quad . \tag{3.9}$$

If one wishes to deal with a finite number of annihilation and creation generators a_s^*, a_s and b_s^*, b_s respectively, there is enough to restrict summations with respect to $\{r\}$,t in (3.6), (3.8) to a finite number N, say. The most general element of the CAR algebra (3.6)-(3.9) is of the form

$$:F(b^*,b): = \sum_{nm} (f_{nm}, b^{*n} b^m) = \sum_{nm} \sum_{\{r\}} \sum_{\{s\}} \cdot$$

$$\cdot\, f_{r_1 \ldots r_n\, s_1 \ldots s_m} \cdot b^*_{r_1} \ldots b^*_{r_n}\, b_{s_1} \ldots b_{s_m} \quad , \tag{3.10}$$

where f_{nm} is the n+m-antisymmetric tensor. One can easily check that

$$:F(b^*,b):\Omega_B = \sum_{nm} (\overset{c}{f}_{nm}, a^{*n} a^m)\,\Omega_B = :\overset{c}{F}(a^*,a):\Omega_B \quad , \tag{3.11}$$

where

$$\overset{c}{f}_{r_1 \ldots r_n\, s_1 \ldots s_m} = f_{r_1 \ldots r_n\, s_1 \ldots s_m} \cdot \varepsilon_{r_1 \ldots r_n}\, \varepsilon_{s_m \ldots s_1} \quad . \tag{3.12}$$

Furthermore, if to take into account a few symmetry arguments, concerning especially the decomposition of n-point tensors into irreducible parts with respect to the symmetry group, one can prove [5] the following <u>projection theorem:</u>

$$:F(b^*,b):F_F = 1_F:\overset{c}{F}(a^*,a):F_F \tag{3.13}$$

being the identity on F_F. If specialized, we find that on F_F, the projected Boson generators are in fact Fermion generators:

$$1_F\ a(f)\ 1_F = b(f)$$
$$1_F\ a(f)^* 1_F = b(f)^* \quad . \tag{3.14}$$

Formulas (3.13), (3.14) provide an elegant way of changing the symmetry properties of any theory under consideration, where expansions into series of normal-ordered products of Fock generators are admitted.

We see thus at once that, if physics in any way makes reasonable the reduction of interests concerning the Boson system to $F_F = 1_F\ F_B$, then the approximation of it by the corresponding (associated) Fermion system is justified.

4. SELECTED APPLICATION: ISOTROPIC HEISENBERG LATTICE

As a special example of the projection theorem (3.13), one can study a Boson theory, whose weakly excited (low temperature) limit well approximates properties of the Heisenberg ferromagnet in low temperatures.

Namely, if to reenumerate the set of generators: $(k) \to (k\alpha)$, $k = 1,\dots,N$, $\alpha = 1,\dots,n$, we can start from the Hamiltonian

$$H_B = G_o - \mu \sum_{k}^{N} \vec{\kappa}\cdot\vec{s}_k - (1/2) \sum_{k,h=1}^{N} J_{kh}\, \vec{s}_k\, \vec{s}_h \,, \tag{4.1}$$

where

$$\vec{s}_k = (s_k^x, s_k^y, s_k^z), \qquad \text{and} \qquad s_k^+ = \sum_{\alpha}^{n} a_{k\alpha}^*,\ s_k^- = \sum_{\alpha}^{n} a_{k\alpha} \,,$$

$$s_k^2 = \{-(n/2) + \sum_{\alpha}^{n} a_{k\alpha}^*\, a_{k\alpha}\} \,.$$

By applying the projector

$$P_o = :\exp(-\sum_{k\alpha} a_{k\alpha}^*\, a_{k\alpha}): + \sum_{k=1}^{N} P_o^k \tag{4.2}$$

with

$$P_o^k = 1_F^k - :\exp(-\sum_{\alpha}^{n} a_{k\alpha}^*\, a_{k\alpha}):,$$

where 1_F^k is given by (3.8) if specialized to the total number n of Boson generators belonging to the k-th from N different collections of them ($a_r^* \to a_{k\alpha}^*$, summation with respect to α), we get

$$P_o\, F_B = F_o \tag{4.3}$$

i.e. the Hilbert space of the spin states (finite spin approximation), and further

$$P_o\, H_B\, P_o = H \,,$$

where $H = H(\vec{s}_k \to \vec{S}_k)$ is the Heisenberg ferromagnet

Hamiltonian, and $\vec{S}_k$ the spin operator at the s-th site of the lattice: $\vec{S}_k = P_o \vec{s}_k P_o$. For n = 1 we get spin-1/2 lattice, while in other cases F_o can be decomposed into subspaces corresponding to the irreducible representations of the SU(2): for n = 2, we get spin 1 and spin 0 examples.

From the physical point of view the above procedure is based on the assumption that the ground state and the first excited level of each single degree of freedom of the Boson system are of importance (spin 1/2 approximation behind the received finally finite spin approximation of the Boson theory). In case when $\alpha = 1,\ldots,n$ one can interpret (4.4) as a kind of a condensation of Bosonic degrees of freedom around the lattice sites, so that in the original Heisenberg lattice, one more lattice (of the condensed magnon gas) appears.

5. THE CORRESPONDENCE PRINCIPLE IN Q.F.T.: QUANTIZATION OF SPINOR FIELDS WITH NO USE OF ANTICOMMUTING C-NUMBERS

Under the Haag-LSZ assumptions, the most general element of the scalar Boson field algebra can be written in the form (compare Klauder's lecture)

$$:F(\phi): = \sum_n (f_n, :\phi^n:) \ , \qquad (5.1)$$

where brackets denote integrations with respect to Minkowski space-time variables, $:\phi^n:$ is a shorthand notation for a normal-ordered product of free (asymptotic) fields taken at different space-time points. Let $\bar{\alpha}(k)$, $\alpha(k)$, $k \in R^3$, denote Fourier amplitudes of the classical scalar field $\overset{c}{\phi}(x)$. On the basis of coherent state techniques, one can employ so-called functional representation of the CCR algebra [4],what we symbolize by

$$:F(\phi):(\bar{\alpha},\alpha) = \sum_n (f_n, \overset{c}{\phi}{}^n) \exp(\bar{\alpha},\alpha) \ , \tag{5.2}$$

and on the r.h.s. of (5.2) the classical free fields $\overset{c}{\phi}(x)$ appear. In the functional representation, $\exp(\bar{\alpha},\alpha) = 1_B(\bar{\alpha},\alpha)$, and is the operator unit (the Fock space transforms in that case into the Bargman space). Obviously $F(\overset{c}{\phi}) = \sum_n (f_n, \overset{c}{\phi}{}^n)$ can appear here as a coherent state expectation value $\langle :F(\phi): \rangle$ of the operator expression, however the use of functional representation has a great advantage of providing the 1-1 map between the classical and quantum level of a given Boson theory, with no polynomial limitations.

Using the functional representations [4,5] of the canonical relations (CCR and CAR) algebras one can prove the following correspondence rule: Let us extend the Haag-LSZ expansion theorem onto the case of Dirac fields:

$$:\Omega(\psi,\bar{\psi}): = \sum_{nm} (\omega_{nm}, :\psi^n \bar{\psi}^m:) \ . \tag{5.3}$$

Then:

(i) the subsidiary Boson level of the starting Fermion theory is given, where

$$1_F :\overset{c}{\Omega}(\overset{B}{\psi},\overset{B}{\bar{\psi}}): 1_F = :\Omega(\psi,\bar{\psi}): \tag{5.4}$$

is an identity in F_F (the spinor $\overset{B}{\psi},\overset{\underline{B}}{\psi}$ as obeying the commutation rules should violate assumptions of spin-statistics theorem);

(ii) the unrestricted (by projections 1_F) Boson level $:\overset{c}{\Omega}(\overset{B}{\psi},\overset{\underline{B}}{\bar{\psi}}):$ admits a straightforward classical map

$$<:\overset{c}{\Omega}(\overset{B}{\psi},\overset{\underline{B}}{\bar{\psi}}):> \; = \; \overset{c}{\Omega}(\overset{c}{\psi},\overset{\underline{c}}{\psi}) \quad , \tag{5.5}$$

where $\overset{c}{\psi},\overset{\underline{c}}{\psi}$ are classical spinor fields (commuting ring).

The converse procedure can stand for a <u>quantization rule</u> of a given classical spinor system.

More details, as well as considerations concerning the map of the algebraic structure, can be found in [5].

REFERENCES

1. S. Yankielowicz, Nonperturbative approach to quantum field theories, SLAC Stanford preprint, (1976), and lecture given at 17th Scottish Universities Summer School in Physics.

2. J.R. Klauder, Ann. Phys. (N.Y.), <u>11</u> (1960) 123.
J.R. Klauder, Classical concepts in quantum contexts, chap. III, lecture given at this school.

3. T.H.R. Skyrme, Proc. Roy. Soc. <u>A262</u> (1961) 237.
R.F. Streater, I.F. Wilde, Nucl. Phys. <u>B24</u> (1970) 561.
A.J. Kàlnay, Progr. Theor. Phys. <u>54</u> (1975) 1848.
S. Coleman, Phys. Rev. <u>D11</u> (1975) 2088.

S.Mandelstam, Phys. Rev. D11 (1975) 3026.
J.A. Swieca, Solitons and confinement, Lecture notes, PUC Rio de Janeiro preprint, (1976).
B. Schroer, Quantum field theory of kinks in two-dimensional space-time, Cargèse lecture notes, (1976).
A. Luther, Eigenvalue spectrum of interacting massive fermions in one-dimension, Nordita 76/8.
M. Lüscher, Dynamical charges in the quantized renormalized massive Thirring model, DESY Hamburg preprint, (1976).

4. J.Rzewnski, Field theory part II, Functional formulation of the S-matrix theory, Iliffe Books Ltd, London, PWN Warsaw, 1969.
J.Rzewnski, Rep. Math. Phys. 1 (1970) 1.
J.Rzewnski, Rep. Math. Phys. 1 (1971) 195.

5. P. Garbaczewski, J.Rzewnski, Rep. Math. Phys. 6 (1974) 431.
P. Garbaczewski, Rep. Math. Phys. 7 (1975) 321.
P. Garbaczewski, Comm. Math. Phys. 43 (1975) 131.
P. Garbaczewski,Bosonization of fermions in Heisenberg ferromagnet, subm. for publ.;also in: Theoretical Physics, Memorial book an occasion of J.Rzewnski's 60th birthday, University of Wroclaw, 1976.
P. Garbaczewski, Z. Popowicz, Rep. Math. Phys. 11 (1977) 57.
P. Garbaczewski, The method of Boson expansions in quantum theory, University of Wroclaw preprints No. 375, 379 (1976),subm. for publ..
P. Garbaczewski, Quantization of spinor fields, University of Wroclaw preprint No. 398 (1977), subm. for publ..
P. Garbaczewski, Z. Popowicz, Plane pendulum in q.f.t.: Ultralocal quantization of Sine-Gordon 1-solitons, subm. for publ..

Acta Physica Austriaca, Suppl. XVIII, 463–469 (1977)

UNITARY SPIN, COLOUR AND UNIFIED THEORY+

by

J. RAYSKI
Institute for Theoretical Physics
University of Krakow, Poland

In contradistinction to ordinary spin, the unitary spin and colour have not acquired a clear, geometrical interpretation. Besides, the usual models based on SU(3) or SU(4) symmetry (with three colours) do not yield a correct value for the asymptotic ratio R(hadr./muon) = 2 or 3.3, whereas the experimental value is about 5.6. (Salam and Pati introduced an SU(4)xSU(4) symmetry which yields 4.4; still too small).

The situation is different within a unified theory of weak, electromagnetic, and gravitational interactions suggested by the present author [1] and also developed (independently) by Scherk [2] et al. The basic idea is

+ Seminar given at XVI.Internationale Universitätswochen für Kernphysik,Schladming,Austria,February 24-March 5,1977.

to consider a manydimensional space-time with some additional compact dimensions. Alternatively, it is possible to use the fibrebundle formalism.

The generalized space-time with four open and four closed (compact) dimensions may be denoted as follows

$$\underbrace{\underbrace{-\;+\;+\;+}_{\mu,\nu}\qquad\underbrace{+\;+\;+\;+}_{\alpha,\beta}}_{a,b,\ldots}$$

It was shown that a dependence of any field quantity upon the subsidiary coordinates x^{α} may be neglected unless unreasonably high concentrations of energy are available.

Splitting the metric tensor as follows,

$$(\gamma_{ab}) = \left[\begin{array}{c|c} g_{\mu\nu} + g_{\alpha\beta}\, f^{\alpha}_{\mu}\, f^{\beta}_{\nu} & g_{\alpha\beta}\, f^{\beta}_{\mu} \\ \hline g_{\alpha\beta}\, f^{\beta}_{\nu} & g_{\alpha\beta} \end{array}\right] , \qquad (1)$$

the quantities $g_{\mu\nu}$ are interpretable as ordinary metric tensor components whereas f^{α}_{μ} as the gauge fields. Indeed, the scalar curvature in 8-dimensions splits into a scalar curvature in 4 ordinary dimensions and a sum of Lagrangians for the gauge fields

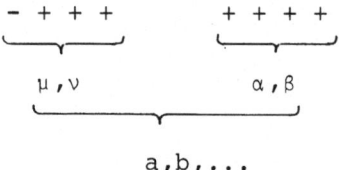

$$R_{(8)} \rightarrow R_{(4)} + \frac{1}{4} f^{\alpha}_{\mu\nu} f^{\alpha}_{\mu'\nu'} g^{\mu\mu'} g^{\nu\nu'} . \qquad (2)$$

The components $g_{\alpha\beta}$ have to exhibit the symmetry of the closed subspace, being also the symmetry of weak interactions. It is broken O(5) symmetry

$$O(5) \rightarrow O(3) \times O(2) \rightarrow SU(2) \times U(1) \qquad (3)$$

in agreement with Weinberg-Salam model of weak and electromagnetic interactions.

The sources of the generalized metric field (1) are leptons and quarks. According to a well known theorem the number of spinor components in an n-dimensional space should be $2^{E(n/2)}$. Thus, for n = 8 we should consider 16-component spinors. This may be achieved in the lepton case by introducing one heavy lepton N. According to Konopinski's suggestion the positively charged muon and the right-handed muonic neutrino are leptons rather than antileptons and we can form a 16-component spinor out from the following multiplets,

$$\begin{bmatrix} \nu_e \\ e^- \end{bmatrix}_L \quad \begin{bmatrix} N \\ e^- \end{bmatrix}_R \quad \begin{bmatrix} \mu^+ \\ \bar{\nu}_\mu \end{bmatrix}_R \quad \begin{bmatrix} \mu^+ \\ N \end{bmatrix}_L . \qquad (4)$$

Obviously, they constitute a natural generalization of the (highly asymmetric) scheme of Weinberg with left-handed doublets and right-handed singlets $(e^-)_R$, $(\mu)_R$.

It is seen that (i) muonic leptons are not ambarras de richesse but fit naturally into the scheme, (ii) parity violation is due entirely to the big mass difference between ordinary neutrinos ν and heavy neutrinos N. Consequently, it may be regarded as a spontaneously broken symmetry and explained in terms of Fayet [4] mechanism involving two Higgs fields interchanging under parity transformation.

A dynamics of the unified gravitational, electromagnetic, and weak interactions follows uniquely from a Lagrange formalism with a Lagrangian consisting of the scalar curvature $R_{(8)}$ and the Lagrangians for the leptonic fields (4) constituting the sources of the unified field. Partial derivatives have only to be replaced by covariant derivatives while a dependence upon the redundant coordinates $x^4,\ldots x^7$ may be neglected for reasons explained in reference [1].

It is not necessary to perform detailed calculations in order to see that this formalism will lead to equally good results as the Weinberg-Salam model does, provided the heavy neutrino N is really very heavy.

Question arises whether the above scheme may be extended so as to include quarks. The discovery (indirect) of charm led to the SU(4) symmetry. Four quarks, each describable by a 4-component spinor, form a 16-component object altogether. This fits well with the assumption of an 8-dimensional world. However, taking colour into account, the number of quark components increases considerably. The point is whether colour is also explicable in a natural way within the same geometrical scheme. It seems to be the case:

Quarks are different from leptons and have to be described differently. The next natural possibility, besides the spinor ψ_ρ representing leptons, is a field quantity endowed with a vector and a spinor index $\psi_{a\rho}$ where a = 0,...7 and ρ = 1,...16. Assume quarks to be represented by such an object. For a = 0,...3 it is a vector index μ in the ordinary space-time so that the corresponding field components describe a mixture of spin $\frac{1}{2}$ and $\frac{3}{2}$. Due to the supplementary condition

$$\gamma^{a}_{\rho\sigma} \psi_{a\sigma} = 0 \tag{5}$$

eliminating time-like components (for particle at rest) the number of independent components is not 8 x 16 but 7 x 16 splitting into 7 x 8 particles and 7 x 8 antiparticles. The 56 particle states split into

$$(3+4)\times 8 = 4\times 4 + 4\times 2 + 4\times 4 \times 2 \tag{6}$$

where 4x4 denotes a unitary spin quartet of particles with (ordinary) spin 3/2; while 4x2 denotes another unitary spin quartet of particles with spin 1/2; and 4x4x2 is interpretable as a unitary spin quartet of particles with spin 1/2 but endowed also with another characteristics that may be called "colour" (an SU(4) quartet of colours). Thus, our model predicts one colour singlet of particles with spin 3/2 as well as a colour singlet and a colour quartet of particles with spin 1/2.

If spin 3/2 quarks are considerably heavier than spin 1/2 quarks, then they do not contribute and the

Drell ratio R is

$$R = (1+4)[(\frac{2}{3})^2 + (\frac{2}{3})^2 + (\frac{1}{3})^2 + (\frac{1}{3})^2] = 5.6 \qquad (7)$$

in agreement with the present experimental evidence (fig. 1). However, it is to be expected that for still higher energies the spin 3/2 quarks would also come into play so that R will increase again up to about 7.8. These are the predictions of our model of unitary spin and colour.

REFERENCES

1. J.Rayski, Acta Phys. Polon. 27, 89 (1965).
   " Preprint TPJU 15 (1975).
   " Preprint Dublin Inst. Ad. Studies TP-76-31 (1976).
   " Preprint TPJU 2 (1977).
   " Nuovo Cimento (in course of publication).

2. E. Cremer, J. Scherk, Nucl. Phys. B 1o3, 399 (1976).

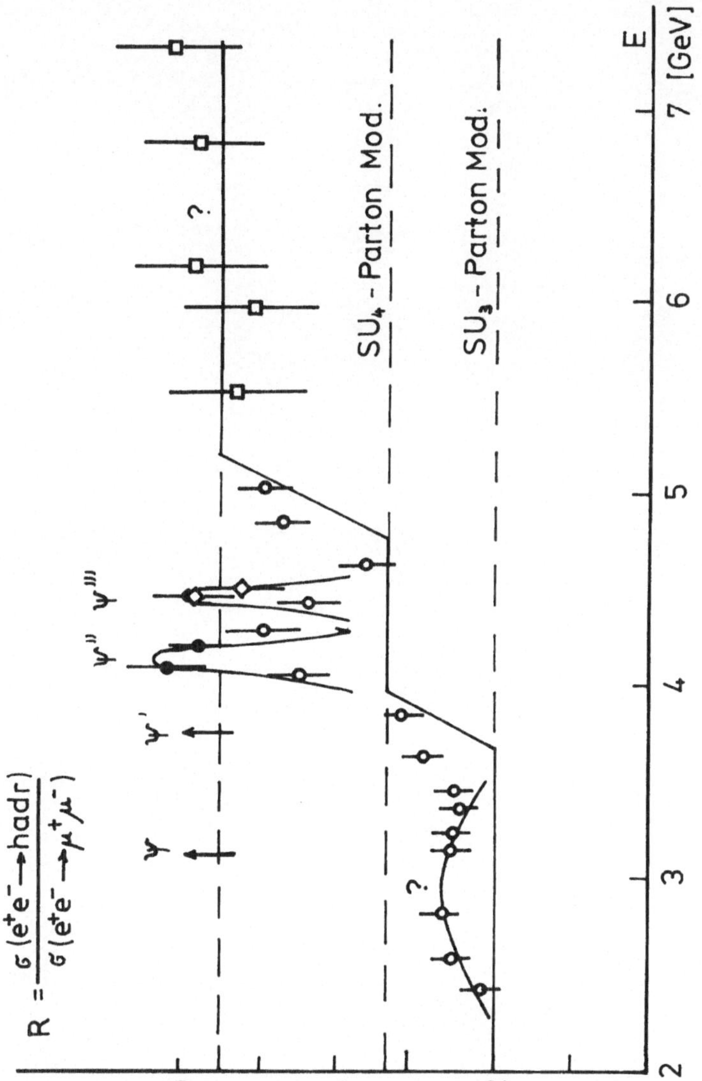

R = σ(e⁺e⁻ → hadr) / σ(e⁺e⁻ → μ⁺μ⁻)
ψ
ψ'
ψ''
ψ'''
?
?
SU₄ - Parton Mod.
SU₃ - Parton Mod.
E [GeV]
2
3
4
5
6
7
2
4
6

Fig.1

Acta Physica Austriaca, Suppl. XVIII, 471-541 (1977)

METHODS OF NONRELATIVISTIC QUANTUM FIELD THEORY+

by

K.W. KEHR
Institut für Festkörperforschung
der Kernforschungsanlage Jülich, Germany

1. INTRODUCTION

The methods of nonrelativistic quantum field theory are used to deal with interacting many-body systems, especially at low temperatures, where the quantum nature of the particles is important. The systems typically contain of the order 10^{23} particles, but also nuclei with typically 100 particles are included. The complete solution of the Schrödinger equation of a given interacting many-body system is normally impossible, but also uninteresting. One is mainly interested in quantities such as

a) ground state energy

b) excited states

   b') "quasiparticles", i.e. modified single-particle excitations

   b") collective excitations

+Lecture given at XVI.Internationale Universitätswochen für Kernphysik, Schladming, Austria, February 24-March 5, 1977.

c) thermodynamic behaviour, e.g. the specific heat
d) transport properties.

In the last decade, the behaviour of many-particle systems near critical points of phase transitions attracted much interest. Also the theoretical understanding of critical behaviour is a task for the nonrelativistic quantum field theory. Since this theme will be treated in the lectures of Professor Wagner and Professor Jona-Lasinio, it will not be touched here.

The general method of nonrelativistic quantum field theory is the use of "Green's functions" or propagators. They are defined with particle creation and destruction operators, which obey appropriate commutation relations. In this way the statistics of the particles is taken into account. The Green's functions play a double role:

a) they are used in perturbation theory;
b) they contain information on the quantities mentioned above: directly on the excitations, indirectly on the ground state energy, density of states, etc..

Many of the methods used in the many-body problem are perturbative. Additional definitions are necessary for systems with "broken symmetry" such as superconductors. Also non-perturbative methods have been applied, e.g. sum rules, or Ward identities. These methods will not be treated here.

Since the audience is mainly composed of high-energy physicists, only a survey will be given on how one works with Green's functions in the many-body problem and what can be achieved with them. Some typical methods will be presented, and some typical results will be mentioned. No scholarly review on Green's functions in the many-body problem is intended. References will be made to some text-books [1-6] and some original papers, without minding about priorities or completeness.

A historical remark seems in order. The great time of the many-body problem was the end of the fifties and the beginning of the sixties. At that time the power of the methods was discovered, and tested on many problems. In the later sixties more complicated problems such as the Kondo problem were investigated. With the advent of the Wilson theory the main effort of many theorists went into the elucidation of critical phenomena. Thus the many-body problem has become a more quiet field. One exception is the recent development of the theory of superfluid He^3 which has attracted much interest.

2. INTRODUCTORY REMARKS ABOUT SYSTEMS

In these lectures we will consider many-body systems consisting of Fermions or Bosons, which interact with two-particle interactions.

Typical examples for Fermion systems are ^3He, electrons, and nuclear matter. In the case of electrons or protons, the long-range Coulomb interaction brings additional complications. Here we will consider neutral isotropic Fermi systems with short-range interactions, e.g. ^3He at not too low temperatures. Fig. 1 shows the momentum distribution $n(k)$ and excitation spectrum $\varepsilon(k)$. For the non-interacting Fermion system at zero temperature $n(k)$ is a step function at the Fermi surface, given by $k_F = (3\pi^2 n)^{1/3}$ where n is the particle density (two spin orientations). The non-interacting particles have an excitation energy $\varepsilon_o(k) = k^2/2m$ ($\hbar$ put equal to one) while the interacting system has well defined single-particle excitations only near the Fermi surface, for an added particle or hole

$$\varepsilon(k) = \varepsilon_F + (k_F/m^*)(k-k_F), \ k \approx k_F \quad . \tag{1}$$

ε_F is the Fermi energy and m^* is an effective mass. It is one of the tasks of the theory to show how the interactions with other particles modify the single-particle energies and lead to a finite lifetime of the excitations. In addition, the interaction of the modified single-particle excitations might allow for the existence of so-called "collective excitations". Due to the interactions, the thermodynamic behaviour is altered compared to the non-interacting system. One would like to relate the thermodynamic quantities, e.g. specific heat, to the modified excitation spectrum of the system. Also the transport properties, e.g. viscosity, will be affected.

Dramatic effects appear in the case of a net attractive interaction of the particles. Then the Fermi surface is unstable against the formation of bound particle pairs, or "Cooper pairs". The aim of many-body theory is the derivation of the new ground state, the new momentum distribution n(k) and excitation spectrum $\varepsilon(k)$ for this case.

There is only one common representative of a condensed Boson system, namely superfluid ^{4}He, but more exotic representatives attract much interest as well, e.g. pion condensates, which are discussed in the lectures of Professor Weise. Fig. 2 shows the momentum distribution n(k) and excitation spectrum of a Boson system at T = O. In the non-interacting system at T = O all N particles condense in the lowest momentum state k = O. Due to the interactions, there are particles outside the Bose condensate even at zero temperature, but there should be still a macroscopic number $N_o < N$ of particles in the condensate. The excitation spectrum of the interacting system shows phonon-like behaviour for small momenta, $\varepsilon(k) = ck$, where c is the sound velocity, and the roton dip, in the case of ^4He, for larger momenta. As will be discussed, one cannot distinguish between single-particle and collective excitations in the case of a condensed Bose system. Again, theory should derive the momentum distribution and the excitation spectrum, including the lifetimes, from the particle interactions, and deduce the thermodynamic behaviour and the transport properties.

It should be mentioned that there are two phenomenological theories for Fermion and Boson systems, both due to Landau, which describe very successfully the low-temperature properties of both systems. These theories are more precisely described as "microscopic models". One important task of the many-body theory is the justification of these models.

There is a wealth of other many-body systems, where field-theoretical methods have been applied, for example

a) phonons

b) coupled electron-phonon systems

c) spin systems
d) impurities.

These systems will be completely neglected in these lectures.

3. GREEN'S FUNCTIONS AT ZERO TEMPERATURE

3.1 Definition of Green's functions

To be specific, the two point Green's function for Fermions at zero temperature will be defined:

$$g_{\alpha\beta}(\underline{r}t,\underline{r}'t') = \frac{1}{\langle\psi_o|\psi_o\rangle} \langle\psi_o|T\psi_\alpha(\underline{r}t)\psi_\beta^+(\underline{r}'t')|\psi_o\rangle \ . \qquad (2)$$

$|\psi_o\rangle$ is the ground state of the system at zero temperature, derived from the hamiltonian given below. T is the time ordering or chronological operator of Dyson

$$T\psi_\alpha(\underline{r}t)\psi_\beta^+(\underline{r}'t') = \begin{cases} \psi_\alpha(\underline{r}t)\psi_\beta^+(\underline{r}'t') & t > t' \\ -\psi_\beta^+(\underline{r}'t')\psi_\alpha(\underline{r}t) & t < t' \ . \end{cases} \qquad (3)$$

$\psi_\alpha(\underline{r}t)$ is a Fermion field operator, time dependent in the Heisenberg picture and α is a spin index. The field operators obey the anticommutation relation at equal times

$$\{\psi_\alpha(\underline{r}),\ \psi_\beta(\underline{r}')\}_+ = \delta_{\alpha\beta}\ \delta(\underline{r}-\underline{r}') \ . \qquad (4)$$

We assume N particles enclosed in a box of volume Ω and impose periodic boundary conditions. Operators for plane-wave states $c_{k\alpha}$, $c^{+}_{k\alpha}$ are introduced, e.g. by

$$\psi_\alpha(\underline{r}) = \Omega^{-1/2} \sum_{\underline{k}} \exp(-i\underline{k}\cdot\underline{r})\, c_{\underline{k}\alpha} \; . \tag{5}$$

In the case of spinless Bosons, the field operators obey the commutation relation

$$[\psi(\underline{r}), \psi^{+}(\underline{r}')]_{-} = \delta(\underline{r}-\underline{r}') \quad . \tag{6}$$

Operators for plane-wave states of Bosons are designated by $a_{\underline{k}}$, $a^{+}_{\underline{k}}$. The hamiltonian of a system of Fermions interacting with two-body potentials is given by

$$H = -(2m)^{-1} \sum_{\alpha} \int d^3r \; \psi^{+}_{\alpha}(\underline{r}) \nabla^2 \psi_\alpha(\underline{r})$$
$$+ \frac{1}{2} \sum_{\substack{\alpha\alpha' \\ \beta\beta'}} \int d^3r\, d^3r' \psi^{+}_{\alpha}(\underline{r}) \psi^{+}_{\beta}(\underline{r}') V_{\alpha\alpha',\beta\beta'}(\underline{r},\underline{r}') \psi_{\beta'}(\underline{r}') \psi_{\alpha'}(\underline{r}) . \tag{7}$$

The correct order of the operators has to be observed. In the case of spin-independent interactions, which will be used exclusively,

$$V_{\alpha\alpha',\beta\beta'}(\underline{r},\underline{r}') = V(\underline{r},\underline{r}')\delta_{\alpha\alpha'}\,\delta_{\beta\beta'} \quad .$$

For Bosons the spin indices have to be omitted.

The <u>physical meaning</u> of the two-point Green's function is obvious from its definition. It represents, for $t > t'$, the matrix element or amplitude of the following process: a particle is added to the ground state of the system at $\underline{r}'$ at time t', propagates to $\underline{r}$ at time t, where it is taken out of the system. If $t < t'$, then a hole is created at time t, and taken out at time t'.

Time translation invariance holds since H is time-independent. If only isotropic homogeneous systems are considered (liquid ^{3}He, ^{4}He), translation invariance in space holds as well. Then the two-point functions depend on $\underline{r}-\underline{r}'$, $t-t'$, and the Fourier transform can be defined by

$$g_{\alpha\beta}(\underline{r}-\underline{r}',t-t')=\int\frac{d^3k\,d\omega}{(2\pi)^4}\exp[i\underline{k}\cdot(\underline{r}-\underline{r}')-i\omega(t-t')]\,g_{\alpha\beta}(\underline{k},\omega) \quad . \tag{8}$$

Inserting $c_{\underline{k}\alpha}$ we have

$$g_{\alpha\beta}(\underline{k},t-t') = \frac{1}{\langle\psi_o|\psi_o\rangle}\langle\psi_o|Tc_{\underline{k}\alpha}(t)c^+_{k\beta}(t')|\psi_o\rangle \quad . \tag{9}$$

For spin-independent interactions, and the hamiltonian invariant under rotations and reflections

$$g_{\alpha\beta}(\underline{k},t-t') = \delta_{\alpha\beta}\quad g(|k|,\, t-t') \quad . \tag{10}$$

Ground state expectation values of single-particle operators are easily expressed in terms of the two-point functions. Also the ground-state energy can be derived from the two-point functions. The formulas will not be reproduced here; they can be found in any of the textbooks, e.g. in Fetter and Walecka [1].

3.2 Lehmann representation

After transformation to frequency space, the Green's functions have a complicated analytic structure, due to the presence of the time ordering operator in their definition. Thus it is convenient to relate them to analytically simpler functions. These functions will also be important for the introduction of "quasiparticles". We decompose

$$g_{\alpha\beta}(\underline{k},t) = \theta(t)g^{>}_{\alpha\beta}(\underline{k},t) + \theta(-t)g^{<}_{\alpha\beta}(\underline{k},t) \tag{11}$$

with the "Wightman functions"

$$g^{>}_{\alpha\beta}(\underline{k},t) = \langle\psi_o|c_{\underline{k}\alpha}(t)c^{+}_{\underline{k}\beta}|\psi_o\rangle ,$$

$$g^{<}_{\alpha\beta}(\underline{k},t) = -\langle\psi_o|c^{+}_{\underline{k}\beta}\ c_{\underline{k}\alpha}(t)|\psi_o\rangle . \tag{12}$$

We have omitted the factor $\langle\psi_o|\psi_o\rangle^{-1}$ for simplicity. Using the Fourier transformation of the step function

$$\theta(\pm t) = \mp \frac{1}{2\pi i}\int d\omega\ \frac{\exp(-i\omega t)}{\omega\pm i\delta} , \qquad \delta \to 0 , \tag{13}$$

we have

$$g_{\alpha\beta}(\underline{k},\omega) = i\int\frac{d\omega'}{2\pi}\ \{\frac{g^{>}_{\alpha\beta}(\underline{k},\omega')}{\omega-\omega'+i\delta} - \frac{g^{<}_{\alpha\beta}(\underline{k},\omega')}{\omega-\omega'-i\delta}\} \tag{14}$$

where e.g.

$$g^{>}_{\alpha\beta}(\underline{k},\omega) = \int dt\ e^{i\omega t}\ \langle\psi_o|c_{\underline{k}\alpha}(t)\ c^{+}_{\underline{k}\beta}|\psi_o\rangle . \tag{15}$$

By using

$$c_{\underline{k}\alpha}(t) = \exp(itH)\; c_{\underline{k}\alpha}\; \exp(-itH)$$

and

$$H|\psi_o\rangle = E|\psi_o\rangle \quad , \quad H|\psi_n\rangle = E_n|\psi_n\rangle$$

we find

$$g^{>}_{\alpha\beta}(\underline{k},\omega) = 2\pi\sum_n \delta(\omega - E_n + E)\langle\psi_o|c_{\underline{k}\alpha}|\psi_n\rangle\langle\psi_n|c^{+}_{\underline{k}\beta}|\psi_o\rangle \tag{16}$$

and similarly

$$g^{<}_{\alpha\beta}(\underline{k},\omega) = -\,2\pi\sum_n \delta(\omega + E_n - E)\langle\psi_o|c^{+}_{\underline{k}\beta}|\psi_n\rangle\langle\psi_n|c_{\underline{k}\alpha}|\psi_o\rangle\;. \tag{17}$$

It has to be observed that the intermediate states of $g^{>}(\underline{k},\omega)$ have N + 1 particles, and of $g^{<}(\underline{k},\omega)$ have N - 1 particles. Hence we can write

$$\omega - E_n\,(N+1) + E(N) = \omega - \varepsilon_n\,(N+1) - \mu \tag{18}$$

where

$$\varepsilon_n(N+1) = E_n(N+1) - E(N+1)$$

is the excitation energy of the N + 1 particle system, counted from the ground state, and

$$\mu = E\,(N+1) - E(N)$$

the chemical potential. For large N, μ is the same for N or N - 1 particles. Thus $g^{>}(\underline{k},\omega)$ has poles at the exact excitation energies of the N + 1 particle system, counted from μ. Similarly, $g^{<}(\underline{k},\omega)$ has poles at μ minus the exact excitation energies of the N - 1 particle system, since

$$\omega + E_n(N-1) - E(N) = \omega + E_n(N-1) - \mu \ . \tag{19}$$

We thus arrive at the following representation of the Green's function, due to Lehmann

$$g_{\alpha\beta}(\underline{k},\omega) = i\sum_n \left\{\frac{<0|c_{\underline{k}\alpha}|n><n|c^{+}_{\underline{k}\beta}|0>}{\omega - E_n(N+1) - \mu + i\delta} + \frac{<0|c^{+}_{\underline{k}\beta}|n><n|c_{\underline{k}\alpha}|0>}{\omega + E_n(N-1) - \mu - i\delta}\right\} . \tag{20}$$

Since excitation energies are positive by definition, the location of the poles of $g_{\alpha\beta}(\underline{k},\omega)$ in the complex ω-plane is as given in Fig. 3.

It is instructive to derive the Green's function for the non-interacting Fermion system. The intermediate states must be plane-wave states of the non-interacting system. Addition of a particle is only possible for $k \geq k_F$, destruction for $k < k_F$. One finds

$$g_{\alpha\beta}(\underline{k},\omega) = i \left\{\frac{\theta(k-k_F)}{\omega - E^{o}_{\underline{k}} - \mu + i\delta} + \frac{\theta(k_F-k)}{\omega + E^{o}_{\underline{k}} - \mu - i\delta}\right\} . \tag{21}$$

The energy $E_{\underline{k}}^{o}$ corresponds to a particle-hole pair, a particle outside the Fermi sphere and a hole at k_F in the first term, and a hole within the Fermi sphere and a particle at k_F in the second term.

3.3 Quasiparticles

Let us be reminded that the Wightman functions have poles at the exact excitation energies E_n. For large particle number N, $E_n(N+1) \approx E_n(N) \approx E_n(N-1)$ will be densely spaced in this limit, hence the poles of $g^{\gtrless}(\underline{k},\omega)$ are densely spaced for large N.

Fig. 4 gives a possible picture of the pole structure of $g^{>}(\underline{k},\omega)$, both for a non-interacting and an interacting Fermion system. When the states are densely spaced, one can introduce a continuous function $g^{>}(\underline{k},\omega)$ in this limit, which is proportional to the square of the matrix elements, and the density of states. This is indicated in Fig. 4 by the envelope function. The same can be done for $g^{<}(\underline{k},\omega)$. In the Lehmann representation and in Fig. 3 the poles have to be replaced by branch lines.

Let us assume that a good representation of $g^{>}(\underline{k},\omega)$ near some energy $E(\underline{k})$ is given by

$$g^{>}(\underline{k},\omega) \approx \frac{2z(\underline{k})\gamma(\underline{k})}{[\omega-\mu-E(\underline{k})]^2 + \gamma^2(k)} + \text{background} \qquad (22)$$

where $\gamma(\underline{k}) \ll E(\underline{k})$ and where spin indices have been omitted. A similar expression can be given for $g^{<}(\underline{k},\omega)$. The resulting Green's function is

$$g(\underline{k},\omega) = \frac{z(\underline{k})}{\omega-\mu-E(\underline{k})+i\gamma(\underline{k})} + \frac{z(\underline{k})}{\omega-\mu+E(\underline{k})-i\gamma(\underline{k})} + \text{rest} \tag{23}$$

and after the transformation to time

$$g(\underline{k},t) \approx \begin{cases} z(\underline{k})\exp\{-i[\mu+E(\underline{k})]t - \gamma(\underline{k})t\} & t > t' \\ z(\underline{k})\exp\{-i[\mu-E(\underline{k})]t - \gamma(\underline{k})t\} & t < t' \end{cases} \tag{24}$$

These expressions should be compared with the non-interacting system, e.g. with Eq. (21). For $t > t'$ we have the propagation of a "quasi-particle" with energy $\mu + E(\underline{k})$ and lifetime $\gamma(\underline{k})$, for $t < t'$ we have the propagation of a "quasihole" with energy $\mu - E(\underline{k})$ and lifetime $\gamma(\underline{k})$. Of course, $g^>(\underline{k},\omega)$, $g^<(\underline{k},\omega)$ have to be calculated for model or realistic systems, in order to show that quasiparticles or quasiholes really occur. It has turned out that the quasiparticles <u>are</u> the important elementary excitations in many-body systems, which determine e.g. the low-temperature behaviour of an interacting Fermion system.

3.4 General properties of Green's functions

Let us very briefly indicate two important general properties of the two-point functions. First, there exist "dispersion relations" which relate the imaginary and real parts of the Green's functions. They are derived from the Lehmann representation by using

$$(\omega-\omega' \pm i\delta)^{-1} = P(\omega-\omega')^{-1} \mp i\pi\,\delta(\omega-\omega')$$

where P means principal value. We have

$$g_{\alpha\beta}(\underline{k},\omega) = iP\int\frac{d\omega'}{2\pi}\frac{g^{>}_{\alpha\beta}(\underline{k},\omega') - g^{<}_{\alpha\beta}(\underline{k},\omega')}{\omega-\omega'}$$

$$+\frac{1}{2}[g^{>}_{\alpha\beta}(\underline{k},\omega) + g^{<}_{\alpha\beta}(\underline{k},\omega)] \quad . \tag{25}$$

It is easy to see that $g^{>}$, $g^{<}$ are real for diagonal Green's functions $\propto \delta_{\alpha\beta}$, cf. Eq. (10). The relations between the imaginary and real parts are then obvious from Eq. (25).

Second, the behaviour of $g(\underline{k},\omega)$ for large ω is governed by "sum rules". If $g^{\gtrless}(\underline{k},\omega)$ vanish for large argument $|\omega|$,

$$g_{\alpha\beta}(\underline{k},\omega) \xrightarrow[\omega\ \text{large}]{} \frac{i}{\omega}P\int\frac{d\omega'}{2\pi}\frac{g^{>}_{\alpha\beta}(\underline{k},\omega') - g^{<}_{\alpha\beta}(\underline{k},\omega')}{1-(\omega'/\omega)}$$

$$\longrightarrow \frac{i}{\omega}\delta_{\alpha\beta} + O(\omega^{-2}) \quad , \tag{26}$$

since

$$\int\frac{d\omega'}{2\pi}[g^{>}_{\alpha\beta}(\underline{k},\omega') - g^{<}_{\alpha\beta}(\underline{k},\omega')] = \delta_{\alpha\beta} \tag{27}$$

represents the equal time anti-commutator between $c_{\underline{k}\alpha}$ and $c^{+}_{\underline{k}\beta}$. This procedure can be continued to higher powers of ω.

4. FEYNMAN DIAGRAMS

The perturbation expansion of the Green's functions and other physical quantities can be expressed in terms of Feynman diagrams. The non-interacting two-point functions appear as elements in this expansion. With their help, resummation procedures can be carried through in a lucid way. This is one of the reasons for the introduction of Green's functions. Here we will indicate the essential steps which lead to Feynman diagrams without writing down explicit rules.

4.1 Interaction picture

It is convenient to split up $H = H_o + H_1$ where H_o represents a one-particle problem, and introduce the interaction picture where operators are time-dependent according to

$$A_I(t) = \exp(it H_o)\, A \exp(-it H_o)$$

and states according to

$$|\psi_I(t)\rangle = U(t,t_o)\,|\psi_I(t_o)\rangle \quad .$$

The time development operator $U(t,t_o)$ obeys

$$i \frac{\partial}{\partial t} U(t,t_o) = H_{1I}(t)\, U(t,t_o) \tag{28}$$

with the initial condition $U(t_o,t_o) = 1$. The solution of this equation is given by

$$U(t,t_o) = \sum_{n=0}^{\infty} \frac{(-i)^n}{n!} \int_{t_o}^{t} dt_1 \ldots \int_{t_o}^{t} dt_n$$

$$\times\ T[H_{1I}(t_1)\ \ldots\ldots\ldots\ldots\ldots\ H_{1I}(t_n)] \qquad (29)$$

where T is the time ordering operator. We introduce an adiabatic switching-on of the interaction,

$$H = H_o + H_1 \exp(-\delta|t|) \ .$$

U then depends on δ. Meaningful results should be independent of δ, such that one can let $\delta \to 0$ in the final results. We consider a finite δ and an early time t_o where the interaction is not yet switched on. We have

$$|\psi_I(t_o)\rangle = |\phi_o\rangle \ , \qquad \text{stationary eigenstate of } H_o \ .$$

Then we let t_o go to $-\infty$. We require coincidence of the pictures at $t = 0$. Hence

$$|\psi_H\rangle = |\psi_I(0)\rangle = U_\delta(0,-\infty)|\phi_o\rangle \ .$$

The limit $\delta \to 0$ of this relation is ill defined. A more detailed examination (see e.g. Ref. 2) shows that diverging phases appear. However, the following quantity might be well defined, since phases can cancel,

$$\lim_{\delta\to 0} \frac{U_\delta(0,-\infty)|\phi_o\rangle}{\langle\phi_o|U_\delta(0,-\infty)|\phi_o\rangle} \equiv \frac{|\psi_o\rangle}{\langle\phi_o|\psi_o\rangle} \ . \qquad (30)$$

There is an important theorem proved by Gell-Mann and Low. It states that this quantity is an eigenstate of H, if it exists to all orders in perturbation theory. It should be noted that the procedure implied in the Gell-Mann and

Low theorem generates the eigenstate adiabatically from the non-interacting ground state. Thus this state can be the true ground state, but not necessarily. For example, the true ground state might not have a convergent series expansion in the interaction. This difficulty appears for example in superconductors. For the moment we will restrict ourselves to "normal systems" where the theorem holds.

Using the Gell-Mann and Low theorem it is easy to rewrite the two-point function, originally defined in the Heisenberg picture, in the interaction picture:

$$g_{\alpha\beta}(\underline{r}t,\underline{r}'t') = \frac{\langle\phi_o|TU_\delta(\infty,-\infty)\psi_{\alpha I}(\underline{r}t)\psi^+_{\beta I}(\underline{r}'t')|\phi_o\rangle}{\langle\phi_o|U_\delta(\infty,-\infty)|\phi_o\rangle} \ . \quad (31)$$

The time ordering operator T refers to the terms of $U_\delta(\infty,-\infty)$ and $\psi_{\alpha I}(\underline{r}t)$, $\psi^+_\beta(\underline{r}'t')$ together. The limit $\delta\to 0$ has to be taken in the final expressions. We have a perturbation expansion of the Green's function when we insert the expansion of the time development operator Eq.(29). Its form is unwieldy, but it can be brought into a more convenient form using the following theorem.

4.2 Wick's theorem

Wick's theorem relates T-products to N-products and contractions. Before we can state it, N-products and contractions have to be defined.

A N-product N(AB.......YZ) of operators AB......YZ is ordered in such a way that all destruction operators (defined with respect to the non-interacting ground state) appear at the right. A sign convention is included in the definition such that a - sign is taken for each permutation of two Fermion operators. It will be necessary to redefine or split up operators, e.g. for Fermions $c^+_{\underline{k}\alpha}$ acts as a destruction operator on the ground state for $k < k_F$. Hence we perform the canonical transformation

$$b_{\underline{k}\alpha} = c^+_{\underline{k}\alpha} \qquad k \leq k_F \quad ,$$

$$b_{\underline{k}\alpha} = c_{\underline{k}\alpha} \qquad k > k_F \quad ;$$

and define the N-product with all b's to the right. For the moment we restrict the discussion to Fermion systems. The definition of the N-products is useful, since its ground state expectation value vanishes,

$$<\phi_o|N(AB............YZ)|\phi_o> = 0 \ . \qquad (32)$$

The contraction AB of two operators A, B is defined as

$$\overbrace{AB} = T\ (AB) - N\ (AB) \quad .$$

We have from Eq. (32)

$$<\phi_o|T\ (AB)|\phi_o> \ = \ \overbrace{AB} \quad .$$

Using a pair of creation and destruction operators, we find the Green's function of the non-interacting system

$$g^o_{\alpha\beta}\ (\underline{r}t,\underline{r}'t') = \overbrace{\psi_{\alpha I}\ (\underline{r}t)\ \psi^+_{\beta I}}\ (\underline{r}'t')$$

$$= \langle\phi_o|T\psi_{\alpha I}(\underline{r}t)\psi^+_{\beta I}(\underline{r}'t')|\phi_o\rangle \ . \tag{33}$$

We only need Wick's theorem applied to ground state expectation values. Then all N-products vanish in view of Eq. (32) and there remains

$$\langle\phi_o|T(ABCD............YZ)|\phi_o\rangle \tag{34}$$

= sum of all possible pair contractions

$$= \overset{\sqcap}{AB}\ \ \overset{\sqcap}{CD}.......\overset{\sqcap}{YZ} \quad - \quad \overset{\sqcap}{AC}\ \overset{\sqcap}{BD}.........\overset{\sqcap}{YZ}$$

$$+ \ ....$$

4.3 Perturbation expansion of $g_{\alpha\beta}(\underline{r}t,\underline{r}'t')$

In section 4.1 we found a perturbation expansion of the Green's function, which consists of a series of expectation values of T-products. Hence Wick's theorem in the form of Eq. (34) applies, and we have a series of sums over all possible contractions, which are given by the non-interacting Green's functions. As an example, we consider the first-order term, in the interaction, of the numerator of $g(\underline{r},t)$. For simplicity we omit spin indices. We have

$$(-i/2)\langle\phi_o|T\int dt_1\int d^3r_1d^3r_2V(\underline{r}_1,\underline{r}_2)\psi^+_I(\underline{r}_1t_1)\psi^+_I(\underline{r}_2t_2)$$

$$\times\ \psi_I(\underline{r}_2t_2)\psi_I(\underline{r}_1t_1)\psi_I(\underline{r}t)\psi_I(r't')|\phi_o\rangle \quad . \tag{35}$$

There are 6 contractions, shown graphically in Fig. 5. The graphical representation of the terms of the perturbation expansion constitutes the Feynman diagrams. There are rules for the evaluation of the contributions of each diagram, which we will not state explicitly. These rules are mainly bookkeeping rules for the correct factors, including combinatorial factors. The rules can be formulated in coordinate or momentum space. The introduction of symmetrized interactions and reduction to point vertices lead to modified rules. It is perhaps the best to write down ones own set of conventions and rules, and to use them consistently.

One important point can already be seen in the first-order diagrams. There appear disconnected parts, i.e. in the two first diagrams of Fig. 5 there are parts which are not linked to the external points. The same diagrams appear in the denominator of the Green's function, and cancel the contributions of the numerator. It can be shown that this cancellation occurs in all orders of the perturbation series. This is the content of the "linked-cluster theorem". Its consequence is that the two-point Green's function is given as the sum of all connected Feynman diagrams linked to two external points. Written symbolically, were c means connected,

$$g_{\alpha\beta}(\underline{r}t,\underline{r}'t') = \langle\phi_o|TU_\delta(\infty,-\infty)\psi_{\alpha I}(\underline{r}t)\psi^+_{\beta I}(\underline{r}'t')|\phi_o\rangle_c \quad .(36)$$

5. PERTURBATION METHODS FOR THE SELF-ENERGY

5.1 Self-energy and Dyson equation

The infinite sum of Feynman diagrams which contribute to g is simplified by introducing the concept of the "self-energy". We consider a general diagram of g in Fig. 6(a). We can identify parts where several lines run from the left to the right, and parts where there is only one line. We define as the self-energy, Σ, the sum of all diagrams, where one line can be attached at both ends, and which do not separate into two pieces, if one internal line is cut, cf. Fig. 6(b). Other names are "mass operator", "proper self-energy", or "1-irreducible self-energy".

We represent the Green's function $g(\underline{r}t,\underline{r}'t')$ by a thick line. Resummation of all different contributions to each individual self-energy insertion leads to an equation, which is depicted graphically in Fig. 7. This equation can be brought into the compact form

$$g_{\alpha\beta}(\underline{r}t,\underline{r}'t') = g^{o}_{\alpha\beta}(\underline{r}t,\underline{r}'t') + \sum_{\gamma\delta} \int dt_1 dt_2 d^3r_1 d^3r_2$$

$$g^{o}_{\alpha\gamma}(\underline{r}t,\underline{r}_1t_1)\Sigma_{\gamma\delta}(\underline{r}_1t_1,\underline{r}_2t_2)g_{\delta\beta}(\underline{r}_2t_2,\underline{r}'t') \qquad (37)$$

which is represented in Fig. 8. Iteration of this "Dyson equation" leads to the equation depicted in Fig. 7. One sees why the introduction of the self-energy is important, because any approximation for Σ already represents an infinite sum when inserted into the Green's function.

The Dyson equation becomes a system of algebraic equations in momentum-frequency space, for homogeneous systems. If

$g_{\alpha\beta} = \delta_{\alpha\beta}\, g$, then also $\Sigma_{\gamma\delta} = \delta_{\gamma\delta}\, \Sigma$

and then

$$g(\underline{k},\omega) = g_o(\underline{k},\omega) + g_o(\underline{k},\omega)\Sigma(\underline{k},\omega)g(\underline{k},\omega) \ . \tag{38}$$

We solve this equation for g and use the non-interacting Green's function derived in Eq. (21);

$$g(\underline{k},\omega) = \frac{i}{\omega - \mu \mp E_o(\underline{k}) - i\Sigma(\underline{k},\omega)} \tag{39}$$

where - refers to $k > k_F$ and + to $k < k_F$.

One can ask for the zeroes of the denominator of $g(\underline{k},\omega)$. Let us assume that there are poles of the following structure:

a) for $k > k_F$

$$\omega = \mu + E(\underline{k}) - i\gamma(\underline{k})$$

where

$$E(\underline{k}) \approx E_o(\underline{k}) + \text{Re}\,\{i\Sigma(\underline{k},E(\underline{k}))\}$$

$$\gamma(\underline{k}) \approx -\text{Im}\,\{i\Sigma(\underline{k},E(\underline{k}))\} \quad ,$$

b) for $k < k_F$

$$\omega = \mu - E(\underline{k}) + i\gamma(\underline{k})$$

where

$$E(\underline{k}) \approx E_o(\underline{k}) - \text{Re}\{i\Sigma(\underline{k},E(\underline{k}))\}$$

$$\gamma(\underline{k}) \approx \text{Im}\{i\Sigma(\underline{k},E(\underline{k}))\} \quad . \tag{40}$$

If there are solutions of this structure with $\gamma(\underline{k}) << E(\underline{k})$, then there exist quasiparticles and quasiholes, as discussed in Section 3.3. The Fermi surface is then determined by the sign change of $\text{Im}\{i\Sigma(\underline{k},E(\underline{k}))\}$.

5.2 Approximations for self-energy

The self-energy has been defined as the sum of all 1-irreducible diagrams, i.e. diagrams which can not be cut into two pieces by just cutting one line. The simplest ones are given in Fig. 9. It is clear that an approximation is obtained by taking a selected subset of the diagrams into account. E.g. one obtains the "Hartree-Fock" approximation by choosing the two first diagrams of Fig. 9, which are the only ones in first order in V. The result for Σ consists of the "direct" and the "exchange" term

$$i\Sigma_{HF}(\underline{k}) = nV(\underline{k}=0) - n\int\frac{d^3k'}{(2\pi)^3} V(\underline{k}-\underline{k}')\theta(k_F-k') \tag{41}$$

where

$$V(\underline{r}-\underline{r}') = \int\frac{d^3k}{(2\pi)^3} \exp[-i\underline{k}\cdot(\underline{r}-\underline{r}')]V(\underline{k})$$

and n is the particle density. $i\Sigma_{HF}(\underline{k})$ is independent of ω and real since V is instantaneous. For the same reason the particle density appears,

$$n = g_o\ (\underline{r}t,\underline{r}t+\delta) = \langle\phi_o|\psi^+(\underline{r}t)\psi(\underline{r}t)|\phi_o\rangle \ . \qquad (42)$$

There are two possibilities to go beyond the first-order approximation:

a) Self-consistent approximations

One is led to a self-consistent approximation, if one tries to introduce into the internal lines the self-energy in the approximation considered, cf. Fig. 10. In general, we imagine all self-energy contributions to internal lines in the diagrams of Σ to be summed up. In this way we obtain diagrams built up with interaction Green's functions g, where the internal lines do no contain any explicit self-energy insertions. They are called "skeleton diagrams" and the first few are given in Fig. 11. Σ is now a functional of g, and the Dyson equation is an equation for a self-consistent determination of g,

$$g = [g_o^{-1} - \Sigma\{g\}]^{-1} \ . \qquad (43)$$

A self-consistent approximation is obtained by selecting a subset of $\Sigma\{g\}$, e.g. the self-consistent Hartree-Fock approximation by taking the first two diagrams of Fig.11. For a homogeneous system, the self-consistent Hartree-Fock approximation leads to the same result as in Eq.(41). A non-trivial result is obtained for the case of an additional external potential, e.g. an atomic potential. One recovers the self-consistent Hartree-Fock equations, which are used in atomic or nuclear calculations. So far nothing new has been derived. The strength of the self-consistent scheme lies more in the derivation of general results.

b) Infinite subsets of diagrams

This will be illustrated by the treatment of hard-core systems. Physical examples are ^{3}He or nuclear matter. However, the following approximation applies to a low-density system. A two-body potential which approaches a hard-core potential for $V_0 \to \infty$ is sketched in Fig.12. The contributions of individual diagrams diverge in that limit. One knows from scattering theory that the scattering amplitude is finite even when the potential diverges, and thereby the individual terms of the Born series for the scattering amplitude diverge. It can be shown that the sum of all ladder diagrams, depicted in Fig. 13, is equivalent to the Born series for the scattering amplitude. The important point is to avoid the evaluation of individual ladder diagrams. This is done by deriving an integral equation for the whole series, which is known as the Bethe-Salpeter equation [7] in the ladder approximation. It still contains the diverging interaction. Eliminating the interaction in favor of the scattering amplitude one obtains the Galitskii [8] integral equation, which remains well defined for a hard-core system. It can be shown that the ladder diagrams give the leading contribution for a low-density hard-core system.

Also the combination of a self-consistent determination of the Green's functions with infinite subsets of diagrams has been investigated. We refer to the text-books, e.g. Ref. 1, for detailed references.

5.3 Some results

We will give a brief survey of some results which can be derived from Galitskii's integral equation, and then indicate the difficulties which appear in the case of ^{3}He. The imaginary part of the self-energy is found to be, for $k \approx k_F$

$$\mathrm{Im}\ \{i\Sigma\} = \frac{k_F^2}{\pi m}\ (k_F a)^2 \left(\frac{k_F - k}{k_F}\right)^2\ \mathrm{sign}\ (k_F - k) \tag{44}$$

where a is the scattering length for s-wave scattering. Only this quantity appears for a low-density hard-core gas. Note that the imaginary part is porportional to a^2, as is expected for a transition rate. Note also that it is proportional to $(k_F - k)^2$. This dependence follows from simple phase-space considerations. There is a sign change of the imaginary part of the self-energy at the Fermi surface, in accordance with the discussion in Section 5.1. The real part of the self-energy is given near k_F by

$$\mathrm{Re}\ \{i\Sigma\} = \frac{k_F}{m^*}\ (k - k_F) \tag{45}$$

where

$$\frac{m^*}{m} = 1 + \frac{8}{15\pi^2}\ (7\ \ln 2 - 1)\ (k_F a)^2 + \ldots . \tag{46}$$

The Fermi energy is given by

$$\varepsilon_F = \mu = \frac{k_F^2}{2m}\ [1 + \frac{4}{3\pi}(k_F a) + \frac{4}{15\pi^2}(11 - 2\ \ln 2)(k_F a)^2] . \tag{47}$$

Since the imaginary part vanishes faster than the real part for $k \to k_F$, we can identify quasiparticle and quasihole excitations with energy $E(\underline{k})$ and lifetimes $\gamma(\underline{k})$, in accordance with Eq.(40).

Can the expressions derived so far be applied to ^3He? This system is not a dilute Fermion gas, and the expansion parameter $k_F a$ is not small. m^*/m which determines the linear term of the specific heat $c_V(T)$ (cf. Section 6.4) is about 2.9. The next-order term of c_V, being $\propto T^3 \ln T$, was found surprisingly large. Berk and Schrieffer [9], and Doniach and Engelsberg [10] recognized that the spin degrees of freedom of the ^3He atoms have to be taken into account, since ^3He is an almost ferromagnetic system. There are "paramagnons" which contribute to m^*, and to the $T^3 \ln T$-term of the specific heat. The theory of these contributions has been elaborated by Brenig, Mikeska, and Riedel [11]. Amit, Kane, and Wagner [12] performed a non-perturbative calculation of m^* and of other quantities, which contribute to the $T^3 \ln T$-term.

6. FINITE TEMPERATURE GREEN'S FUNCTIONS

6.1 Definition of two-point function at $T \neq 0$

Green's functions can be defined for finite temperature by replacing ground-state expectation values with thermal averages. In this way the statistical-mechanical many-body problem at finite temperature is treated. We will see how this work out in examples.

The basic statistical-mechanical quantity is the statistical operator ρ,

$$\rho = Z^{-1} \exp(-\beta H) \tag{48}$$

where $\beta = 1/T$ is the reciprocal temperature ($k_B = 1$) and Z the partition function

$$Z = \text{trace}\,\{\exp(-\beta H)\} \quad . \tag{49}$$

The chemical potential has been included in the hamiltonian,

$$H - \mu N \to H \quad .$$

μ can be fixed in the final expressions by prescribing the mean value of the particle number, $\langle N \rangle$. From Z follows the grand potential Y,

$$Y = -\beta^{-1} \ell n\, Z \quad . \tag{50}$$

Y is related to the Helmholtz free energy F by $Y = F - \mu N$.

Instead of real times, imaginary times τ (cf.Fig.14) are introduced for the time dependence of operators,

$$A_H(\tau) = \exp(\tau H)\, A \exp(-\tau H) \quad . \tag{51}$$

Note that $[A(\tau)]^+ \neq A^+(\tau)$.

The finite-temperature two-point function is defined by

$$G_{\alpha\beta}(\underline{r}\tau, \underline{r}'\tau') = \text{trace}\,\{\rho T_\tau \psi_{\alpha H}(\underline{r}\tau)\psi^+_{\beta H}(\underline{r}'\tau')\} \quad . \tag{52}$$

The time ordering operator T_τ orders with respect to increasing τ, i.e. the operator with the largest τ appears at the left, cf. also Fig. 14. We continue to consider Fermions and include the same sign convention as at T = O in the definition of T_τ. The arguments of G will be restricted to the interval $0 \leq \tau, \tau' \leq \beta$. Inside this interval the absolute convergence of the expression (52) is guaranteed.

The $T \neq 0$ two-point function can be related to single-particle thermal averages. We will show this for the particle number $N(T,\mu,\Omega)$ which is given by

$$N = -\sum_\alpha \int d^3r\; G_{\alpha\alpha}(\underline{r}\tau, \underline{r}\tau + \delta) \;, \quad \delta \to 0 \;. \tag{53}$$

We have

$$\sum_\alpha G_{\alpha\alpha}(\underline{r}\tau, \underline{r}\tau + \delta) = -\sum_\alpha \text{trace} \; \{\rho\psi_\alpha^+(\underline{r}\tau)\psi_\alpha(\underline{r}\tau)\}$$

and because of the cyclic invariance of the trace

$$= -Z^{-1} \sum_\alpha \text{trace}\; \{e^{-\beta H} \; \psi_\alpha^+(\underline{r})\psi_\alpha(\underline{r})\} = -\langle n(\underline{r})\rangle \;.$$

Other single-particle thermal averages can be expressed in a similar way. There also exist expressions for the mean energy and the grand potential.

It is illustrative to write down the non-interacting $T \neq 0$ two-point function, which is defined with a single-particle hamiltonian H_0. For a homogeneous isotropic system in k-space

$$G^{o}_{\alpha\beta}(\underline{k},\tau-\tau') = Z_o^{-1}\ \mathrm{trace}\,\{e^{-\beta H_o}\, T_\tau c_{\underline{k}\alpha}(\tau) c^{+}_{\underline{k}\alpha}(\tau')\} \tag{54}$$

where

$$H_o = \sum_{\underline{k}\alpha} (\varepsilon^{o}_{\underline{k}} - \mu)\ c^{+}_{\underline{k}\alpha}\ c_{\underline{k}\alpha} \quad .$$

The time dependence of $c_{\underline{k}\alpha}$, $c^{+}_{\underline{k}\beta}$ is here given by

$$c_{\underline{k}\alpha}(\tau) = c_{\underline{k}\alpha}\ \exp\,[-\tau(\varepsilon^{o}_{\underline{k}} - \mu)]$$

$$c^{+}_{\underline{k}\beta}(\tau') = c^{+}_{\underline{k}\beta}\ \exp\,[\tau'(\varepsilon^{o}_{\underline{k}} - \mu)] \ .$$

Hence

$$G^{o}_{\alpha\beta}(\underline{k},\tau-\tau') = \begin{cases} \exp\,[-(\tau-\tau')(\varepsilon^{o}_{\underline{k}}-\mu)]\langle c_{\underline{k}\alpha} c^{+}_{\underline{k}\beta}\rangle_o & \tau>\tau' \\ -\exp\,[(\tau-\tau')(\varepsilon^{o}_{\underline{k}}-\mu)]\langle c^{+}_{\underline{k}\beta} c_{\underline{k}\alpha}\rangle_o & \tau<\tau' \ . \end{cases} \tag{55}$$

One sees that the mean occupation numbers at finite temperature appear.

6.2 The boundary condition

The boundary condition is an extremely important property of the $T \neq 0$ Green's functions. It has the following form for Fermions

$$G_{\alpha\beta}(\underline{r}\tau, r'0) = -\,G_{\alpha\beta}(\underline{r}\tau, r'\beta) \quad . \tag{56}$$

It is easily proved by using the cyclic invariance of the trace,

$$\text{trace}\,\{e^{-\beta H}\, e^{\tau H}\, \psi_\alpha(\underline{r})\, e^{-\tau H}\, \psi_\beta(\underline{r}')\}$$

$$= \text{trace}\,\{e^{-\beta H}\, e^{\tau H}\, \psi_\alpha(\underline{r})\, e^{-\tau H}\, e^{-\beta H}\, e^{\beta H}\, \psi_\beta(\underline{r}')\}$$

$$= \text{trace}\,\{e^{-\beta H}\, \psi_\beta(\underline{r}'\beta)\, \psi_\alpha(\underline{r})\} \quad .$$

Similarly

$$G_{\alpha\beta}(\underline{r}0, \underline{r}'\tau') = -G_{\alpha\beta}(\underline{r}\beta, \underline{r}'\tau') \quad . \tag{57}$$

For illustration, the boundary condition is shown to hold for the non-interacting two-point function, where the spin indices are omitted for simplicity. Starting from eq. (55) one sees that the boundary condition requires

$$\langle c_{\underline{k}}\, c^{+}_{\underline{k}} \rangle_o = \exp\,[\beta(\varepsilon^o_{\underline{k}} - \mu)]\, \langle c^{+}_{\underline{k}}\, c_{\underline{k}} \rangle_o \quad .$$

Using the anticommutation relation the Fermi distribution of the non-interacting system is found

$$\langle c^{+}_{\underline{k}}\, c_{\underline{k}} \rangle_o = \{\exp\,[\beta(\varepsilon^o_{\underline{k}} - \mu)]\, -1\}^{-1} \quad . \tag{58}$$

Conversely, if $\langle c^{+}_{\underline{k}}\, c_{\underline{k}} \rangle_o$ is the Fermi distribution of the free system, then the boundary condition is fulfilled.

The boundary condition can be written in a slightly different form in the case of time translation invariance

$$G_{\alpha\beta}(\underline{r},\underline{r}',\tau-\tau') = -G_{\alpha\beta}(\underline{r},\underline{r}',\tau-\tau' \pm \beta) \quad , \tag{59}$$

+ when $\tau - \tau' < 0$

- when $\tau - \tau' > 0$.

$G_{\alpha\beta}(\tau-\tau')$ is defined in the interval $-\beta \leq \tau - \tau' \leq \beta$. In this interval, it can be expressed as a Fourier series

$$G_{\alpha\beta}(\underline{r},\underline{r}',\tau-\tau') \doteq \frac{1}{\beta}\sum_{n} G_{\alpha\beta}(\underline{r},\underline{r}',\omega_n) \exp[-\omega_n(\tau-\tau')] \ . \tag{60}$$

In order to fulfill the boundary condition, ω_n has to take on the following discrete values

$$\omega_n = (1 + 2n)\,(i\pi/\beta) \quad , \qquad n \quad \text{integer} \quad . \tag{61}$$

Up to now Fermions have been considered. For Bosons a similar boundary condition without sign change holds, and the discrete frequencies of the Fourier series are given by

$$\omega_n = 2n\,(i\pi/\beta) \quad , \qquad n \quad \text{integer} \quad . \tag{62}$$

We return to the Fermion case. The Fourier coefficients are given by

$$G_{\alpha\beta}(\underline{r},\underline{r}',\omega_n) = \int_0^\beta d(\tau-\tau')G_{\alpha\beta}(\underline{r},\underline{r}',\tau-\tau') \exp[\omega_n(\tau-\tau')] \quad . \tag{63}$$

It is an easy exercise to calculate the Fourier coefficients for the non-interacting Green's function starting from Eq. (55). One finds

$$G^{o}_{\alpha\beta}(\underline{k},\omega_n) = -[\omega_n - \varepsilon^{o}_{\underline{k}} + \mu]^{-1} \quad . \tag{64}$$

One can also define finite-temperature Green's functions for real times, $G(\underline{r}t,\underline{r}'t')$. The analytic continuation of the Fourier transform of these functions in the complex ω-plane coincides with the analytic continuation of the Fourier coefficients given above. For more details see the textbooks, e.g. Refs.[1],[3] or [4].

6.3 Perturbation theory

The perturbation theory for finite temperature Green's functions and other quantities is completely analogous to that at zero temperature. What follows is just a transcription of all procedures and definitions of the T = O case to finite T. We split up $H=H_o+H_1$, and define time dependence in the interaction picture by

$$A_I(\tau) = \exp(\tau H_o)\, A \exp(-\tau H_o) \quad . \tag{65}$$

The time development operator obeys

$$\frac{\partial}{\partial\tau} U(\tau,\tau') = - H_{1I}(\tau) U(\tau,\tau') \tag{66}$$

with the initial condition $U(\tau,\tau) = 1$. $U(\tau,\tau')$ is no more unitary, but still satisfies the group property. The solution of the differential equation is given by

$$U(\tau,\tau') = \sum_{n=0}^{\infty} \frac{(-1)^n}{n!} \int_{\tau'}^{\tau} d\tau_1 \ldots\ldots \int_{\tau'}^{\tau} d\tau_n$$

$$\times\, T_\tau\, [H_{1I}(\tau_1) \ldots\ldots H_{1I}(\tau_n)] \quad . \tag{67}$$

The following relation holds for $U(\tau,0)$:

$$\exp(-\tau H) = \exp(-\tau H_o) U(\tau,0) \quad . \tag{68}$$

We set $\tau = \beta = 1/T$ and form the trace,

$$Z = \text{trace}\,\{\exp(-\beta H)\} = \text{trace}\,\{\exp(-\beta H_o) U(\beta,0)\} \quad . \tag{69}$$

Thus a perturbation expansion of the partition function Z is obtained by using the expansion of $U(\beta,0)$ given in Eq.(67).

The $T \neq 0$ Green's functions were originally defined in the Heisenberg picture. After transforming to the interaction picture the following representation of the two-point function is obtained

$$G_{\alpha\beta}(\underline{r}\tau,\underline{r}'\tau') = \frac{\text{trace}\{e^{-\beta H_o}\, T_\tau U(\beta,0)\psi_{\alpha I}(\underline{r}\tau)\psi^+_{\beta I}(\underline{r}'\tau')\}}{\text{trace}\,\{e^{-\beta H_o}\, U(\beta,0)\}} \quad . \tag{70}$$

Also the perturbation expansion of $G_{\alpha\beta}(\underline{r}\tau,\underline{r}'\tau')$ is obtained by using the expansion of $U(\beta,0)$.

The form of the perturbation expansion of $G_{\alpha\beta}$ is not yet convenient. At $T = 0$ the expressions were simplified with the help of Wick's theorem. It can not be applied directly at finite temperatures, since the thermal average of a N product generally does not vanish. However, the following generalization holds

$$\langle T_\tau \ (AB \ldots\ldots\ldots\ldots YZ)\rangle_o \tag{71}$$
$$= \text{sum over all possible pair contractions} \ .$$

Here $\langle\ \rangle_o$ denotes the thermal average formed with the statistical operator ρ_o of the noninteracting system, $\rho_o = Z_o^{-1}\exp(-\beta H_o)$. A contraction is defined as

$$\overbrace{AB} = \langle T_\tau(AB)\rangle_o \quad .$$

Especially

$$\overbrace{\psi_{\alpha I}(\underline{r}\tau)\,\psi^+_{\beta I}}(\underline{r}'\tau') = G^o_{\alpha\beta}(\underline{r}\tau,\underline{r}'\tau') \quad . \tag{72}$$

Due to the validity of the generalized Wick theorem at $T \neq 0$, all diagrammatic expansions at $T = 0$ can be transferred to the case $T \neq 0$. Time integrals from $-\infty$ to $+\infty$ have to be replaced by integrals from 0 to $-i\beta$, or frequency integrals by sums over the discrete frequencies. Also the linked-cluster theorem is valid at finite temperature. Thus the $T \neq 0$ two-point function $G_{\alpha\beta}(\underline{r}\tau,\underline{r}'\tau')$ is the sum of all connected diagrams, linked to two external points. Application of the linked-cluster theorem to the partition function shows that $\ln(Z)$ is given by the sum of all connected diagrams without external points.

Analogous to the $T = 0$ case, the series for the Green's function can be partially summed up by the introduction of the self-energy, $\Sigma_{\alpha\beta}(\underline{k},\omega_n)$. In the same manner as for T = 0 the dyson equation for the two-point function is obtained, as represented in Fig. 8. For homogeneous systems the following form in frequency and momentum space is found (cf. Eq. 38)

$$G_{\alpha\beta}(\underline{k},\omega_n)=G^{o}_{\alpha\beta}(\underline{k},\omega_n)+\sum_{\gamma\delta} G^{o}_{\alpha\gamma}(\underline{k},\omega_n)\Sigma_{\gamma\delta}(\underline{k},\omega_n)G_{\delta\beta}(\underline{k},\omega_n). \quad (73)$$

When $G_{\alpha\beta}$ and $\Sigma_{\alpha\beta} \propto \delta_{\alpha\beta}$, then

$$G(\underline{k},\omega_n) = -[\omega_n - \varepsilon^{o}_{\underline{k}} + \mu + \Sigma(\underline{k},\omega_n)]^{-1} \quad . \quad (74)$$

Approximations for the self-energy are obtained in the same way as at T = 0. Also resummation of the internal lines is possible such that Σ is given as the sum of all skeleton diagrams, where the internal lines are interacting Green's functions, cf. Fig. 11.

We briefly mention a few applications of the finite-temperature formalism, before we discuss one specific application in the next section. The electron gas can be treated at low and high temperatures. At high temperature and low density the classical Debye-Hückel equation of state is found. One can also treat the imperfect Boson gas above the condensation temperature T_c and derive the shift of this temperature. More interesting are the applications to superconductivity and superfluidity at finite temperatures. The treatment of these systems requires the additional concept of "broken symmetry" which will be introduced in the next chapter.

6.4 Low-temperature behaviour of many-body systems

One important problem is the derivation of the thermodynamic behaviour of many-body systems from basic principles. How can the low-temperature behaviour of an interacting system be calculated? There exists an intuitive derivation of the low-temperature behaviour using the concept of quasiparticles. In this picture, the low-lying quasiparticles are used to calculate the thermodynamic potential and related quantities, analogous to the derivations for non-interacting systems. The task of the many-body theory is the derivation and justification of these results.

The starting point of the formal derivations is the representation of the grand potential as a functional of the <u>interacting</u> two-point function, and mass operator. One obtains with resummation techniques (De Dominicis and Martin [13]) or with other arguments (Abrikosov, Gorkov, and Dzyaloshinksi [4])

$$Y - Y_o = -\frac{2\Omega}{\beta}\sum_n \int \frac{d^3k}{(2\pi)^3}\{\ell n[1-G^o(\underline{k},\omega_n)\Sigma(\underline{k},\omega_n)] + \Sigma(\underline{k},\omega_n)G(\underline{k},\omega_n)\} + \phi[G] \qquad (75)$$

where $\phi[G]$ is the sum of all connected closed skeleton diagrams, such that

$$\delta\phi[G]/\delta G = \Sigma.$$

It is easy to see that Y is a "stationary functional", stationary with respect to variations of Σ, provided that G satisfies the Dyson equation. One important point is that Y[G] is an intermediate step in a renormalization program [13]. In the perturbation theory with non-interacting Green's functions, $Y[G^{O}]$ is essentially a functional of the two-particle interactions; the internal lines do not contain information on the interacting system. Y[G], however, is a functional of the two-particle interactions, and of lines, which contain information on the single-particle excitations of the interacting system. In using Y[G] one can investigate how the quasiparticles determine the thermodynamics of the system.

It has been shown by Luttinger [14] to all orders in perturbation theory that the leading term of the specific heat at low temperature is given by the $T = 0$ two-point function, and is explicitly

$$c_V = \Omega \frac{k_B^2 \, m^* \, k_F}{3\hbar^2} T \tag{76}$$

where m^* is defined by

$$\left. \frac{dE(\underline{k})}{dk} \right|_{k_F} = \frac{\hbar^2 k_F}{m^*} \; .$$

A similar, but more complicated proof of the Debye formula for $c_V(T)$ of a condensed interacting Boson system has been given by Götze and Wagner [15] and supplemented by the author [16]. For low temperature, to all orders in perturbation theory

$$c_V = \Omega \frac{2\pi^2 \, k_B^4}{15\hbar^3 c^3} T^3 \tag{77}$$

where c is the velocity of sound.

The higher-order terms of the low-temperature expansion of Y have not yet been obtained in a satisfactory manner. The expressions become very complicated. The derivation of these terms would be necessary for a complete evaluation of the $T^3 \ln T$-term of $c_V(T)$ in Fermion systems.

7. SYSTEMS WITH BROKEN SYMMETRY

Let us consider a non-interacting Boson system. At T = 0 all particles condense in the ground state, such that

$$a_o|\phi_o;\ N> = \sqrt{N}|\phi_o;\ N-1> \quad . \tag{78}$$

It is assumed that also an interacting Boson system has a condensate at T = 0 and at low temperatures, although no general proof exists,

$$a_o|\psi_o;N> = \sqrt{N_o}|\psi_o;\ N-1> \tag{79}$$

where $|\psi_o;N>$ is the ground state of the interacting Boson system with N particles. The number of particles in the condensate N_o is smaller than N, but still a macroscopic quantity, i.e. proportional to the volume for large N. In view of Eq.(79) Wick's theorem about normal products is useless. We perform the following canonical transformation

$$a_o = \sqrt{N_o} + a_o' \ . \tag{80}$$

For an exact eigenstate of H

$$<\psi_o| a_o|\psi_o> = 0 \quad . \tag{81}$$

This relation is a consequence of the fact that $|\psi_o>$ is an eigenstate of H and the particle number operator $\hat{N}$. A gauge transformation yields

$$\exp (i\alpha\hat{N})|\psi_o> = \exp (i\alpha N)|\psi_o> \quad . \tag{82}$$

Eq. (81) is easily proved by forming the expectation value of a_o with the vectors of Eq.(82). We would like to have a state where the expectation value of a_o is $\sqrt{N_o}$ rather than zero. Such a state can be constructed from the states of the hamiltonian, which are degenerate with respect to the phases, in form of a wave packet where the mean value of the phase is fixed. This will be called a state with "broken symmetry", since the gauge symmetry is broken then.

There is a close analogy to the ground state of an isotropic Heisenberg ferromagnet. The hamiltonian is invariant against rotations in spin space. The ferromagnetic ground state has a distinguished direction since the mean value of the magnetization is unequal zero. Thus the ground state does not have the symmetry of the hamiltonian.

States with broken symmetry can also be obtained in the following way, according to Bogoliubov. A small field is added to the hamiltonian, which breaks the symmetry of it, and which ensures that the appropriate mean values are unequal zero. This procedure is very suggestive for isotropic Heisenberg magnets, where the additional magnetic field fixes the direction of the magnetization. Then the field is let go to zero. If a non-zero mean value remains, one has a state with broken symmetry. For Boson systems the additional field has the form

$$h_\nu = \nu a_o + \nu^* a_o^+$$

which however cannot be physically realized. Bogoliubov showed for the ideal Boson system

$$\lim_{\nu\to 0} \lim_{\Omega\to\infty} \langle\nu|a_o|\nu\rangle \neq 0 \qquad (83)$$

where $|\nu\rangle$ are eigenstates of $H + h_\nu$. The mean values formed with the additional field terms will be called "quasi-averages". There are mathematical problems connected with these procedures, and with the concept of broken symmetry. We refer to Wagner [17] for a more detailed account and further references.

Here we will use the concept of broken symmetry in a rather naive way. We take it for granted that "anomalous averages" appear in condensed systems, e.g. in Boson system $\langle|a_o|\rangle \neq 0$. The application to superconductors will follow in the next chapter. A few additional remarks concerning Boson systems will be made.

Replacing $a_o \to \sqrt{N_o} + a_o'$ we obtain a hamiltonian $H(N_o)$. Now the definition of Green's functions and the diagrammatic perturbation theory become possible. The formalism is complicated due to the presence of the factors $\sqrt{N_o}$. The lowest-order or Bogoliubov approximation consists in taking into account the leading terms in N_o of $H(N_o)$. It gives already a linear excitation spectrum for small k, $E(k) = ck$, where c is the sound velocity. Consistency problems arise in higher-order approximations, but progress has been made over the years. Also the hard-core problem has been treated.

We would like to mention a general theorem for the excitation spectrum, due to Hugenholtz and Pines [18]. It states that there is no gap in the excitation spectrum $E(k)$ of the interacting condensed Boson system at $k = 0$. The theorem has been derived to all orders in perturbation theory, and also from Ward identities at zero and finite temperatures.

Finally we will give a simple but nonrigorous argument that the spectrum of single-particle excitations and density oscillations coincide in a Boson system with a condensate. Density oscillations should be excited by applying $\rho_{\underline{k}}$ to the ground state, where

$$\rho_{\underline{k}} = \sum_{\underline{k}'} a^+_{\underline{k}'} a_{\underline{k}'+\underline{k}} \ .$$

We have

$$\rho_{\underline{k}}|\psi_o\rangle = \sum_{\underline{k}'} a^+_{\underline{k}'} a_{\underline{k}'+\underline{k}}|\psi_o\rangle$$

$$= \sqrt{N_o}\,(a_{\underline{k}} + a^+_{-\underline{k}})|\psi_o\rangle \ + \ \text{rest} \qquad (84)$$

where e.g. $a^+_{-\underline{k}}$ creates a single-particle excitation with momentum $-\underline{k}$.

8. SUPERCONDUCTIVITY

8.1 Introduction

Some metals such as Pb, Hg exhibit the phenomenon of superconductivity below a critical temperature T_c of a few Kelvin. Below T_c these metals show infinite conductivity, perfect diamagnetism, and the Meissner effect, i.e. $B \equiv 0$ in the interior of the material. London developed a phenomenological theory of superconductivity which consists of the Maxwell together with supplementary equations. The development of a microscopic theory of superconductivity took a long time, with several preparatory steps.

It became clear in the course of time that one of the tasks of the microscopic theory was the derivation of an "energy gap" in the quasiparticle spectrum. A rough argument can be given in the following way. If the quasiparticle spectrum has the form indicated in Fig. 15, then there is a critical velocity for the flow properties of the system. It can be deduced from Galilean invariance that excitation of quasiparticles is not possible for $v < v_c$. This is essentially Landau's argument for the existence of a critical velocity in superfluid helium. It can serve as a plausibility argument for superconductors, namely that the existence of a gap implies flow without resistivity for small velocities. Fröhlich recognized the importance of the phonon-mediated attractive interaction between electrons: the exchange of phonons between electrons leads to an attraction which, however, has to overcome the Coulomb repulsion. The observation of an isotope effect in the transition temperature, $T_c \propto M^{-1/2}$ where M is the mass of the metal atoms, supported the propositions of Fröhlich. Cooper showed that the Fermi surface is unstable against formation of bound electron pairs ("Cooper pairs") in the case of a net attractive interaction. Bardeen, Cooper, and Schrieffer [19] (BCS) finally derived the new ground state of a superconductor, starting from a model hamiltonian. They could show that the superconducting ground state is a distinctive many-body state, which consists of a coherent superposition of Cooper pairs. BCS also derived the excitation spectrum, the thermodynamics, and many other quantities. In their formulation wave functions were used, but the theory is well suited for the application of field-theoretical methods, as was noticed soon after BCS.

8.2 Temperature-dependent gap

We will derive the temperature-dependent gap using finite temperature Green's functions, in order to demonstrate how this formalism is applied to a nontrivial example. The derivation was given by Gorkov [20] who formulated the BCS theory in terms of $T \neq 0$ Green's functions.

We start with the following reduced hamiltonian (BCS)

$$H = \sum_{\underline{k}} (\varepsilon^{o}_{\underline{k}} - \mu)[c^{+}_{\underline{k}\uparrow} c_{\underline{k}\uparrow} + c^{+}_{\underline{k}\downarrow} c_{\underline{k}\downarrow}]$$

$$- \frac{g}{\Omega} \sum_{\underline{kk}'} c^{+}_{\underline{k}'\uparrow} c^{+}_{-\underline{k}'\downarrow} c_{-\underline{k}\downarrow} c_{\underline{k}\uparrow} \quad . \tag{85}$$

Several simplifications have been made in the derivation of this hamiltonian. First, the phonon-mediated attractive interaction between two electrons has been represented by an instantaneous $\underline{k}$-independent interaction of strength g. Second, the interaction has been restricted to electron pairs of opposite spin, since Cooper pairs should be formed with opposite spins; equal spins would experience an additional repulsion from the Pauli principle. Third, only pairs with total momentum zero have been included.

Starting from the hamiltonian Eq.(85) we could try to calculate two-point functions using diagrammatic techniques. However, the existence of a new ground state with broken symmetry has to be taken into account. Instead of extending the formalism to include this complication, we choose a more direct approach. This is the "method of the equations of motion" which is equivalent to a generalized Hartree-Fock treatment. We use imaginary times, as introduced in Eq. (51), and calculate the time dependence of $c_{\underline{k}\uparrow}$; $c^{+}_{-\underline{k}\downarrow}$,

$$\frac{\partial}{\partial\tau} c_{\underline{k}\uparrow} = -(\varepsilon^{o}_{\underline{k}}-\mu)c_{\underline{k}\uparrow} + \frac{g}{\Omega}\sum_{\underline{k}'} c^{+}_{-\underline{k}\downarrow}\, c_{-\underline{k}'\downarrow}\, c_{\underline{k}'\uparrow} \quad ,$$

$$\frac{\partial}{\partial\tau} c^{+}_{-\underline{k}\downarrow} = (\varepsilon^{o}_{\underline{k}}-\mu)\, c^{+}_{-\underline{k}\downarrow} + \frac{g}{\Omega}\sum_{\underline{k}'} c^{+}_{\underline{k}'\uparrow}\, c^{+}_{-\underline{k}'\downarrow}\, c_{\underline{k}\uparrow} \quad . \qquad (86)$$

We linearize these equations of motion by replacing operators by their expectation values. We assume that there is broken symmetry, such that the following mean values exist

$$\Delta = \frac{g}{\Omega}\sum_{\underline{k}'} \langle c_{-\underline{k}'\downarrow}\, c_{\underline{k}'\uparrow}\rangle \quad ,$$

$$\Delta^{*} = \frac{g}{\Omega}\sum_{\underline{k}'} \langle c^{+}_{\underline{k}'\uparrow}\, c^{+}_{-\underline{k}'\downarrow}\rangle \quad . \qquad (87)$$

We can view the broken symmetry in a vague way as associated with a condensate of Cooper pairs. Then the analogy to the Boson case is obvious. After linearization, the equations have the form

$$\frac{\partial}{\partial\tau} c_{\underline{k}\uparrow} = -(\varepsilon^o_{\underline{k}} - \mu)\, c_{\underline{k}\uparrow} + \Delta\, c^+_{-\underline{k}\downarrow}$$

$$\frac{\partial}{\partial\tau} c^+_{-\underline{k}\downarrow} = (\varepsilon^o_{\underline{k}} - \mu)\, c^+_{-\underline{k}\downarrow} + \Delta^*\, c_{\underline{k}\uparrow} \quad . \tag{88}$$

These equations could be decoupled by introducing linear combinations of the $c_{\underline{k}\uparrow}$ and $c^+_{-\underline{k}\downarrow}$, the so-called Bogoliubov transformation. Here we will calculate Green's functions. We consider the two-point function

$$G(\underline{k},\tau) = \langle T_\tau\; c_{\underline{k}\uparrow}(\tau)\, c^+_{\underline{k}\uparrow} \rangle$$

which is defined in the time interval $-\beta \leq \tau \leq \beta$. An equation of motion is obtained by differentiating with respect to τ. Hereby the differentiation of the step functions $\theta(\tau)$, $\theta(-\tau)$ has to be included, which are implicit in the definition of $G(\underline{k},\tau)$. The term due to the step functions is simplified by the anticommutation relation. We have

$$\frac{\partial}{\partial\tau} G(\underline{k},\tau) = \delta(\tau) + \langle T_\tau\; \frac{\partial c_{\underline{k}\uparrow}(\tau)}{\partial\tau}\; c^+_{\underline{k}\uparrow} \rangle \quad . \tag{89}$$

Inserting the linearized equation of motion we obtain

$$\left(\frac{\partial}{\partial\tau} + \varepsilon^o_{\underline{k}} - \mu\right) G(\underline{k},\tau) - \Delta F^*(\underline{k},\tau) = \delta(\tau) \tag{90}$$

where we have defined an "anomalous" Green's function

$$F^*(\underline{k},\tau) = \langle T_\tau \; c^+_{-\underline{k}\downarrow}(\tau) \; c^+_{\underline{k}\uparrow} \rangle \; . \tag{91}$$

$F^*(\underline{k},\tau)$ obeys the following equation of motion

$$(\frac{\partial}{\partial\tau} - \varepsilon^o_{\underline{k}} + \mu) F^*(\underline{k},\tau) - \Delta G(\underline{k},\tau) = 0 \quad . \tag{92}$$

After introducing the Fourier coefficients the following system of equations is obtained

$$(-\omega_n + \varepsilon^o_{\underline{k}} - \mu) G(\underline{k},\omega_n) - \Delta F^*(\underline{k},\omega_n) = 1 \; ,$$

$$(-\omega_n - \varepsilon^o_{\underline{k}} + \mu) F^*(\underline{k},\omega_n) - \Delta^* G(\underline{k},\omega_n) = 0 \; . \tag{93}$$

It is easily solved

$$G(\underline{k},\omega_n) = \frac{\omega_n + \varepsilon^o_{\underline{k}} - \mu}{-\omega_n^2 + (\varepsilon^o_{\underline{k}}-\mu)^2 + \Delta\Delta^*} \; ,$$

$$F^*(\underline{k},\omega_n) = - \frac{\Delta^*}{-\omega_n^2 + (\varepsilon^o_{\underline{k}}-\mu)^2 + \Delta\Delta^*} \quad . \tag{94}$$

We now identify

$$\Delta^* = - \frac{g}{\Omega} \sum_{\underline{k}} F^*(\underline{k},\tau = -0) \tag{95}$$

and assume Δ to be real. Then the second equation of (94) is a self-consistency equation for Δ,

$$\Delta = \frac{g}{\Omega\beta} \sum_{\omega_n} \sum_{\underline{k}} e^{\omega_n \delta} \frac{\Delta}{-\omega_n^2 + (\varepsilon_{\underline{k}}^o - \mu)^2 + \Delta\Delta^*} , \quad \delta \to 0 . \quad (96)$$

Here the sum over the discrete frequencies appears. This sum will be converted into an integral over real frequencies by a standard procedure. We observe that $(\beta/2)$ $\tanh(\beta\omega/2)$ has poles at the Fermion frequencies $\omega_n = (2n+1)(i\pi/\beta)$, and residues 1. Thus

$$\Delta = \frac{g}{4\pi i\Omega} \sum_{\underline{k}} \int_C d\omega \ \tanh(\beta\omega/2) \frac{\Delta}{-\omega^2 + (\varepsilon_{\underline{k}}^o - \mu)^2 + \Delta^2} . \quad (97)$$

The contour C is indicated in Fig. 16. It will be deformed into the contour C'. Then the integral can be performed. We introduce

$$E(\underline{k}) = [(\varepsilon_{\underline{k}}^o - \mu)^2 + \Delta^2]^{1/2} . \quad (98)$$

When we would have introduced new operators $b_{\underline{k}}^+$, $b_{\underline{k}}$ in order to decouple the equation of motion for $c_{\underline{k}}^+$, $c_{\underline{k}}$, then the $E(\underline{k})$ would appear in the hamiltonian in the form

$$\sum_{\underline{k}} E(\underline{k}) \, b_{\underline{k}}^+ b_{\underline{k}} ,$$

i.e. $E(\underline{k})$ is the excitation energy of new quasiparticles, and Δ has the interpretation of a gap in the excitation spectrum, cf. also Fig. 15. The energy 2Δ is necessary to break up a Cooper pair.

The equation for the gap $\Delta(T)$ has the form

$$1 = \frac{g}{2\Omega} \sum_{\underline{k}} \tanh\,(\beta E(\underline{k})/2)\, \frac{1}{E(\underline{k})} \tag{99}$$

where $\Delta(T)$ is contained in $E(\underline{k})$. This equation can be easily solved

a) at $T = 0$, where the tanh can be replaced by 1. The solution is

$$\Delta = 2\omega_D \exp\,(-1/g\nu_F) \tag{100}$$

where ν_F is the density of states at the Fermi surface. A cutoff ω_D has been used for the integration. Note the nonanalytic dependence on the coupling constant g.

b) For $T \to T_c$, the highest temperature where a nontrivial solution exists, with $\Delta(T \to T_c) \to 0$. The result is

$$k_B\,T_c = 1.14\,\omega_D \exp\,(-1/g\nu_F)\ ,$$

$$\Delta = 1.76\,k_B\,T_c\ . \tag{101}$$

For arbitrary temperature, the gap equation has to be solved numerically.

The equation for $\Delta(T)$ has already been given by

BCS. The derivation with finite temperature Green's functions is shorter, and more "mechanical". The thermodynamics is contained in the formalism when the boundary condition is used, i.e. when the discrete frequencies ω_n are employed, as we have seen in the derivation of $\Delta(T)$.

8.3 Superfluid ^{3}He

Concluding these lectures, a short outlook will be given on a quite active field of many-body theory. This is the theory of superfluid ^{3}He. After the advent of the BCS theory the question was raised whether pairing is also possible in the Fermion liquid ^{3}He. The interaction between two ^{3}He atoms consists of a strong repulsive interaction, represented by a hard core, and a weaker attractive interaction. Pairing in states with angular momentum $\ell = 0$ is impossible because of the repulsive core. The next possibility is pairing in $\ell = 1$ states. Then the spin states of the Cooper pairs have to be triplett states. One has the possibility to combine either

$$|\uparrow\uparrow\rangle \quad \text{and} \quad |\downarrow\downarrow\rangle \tag{A}$$

or

$$|\uparrow\uparrow\rangle, \ |\uparrow\downarrow + \downarrow\uparrow\rangle \ , \quad \text{and} \quad |\downarrow\downarrow\rangle \ . \tag{B}$$

The possibility A was pursued by Anderson and Morel, B by Balian and Werthamer, in theoretical papers. Only in the last few years the experiments could be performed in the temperature region below 3 mK. The superfluid phase transition was indeed found. Fig. 17 indicates the phase diagram of this system in this temperature range. As a surprise, two phases were found, which could be identified with the two theoretical possibilities A and B. We refer to the reviews of Leggett [21] and Wheatley [22] for a more detailed account of the theory and the experiments on this fascinating system.

REFERENCES

a) Textbooks

1. A.L. Fetter and J.D. Walecka, Quantum Theory of Many-Particle Systems, McGraw Hill, New York (1971).
2. P. Nozières, Theory of Interacting Fermi Systems, Benjamin, New York (1964).
3. L.P. Kadanoff and G. Baym, Quantum Statistical Mechanics, Benjamin, New York (1962).
4. A.A. Abrikosov, L.P. Gorkov, and I.E. Dzyaloshinski, Methods of Quantum Field Theory in Statistical Physics, Prentice Hall, Englewood Cliffs (1965).
5. G.E. Brown, Many-Body Problems, North Holland, Amsterdam (1972).
6. D. Pines, The Many-Body Problem (Lecture Notes and Reprints), Benjamin, New York (1961).

b) Original Papers

7. E.E. Salpeter and H.A. Bethe, Phys. Rev. 84 (1951) 1232.
8. V.M. Galitskii, Sov. Phys.-JETP 7 (1958) 104.

9. N.F. Berk and J.R. Schrieffer, Phys. Rev. Lett. 17 (1966) 433.

10. S. Doniach and S. Engelsberg, Phys. Rev. Lett. 17 (1966) 750.

11. W. Brenig, H.J. Mikeska, and E. Riedel, Z. Physik 206 (1967) 439.

12. D.J. Amit, J.W. Kane, and H. Wagner, Phys. Rev. 175 (1968) 326.

13. C. De Dominicis and P.C. Martin, J. Math. Phys. 5 (1964) 31.

14. J.M. Luttinger, Phys. Rev. 119 (1960) 1153.

15. W. Götze and H. Wagner, Physica 31 (1965) 475.

16. K. Kehr, Z. Physik 221 (1969) 291.

17. H. Wagner, Z. Physik 195 (1966) 273.

18. N. Hugenholtz and D. Pines, Phys. Rev. 116 (1959) 489.

19. J. Bardeen, L.N. Cooper, and J. R. Schrieffer, Phys. Rev. 108 (1957) 1175.

20. L.P. Gorkov, Sov.Phys.-JETP 7 (1958) 505.

21. A.J. Leggett, Rev. Mod. Phys. 47 (1975) 331.

22. J. Wheatley, Rev. Mod. Phys. 47 (1975) 415.

FIGURE CAPTIONS

Fig. 1: Momentum distribution n(k) and single-particle excitation energy $\varepsilon(k)$ for a Fermion system at T = 0. Dashed lines: non-interacting system. Thick lines: interacting system. The spin has been disregarded.

<u>Fig. 2:</u> Momentum distribution $n(k)$ and excitation spectrum $\varepsilon(k)$ for a Boson system at $T = 0$. Dashed line: excitation spectrum of non-interacting system. $n(k)$ is discussed in the text.

<u>Fig. 3:</u> Location of the poles (x) of the Green's function in the complex ω-plane according to the Lehmann representation Eq. (20).

<u>Fig. 4:</u> Poles of $g^{>}(k,\omega)$ for fixed k as a function of ω. (a) Free system: only one plane-wave matrix element is unequal zero for a given k. (b) interacting system: several states in the neighborhood of $\varepsilon(k)$ may have matrix elements unequal zero. The height of the vertical lines indicates the square of the matrix elements.

<u>Fig. 5:</u> All Feynman diagrams of the numerator of the two-point function which are of first order in the interaction. Lines: non-interacting two-point function $g_o(\underline{r}t,\underline{r}'t')$. Dashed line: interaction $V(\underline{r}_1,\underline{r}_2)$.

<u>Fig. 6:</u> (a) General Feynman diagram which contributes to the two-point function. (b) Self-energy diagrams.

<u>Fig. 7:</u> Equation for the two-point function where all self-energies are isolated.

<u>Fig. 8:</u> Dyson equation for the two-point function.

<u>Fig. 9:</u> Expansion of self-energy in 1-irreducible diagrams.

<u>Fig.10:</u> Resummation of self-energy contributions to the internal lines.

<u>Fig.11:</u> Expansion of the self-energy in skeleton diagrams.

<u>Fig.12:</u> Ladder approximation of the self-energy.

Fig. 13: Approximation of a hard-core potential by a soft-core potential of strength V_o.

Fig. 14: Introduction of imaginary times τ. The two-point function is defined in the interval designated by the thick line.

Fig. 15: Quasiparticle spectrum with gap, and determination of the critical velocity v_c.

Fig. 16: Deformation of integration contours in the complex ω-plane. The crosses indicate the position of the discrete frequencies Eq.(16).

Fig. 17: Schematic phase diagram of ^{3}He at very low temperatures after Wheatley [22]. A and B are superfluid phases.

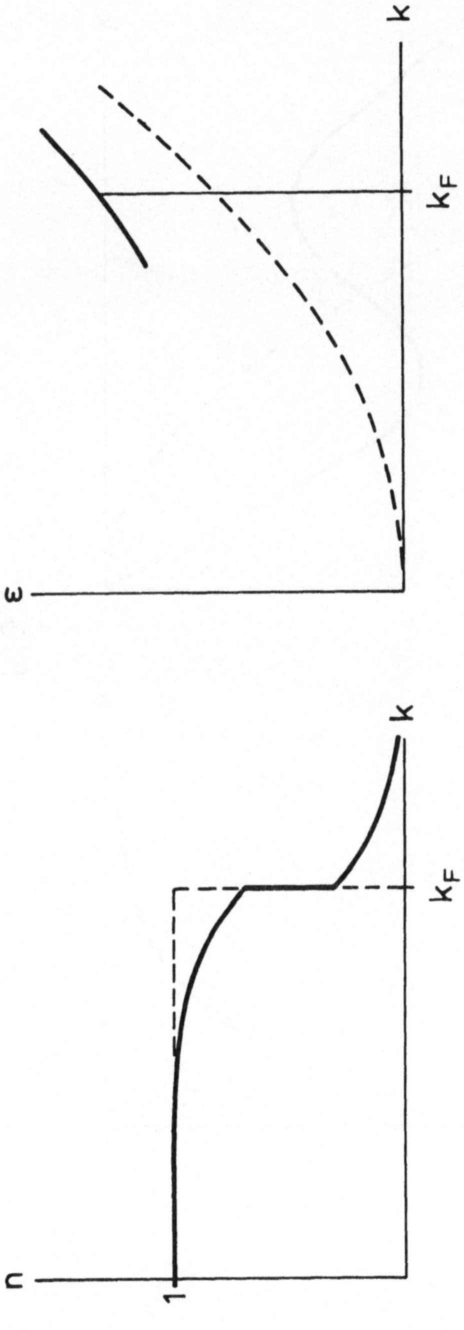

FIG.1

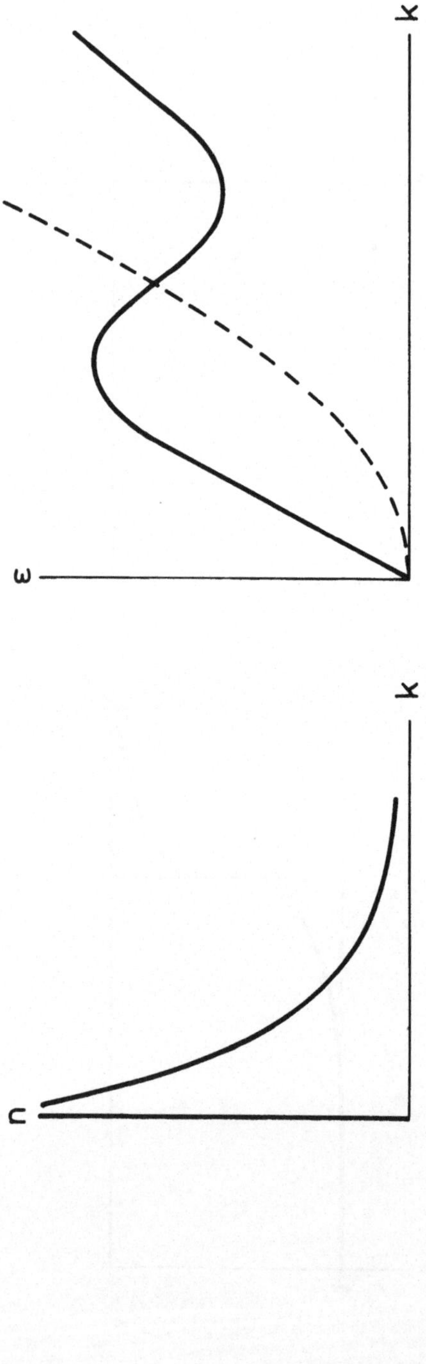

FIG.2

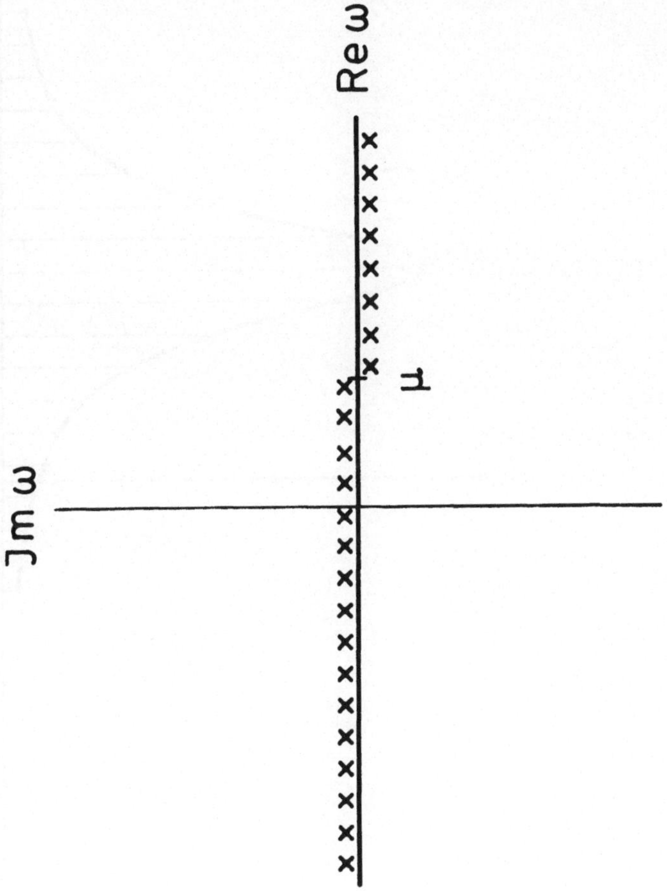

FIG.3

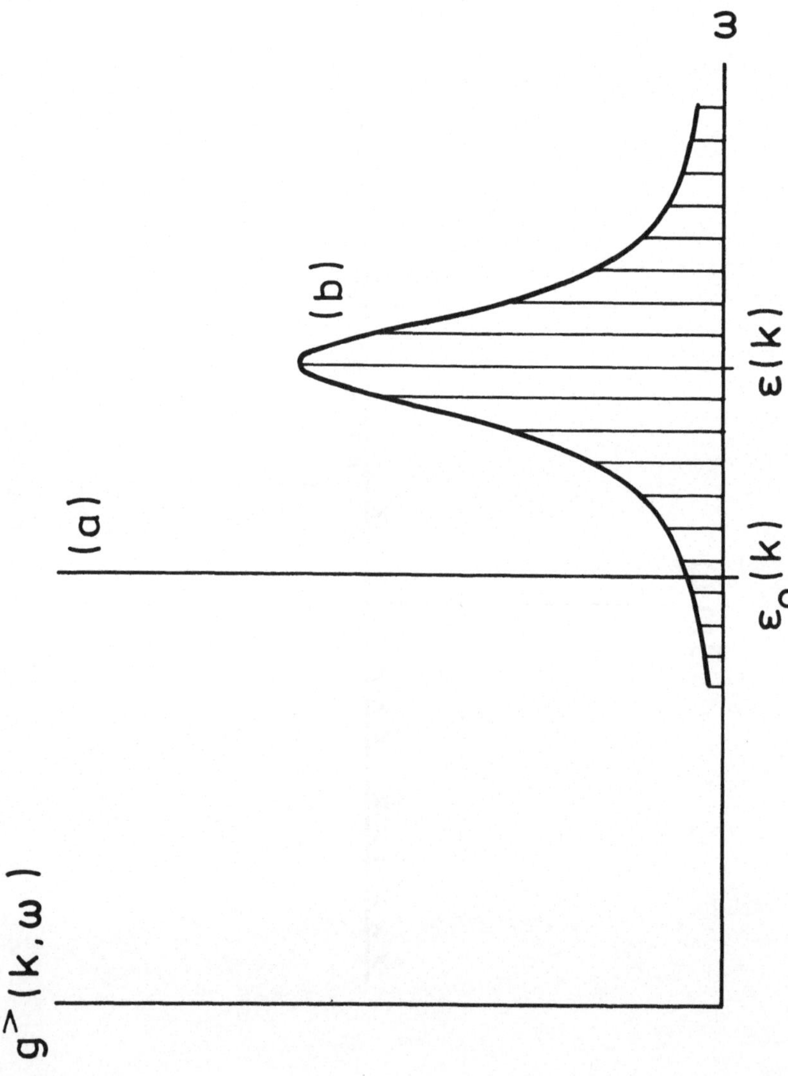

FIG.4

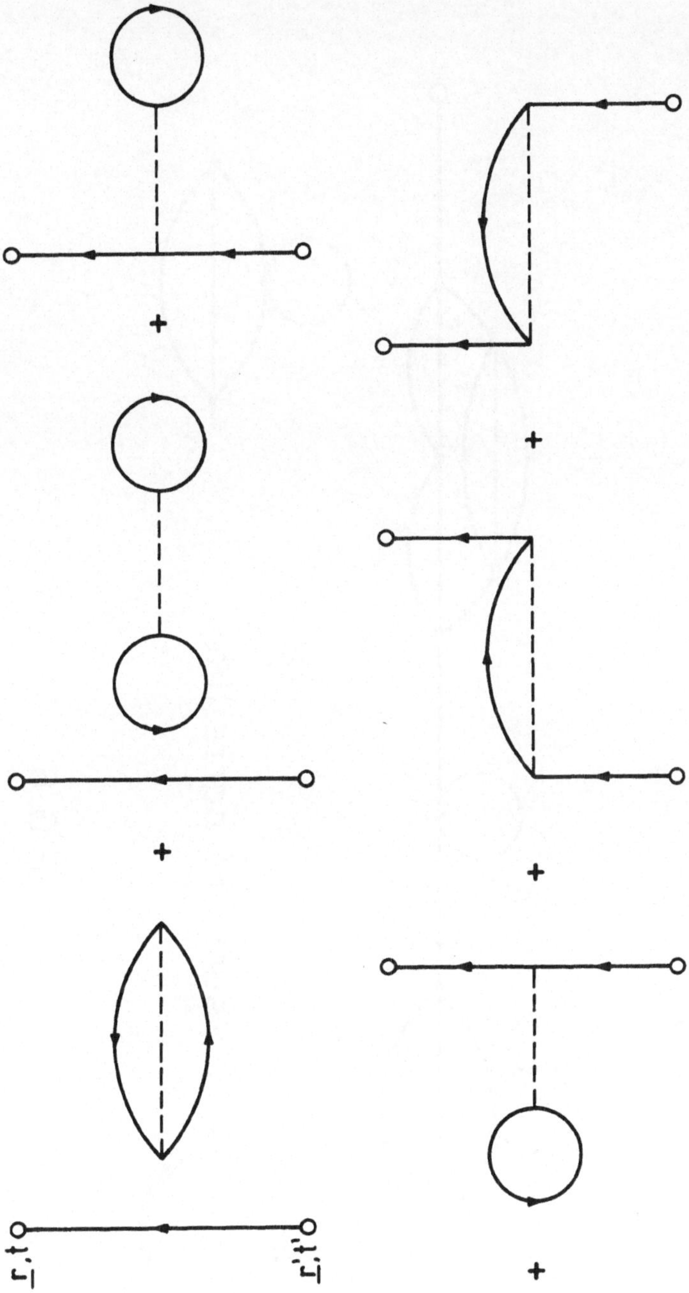

FIG.5

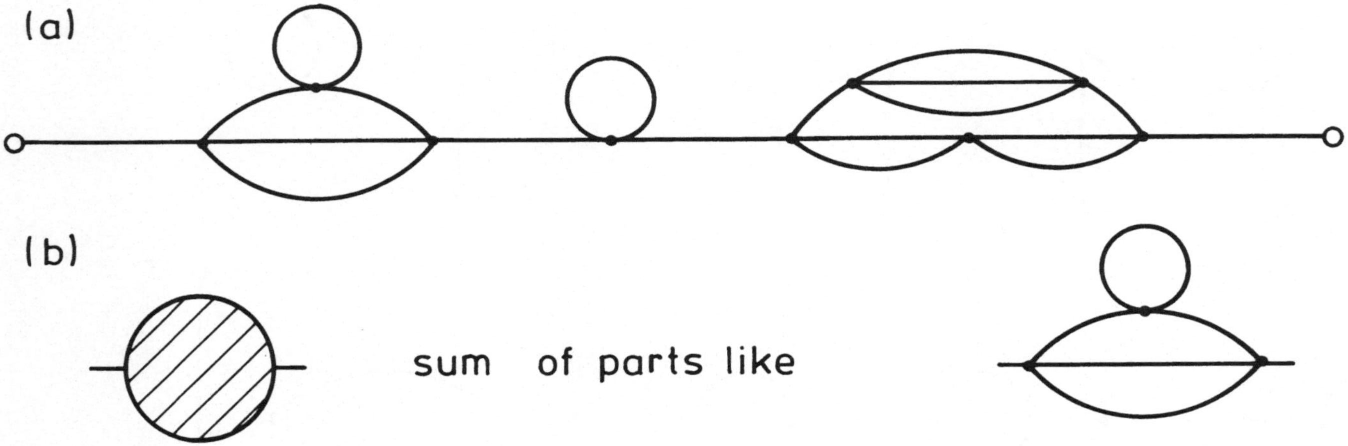

FIG. 6

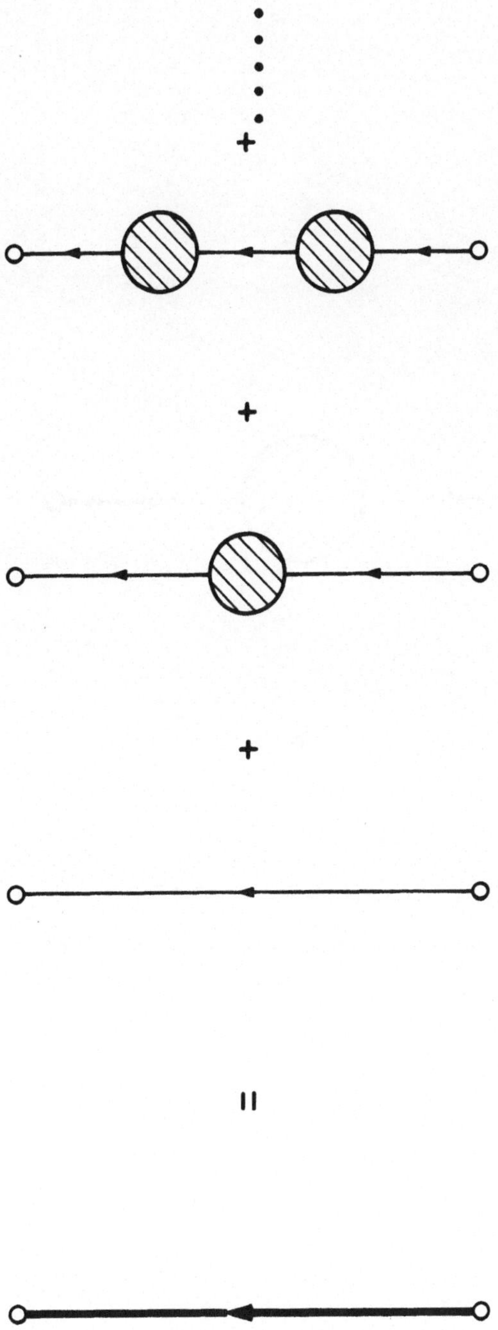

FIG.7

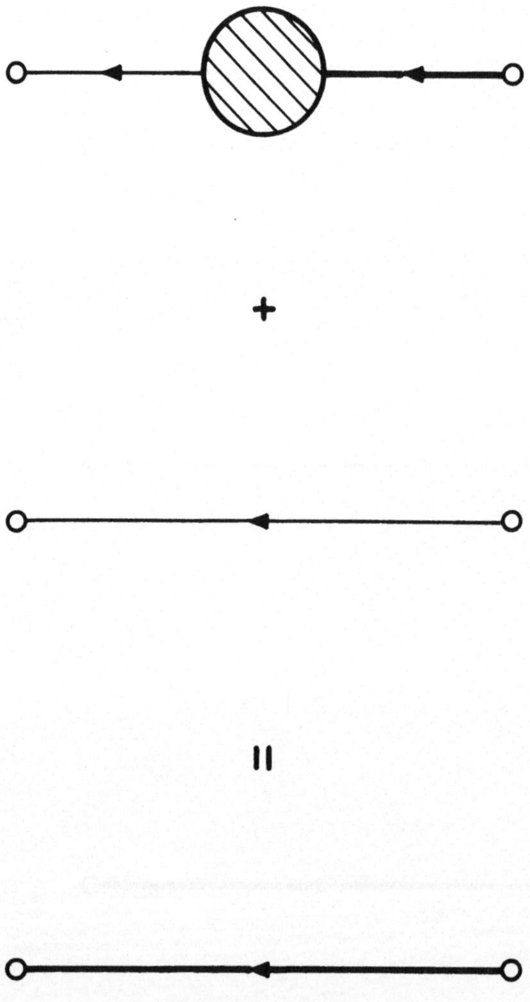

FIG. 8

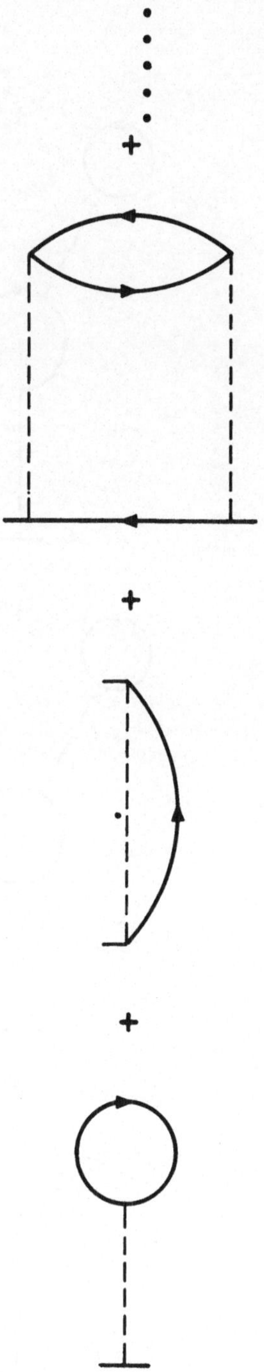

FIG.9

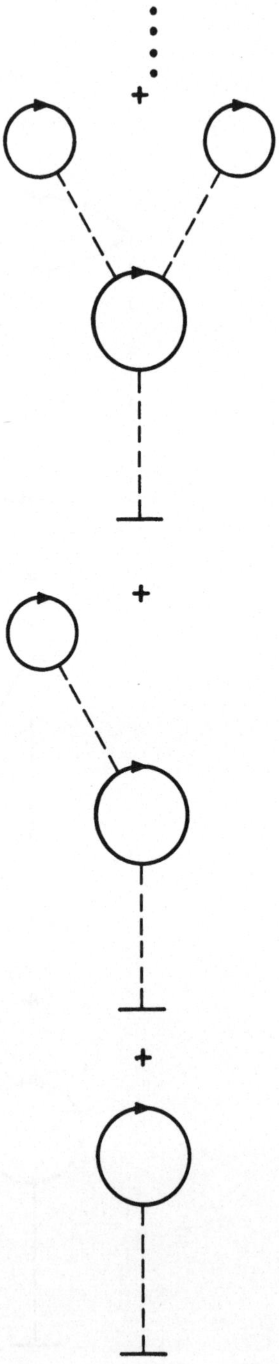

FIG.10

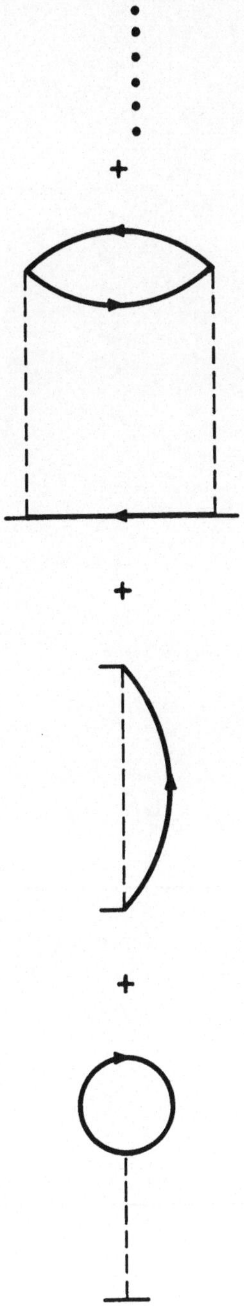

FIG.11

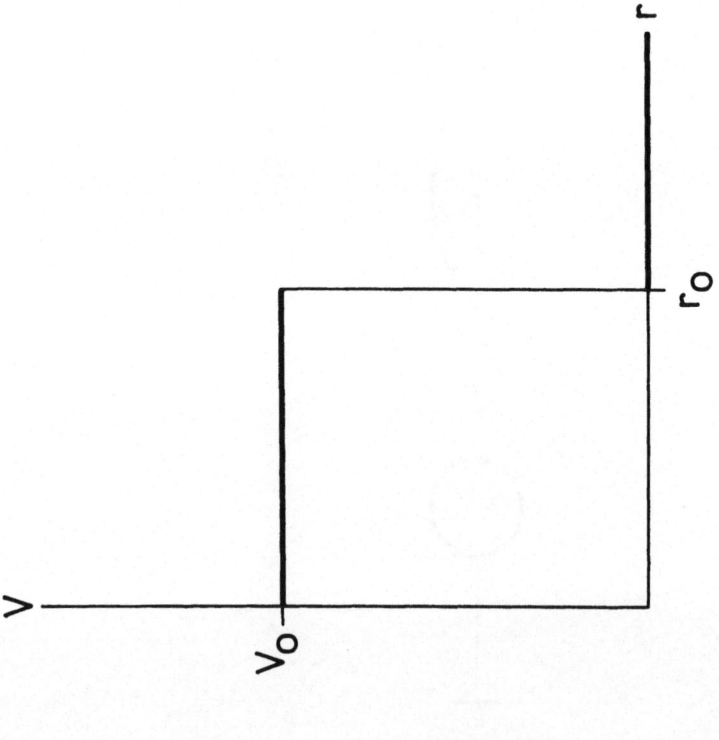

FIG.12

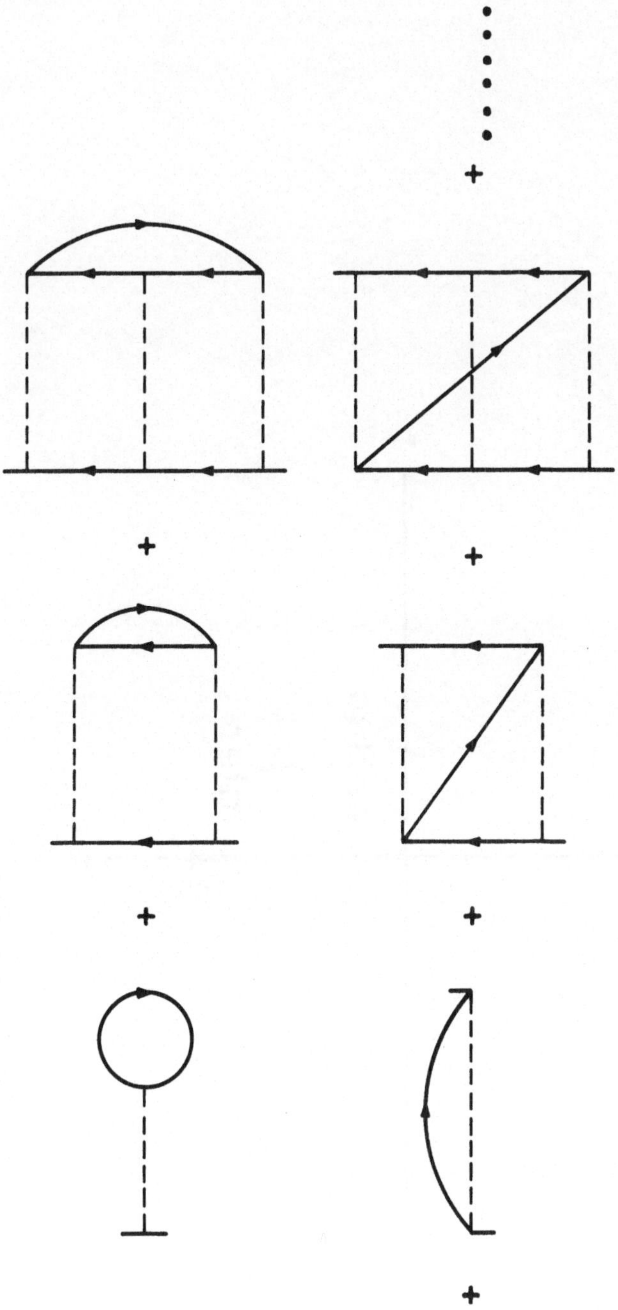

FIG.13

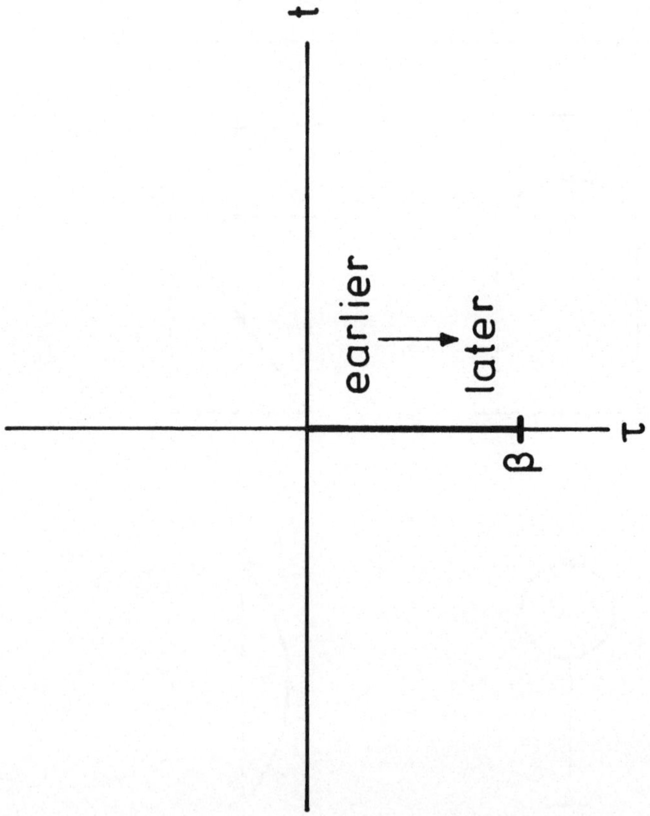

FIG.14

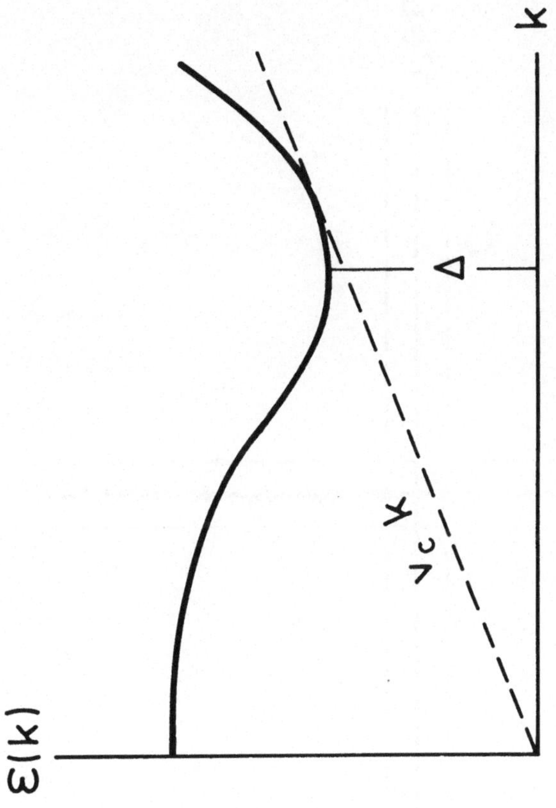

FIG.15

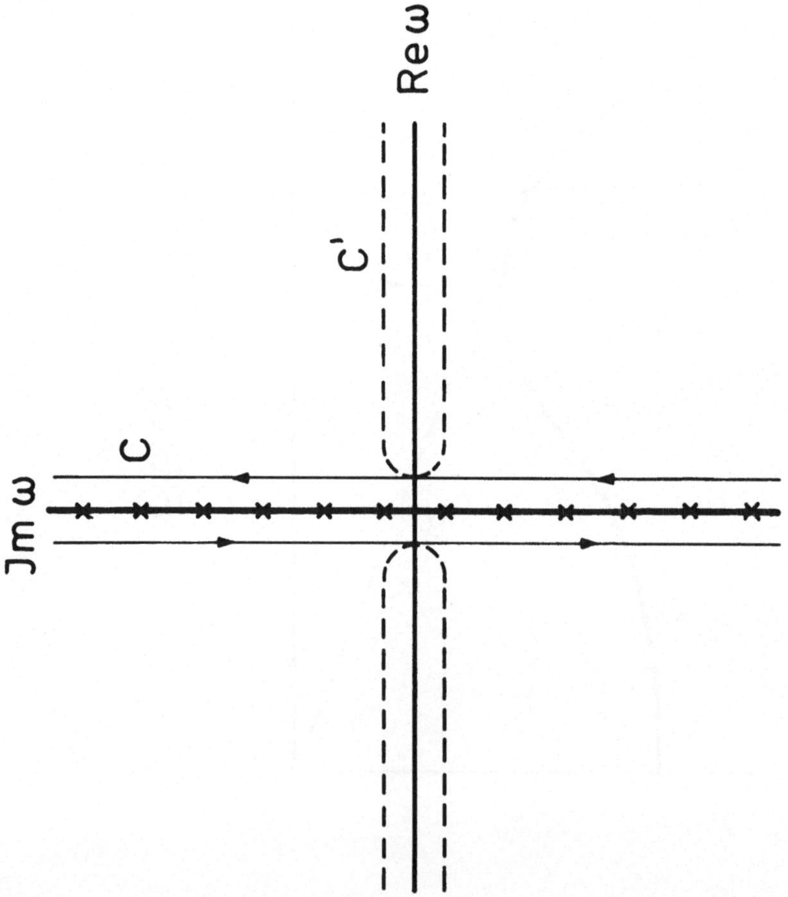

FIG.16

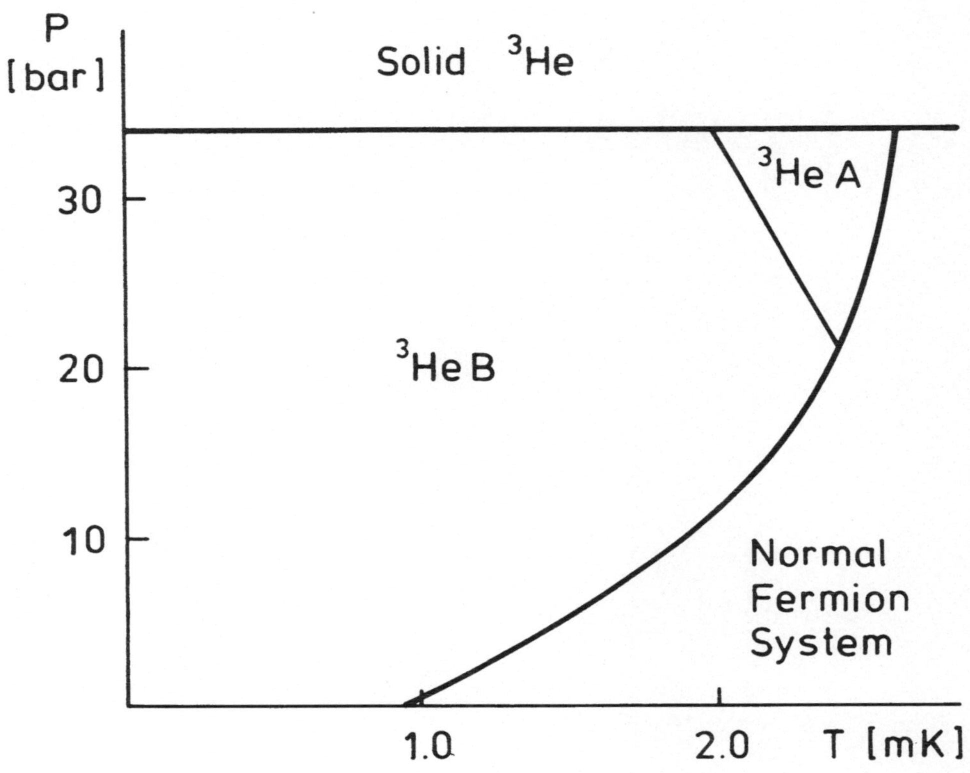
P
[bar]
Solid 3He
30
20
10
3He A
3He B
Normal
Fermion
System
1.0
2.0
T [mK]

FIG.17

Acta Physica Austriaca, Suppl. XVIII, 543–566 (1977)

RENORMALIZATION GROUP APPROACH TO CRITICAL PHENOMENA+

by

H. WAGNER
Sektion Physik
Universität München

I. INTRODUCTION

In this lecture we sketch the application of renormalization group (RG) methods to the study of continuous phase transitions. The basic ideas behind these methods have first been formulated in Lagrangian quantum field theory in an attempt to improve perturbation expansions and to investigate the asymptotic behaviour of Green functions at large momenta. After the pioneering work of Wilson [1] the RG methods quickly conquered the field of statistical mechanics and led to a breakthrough in our understanding of critical phenomena.

We shall first briefly review some current concepts in the theory of phase transitions. In the second part we outline the application of modern renormalized

---

+Lecture given at XVI.Internationale Universitätswochen für Kernphysik,Schladming,Austria,February 24-March 5,1977.

perturbation theory to static critical phenomena. The third part is concerned with an extension of RG methods to deal with dynamic critical effects.

II. CONTINUOUS PHASE TRANSITIONS

Consider a ferromagnet. Below the critical (Curie-) temperature T_c there is a spontaneous magnetization $\phi_o(T)$. This $\phi_o(T)$ is an example of an order parameter and characterizes thermodynamically the ordered phases below T_c. As the temperature is increased towards T_c, $\phi_o(T)$ decreases and vanishes continuously at T_c. Near T_c there are large fluctuations in the magnetization within subvolumes of the sample. In this critical region one also observes anomalies in some thermodynamic quantities. For example, the magnetic susceptibility χ is found to diverge like $(T-T_c)^{-\gamma}$, with γ denoting a critical exponent. One of the most remarkable features of these critical singularities is their apparent universal character. For many ferromagnetic materials, the value of the exponent γ is found to be near to 1.3. Furthermore, the isothermal compressibility of CO_2, for instance, diverges near its gas-liquid critical point in the same way as the magnetic susceptibility of Fe near the Curie point, with about the same value of γ.

These singularities are believed to be ultimately caused by the large fluctuations of the local order parameter. In order to elaborate on this vital point in more detail, we start with statistical mechanics. For a ferromagnet, the fundamental connection between thermodynamics and statistical mechanics reads (k_B=1):

$$G_N(TH) = - T \ \ell n \ Z_N(TH) \ .$$

This expression relates the Gibbs free energy $G_N(TH)$ with the partition function

$$Z_N(TH) = \sum_{\text{states}} e^{-\mathcal{H}_N(H)/T} \ ,$$

where H denotes the external magnetic field and N is the number of spins in the sample. In order to proceed we have to specify a model, since the Hamiltonian $\mathcal{H}_N$ enters explicitly in Z_N. The simplest model of a ferromagnet with some realistic features is the Ising model, where we have N spins $S_R = \pm 1$ on a lattice $\{R\}$ with a nearest neighbour (n.n.) coupling $J > O$,

$$\mathcal{H}_N(H) = J \sum_{\text{n.n.}} S_R S_{R'} - H \sum_R S_R \ .$$

All the thermodynamic information about the ferromagnet can be derived from $G_N(TH)$. For instance, the zero-field susceptibility per spin is given by

$$\chi(T) = \lim_{N\to\infty} \frac{T}{N} \left. \frac{\partial^2 \ell n Z_N(TH)}{\partial^2 H} \right|_{H=O} \ .$$

This also yields

$$T\chi(T) = \sum_R [\langle S_R S_O \rangle - \langle S_R \rangle \langle S_O \rangle]$$

$$\equiv \sum_R \Delta(RT) \ .$$

Empirically we know that $\chi(T)$ diverges as $T \to T_c$. Since $|\Delta(RT)| \leq 1$ for all R and T, the only way in which the above sum may become divergent is by having the number of non-vanishing terms in this sum increase indefinitely as $T \to T_c$. In other words, the spins have to be correlated, $\langle S_R S_o \rangle \neq \langle S_R \rangle \langle S_o \rangle$, over increasing distances $|R|$ as we approach T_c.

Away from the critical point we may assume that there is some finite correlation length $\xi(T)$ such that $\Delta(RT)$ vanishes rapidly for $|R| >> \xi(T)$. (A possible definition of ξ is

$$\xi^2 = \left| \frac{\sum_R R^2 \Delta(RT)}{\sum_R \Delta(RT)} \right| .)$$

Then

$$T\chi(T) \approx \sum_{|R|<\xi} \Delta(RT) ,$$

and we expect $\xi \to \infty$ as $T \to T_c$. From the theory of thermodynamic fluctuations we then conclude that the divergence of ξ leads to divergent fluctuations in the magnetization. This argument is indeed well supported by soluble models and also, empirically, by neutron and light scattering experiments.

The concept of a single, dominant correlation length is the basis of the phenomenological scaling theory from which relations between various critical exponents may be derived ("scaling laws"). In the simplest form (H = 0) one writes, for instance,

$$\Delta(RT) \approx \frac{1}{R^{d-2+\eta}} D_{\gtrless}\left(\frac{R}{\xi}\right) . \tag{2.1}$$

Here, d denotes the dimensionality of the system and η is a critical exponent characterizing the decay of correlations at T_c. The correlation length is supposed to follow a power law ($\tau = (T-T_c)/T_c$)

$$\xi \sim |\tau|^{-\nu}$$

and the "shape function" $D_{\gtrless}(x)$ (refering to $\tau \gtrless 0$) is assumed to be finite at T_c, $D(x) \to D_o \neq 0$, $x \to 0$. One then finds by standard arguments that $\gamma = \nu(2-\eta)$, which is one of the scaling laws.

Summary: There is experimental as well as theoretical support for the assumption that critical singularities follow power laws with exponents obeying the scaling relations. Furthermore there is evidence in favour of the so called universality hypothesis according to which the values of critical exponents depend only on the following factors:

1) the dimensionality d of the system ;
2) the dimensionality n of the order parameter (liquid-vapour, Ising: n=1; superfluid He^4: n=2; (OP-complex); Isotropic Heisenberg magnets: n=3)
3) the nature and the rate of decrease of long-ranged interactions, provided this decrease is sufficiently slow (e.g. like $R^{-(d+\sigma)}$).

Literature:

M.E. Fisher: Rep. Progr. Physics 30 (1967) 615.

L.P. Kadanoff et al.: Rev. Mod. Phys. 39 (1967) 395.

H.E. Stanley: Introduction to Phase transitions and Critical Phenomena (Oxford Univ.Press, 1974).

III. RENORMALIZATION GROUP METHODS FOR STATIC CRITICAL PHENOMENA

Renormalization group methods are currently applied to the study of phase transitions in the form of

1) Wilsons's recursion formalism [1-4]
2) Block spin methods [5]
3) Renormalized perturbation theory [6-7].

The first two formulations are more adapted to the language of statistical mechanics. However, a brief discussion of the third one presumably serves best the intention of this meeting since it exhibits the links between statistical mechanics and (Euclidean) field theory in the most direct way. For a comparison between 1) and 3) see the lecture by Jona-Lasinio in this volume.

We start with a path-integral expression for the partition function

$$Z = \text{const.} \int e^{-H\{\phi\}} \, d\{\phi\} \quad , \tag{3.1}$$

with a Hamiltonian (T absorbed) of the Landau-Ginzburg form

$$H = \int [\tfrac{1}{2}(\nabla\phi(x))^2 + \tfrac{1}{2}m_o^2\phi^2(x) + \tfrac{1}{4!}g_o\phi^4(x)] \, d^dx \ . \tag{3.2}$$

$\phi(x)$, the (scalar) order parameter field, may be considered as an Ising spin averaged over cells of the size Λ^{-1} >> lattice distance. The sum of the quadratic and the quartic term in (3.2) is interpreted as the free energy of a single cell, with m_o^2 as a linear measure of temperature and $g_o > 0$. The gradient term describes the interaction between cells.

There are arguments which indicate that the critical behaviour of correlation functions obtained from this statistical Landau theory is the same as in the Ising model, see e.g. [6]. Correlation functions, e.g.

$$\Delta_2(x-x') = \langle \phi(x)\phi(x') \rangle$$

$$= \frac{1}{Z}\int \phi(x)\phi(x')e^{-H\{\phi\}}d\{\phi\}$$

may be computed by a perturbative expansion in powers of g_o. Individual terms in the perturbation series can be represented by diagrams. Up to second order in g_o we have for instance

$\Delta_2(x-x')$:

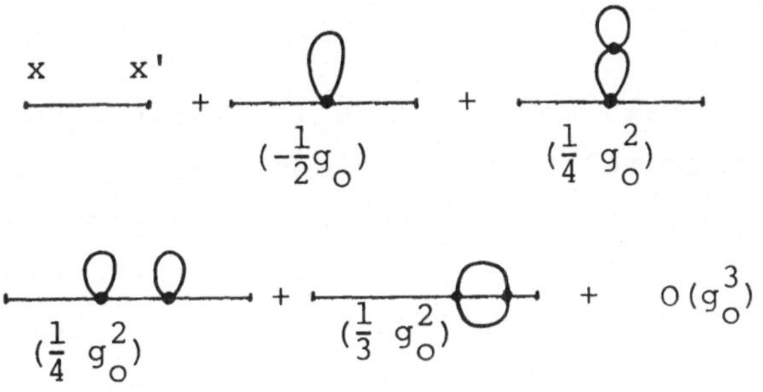

with the propagator

$$: \Delta_2^{(o)}(x-x') = \int \hat{\Delta}_2^{(o)}(q)e^{iq\cdot x}\frac{d^dq}{(2\pi)^d},$$

$$\hat{\Delta}_2^{(o)}(q) = [q^2 + m_o^2]^{-1}.$$

The formally divergent q-integrals over internal loops are cut off by Λ. The cut-off arises naturally because of the averaging implicit in (1) and (2). Λ^{-1} also provides the basic length scale and serves to define the dimensions of ϕ, m_o^2, g_o. Since H is dimensionless we have

$$\phi(x) \sim \Lambda^{(d-2)/2} \quad , \quad m_o^2 \sim \Lambda^2 \quad , \qquad g_o \sim \Lambda^{4-d} \quad .$$

In the following we shall mainly consider the one-particle irreducible parts (vertex functions) $\Gamma_N(x_1 \ldots x_N)$ of $\Delta_N(x_1 \ldots x_N)$. Example:

$$\hat{\Gamma}_2(q) = \hat{\Delta}_2^{-1}(q)$$

$= q^2 + m_o^2 +$ 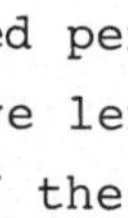 $+$ 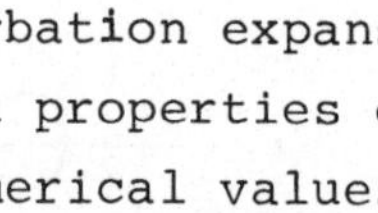 $+ \; O(g_o^3)$.

It would surely be too naive to expect meaningful results for correlation functions near T_c from this unsophisticated perturbation expansion. For instance, the long-wave length properties of $\hat{\Gamma}_N$ should be independent of the numerical values of g_o and Λ ("universality") which enter, of course, quite explicitly into the perturbation series.

As a heuristic guide and in order to provide a poor man's motivation for renormalization, let us look at the form of $\hat{\Gamma}_2(q)$ as postulated by the scaling hypothesis (2.1). At T_c (i.e. $m_o^2 = m_{oc}^2$) we have for $q \ll \Lambda$

$$\hat{\Gamma}_2(q) \approx q^2 \left(\frac{q}{\Lambda}\right)^{\eta} \quad \left[1 + \text{const.} \left(\frac{q}{\Lambda}\right)^{x}\right] \quad . \tag{3.3}$$

asymptotic part — correction ($x > 0$)

Once a power law behaviour of $\hat{\Gamma}_2(q)$ is assumed, the above form is dictated by dimensional arguments, since dim $\hat{\Gamma}_2$ = length^{-2}.

In the case of phase transitions we are mainly interested in the leading asymptotic term in $\hat{\Gamma}_2(q)$ for $q \to 0$. In order to get rid of the correction, it is tempting to take the limit $\Lambda \to \infty$. However, from (3.3) it is obvious that this limit cannot be taken in a straight forward manner. The reason lies in the two-fold nature of Λ, which is clearly displayed by the expression (3.3):

1) Λ serves as a <u>cut-off</u> which enters into the correction to the asymptotic behaviour;
2) at T_c, Λ^{-1} is the only <u>length-scale</u> available to appear in the asymptotic part of $\hat{\Gamma}_2$, as required for dimensional reasons if $\eta \neq 0$.

It is clear that we have to separate these two aspects of Λ by introducing another (arbitrary) length scale μ^{-1}, say, before we can execute the limit $\Lambda \to \infty$. For this purpose, we multiply (3.3) by $(\mu/\Lambda)^{\eta}$ to obtain

$$\left(\frac{\mu}{\Lambda}\right)^{\eta} \hat{\Gamma}_2(q) \approx q^2 \left(\frac{q}{\mu}\right)^{-\eta} \quad \left[1 + \text{const.} \left(\frac{q}{\Lambda}\right)^{x}\right] \quad .$$

Now, the limit $\Lambda \to \infty$ for fixed μ properly extracts the asymptotic part we are looking for.

This argument suggests that the "renormalized" vertex function

$$\hat{\Gamma}_{2R}(q) \equiv \hat{Z}_{\phi}\, \hat{\Gamma}_2(q) \quad , \qquad \hat{Z}_{\phi} = \left(\frac{\mu}{\Lambda}\right)^{\eta} \quad ,$$

or, equivalently,

$$\langle \phi_R(x)\phi_R(x')\rangle \equiv \langle \hat{Z}_{\phi}^{-\frac{1}{2}}\phi(x)\ \hat{Z}_{\phi}^{-\frac{1}{2}}\ \phi(x')\rangle$$

should have a meaningful limit for $\Lambda \to \infty$.

This consideration also tells us that we have to reparameterize our ϕ^4-model by renormalizing not only the field ϕ but also the other parameters of the model which carry dimensions. At this point the contact with the renormalization problem of quantum field theory becomes obvious; our model as given by (3.1) and (3.2) corresponds to a renormalizable field theory if $d \leq 4$ (super-renormalizable, if $d < 4$).

Following the standard procedures of renormalized perturbation theory we introduce the quantities ϕ_R, τ, g by

$$\phi_R = Z_{\phi}^{-\frac{1}{2}}\phi, \quad \tau = Z_{\tau}\ \tau_o \ , \quad (\tau_o \equiv m_o^2 - m_{oc}^2)$$

$$g \equiv u\mu^{4-d} = Z_g\ g_o \quad , \quad (g_o \equiv u_o\ \Lambda^{4-d}) \quad . \qquad (3.4)$$

After insertion of (3.4) into the "bare" Hamiltonian (3.2) we arrive at the renormalized H_R:

$$H_R = \int d^dx \, [\tfrac{1}{2}(\nabla\phi_R)^2 + \tfrac{1}{2}\tau\phi_R^2 + \tfrac{1}{4!}\,\phi_R^4 \tag{3.5}$$

$$+ \text{ counter terms}]$$

with

"Counter-terms" =

$$\tfrac{1}{2}\,(Z_\phi - 1)(\nabla\phi_R)^2 + m_{oc}^2 Z_\phi \phi_R^2 + \tau\,(Z_\tau Z_\phi - 1)\phi_R^2$$

$$+ \frac{1}{4!}\, g\,(Z_g\, Z_\phi^2 - 1)\,\phi_R^4 \quad . \tag{3.6}$$

The correlation functions $\Delta_{NR}(x_1 \ldots x_N) = \langle\phi_R(x_1)\ldots\phi_R(x_N)\rangle$ (or the $\Gamma_{NR}(x_1\ldots x_n)$) are now calculated from (3.5), using $[(g/4!)\phi_R^4$ + counter-terms] as a perturbation. The m_{oc}^2 and the dimensionless Z-factors will be determined by normalization conditions for renormalized vertex functions. Specifically, m_{oc}^2 has to be defined such that $\tau = 0$ corresponds to $T = T_c$ i.e. $\xi = \infty$. In this way the counter-terms produce the required substractions to remove the cut-off dependence of Feynman integrals for $\Lambda \to \infty$. The connection between the bare and the renormalized vertex functions reads

$$\hat{\Gamma}_{NR}(\{q\};\tau u\mu;\Lambda) = Z_\phi^{\frac{N}{2}} \;\; \hat{\Gamma}_N(\{q\};\; m_o^2\; u_o;\Lambda) \tag{3.7}$$

(an overall $\delta(\Sigma q)$ is divided out).

As for the normalization conditions we may choose

$$\hat{\Gamma}_2\Big|_{\substack{q=0\\ \tau=0}} = 0 \;, \qquad \hat{\Gamma}_4\Big|_{\substack{q_i=0\\ \tau=\mu^2}} = g = u\mu^{4-d} \;,$$

$$\partial_\tau \hat{\Gamma}_2 \Big|_{\substack{q=0 \\ \tau=\mu^2}} = 1 \ , \qquad \partial_{q^2} \hat{\Gamma}_2 \Big|_{\substack{q=0 \\ \tau=\mu^2}} = 1 \tag{3.8}$$

from which we obtain m_{oc}^2/Λ^2 and the Z-factors as functions of μ/Λ and u.

By construction, the renormalized vertex functions $\hat{\Gamma}_{NR}$ are finite in the limit $\Lambda \to \infty$ (τ, u, μ fixed) in every order of the perturbation expansion in powers of u, provided $d \leq 4$.

The $\hat{\Gamma}_{NR}$ still depend on the arbitrary length-scale parameter μ whose numerical value should be insignificant. A variation of μ with fixed bare parameters m_o^2, u_o, Λ must be compensated by corresponding variations of τ,u. (In the context of statistical mechanics it is the bare theory which is the "physical" one, with the cut-off substituted for the inverse lattice spacing.) It is well-known that this compensation is expressed in the renormalization group equations (RGE)

$$[\mu\partial_\mu + \kappa\tau\partial_\tau + \beta\partial_\mu - \frac{N}{2}\gamma_\phi]\hat{\Gamma}_{NR} = 0 \ , \tag{3.9}$$

with

$$\kappa = \kappa(u\ \tfrac{\mu}{\Lambda}) \equiv \tfrac{\mu}{\tau}\partial_\mu\tau|_o \xrightarrow{\Lambda\to\infty} \kappa(u) \ ,$$

$$\beta = \beta(u\ \tfrac{\mu}{\Lambda}) \equiv \mu\partial_\mu u|_o \xrightarrow{\Lambda\to\infty} \beta(u) \ ,$$

$$\gamma_\phi = \gamma_\phi(u,\tfrac{\mu}{\Lambda}) \equiv \frac{\mu}{Z_\phi}\partial_\mu Z_\phi|_o \xrightarrow{\Lambda\to\infty} \gamma_\phi(u) \ . \tag{3.10}$$

The RGE (3.9) are simply obtained from (3.7) by differentiating (3.7) with respect to μ for fixed bare parameters $(\mu\partial_\mu \ldots|_o)$.

In order to see how the scaling relations emerge from the RGE (3.9) we now assume that there is a "fixed point" u^*,

$$\beta(u^*) = 0 . \tag{3.11}$$

Setting $u = u^*$ in (3.9), the RGE reduce to Euler-type relations for homogeneous functions. To simplify the discussion we put $N = 2$. In terms of dimensionless variables we have in general

$$\hat{\Gamma}_2(q) = q^2 \, f(\frac{g}{\mu}, \frac{\tau}{\mu^2}, u) \quad . \tag{3.12}$$

Then, for $u = u^*$, the RGE (3.9) imply that

$$f(\frac{g}{\mu}, \frac{\tau}{\mu^2}, u) = (\frac{g}{\mu})^{-\eta} \psi \; (\frac{g}{\mu}(\frac{\tau}{\mu^2})^{-\nu}) \; , \tag{3.13}$$

where $\psi(x)$ remains unspecified, but

$$\eta = \gamma_\phi(u^*) \; , \qquad \nu = - \, [\kappa(u^*) - 2]^{-1} \; . \tag{3.14}$$

With the identification (3.14), the eq. (3.13) is a scaling relation of the form postulated in II. At this point it seems as if we would be able to derive scaling only for a special value of the coupling constant, $u = u^*$. This would be quite unsatisfactory in view of the empirically well-established universality of critical phenomena.

Fortunately, the concept of universality <u>is</u> embodied in this formalism. This can be seen as follows:

The RGE (3.9) may be integrated by the method of characteristics. To save writing we take $\tau = 0$. We introduce the scale variable ℓ by $\bar{\mu}(\ell) = \mu\ell$ and define $\bar{u}(\ell)$ through

$$\ell\frac{d\bar{u}}{d\ell} = \beta(\bar{u}) \ , \qquad \bar{u}(\ell=1) = u \ . \tag{3.15}$$

From (3.9) with (3.11) we then obtain

$$f(\frac{q}{\mu},0,u) = f(\ell\frac{q}{\mu},0,\bar{u}(\ell)) \exp[-\int_u^{\bar{u}(\ell)} \frac{\gamma_\phi(y)}{\beta(y)} dy] \ . \tag{3.16}$$

If $\beta(u)$ has a zero, u^*, and is regular in its neighbourhood, $\beta(u) = (u-u^*)x + O((u-u^*)^2)$, $x = \beta'(u^*)$, then

$$\bar{u}(\ell) = u^* + (u-u^*)\ell^x + O((u-u^*)^2) \ .$$

Thus $\bar{u}(\ell) \to u^*$ for $\ell \to 0$, provided $x > 0$ i.e. provided the fixed point u^* is "infrared stable". Furthermore,

$$\exp[-\int_u^{\bar{u}(\ell)} \frac{\gamma_\phi(y)}{\beta(y)} dy] \xrightarrow{\ell\to 0} \ell^{-\eta} [1+O((u-u^*)\ell^x)]. \tag{3.17}$$

Setting $\ell^{-1} = q/\mu$ in (3.16) we arrive at

$$f(\frac{q}{\mu},0,u) = (\frac{q}{\mu})^{-1} f(1,0,u^*)[1+O((u-u^*)q^x)] \ , \tag{3.18}$$

for $q/\mu \to 0$. In other words, the (infrared stable) fixed point describes the asymptotic behaviour of the vertex-functions for $q/\mu \ll 1$. Deviations of u from u^* merely produce correction to the leading asymptotic term.

This conclusion remains valid if $\tau > 0$, as can be easily checked. Therefore we may write

$$f(\frac{q}{\mu}, \frac{\tau}{\mu^2}, u) = f(\frac{q}{\mu}, \frac{\tau}{\mu^2}, u^*) \, [1+O((u-u^*)q^x)] \,, \qquad (3.19)$$

with $f(q/\mu, \tau/\mu^2, u^*)$ given by (3.13) and $\psi(y) = f(1, y^{-\nu}, u^*)$.

If one calculates $\beta(u)$ in perturbation theory one obtains

$$\beta(u) = -\varepsilon u + a u^2 + O(u^3) \,,$$

$$\varepsilon \equiv 4-d \,,$$

$$a > 0 \,. \qquad (3.20)$$

Hence $u^* \sim \varepsilon$ is an infrared stable fixed point ($x \sim \varepsilon$) if $d < 4$. (The other fixed point, $u^* = 0$, is ultraviolet stable, $\bar{u} \to 0$, $\ell \to \infty$, and describes the behaviour of correlations for $q/\mu >> 1$, which is of minor interest in the case of phase transitions.)

Two important conclusions can be drawn from (3.20):

1) For $\varepsilon < 0$ the long-wavelength properties of correlation functions are governed by the trivial Gaussian fixed point $u^* = 0$. The quartic term in (3.1) is "irrelevant" for critical phenomena; the critical exponents attain their classical values. (At $d = 4$ logarithmic correction occurs.)

2) For $0 < \varepsilon << 1$ critical exponents (and scaling functions, like in (3.13)) may be calculated within

an ε-expansion [8]. For illustration, we quote the results:

$$\eta = \frac{1}{56}\varepsilon^2 + O(\varepsilon^3) \quad ,$$

$$\nu = \frac{1}{2} + \frac{1}{12}\varepsilon + \frac{7}{162}\varepsilon^2 + O(\varepsilon^3) \quad .$$

If one neglects terms $O(\varepsilon^3)$ (which have also been computed) and boldly sets $\varepsilon = 1$ (d=3), one finds $\eta \approx 0.018$, $\nu \approx 0.627$ which should be compared with the series-results for the d=3 Ising model: $\eta = 0.041 \pm 0.01$, $\nu = 0.638 \pm 0.002$.

Above we mentioned that in statistical mechanics one is ultimately interested in the form of the bare correlation functions. The bare theory can be reconstructed from the renormalized one. This leads, in essence, to a replacement of μ by Λ in scaling expressions like (3.13).

IV. RENORMALIZATION GROUP METHODS IN CRITICAL DYNAMICS

In critical region, anomalies also occur in relaxation rates and transport coefficients. For instance, the spin diffusion coefficient in a ferromagnet and the thermal conductivity in a fluid both vanish at T_c. These are only two examples of a whole variety of transport quantities showing "critical slowing down". The dynamic anomalies may be described by power laws in $T-T_c$. However, the corresponding exponents turn out to be less universal than the static exponents and depend on properties of the materials which are irrelevant for the static critical behaviour.

A simple explanation of critical slowing down is offered by the socalled conventional theory. Consider a ferromagnet. The diffusion equation for the local magnetization reads

$$\partial_t M = D \nabla^2 M .$$

The spin diffusion coefficient D is connected with the susceptibility by Einstein's relation

$$D = \frac{\lambda}{\chi}$$

where λ is a kinetic (Onsager) coefficient. With reference to kinetic theory one argues that λ is determined by short-ranged spin-spin correlations which are supposed to be insensitive to critical fluctuations. Therefore one expects λ to remain finite at T_c and thus $D \sim \chi^{-1} \sim (T-T_c)^{\gamma}$. In this way we indeed find that $D \to 0$ as $T \to T_c$. However, the prediction $D \sim \chi^{-1}$ is incorrect in most cases. The reason is, as shown within the more refined but still phenomenological "mode-coupling" theory that near T_c correlations of the range ξ do contribute nonlinearly to kinetic coefficients, leading to a singularity in λ.

In the past few years, successful efforts have been made to extend RG methods to the study of critical dynamics. In this last part of the lecture we outline a formulation of critical dynamics which is well suited for an application of renormalized perturbation theory.

As usual in this field, we start with stochastic models described by generalized Langevin equations. These stochastic equations govern the temporal behaviour of appropriately chosen gross variables. The latter are assumed to provide a closed macroscopic description of a thermal system near a critical point. The choice of the set of gross variables depends on the nature of the model system. In principle, one should include all "slow" variables, as for instance the local order parameter and the densities of conserved quantities. Simplifications arise if some of the gross variables decouple because of symmetry requirements.

As an example, we consider a model of a ferromagnet with a three-component order parameter $\phi(xt) = \{\phi_\alpha(xt) | \alpha = 1,2,3\}$ and specified by the stochastic equation ($\dot{\phi} = \partial_t \phi$)

$$\dot{\phi} = - L_o \frac{\delta H}{\delta \phi} + \lambda_o f_o \quad \phi \times \frac{\delta H}{\delta \phi} + \sigma$$

$$\equiv v(\phi) + \sigma, \tag{4.1}$$

with

$$H = \int [\tfrac{1}{2} \nabla_j \phi_\alpha \nabla_j \phi_\alpha + \frac{m_o^2}{2} \phi_\alpha \phi_\alpha + \frac{g_o}{4!} (\phi_\alpha \phi_\alpha)^2] d^d x \ . \tag{4.2}$$

The meaning of the various terms on the right-hand side of (4.1) is as follows:

1) $- L_o \ \delta H / \delta \phi$ is a dissipative term,

$$L_o = \lambda_o (i\nabla)^a, \ a = \begin{cases} 0, & \phi \text{ not conserved,} \\ 2, & \phi \text{ conserved.} \end{cases} \tag{4.3}$$

λ_o denotes the bare Onsager coefficient.

2) $\lambda_o f_o \quad \phi \times \delta H/\delta\phi$ is a convective term, which describes the Larmor precession of the local magnetization in the local magnetic field $\delta H/\delta\phi$. If the coupling constant $f_o = 0$ then (4.1) reduces to the time-dependent Ginzburg-Landau model.

3) The third term, $\sigma(=\{\sigma_\alpha(x,t)\})$, denotes a random force and describes the effect of the "fast" variables not included in the set of gross variables. σ is assumed to be Gaussian distributed with a white-noise spectrum, i.e.

$$\omega\{\sigma\} \sim \exp\left[-\frac{1}{4}\int \sigma(x,t)\cdot(L_o^{-1}\sigma)(x,t)d^dx\,dt\right], \qquad (4.4)$$

where $\omega\{\sigma\}$ denotes the path probability distribution for the process σ.

The Onsager coefficient λ_o provides the basic time scale of the model. We also introduce a momentum cut-off Λ which defines the basic length scale. Finally, we assert that the thermal equilibrium distribution is given by $\exp(-H)$. This can be verified by going over to the Fokker-Planck equation associated with (4.1) and (4.4).

The RG methods may now be applied in analogy with the static case by successive elimination of short-wavelength fluctuations, either within a perturbative expansion for correlation functions [9-11] or directly in the stochastic equation (4.1) [12-14]. Here, we shall outline a different approach [15,16] in which one transforms (4.1) into a path integral. This leads to a Lagrangian-type version of critical dynamics which allows us to make straight forward use of the tools of renormalized perturbation theory, as in the static case. (An alternative Lagrangian formulation of critical dynamics which does not include convective mode-coupling terms has been given in ref. [17].)

As a first step we write

$$\omega\{\sigma\}d\{\sigma\} = e^{G\{\phi\}}d\{\phi\} , \tag{4.5}$$

by substituting $\sigma = \dot{\phi} - v(\phi)$ in (4.4). Being careful to include the functional Jacobian which arises from transformation $\sigma \to \phi$ one then finds [18]

$$G = -\frac{1}{4}\int [(\dot{\phi}-v)\cdot L_o^{-1}(\dot{\phi}-v) + 2\frac{\delta v}{\delta\phi}]d^dx\, dt . \tag{4.6}$$

It is now possible to generate a perturbation expansion of correlation functions $\langle\phi(x_1,t_1)\ldots\phi(x_n,t_n)\rangle$ using the weight exp $G\{\phi\}$.Presumably, this would not be a convenient procedure, mainly for two reasons:

1) G contains coupling terms up to 6th order in ϕ (if H is of the form (4.2)) of different structure. This would complicate diagrammatics.

2) If ϕ is a conserved order parameter, then $L_o^{-1} \sim q^{-2}$ leads to superficial infrared divergences in the vertices which could cause technical troubles.

A more convenient form of the functional weight is obtained after linearizing G with the help of a Gaussian transformation [15],

$$e^{G\{\phi\}} \sim \int e^{J\{\phi,\tilde{\phi}\}} \, d\{\phi, i\tilde{\phi}\} \,, \tag{4.7}$$

where

$$J = \int [\tilde{\phi}\cdot L_o\tilde{\phi} - \tilde{\phi}(\dot{\phi}-v(\phi)) - \frac{1}{2}\frac{\delta v}{\delta \phi}] \, d^dx \, dt \quad . \tag{4.8}$$

Now, we may calculate correlation functions

$$\langle\phi\ldots\tilde{\phi}\ldots\rangle \sim \int \phi\ldots\tilde{\phi}\ldots e^{J\{\phi,\tilde{\phi}\}} \, d\{\phi, i\tilde{\phi}\} \tag{4.9}$$

by a perturbation expansion in terms of the non-linear coupling in $\tilde{\phi}\cdot v(\phi)$. The structure of this expansion is only slightly more complicated than in the corresponding static case, due to the presence here of a second field $\tilde{\phi}$. This field has been formally introduced above as a dummy variable. It turns out however, that $\tilde{\phi}$ does have an important physical significance. One finds that mixed correlation functions e.g. $\langle\phi(x,t)\tilde{\phi}(x',t')\rangle$ describe the response of the system to an external field. Thus, response functions are obtained simultaneously with the normal correlation functions e.g. $\langle\phi(x,t)\phi(x',t')\rangle$ at the same level of approximation.

We are ultimately interested in the low-frequency, long-wavelength behaviour of correlation and response functions from which we may deduce the transport coefficients. By the same arguments as given in the static case, the perturbation expansion will be renormalized in order to extract the leading asymptotic form of the correlation and response functions.

Power counting indicates that the present model is renormalizable in the field theoretic sense, with the additional reparameterizations

$$\tilde{\phi}_R = Z_{\tilde{\phi}}^{-\frac{1}{2}}\tilde{\phi} , \quad \lambda = Z_\lambda \lambda_o , \quad f \equiv v\mu^{\frac{6-d}{2}} = Z_f f_o \qquad (4.10)$$

supplementing (3.4). The new Z-factors have to be specified by additional normalization conditions. In the standard way, we then arrive at RG equations for the various vertex functions. As an example, we obtain for the vertex function $\hat{\Gamma}_{11}$ associated with the response function $<\phi\tilde{\phi}>$:

$$[\mu\partial_\mu + \kappa\tau\partial_\tau + \beta_u\partial_u + \beta_v\partial_v + \zeta\lambda\partial_\lambda - \frac{1}{2}\gamma_\phi - \frac{1}{2}\tilde{\gamma}_\phi]\hat{\Gamma}_{11} = 0 . \qquad (4.11)$$

The coefficient functions in (4.11) are defined analogously to (3.10) and depend, in the limit $\Lambda\to\infty$, on u and v only.

If there is an infrared-stable fixed point

$$\beta_u(u^*,v^*) = \beta_v(u^*,v^*) = 0 , \qquad (4.12)$$

then the asymptotic part of $\hat{\Gamma}_{11}$ exhibits the "dynamic scaling" form

$$\hat{\Gamma}_{11}^{as}(q,\omega) = \lambda q^{2+a} (\frac{q}{\mu})^y \Phi(\frac{\omega}{\omega(q)}, q\xi) , \qquad (4.13)$$

$$\omega(q) \sim q^Z$$

with $Z = 2+a+\zeta^*$ and $y = Z-2-a+(\eta+\tilde{\eta})/2$ where

$\zeta^* = \zeta(u^*,v^*)$ and $\tilde{\eta} = \tilde{\gamma}_\phi(u^*,v^*)$. In some cases the exponent Z can be expressed exactly in terms of static exponents. For a = 2 (ϕ conserved) one has $Z = 4 - \eta$ if $f_o = 0$ and $Z = (d + 2 - \eta)/2$ if $f_o \neq 0$. For a = 0, $f_o \neq 0$, one finds $Z = 4 + (\eta + \tilde{\eta})/2$, where $\tilde{\eta}$ is now a genuine dynamic exponent, being of $O(\varepsilon^2)$ in the ε-expansion. Corrections to scaling arise from $u \neq u^*$, $v \neq v^*$ as in the static case.

With the use of a short-distance expansion one also obtains some information about the scaling function Φ in (4.13). In the range $\xi^{-1} << q << \mu(\sim\Lambda)$ one finds ($\Omega \equiv \omega/\omega(q)$)

$$\Phi(\Omega,q\xi) = \Phi_o(\Omega) + \Phi_1(\Omega)(q\xi)^{-\frac{1}{\nu}}$$

$$+ \Phi_2(\Omega)(q\xi)^{-\frac{1-\alpha}{\nu}} + (\text{terms} \sim (q\xi)^{-\frac{2}{\nu}}) . \qquad (4.14)$$

Here, α is the static critical exponent for the specific heat. The functions $\Phi_i(\Omega)$ remain unspecified. The powers of $q\xi$ turn out to be identical to those appearing in the corresponding short-distance expansion of static correlation functions. Finally, let us mention that one may demonstrate, that deviations from the white-noise spectrum of the random force are irrelevant for dynamic behaviour.

REFERENCES

1. For a review see: K.G. Wilson, J. Kogut, Phys. Reports <u>12C</u> (1974) 75.

2. S.K. Ma, Modern Theory of Critical Phenomena, Benjamin, London 1976.

3. F. Wegner, in:Phase Transitions and Critical Phenomena, Vol. 6, C. Domb and M.S. Green eds., Academic Press, New York, 1976.

4. M.N. Barber, Phys. Reports 29C (1977) 1.

5. Th. Niemeijer, J.M.J. van Leeuwen, in: Domb + Green, Vol. 6.

6. E. Brézin, J.C.Le Guillon, J. Zinn-Justin, in: Domb + Green, Vol. 6 .

7. C.Di Castro, G.Jona-Lasinio, in: Domb + Green, Vol.6.

8. D.J.Wallace, in: Domb + Green, Vol. 6.

9. B.I.Halperin, P.C. Hohenberg, S.Ma: Phys. Rev. B10 (1974) 139.

10. P.C. Hohenberg, B.I. Halperin, to appear in Rev. Mod. Phys. 49 (July 1977).

11. S. Ma, G.F. Mazenko, Phys. Rev. B11 (1975) 4077.

12. Y. Kuramoto, Progs. Theor. Phys. 51 (1974) 1712.

13. K. Kawasaki, J. Gunton, preprint.

14. K. Kawasaki, preprint.

15. H.K.Janssen, Z. Physik B23 (1975) 377.

16. R. Bausch, H.K. Janssen, H. Wagner, Z. Physik B24 (1976) 113.

17. C. De Dominicis, E. Brézin, J. Zinn-Justin, Phys. Rev. B12 (1975) 4945.

18. R. Graham, in Springer Tracts in Modern Physics, Vol. 66, Berlin, Heidelberg, New York, 1973.

Acta Physica Austriaca, Suppl. XVIII, 567–588 (1977)

CRITICAL BEHAVIOUR AND NONLINEAR EFFECTS IN UNIAXIAL FERROELECTRICS+

by

G. MEISSNER and H. MAIS
Fachrichtung Theor. Physik, Univ. des Saarlandes
Saarbrücken, Germany

Based on the field theoretical method of the renormalization group theory for critical phenomena, we present calculations of the critical temperature dependence of static and dynamic properties for pure and chemically disordered uniaxial ferroelectrics. In particular, the anomalous temperature dependence of the Debye-Waller factor, of the static limit of the critical two-phonon scattering amplitudes (i.e. of the non-linear expansion coefficient of the electric field in powers of the polarization) and of the zero wave-vector dynamic electric susceptibility (dielectric dispersion) is examined. Experimental implications of the results for these pure and random

+ Seminar given at XVI. Internationale Universitätswochen für Kernphysik, Schladming, Austria, February 24 - March 3, 1977. Supported in part by the Deutsche Forschungsgemeinschaft under SFB 130, Ferroelektrika.

ferroelectric systems with marginal dimensionality $d^* = 3$ are discussed, since they may allow for carefully checking and verifying exactly solvable predictions of the renormalization group theory at marginal dimensionalities.

1. INTRODUCTION

The study of critical phenomena concerns the behaviour of many-particle systems whose order parameter correlation length is very large in comparison with their interparticle spacing. Therefore, many microscopic details of the corresponding many-body Hamiltonian should not be of importance for the static and dynamic properties at or close to the critical point. Crucial in determining the critical behaviour of a system with a given many-body Hamiltonian are, however, the spatial dimensionality d and the component number n of the order parameter in this space, i.e., more generally the symmetry of the order parameter.

The application of renormalization group methods in recent years has then indeed allowed for a proper treatment of static [1] and dynamic [2] critical behaviour of many systems. For comparison with experiments practical calculations based on the renormalization group approach usually have to be performed as an expansion in powers of the parameter $\varepsilon = d^* - d$ since the marginal dimensionality d^* equals four for standard systems with critical phenomena [3]. Important in this context is then, that certain microscopic details may give rise to a modification of the marginal dimensionality, of the effective spatial dimensionality and of the order parameter component number. Extremely interesting are for these reasons uniaxial ferroelectrics where owing to dipolar interactions critical fluctuations of the spontaneous polarization are reduced to a plane perpendicular to the polar axis which in turn results in a reduction of the marginal dimensionality to $d^* = 3$. The ε-expansion thus is no longer required and exactly solvable predictions of the renormalization group theory at the marginal dimensionality three are accessible to experimental study. Due to the uniaxial character the effective order parameter component number $n = 1$.

Because of a far-reaching correspondence between such a three-dimensional dipolar system and a four-dimensional short range system, first noted by Larkin and Khmelnitskii [4], a field theoretical model with short-range interactions in four-dimensional space will finally be investigated in this paper. The physical properties of this model as well as the general formalism for investigating its critical behaviour will be presented in section II. An extension of the formalism in order to be applicable to random, e.g. chemically disordered, uniaxial ferroelectrics is there given, too. This is of general interest and of practical importance, e.g., for a comparison of theoretical and experimental results in partially deuterated triglycine sulfate. There, local fluctuations in the critical temperature might appear due to a random substitution of hydrogen (H) by deuterium (D) in certain hydrogen bonds of the molecules $(NH_2CH_2COOH)_3H_2SO_4$. Applications of the theory to static critical behaviour will be discussed in section III. The anomalous temperature dependence of the Debye-Waller factor and of the static limit of the critical two-phonon scattering amplitude will there be derived from the solution of the renormalization group equation of the free energy. The application to the dynamical critical behaviour will be illustrated in section IV by calculating the critical behaviour of the zero wave-vector electrical susceptibility (dielectric dispersion) above T_c. Section V is devoted to concluding remarks.

2. FIELD THEORETICAL MODEL AND GENERAL FORMALISM

The field-theoretical models generally studied with renormalization group techniques replace the discrete theory by a continuous one described in terms of fields. This semimicroscopic approach is then applicable to a temperature region, where the correlation length is much larger than the lattice spacing, i.e., the distances in the correlation functions to be explored are large compared to the lattice spacing.

With the polarization density P(xt) along the polar axis as the field, we thus write the Lagrangian of the uniaxial ferroelectrics to be discussed:

$$L[P,\psi]=\int d^3xdtP(xt)\left(-\frac{1}{2\Gamma_o}\frac{\partial}{\partial t}\right)P(xt)-V[P]-\int d^3xdt\psi(x)P^2(xt) \quad . \tag{1}$$

The first term of the r.h.s. of (1) accounts for the time dependence of a purely relaxational system with the bare kinetic coefficients Γ_o. The t-integration in (1) is along the imaginary time (t)-temperature (T) axis in the interval (0, -i/T). The functional of the potential energy of such a system exhibiting a second order phase transition of the order disorder type:

$$V[P]=\int d^3x dt P(xt)\{\tfrac{1}{2}(m_o^2-\partial_k^2)\delta(x-x')+\tfrac{1}{2}\alpha_o\partial_3^2\frac{1}{|x-x'|}\}P(x't)$$

$$+\int d^3x dt\,\frac{g_{oc}}{4!}P^4(xt) \qquad (2)$$

couples the long-range interaction of dipoles oriented parallel to the polar x_3-axis and the continuous version of harmonically coupled double-well potentials with the bare coefficients m_o^2 for the second order term and g_{oc} for the fourth order term in P(xt). In the particular case of triglycine sulfate (TGS), the double-well potential may be related to a simulation of certain hydrogen bonds in the glycine groups. From a lattic dynamical point of view $m_o^2 \propto (T/T_{co}-1)$ is related to the zero-wave vector value of the unstable optical mode at the structural transition, where the transition temperature T_{co} results from the anharmonic thermal fluctuations of the non-critical modes. Finally, α_o can be related to the effective charge of the permanent dipoles. In our notation, $\partial_k^2 = \partial_1^2 + \partial_2^2 + \partial_3^2$ with $\partial_k = \partial/\partial x_k$. The non-dipolar part of V in (2) obviously represents the usual ϕ^4-theory in terms of P, studied for instance in non-linear field theory.

The Fourier transform of the coefficient of the term of V bilinear in P(xt):

$$\omega^2(q)=m_o^2+q^2+\alpha_o(q_3/q)^2 \equiv m_o^2+q_1^2+q_2^2+q_3^2+q_4^2 \qquad (3)$$

can either be interpreted as the small wave vector dispersion relation of the optical phonon frequencies in the presence of dipolar forces in a three-dimensional system or, following Larkin et Khmelnitskii [4], as the dispersion relation in a quasi four-dimensional space, by interpreting $q_4 = \sqrt{\alpha_o}\,(q_3/q)$ as the fourth wave vector component. It can be shown that the correspondence of a three-dimensional dipolar system to a four-dimensional short-range system holds in the leading temperature behaviour at the critical point to all orders of perturbation theory. Instead of the dipolar Lagrangian in three dimensions we shall therefore use in this lecture the corresponding short-range Lagrange function in four-dimensional space:

$$L[P,\psi]=\int d^4x dt\{P(xt)\tfrac{1}{2}(-\frac{1}{\Gamma_o}\frac{\partial}{\partial t} - m_o^2 + \partial_\mu^2)P(xt) -$$

$$- \frac{g_{oc}}{4!}P^4(xt) - \psi(x)\,P^2(xt)\} \tag{4}$$

where $\mu = 1, \ldots, 4$ now [5].

The space-dependent function $\psi(x)$ introducing an additional term quadratic in $P(xt)$ can achieve a local change of the transition temperature T_c due to a local change of the optical frequency m_o^2. In TGS, this extra term allows, e.g., for studying systems, where protons have been replaced by deuterium, i.e., deuterized TGS. To take into account randomness of deuterization, averaging over all possible configurations of ψ is required. This averaging will be denoted by $\langle \ldots \rangle_\psi \equiv \int\{\delta\psi\}\omega(\psi)\ldots$, where $\omega(\psi)$ represents the distribution function of ψ.

A convenient starting point for calculating now static and dynamical quantities is the path integral representation of the partition function [1] in an external electric field $E(xt)$ conjugate to $P(xt)$:

$$Z[E,\psi] = \int\{\delta P\} \exp\,(L[P,\psi] + P(1)E(1)) \ . \tag{5}$$

For notational simplicity, $P(1)E(1)$ stands for $\int d^4x dt$ $P(x_1t_1)E(x_1t_1)$, i.e., $1 \equiv x_1t_1$. Also an E-independent normalization factor is ignored in (5). The expansion of a generalized free enthalpy functional defined as:

$$W[E] = -\langle \ell n\, Z[E,\psi]\rangle_\psi = \sum_N \frac{1}{N!} E(1)\ldots E(N)G^{(N)}(1\ldots N) \tag{6}$$

in a functional Taylor series of E can be used to define the correlation functions or connected Green functions:

$$G^{(N)}(1..N) = \delta^N W/\delta E(1)..\delta E(N)_{E=0} = \langle\langle (P(1)..P(N))_c\rangle\rangle \tag{7}$$

as the successive coefficients in this series. The free energy functional is then defined by a functional Legendre transformation of W[E], i.e.:

$$\Gamma[P] = W[E]-E(1)P(1) = \sum_N \frac{1}{N!} P(1)..P(N)\Gamma^{(N)}(1..N) \ . \tag{8}$$

The expansion on the r.h.s. of (8) with respect to P in a manner similar to that of (6) with respect to E can be used to define the proper vertex functions:

$$\Gamma^{(N)}(1..N) = \delta^N\Gamma/\delta P(1)..\delta P(N)_{P=0} \tag{9}$$

again as the successive coefficients in that series. From the definition (8) it follows directly that:

$$E(1) = -\ \delta\Gamma/\delta P(1) \ . \tag{10}$$

The double brakets in the r.h.s. of (7) should indicate both the thermodynamic average according to (5) and the configurational average with respect to ψ. Because of this configurational average the correlation functions $G^{(N)}$ and the vertex functions $\Gamma^{(N)}$ are again translational invariant functions of the space-time variables. Hence, the two-point functions $G^{(2)}(12)$ and $\Gamma^{(2)}(12)$ are functions of the differences $x = x_1-x_2$ and $t = t_1-t_2$ only. Their Fourier transforms thus may be written as:

$$G^{(2)}(q,\Omega)=\int d^4x \int_0^{-i/T} dt\ e^{-i(qx-\Omega t)} G^{(2)}(xt)=[\Gamma^{(2)}(q,\Omega)]^{-1} . \tag{11}$$

Of practical importance in studying phase transitions are the following related functions: the Fourier transform of the expectation value of the polarization fluctuations which according to the fluctuation-dissipation theorem for classical systems:

$$C^{(2)}(q,\omega) = \frac{2}{\omega} \ \mathrm{Im}\ G^{(2)}(q,\Omega = \omega+io) \tag{12}$$

and the static susceptibility:

$$\chi(q) = \int\frac{d\omega}{2\pi} C^{(2)}(q,\omega) = G^{(2)}(q,\Omega = 0) . \tag{13}$$

For explicitly calculating the anomalous temperature dependence of measurable quantities, an actual evaluation of the configuration averages as defined in (6) is needed. This can be achieved by the use of the replica trick or the $n \to 0$ trick [6]. Starting from the identity for the partition function (5):

$$\ell n\, Z[E,\psi] = \frac{\partial}{\partial n} Z^n[E,\psi]_{n=0} =$$

$$= \frac{\partial}{\partial n}\left(\int\{\delta P_1 .. \delta P_n\}\exp\left(\sum_{i=1}^{n} (L[P_i,\psi]+E_i(1)P_i(1))\right)\right)_{n=0}$$

where n is the analytic continuation to zero of an arbitrary positive integer, we find for (6):

$$W[E] = -\frac{\partial}{\partial n}\left(\int\{\delta\vec{p}\}\exp(L_{eff}[\vec{p}]+\vec{E}\cdot\vec{p})\right)_{n=0} =$$

$$= \lim_{n\to 0} \ell n \int\{\delta\vec{p}\}\exp(L_{eff}[\vec{p}]+\vec{E}\cdot\vec{p}) \quad . \tag{14}$$

In (14) the n-component field:

$$\vec{p} \equiv \{P_1(xt),..,P_n(xt)\} \ ,$$

the scalar product:

$$\vec{E}\cdot\vec{p} \equiv \int d^4x\, dt \sum_{i=1}^{n} E_i(xt)\, P_i(xt) \ ,$$

and the effective Lagrangian of the n-component field:

$$L_{eff}[\vec{p}] = \int d^4x dt\, \vec{p}(xt)\left(-\frac{1}{2\Gamma_o}\frac{\partial}{\partial t}\right)\vec{p}(xt) - V_{eff}[\vec{p}], \tag{15}$$

where V_{eff} denotes now the effective potential:

$$V_{eff}[\vec{p}] = \int d^4x dt\{\tfrac{1}{2}m_o^2\vec{p}^2 + \tfrac{1}{2}(\partial_\mu\vec{p})^2 + \frac{g_{oc}}{4!}\sum_{i=1}^{n}\vec{p}_i^4 + \frac{g_{os}}{4!}(\vec{p}^2)^2 +$$

$$+ O\ ((\vec{p}^2)^3)\} \tag{16}$$

with the leading term in a cumulant expansion:

$$g_{os} \propto \langle (\int d^4x dt \vec{p}^2\ \psi_c^2 \rangle_\psi\ /\ \int d^4x dt (\vec{p}^2)^2\ .$$

Cumulants of ψ of higher than second order can be omitted since the additional factors of $\vec{p}$ to sixth order at least are irrelevant at the phase transition.

Expression (14) for W[E] together with (15) and (16) may then be used for an explicit investigation of the critical behaviour of uniaxial ferroelectrics. The limiting case of a chemically pure system is of course recovered by setting $g_{os} \equiv 0$.

3. STATIC CRITICAL BEHAVIOUR

We shall now apply the general formalism of the last section to the static critical behaviour of uniaxial ferroelectrics. The effective Lagrangian of (15) then reduces to:

$$L_{eff}[\vec{p}] = -\ V_{eff}[\vec{p}] \tag{17}$$

where now $\vec{p}$ stands for $\{P_1(x),\ldots,P_n(x)\}$ with $P(x) \equiv \int dt\ P(xt)$ and $P(1)E(1) \equiv \int d^4x_1 P(x_1)E(x_1)$, i.e. $1 \equiv x_1$. We introduce the Fourier transforms of the time-independent static vertex functions:

$$\Gamma^{(N)}(x_1..x_N) = \delta^N\Gamma/\delta P(x_1)..\delta P(x_N)_{P=0}$$

$$= \int d^4q_1..d^4q_N\ e^{-i(q_1x_1+..+q_Nx_N)}\tilde{\Gamma}^{(N)}(q_1..q_{N-1})\delta(q_1+..+q_N). \tag{18}$$

In the expression for the critical part of the Debye-Waller factor $W \propto \langle\langle P^2(y)\rangle\rangle$ as well as in that for the specific heat at zero electric field, $c_{E=0} \propto \int d^4y\cdot \langle\langle P^2(y)P^2(0)\rangle\rangle$, correlations of squares of the polarization P at coinciding points, e.g., y, have to be evaluated. In that case the theory has to be extended [1] by expanding:

$$\Gamma^{(N)}(x_1..x_N) = \sum_L \frac{1}{L!}\tau(y_1)..\tau(y_N)\Gamma^{(L,N)}(x_1..x_N;y_1..y_L) \tag{19}$$

in a functional Taylor series of the space-dependent temperature $\tau(y)$. The correlation functions:

$$\Gamma^{(L,N)}(x_i,y_i) = \delta^L\Gamma^{(N)}(x_1..x_N)/\delta\tau(y_1)..\delta\tau(y_L)_{\tau=0} \tag{20}$$

defined as the successive coefficients in this series are then related to the required quantities as follows:

a) static susceptibility:

$$\int d^4x\ e^{iqx}\ \delta^2\Gamma/\delta P(x)\delta P(0) \to \tilde{\Gamma}^{(0,2)}(q) \to \chi^{-1}(q); \tag{20a}$$

b) static limit of the critical two-phonons scattering amplitude:

$$\int d^4x_1 d^4x_2 d^4x_3 \delta^4\Gamma/\delta P(x_1)\delta P(x_2)\delta P(x_3)\delta P(0) \to \tilde{\Gamma}^{(0,4)}(0,0,0) \to f_2 \tag{20b}$$

which corresponds to the non-linear coefficients between E and P in the equation of state:

$$E = f_o P + f_2 P^3 + \dots ;$$

c) Debye-Waller factor [7]:

$$\delta\Gamma/\delta\tau(y) \to \Gamma^{(1,0)}(y) \to W ; \tag{20c}$$

d) Specific heat:

$$\int d^4y\ \delta^2\Gamma/\delta\tau(y)\delta\tau(0) \to \tilde{\Gamma}^{(2,0)}(0,0) \to C_{E=0} . \tag{20d}$$

The temperature dependence $\tau = T/T_c - 1$ of all these quantities may therefore be deduced from the temperature dependence of the free energy $\Gamma[P,\tau,g_c,g_s]$ which we derive from the solution:

$$\Gamma[P,\tau,g_c,g_s] = \lambda^4 \Gamma[\frac{P(\lambda)}{\lambda}, \frac{\tau(\lambda)}{\lambda^2}, g_c(\lambda), g_s(\lambda), 1] - - \frac{1}{2}\int_1^{\lambda} B(\lambda')\tau^2(\lambda')\frac{d\lambda'}{\lambda'} \tag{21}$$

of the renormalization group (R.G.) equation [1]:

$$\lambda\frac{d\Gamma}{d\lambda} = \{\lambda\frac{\partial}{\partial\lambda} + \lambda\frac{\partial v_i}{\partial\lambda}\frac{\partial}{\partial v_i}\}\Gamma[P,\tau,g_c,g_s,\lambda] = \frac{1}{2}B(g_c,g_s)\tau^2 . \tag{22}$$

The R.G. equation (22) describes to what extent an explicit change of Γ under a dilatation λ of the lattice spacing in the original system can be compensated for by an appropriate implicit change via the coupling constants g_c and g_s together with an appropriate rescaling of the polarization P and the temperature τ. From the chain-rule differentiation of Γ, thus:

$$\frac{\partial v_i}{\partial \lambda} \frac{\partial}{\partial v_i} \equiv \frac{\partial g_c}{\partial \lambda} \frac{\partial}{g_c} + \frac{\partial g_s}{\partial \lambda} \frac{\partial}{\partial g_s} + \frac{\partial}{\partial \lambda} \ell n\ P\ P\frac{\partial}{\partial P} + \frac{\partial}{\partial \lambda} \ell n \tau\ \tau\frac{\partial}{\partial \tau}\ .$$

The first term on the r.h.s. of (21) results from the dimensional analysis in four dimensional space:

$$\Gamma[P(\lambda),\tau(\lambda),g_c(\lambda),g_s(\lambda),\lambda]=\lambda^4\Gamma[\frac{P(\lambda)}{\lambda},\frac{\tau(\lambda)}{\lambda^2},g_c(\lambda),g_s(\lambda),1]\ . \tag{23}$$

In order to further exploit (21) one has to determine from perturbation theory the Gell-Mann-Low functions, i.e., the right hand sides of the characteristic equations [1]:

$$\lambda\frac{\partial}{\partial \lambda}\ g_c = W_1(g_c,g_s)\ , \tag{24a}$$

$$\lambda\frac{\partial}{\partial \lambda}\ g_s = W_2(g_c,g_s)\ , \tag{24b}$$

$$-\lambda\frac{\partial}{\partial \lambda}P = \eta(g_c,g_s)\ , \tag{24c}$$

$$-\lambda\frac{\partial}{\partial \lambda}\ell n\tau = \nu^{-2}(g_c,g_s) - 2\ , \tag{24d}$$

with the boundary conditions, $g_c(\lambda=1)=g_c, g_s(\lambda=1)=g_s$, $P(\lambda=1)=P$ and $\tau(\lambda=1)=\tau$, together with:

$$B(g_c,g_s) = [-2(\nu^{-1}(g_c,g_s)-2) + \lambda\frac{\partial}{\partial\lambda}]A(g_{oc},g_{os},\lambda) \tag{24e}$$

which appears in (22) due to the non-multiplicative character of the vertex functions $\Gamma^{(L,N)}$ for $N = 0$ and $L \leq 2$. In our four-dimensional system the functions W_1, W_2, η, ν and B can be calculated as power series in g_c and g_s without an ε-expansion. Using these power series expansions [8-12] one can integrate Eqs. (24) and infer that the renormalized coupling constants g_c and g_s are driven towards values g_c^* and g_s^* which are fixed points of the system, i.e., simultaneous zeros of the functions W_1 and W_2 as λ goes to zero. With the choice for λ such that $|\tau(\lambda_1)/\lambda_1^2| = 1$, one can moreover show that the coupling constants g_c and g_s go to zero as $\lambda \to 0$. Since this same choice, finally, prevents the expansion of $\Gamma[P(\lambda)/\lambda, \tau(\lambda)/\lambda^2, g_c(\lambda), g_s(\lambda), 1]$ to diverge when expanded into powers of $g_c(\lambda)$ and $g_s(\lambda)$ it suffices to study the asymptotic expansion of (21):

$$\Gamma[P,\tau,g_c,g_s] = \frac{1}{2}\tau(\lambda)P^2 + \frac{1}{4!}\{g_c(\lambda) + g_s(\lambda)\}P^4 -$$

$$- \frac{1}{2}\int_1^\lambda B(\lambda')\tau^2(\lambda')d\lambda'/\lambda' + \dots \tag{25}$$

which has to be supplemented by the results in the limit $\lambda \to 0$ for the anomalous temperature dependence of

$$\tau(\lambda) \propto \begin{cases} |\tau||\ell n|\tau||^{-1/3} , & (g_{os} \equiv 0) \\ |\tau|\exp\{-(\frac{6}{53}|\ell n|\tau||)^{1/2}\} , & \end{cases} \qquad (26a)$$

of:

$$g_c(\lambda) \propto \begin{cases} |\ell n|\tau||^{-1} , & (g_{os} \equiv 0) \\ |\ell n|\tau||^{-1/2} \propto g_{os} \neq 0 , & \end{cases} \qquad (26b)$$

and of:

$$\int_1^{\lambda} B(\lambda')\tau^2(\lambda')d\lambda'/\lambda' \propto \begin{cases} |\tau|^2|\ell n|\tau||^{1/3} & (g_{os} \equiv 0) \\ |\tau|^2|\ell n|\tau||^{1/2}\exp\{-2(\frac{6}{53}|\ell n|\tau||)^{1/2}\} . & \end{cases} \qquad (26c)$$

This then enables us to illustrate the critical temperature behaviour of uniaxial ferroelectrics, e.g. by the quantities listed in Table 1.

	pure ($g_{so} \equiv 0$)	random (e.g. deuterated)
f_2	$\lvert\ell n\lvert\tau\rvert\rvert^{-1}$	$\lvert\ell n\lvert\tau\rvert\rvert^{-1/2}$
W	$\lvert\tau\rvert\lvert\ell n\lvert\tau\rvert\rvert^{1/3}$	$\lvert\tau\rvert\lvert\ell n\lvert\tau\rvert\rvert^{1/2}\exp\{-2(\frac{6}{53}\lvert\ell n\lvert\tau\rvert\rvert)^{1/2}\}$
$C_E = 0$	$\lvert\ell n\lvert\tau\rvert\rvert^{1/3}$	$\lvert\ell n\lvert\tau\rvert\rvert^{1/2}\exp\{-2(\frac{6}{53}\lvert\ell n\lvert\tau\rvert\rvert)^{1/2}\}$
χ^{-1}	$\lvert\tau\rvert\lvert\ell n\lvert\tau\rvert\rvert^{-1/3}$	$\lvert\tau\rvert\exp\{-(\frac{6}{53}\lvert\ell n\lvert\tau\rvert\rvert)^{1/2}\}$

Table 1:

Anomalous temperature dependence of the non-linear expansion coefficient f_2, the Debye-Waller factor W, the specific heat $C_{E=0}$ and the inverse electric susceptibility χ^{-1} in pure and random uniaxial ferroelectrics as TGS for $T \geq T_c$.

The temperature dependence of $C_{E=0}$ and χ^{-1} of pure uniaxial ferroelectrics have first been derived by Larkin and Khmelnitskii [4]. The temperature dependence of f_2 and W and their experimental observability have been investigated in ref.[13]. In the random system, the Debye-Waller factor has first been derived in a recent paper by Meissner [14], the results for $C_{E=0}$ and χ^{-1} are due to Aharony [11] and the temperature dependence of f_2 has recently been obtained by Shalayev [15]. One of the striking results for the critical behaviour of the random ferroelectrics as compared to the pure case is the appearence of the new type of critical singularity described by $\exp\{-(6/53|\ell n|\tau||)^{1/2}\}$. Since the marginal dimensionality $d^* = 3$ in the dipolar uniaxial ferroelectrics this type of singularity should be accessible in the laboratory as well as the modification of the mean field behaviour by powers of logarithmic factors in pure uniaxial ferroelectrics.

It might be worth mentioning that several further predictions like crossover effects of critical amplitude ratios etc. are possible within our theoretical frame. Due to the limitation of time they cannot be discussed here, however. In the remaining part of this seminar we shall rather give a brief illustration of the dynamical critical behaviour of uniaxial ferroelectrics.

4. DYNAMICAL CRITICAL BEHAVIOUR

The inclusion of the purely relaxating time-dependent term $\int d^4x dt \vec{p}(xt)(-(2\Gamma_o)^{-1}\partial/\partial t)\vec{p}(xt)$ in the Lagrangian $L_{eff}[\vec{p}]$ of (15) gives now rise to the presence of finite frequencies in the Fourier transform of the correlation- and vertex-functions. For simplicity and since it is accessible experimentally we shall discuss the critical dynamics of our system in terms of the dielectric dispersion above T_c, i.e., for the zero wave vector inverse susceptibility:

$$\chi^{-1}\ (q = 0,\Omega) = \tilde{\Gamma}^{(2)}\ (q = 0,\Omega)\ , \quad T \geq T_c\ .$$

The analysis follows very closely that of the static case. We therefore derive the temperature dependence of the two-point vertex function $\tilde{\Gamma}^{(2)}(\frac{i\Omega}{\Gamma_o})[\tau, g_c, g_s]$ again from the solution:

$$\tilde{\Gamma}^{(2)}(\frac{i\Omega}{\Gamma_o})[\tau, g_c, g_s] = \tag{27}$$

$$= \lambda^2\tilde{\Gamma}^{(2)}(\frac{i\Omega}{\lambda^2\Gamma_o(\lambda)})[\frac{\tau(\lambda)}{\lambda^2}, g_c(\lambda), g_s(\lambda), 1]\exp(-\int_1^\lambda \eta(\lambda')\frac{d\lambda'}{\lambda'})$$

of the R.G. equation:

$$\lambda\frac{d}{d\lambda}\tilde{\Gamma}^{(2)} = \{\lambda\frac{\partial}{\partial\lambda} + W_1\frac{\partial}{\partial g_c} + W_2\frac{\partial}{\partial g_s} - (\nu^{-1}-2)\tau\frac{\partial}{\partial\tau} - (z-2)\Omega\frac{\partial}{\partial\Omega}\}\tilde{\Gamma}^{(2)} =$$

$$= \eta\tilde{\Gamma}^{(2)}\ . \tag{28}$$

On the r.h.s. of (27) we have used the dimensional analysis:

$$\tilde{\Gamma}^{(2)}\left(\frac{i\Omega}{\Gamma_o}\right)[\tau, g_c, g_s, \lambda] =$$

$$\lambda^2\,\tilde{\Gamma}^{(2)}\left(\frac{i\Omega}{\lambda^2\Gamma_o(\lambda)}\right)\left[\frac{\tau(\lambda)}{\lambda^2},\ g_c(\lambda),\ g_s(\lambda),\ 1\right] \quad . \tag{29}$$

To the Gell-Mann-Low functions (24a-d) we have to add another one:

$$\lambda\frac{\partial}{\partial\lambda}\ell n\ \Gamma_o = -z(g_c, g_s) + 2 \qquad \text{where} \qquad \Gamma_o(\lambda=1)=\Gamma_o$$

since the compensation of the explicit change of $\tilde{\Gamma}^{(2)}$ under a dilatation λ requires an additional implicit change via the kinetic coefficient Γ_o, with the dynamical critical exponent z. The asymptotic expansion of (27) into powers of g_c and g_s can be taken from Grinstein, Ma and Mazenko [16] which in our notation reads:

$$\tilde{\Gamma}^{(2)}\left(\frac{i\Omega}{\Gamma_o}\right)[\tau, g_c, g_s] = \lambda^2\left\{\frac{\tau(\lambda)}{\lambda^2} - \frac{i\Omega}{\lambda^2\Gamma_o(\lambda)}[1-g_s(\lambda)Q(\Omega)]\right\} \tag{30}$$

where

$$Q(\Omega) = \frac{1}{2}\left\{-\ell n\left[\frac{\tau(\lambda)}{\lambda^2} - \frac{i\Omega}{\lambda^2\Gamma_o(\lambda)}\right] + \right.$$

$$\left. + \frac{\tau(\lambda)}{i\Omega/\Gamma_o(\lambda)}\ell n\left[\frac{\left(1+\frac{\lambda^2}{\tau(\lambda)}\right)\left(\frac{\tau(\lambda)}{\lambda^2} - \frac{i\Omega}{\lambda^2\Gamma_o(\lambda)}\right)}{1+\frac{\tau(\lambda)}{\lambda^2} - \frac{i\Omega}{\lambda^2\Gamma_o(\lambda)}}\right]\right\} \quad . \tag{31}$$

The physical implications of these results will be considered in the frequency region: $\omega/\Gamma_o \ll \tau$. With the same choice for λ, i.e., $\tau(\lambda_1)/\lambda_1^2 = 1$ as in statics we can evaluate the temperature dependence of the kinetic coefficient:

$$\Gamma_o(\tau) \propto \begin{cases} \Gamma_o & (g_{os} \equiv 0) \\ \Gamma_o \exp\{-(\frac{6}{53}|\ln|\tau||)^{1/2}\} & ; \end{cases} \tag{32}$$

take $\tau(\lambda)$ from (24) and approximate (31) to lowest order:

$$Q(\Omega) \simeq -\frac{1}{4} \tag{33}$$

which gives for (30):

$$\tilde{\Gamma}^{(2)}(\frac{i\Omega}{\Gamma_o})[\tau,g_c,g_s] \simeq \tau(\lambda) - \frac{i\Omega}{\Gamma_o(\lambda)}[1+\frac{1}{4}g_s(\lambda)] \quad . \tag{34}$$

With (34) we may immediately evaluate the correlation function $C^{(2)}(q=0,\omega)$ from (12) and write in the scaling form:

$$C^{(2)}(q=0,\omega) = \frac{2}{\omega}\,\mathrm{Im}\,G^{(2)}(q=0,\omega) = \frac{\chi(\tau)}{\omega_c(0)}\,f(\frac{\omega}{\omega_c(0)}) \quad . \tag{35}$$

The dissipative characteristic frequency:

$$\omega_c(q=0) = \begin{cases} \Gamma_o\,\chi^{-1}(\tau) & (g_{os} \equiv 0) \\ \Gamma_o\,\chi^{-1}(\tau)[1+g_s(\tau)/4]^{-1} & \end{cases} \tag{36}$$

where from statics $g_s(\tau) \propto |\ln|\tau||^{-1/2}$ and $\chi^{-1}(\tau)$ acquires the temperature dependence as given in Table 1. The shape function:

$$f\left(\frac{\omega}{\omega_c(0)}\right) = 2\,[1 + (\omega/\omega_c(0))^2]^{-1} \quad . \tag{37}$$

A striking feature of these results is the reduction of the characteristic frequency $\omega_c(0)$ due to the random fluctuations (with the factor: $\exp\{-(6/53)\,|\ln|\tau||)^{1/2}\}$) in comparison to the pure system (with the factor: $|\ln|\tau||^{-1/3}$). The resulting enhanced low-frequency correlations in the vicinity of T_c should be accessible experimentally, too.

5. CONCLUSIONS

Our basic theoretical results of the static and dynamic critical behaviour of pure and random uniaxial ferroelectric materials are very important for the general theory of critical fluctuations at second order phase transitions. This results from the fact that their marginal dimensionality $d^* = 3$ or equivalently that the leading logarithmic corrections to the meanfield-like critical behaviour in these dipolar systems are the same as those of a short range system at $d = 4$. Therefore, exact solutions of the renormalization group equations can be compared with experiments. The field theoretical formulation allows for treating static and dynamic behaviour within the same formalism [17]. The field theoretical model obtained by replacing the discrete theory by a continuous one in terms of the order parameter field exhibits interesting similarities between solid state physics and field theory. The fields corresponding to additional degrees of freedom, e.g., due to elasticity, have, of course, to be included in a more complete theory, too. This, however, would exceed the scope of this contribution.

ACKNOWLEDGEMENTS

This manuscript has been performed while one of the authors (G.M.) has been visiting the Max-Planck-Institut für Physik und Astrophysik, München. The kind invitation by Professor W. Götze and the kind hospitality extended by Professor W. Zimmermann are gratefully acknowledged.

REFERENCES

A short account of parts of this work has been given by G. Meissner at the SFB-130 Colloquium, Saarbrücken, July 1976.

1. For the field theoretical renormalization group approach to static critical behaviour see, E. Brézin, J.C. Goullou and J. Zinn-Justin, in Phase Transitions and Critical Phenomena, C. Domb and M.S. Green, eds. (Academic Press, London, 1976), Vol. 6, page 125.

2. For a review of the theory of dynamical critical phenomena see, P.C. Hohenberg and B.I. Halperin, Rev. Mod. Phys., July 1977 issue.

3. K.G. Wilson and J. Kogut, Phys. Rev. 12C (1974) 75.

4. A.I. Larkin and D.E. Khmelnitskii, Z. Eksp. Teor. Fiz. 56 (1969) 2087; (Sov. Phys.-JETP 29 (1969) 1123).

5. The equivalence of describing a purely dissipative system by either a time-dependent Ginzburg-Landau equation with Gauss-distributed Langevin forces or by an effective Lagrangian has first been shown by C. de Dominicis, Nuovo Cimento Lett., 12 (1975) 567.

6. S.F. Edwards and P.W. Anderson, J. Phys. F5 (1975) 965; V.J. Emery, Phys. Rev. B11 (1975) 239.

7. G. Meissner and K. Binder, Phys. Rev. B12 (1976) 3948.

8. G. Grinstein and A. Luther, Phys. Rev. B13 (1976) 1329.

9. T.C. Lubensky, Phys. Rev. B11 (1975) 3573.

10. Y. Yamazaki, Progr. Theor. Phys., to be published.

11. A. Aharony, Phys. Rev. B13 (1976) 2092.

12. D.E. Khmelnitskii, Zh. Eksp. Teor. Fiz. 68 (1975) 1960; (Sov. Phys.-JETP 41 (1976) 981).

13. K. Binder, G. Meissner and H. Mais, Phys. Rev. B13 (1976) 4890.

14. G. Meissner, preprint, to be published.

15. B.N. Shalayev, Zh. Eksp. Teor. Fiz. 72 (1977) 1192.

16. G. Grinstein, S.K. Mai, G.F. Mazenko, Phys. Rev. B15 (1977) 258.

17. The extension of the field theoretical renormalization group approach to critical dynamics of systems with mode-coupling terms has been developed by R. Bausch, H.K. Janssen and H. Wagner, Z. Physik B24 (1976) 113.

Acta Physica Austriaca, Suppl. XVIII, 589-592 (1977)

PROBABILISTIC IDEAS IN FIELD THEORY AND STATISTICAL MECHANICS+

by

G. JONA LASINIO
University of Rome, Italy

SUGGESTED READING

The lectures stressed and illustrated the increasing role of probability theory in unifying a number of problems in quantum mechanics, quantum field theory and classical statistical mechanics. We first explained the formal equivalence of quantum mechanics and quantum field theory at imaginary times (the latter goes under the name of Euclidean field theory) with classical statistical mechanics. This problems has a long history. As a very interesting reading on the early work we recommend

M. Kac, "Probability and related topics in physical sciences", Interscience Publ. New York, N.Y. 1959, especially Ch. IV.

+Lecture given at XVI.Internationale Universitätswochen für Kernphysik,Schladming,Austria,February 24 - March 5, 1977.

In recent times the equivalence with classical statistical mechanics has become a key tool in the development of constructive field theory. Here very good references are

"Constructive quantum field theory" Lecture Notes in Physics No.25, G.Velo and A.S. Wightman editors, Springer Verlag 1973.

B. Simon "The $P(\phi)_2$ euclidean (quantum) field theory" Princeton University Press, 1974.

These books contain an exhaustive bibliography on the development of the subject. For the most recent results a good source, when available, will be also the Proceedings of the 1976 Cargèse Summer School (to be published by Plenum Press).

These problems have however another side connected with the physical interpretation of both quantum mechanics and quantum field theory. It is an aspect of general interest although not yet very developed. We suggest:

E. Nelson, Phys. Rev. 150 (1966) 1079.

E. Nelson, "Dynamical theories of Brownian motion" Princeton University Press, 1967.

F. Guerra, P. Ruggiero, Phys. Rev. Lett. 31 (1973) 1022.

More recently one has realized that the renormalization group approach to critical phenomena has a natural probabilistic interpretation. So far the papers contributing to these ideas have been rather technical and some of them exist only in Russian. The following articles emphasize the general aspects:

G. Jona-Lasinio, Nuovo Cimento 26B (1975) 99.

Ya. G. Sinai, "Some rigorous results in the theory of phase transitions" in "Statistical physics", Proceedings of the 1975 IUPAP conference, Budapest.

G. Jona-Lasinio, "Probabilistic approach to critical behaviour", Lectures given at the 1976 Cargèse Summer School, to be published by Plenum Press.

The relationship with the standard renormalization group in field theory is discussed in:

G. Benettin, C. Di Castro, G. Jona-Lasinio, L. Peliti, A.L. Stella, "On the equivalence of different renormalization groups", in the same Proceedings of the Cargèse Summer School.

In preparation, an elementary introduction to the subject by

M. Cassandro, G. Jona-Lasinio, "Strongly dependent random variables and the theory of the critical point", to be published in Advances in Physics.

Another topic in which probability theory has provided new ideas and insight is spontaneous breakdown of a symmetry. In the lectures we discussed the following paper:

R.L. Dobrushin, S.B. Shlosman, Comm. Math. Phys. 42 (1975) 31.

This article discusses in probabilistic terms and extends well known results of Mermin and Wagner on the absence of spontaneous breakdown of a continuous symmetry in two-dimensional systems.

N.D. Mermin, H. Wagner, Phys. Rev. Lett. 17 (1966) 1133.

N.D. Mermin, J. Math. Phys. 8 (1967) 1061.

The paper by Dobrushin and Shlosman is not easy. For the intuitive physical ideas one may refer to

M.E. Fisher, J. Appl. Phys. 38 (1967) 981.

For a probabilistic treatment of spontaneous breakdown of a symmetry in statistical mechanics we suggest

G. Gallavotti, Riv. Nuovo Cimento 2 (1972) 133.

This article is very clear and although it treats only Ising models, provides a very good background.

Acta Physica Austriaca, Suppl. XVIII, 593–628 (1977)

PARTICLE PHYSICS AND EXOTIC ATOMS+

by

G. BACKENSTOSS
Univ. of Basel, Switzerland

CONTENTS

+Lecture given at XVI.Internationale Universitätswochen für Kernphysik,Schladming,Austria,February 24 - March 5, 1977.

D. STRONG INTERACTIONS

1. Pionic atoms
2. Kaonic atoms
3. Σ-hyperonic atoms
4. Antiprotonic atoms

E. OUTLOOK

A. INTRODUCTION

1. What are exotic atoms?

Any negatively charged particle which lives sufficiently long may replace an atomic electron and form a bound state with the nucleus due to its attractive Coulomb potential. The gross features are already evidenced by the simple Bohr model, the most salient ones being repeated here:

i) The binding energy $E \sim mZ^2$ leading to energies in exotic atoms typically 2-3 orders of magnitude larger than in ordinary atoms i.e. in the region of keV or MeV.

ii) The Bohr radius r_n connected to the principal quantum number $n, r_n \sim \frac{n^2}{mZ}$ and hence 2-3 orders of magnitude smaller than for electrons thus becoming comparable with nuclear radii. This fact leads to the most important consequences.

iii) For a critical quantum number $n_c = \sqrt{m/m_e}$ with m and m_e being the masses of the orbiting particle and the electron respectively the particle is inside the electronic K-shell. For all $n < n_c$, therefore, the exotic atom is essentially hydrogene like.

In table (1) the relevant numbers for particle life times and masses as well as the ground state Bohr orbits are given for the various exotic atoms so far observed.

2. How exotic atoms are formed and detected

Exotic atoms are inevitably formed in highly excited states when negative particles are stopped in material. Although the detailed formation mechanism is not known the observations are rather consistent with the plausible assumption that capture takes place at $n \approx n_c$ (greatest overlap between K-electron and particle wave functions) and a statistical $(2\ell+1)$ distribution over the ℓ-substates. The exotic atom is then deexcited first by Auger transitions and in the lower n-states predominantly by radiative E1-transitions. This X-ray cascade is observed by suitable detectors by which precise energies, intensities and line shapes are measured. With a few exception Ge- and Li-detectors are currently used. At very low energies highly efficient proportional counters of modest energy resolution and for very intense lines produced at meson factories bent crystal spectrometers with very high resolution have been employed.

In order to obtain good spectra a powerful source i.e. particle beams producing a high stop density is of great importance. This is for short living particles like K^--mesons and Σ^--hyperons by no means trivial particularly, since they are produced by orders of magnitudes less than pions. Therefore, mass separators and refined electronic trigger conditions are essential preequisites for high quality spectra allowing detailed studies of exotic atoms.

3. Why do we study exotic atoms?

The atoms consist of the nuclei and the orbiting particles. Hence, we may expect to obtain information on the particles, the nucleus and the interaction between them as described in a self-explanatory way in table (2). A very useful distinction is the range of the various potentials the particle experiences. The long range potentials $\sim 1/r;\ 1/r^2 \ldots$ are of electromagnetic nature and hence well known. Short range potentials are connected with finite size effects and strong interaction. Accordingly, states can be selected in the various regions of potentials and very different aspects of physics can be investigated as indicated in fig. (1).

B. PARTICLE PROPERTIES

1. Masses

X-ray transitions affected only by the Coulomb potential can be used to determine the masses. Energies are directly proportional to the reduced particle-nucleus mass

$$E = -\frac{\mu c^2}{2}\left(\frac{Z\alpha}{n}\right)^2\left[1 + \left(\frac{Z\alpha}{n}\right)\left(\frac{n}{\ell+1/2} - \frac{3}{4}\right) \ldots\right]$$

where the reduced mass $\mu = m/(1+m/A)$ with A being the

nuclear mass, Z the nuclear charge, α the fine structure constant. The angular momentum quantum number ℓ present in bosonic atoms has to be replaced by $j = \ell \pm 1/2$ in fermionic atoms. Radiative corrections, predominantly vacuum polarization, correction due to finite size effect, strong interaction effects, nuclear polarization and electron screening must be considered. As an example these contributions are displayed in table (3) for the pionic 5g-4f transition in $_{53}$I and the 6h-5g transition in $_{81}$Tl from which the pion mass was deduced [1]. The best data for the masses of π^-, K^-, $\bar{p}$ and Σ^- originate from measurements on exotic atoms [2-6] as tabulated in table (4).

2. Magnetic moments

The magnetic moments can be deduced from the fine structure splitting of observed X-ray lines. In the above approximation this energy splitting ΔE is given by

$$\Delta E_{n,\ell} = \frac{\mu c^2}{2} \left(\frac{Z\alpha}{n}\right)^4 \frac{n}{\ell(\ell+1)} \,.$$

This relation is valid for a Dirac particle with a magnetic moment $\mu = g_o I_s = \frac{e\hbar}{2mc}$, where g_o for $I_s = 1/2$ equals very nearly 2 and μ is the Bohr magneton. Hadrons have an anomalous magnetic momentum $\mu_H = (g_o + g_1) \cdot \mu$ where g_1 describes the anomalous part of the magnetic moment. The fine structure splitting becomes then for hadronic fermions

$$\Delta E_{n,\ell} = (g_o + 2\, g_1) \frac{\mu c^2}{2} \left(\frac{Z\alpha}{n}\right)^4 \frac{n}{\ell(\ell+1)} \,.$$

Since ΔE depends on Z^4 measurements are advantageously performed with heavy atoms. Since experimentally transitions between two spin doubletts are observed one has to resolve the 3 transitions, as indicated in figure (2) for a negative magnetic moment. Assuming a statistical weight of all levels $\propto$ (2j+1) one can calculate easily the relative intensities (I) of the three components. For ℓ = 10 corresponding to the lower doublett of the n = 11, ℓ = 10 level and the upper doublett n = 12, ℓ = 11 one finds

$$I_i : I_{II} : I_{III} = 252 : 1 : 230.$$

That means that for sufficiently large ℓ the line (II) with the lowest energy can be neglected and hence the energy difference $E_{III} - E_I$ yields the difference between the two doublett splittings. It should be noted that the intensity ratios are reversed for a positive magnetic moment whereas the energy difference depends only on the magnitude of the magnetic moment. $\mu_{\bar{p}}$ has been determined from the n = 11 $\rightarrow$ n = 10 transitions in Pb and U (see table 4) and has been found to have the same amount than the proton magnetic moment but the negative sign.

The measurement of the magnetic moment of the Σ^- hyperon is far more difficult because of the very poor statistics and the small energy splitting due to the large n necessary, which is less than the experimental resolution. One has to assume an intensity pattern for the two main lines and then try to optimize a fit to the data by variation of the size of the splitting. It was not possible so far to determine uniquely the sign of the magnetic moment since both signs lead to fits which are not substantially different in their quality [4]. In table (4) some data are displayed which signify the present state of our knowledge.

3. Polarizability

The particle moving around the nucleus tends to polarize the nucleus which would result in an increase of the binding energy. But the particle itself moves in a field ($\frac{Ze}{r^2}$) which may reach for the orbits of heavy particles 10^{19} V/cm. Therefore, also the hadron may be polarized leading to an additional energy shift [7]. In principle with the knowledge of the nuclear polarizability one could arrive at a determination of the hadron polarizability. At present for kaons a limit of 0.02 fm^3 could be established [8].

C. ELECTROMAGNETIC INTERACTION
(Muonic Atoms)

The muonic atom represents an ideal system to test the electromagnetic interaction since muon and nucleus can interact only electromagnetically as soon as the weak interaction can be neglected. The basic interaction i.e. the interaction independent of all details of nuclear structure such as distributions of charges, magnetic moments etc. can be studied for muonic states essentially outside the reach of the short range finite size potentials.

1. Vacuum polarization

Although the validity of QED is established to a high degree there exist nevertheless measurements in muonic atoms which are of interest with respect of testing QED. In contrast to the Lamb shift in electronic atoms where vacuum polarization contributes only 2% the vacuum polarization correction in muonic atoms is responsible for the overwhelming amount of radiative correction. This is connected with the larger mass of the muon since the self energy is $\sim \frac{1}{\mu^2}$. In muonic atom states where all other corrections are small and sufficiently known a precise energy measurement can be interpreted in terms of the vacuum polarization correction. In particular the 5g-4f transitions in $_{82}$Pb and the 4f-3d transitions in $_{56}$Ba with energies between 431 and 441 keV have been measured recently [9],[10] with an accuracy of about 15 eV and shown to agree within these errors with the theoretically predicted energies which in their terms are believed to be accurate to about 8 eV. This agreement has been reached after severe doubts to it have been expressed [11] which however have been the consequence of overestimating experimental accuracy as well as of theoretical calculating errors. More about the calculations can be found in the lectures of F.Scheck [12]. Comparison of the experimental energies with calculations [13,14,15] (see table (5)) therefore shows that the total vacuum polarization term is tested to an accuracy of about 0.6% whereas the higher terms are tested to about 25%. In spite of the fact that there are more accurate experimental results to the validity of QED as for example the very precise g-factor of the muon the present result lives on its own right since there is no other experiment which tests the theory at such high electric field strength as it is done in heavy muonic atoms where field strengths of more than 10^{18} V/cm are being accessible.

The most precise measurement so far is the measurement of the Lamb-shift in $\mu\,^4He$ [16]. This shift corresponds to the near red end of the visible spectrum and hence a novel technique using a tunable dye laser has been applied.

The metastable 2 $S_{1/2}$ level in fig. (3) is populated by the muonic cascade to a few percent and a tunable dye laser is used to induce the transition to the $2p_{3/2}$ level. Most of the deexcitation of the μHe atom leads to the 2p level and is signalized by the 2p-1s EI X-ray transition of 8.2 keV which occurs essentially prompt with respect to the μ-stop signal. A μ-stop signal not followed by the 8.2 keV transition is an indication for the trapping of the μ- in the 2s level. Such a signal is used to trigger the laser. If the frequency of the laser corresponds to the Lamb shift the muon is lifted to the $2p_{3/2}$ level from which it promptly undergoes transition to the ground state level by emission of a 8.2 keV photon. Thus these photons are observed in coincidence with the laser trigger as a function of the laser wavelength. A signal has been found for $\lambda = (8117 \pm 5)$ Å corresponding to an energy difference

$$E(2p_{3/2}) - E(2S_{1/2}) = (1.5274 \pm 0.009)\,\mathrm{eV}.$$

The various theoretical contributions to the calculated

shift are given in table (6).

One notices that the only important uncertainty originates from the uncertainty in the finite size of He nucleus. The difference

$$\Delta E_{exp} - \Delta E_{th} = (0.0023 \pm 0.0087)\,eV$$

is well within the error of the theoretical value and thus a sensitive test for the validity of QED. Provided QED is believed to be correct a new and better value for the charge rms-radius of ^4He of $\langle r^2 \rangle^{1/2} = (1.644 \pm 0.005)$ fm can be derived.

Similar measurements of the $\mu^- p$ system would be of even more fundamental interest. Besides a most sensitive test of QED information about the proton structure could be obtained. Unfortunately those experiments are even more complicated as in the case of ^4He since the life-time of the 2s level may be strongly reduced by Stark mixing and the transition energies of interest near 0.2 eV are located in the ultraviolet region.

2. Parity in electromagnetic interaction

So far electromagnetic interactions have been assumed to conserve parity strictly. The existence of neutral currents in weak interactions as evidenced by neutrino experiments has stimulated the interest in theories aiming at a unified description of weak and electromagnetic interactions. The question arises now to which extent are neutral currents parity violating and to which extent can parity violation be observed in processes dominated by electromagnetic interactions.

The muonic atom is a system very suitable to study these questions since it is a bound system where the muonic wave function overlaps the nucleus much more than in electronic atoms. It has been suggested to investigate the MI(EI) mixing in light or medium muonic atoms [17], [18]. The mixing is given by

$$\eta = \frac{\langle\psi_j|V^{PV}|\psi_i\rangle}{|E_i-E_j|}$$

where V^{PV} is a parity violating potential $V^{VP} \ll H_o$ and of short range. Mixing would occur if for example the 2s and 2p levels are nearly degenerate and a parity violating potential would exist. Since the short range finite size effect shifts the 2s level much more upwards than the 2p level and the vacuum polarization shifts the two levels downwards always a Z can be found for which near degeneracy occurs (see fig. (4)). As a result of a parity mixing the interference term which manifests itself by a circular polarization or an asymmetry of the emitted photons with respect to the muon spin in a MI transition of the type 2s-1s must be observed. Whereas the effects expected here seem at present to be unobservable the E2(EI) mixing in nearly degenerated 3d-2p states where E2 transitions of the type 3d-1s are being observed [19] may be slightly more promising [20]. However, the effects estimated to be of the order of 10^{-6} to 10^{-7} require in order to be statistically significant 10^{12} - 10^{14} events which require long running times even at the new meson factories.

D. STRONG INTERACTIONS

Strong interactions are short range and can be observed only where the strong potentials contribute a moderate perturbation to the Coulomb potential, i.e. in the lower states of the cascade before the cascade ceases completely as a consequence of the final absorption of the particle by the nucleus. Three kinds of strong interaction effects occur. a) a shift of the energy level as compared to the energy expected on the basis of purely electromagnetic interaction, b) a line broadening and c) an intensity decr. of the last observable trans.. Effect c) is due to nuclear absorption from the upper level and b) from the lower level of the transition where b) is usually 2-3 orders of magnitude larger. In fig. (5-7) some recent representative spectra of π, K^- and $\bar{p}$-atoms are given [21,22,2] showing the various level broadening and intensity reductions. In table (7) typical data for the strong interaction effects are presented.

Theoretically one describes the hadron nucleus interaction by an optical potential which is added in the wave equation (Dirac or Klein-Gordon) to the Coulomb potential [23,24]. One tries to relate the parameters of this potential to the elementary hadron nucleon interaction. If successful the study of strong interaction effects would yield results on the elementary interaction at zero energies. The difficulties are that also the structure of the nucleus, especially its matter distribution enters. The most simple version of an ansatz is

$$V \simeq A_p \rho_p + A_n \rho_n$$

where A_p and A_n is related to the proper isospin hadron proton and hadron neutron scattering amplitudes and ρ_p and ρ_n are the nuclear proton and neutron distributions respectively. Therefore, one meets the difficulty to disentangle the nuclear effects. On the other hand one also has the possibility to study nuclear structure, e.g. neutron distributions, provided the A's are known.

1. Pionic atoms

In pionic atoms the gross features are understood [23,24]. The optical potential is, due to the strong π-nucleon p-wave interaction and the pion absorption on two nucleons, rather complex necessitating at least 6 independent parameters. Nevertheless, it turns out that to a good approximation each of these parameters is predominantly connected to one observable quantity. For instance, energy shifts being repulsive for s-levels and attractive for p levels are connected to the real part of the corresponding scattering lengths etc. This, of course, facilitates the experimental determination of those parameters and a large amount of data can be described in this way [24]. Unfortunately enough the s-wave π^-p and π^-n scattering lengths are nearly equal but with opposite sign so that higher order effects play a leading role and it becomes difficult to extract reliable data about those elementary interaction constants. Recently also a prediction could be proved [25] that the 2p level energy shift becomes less and less attractive with increasing Z as a consequence of the shrinking of the 2p-orbits thus enhancing s-wave interaction with its repulsive character (see fig.8)).

Interesting problems remain open, particularly in light nuclei where microscopic theories must be applied and the experimental precision is high enough as to shed light on so fundamental processes as the pion absorption on 3 or 4 nuclei as in ^3He or ^4He. Isotope effects on the lightest nuclei such as $_2He$ [21], $_3Li$, $_5B$ may be particularly easily understood. However, it turns out that even there the data can not be sufficiently well explained by a simple superposition of the interaction on the extra neutrons.

2. Kaonic atoms

For kaonic atoms the situation was rather unexpected. It was thought that the K-nucleus potential should be simpler than the π-nucleus potential since p-wave interaction is small and might be neglected for the begin. Further K-absorption can take place on one nucleon only (e.g. $K+N\rightarrow\Lambda(\Sigma)+\pi$) and the interaction takes place mainly on the nuclear surface where the nuclear density is small. In spite of the success experienced on pionic atoms the corresponding treatment for kaonic atoms was a complete failure in so far as even the sign of the real part of the optical potential had to be reversed toward attractive interaction in order to fit the data [26], whereas a potential constructed simply from the KN scattering lengths is repulsive. It has been suggested [27] that this might be connected with the existence of a resonance in the K^-p channel (Y_o^* 1405) 27 MeV below threshold which might be responsible for attraction. It is however an interesting question on its own whether such states found in the elementary system exist also in nuclei. A satisfactory interpretation of kaonic atom data to which various attempts have been done [28,29] may shed light on this question.

3. Σ-hyperonic atoms

A high energy hyperon beam is not suited for the production of Σ^--atoms since the slowing donw time is large compared with the Σ-lifetime. However, the Σ^- produced after K^- capture are useful to form Σ-atoms. They are produced in about 8% of all K^--nuclear absorptions and have a kinetic energy of only 15 MeV thus being able to be captured by the atoms of the same target in which the K^- are stopped before they decay. Thus Σ-spectra with a yield of ∿8% can be observed simultaneously with the kaonic spectra. Only a few intensity reductions have been measured so far in Σ-atoms [30]. The data are not sufficient to study the Σ-nucleus interaction in some detail. However, a comparison of the intensities between kaonic and Σ-hyperonic spectra yields some information about the Σ^- production in nuclei and their escape probability from the nuclei.

4. Antiprotonic atoms

In antiprotonic atoms a sitution similar to that with K^--atoms holds where also an attractive potential is required to fit the data though there are no bound states known so far near threshold. Is this a hint to the narrow bound states recently discussed [31] in the $p\bar{p}$ system? Measurements on $\bar{p}H$ and K^-H where no nuclear effects enter are hence most important. They are being attempted [32] but are very difficult because of the low energies of the X-rays and even more because of the low yield $\lesssim 10^{-3}$ due to the Stark effect the neutral $\bar{p}p$ system experiences in the liquid H_2 and by which s-states may be admixed even for rather high principal quantum number states thus increasing the nuclear absorption from high states. A somewhat easier but less direct way is the measurement of isotope effects (e.g. $\bar{p}^{16}O/\bar{p}^{18}O$) where one hopes to be better able to eliminate nuclear structure effects and to interprete the isotope effect in terms of the additional neutrons. Thus, a separate rough determination of the interaction constants $1/2 \cdot (A_{\bar{p}p} + A_{\bar{p}n}) = 2.0 \pm 0.5 + i(2.0 \pm 0.5)$ fm and $A_{\bar{p}n} = -0.3 \pm 1.4 + i \cdot (0.9 \pm 1.7)$ fm was possible.

Also first attempts to search for the narrow quasistationary bound states [31] of the $\bar{p}p$ system have been made which may be found below threshold (1876 MeV). At least in some reaction channels the width of these states should be rather narrow [33] of the order of some MeV. They are supposed to be located between the threshold and a few hundred MeV below it. The levels should be populated by γ-transitions from atomic states of the $\bar{p}p$ system or from other quasi-nuclear states of higher energy.

In a recent experiment at the CERN-PS the γ-spectrum after $\bar{p}$ absorption in H_2 was measured [32] between 50 and 1000 MeV with a large NaI crystal. The shape of the γ-spectrum could be explained by a good degree by the γ's from π^0-decay. Above that background a few weak lines could be observed whose width was compatible with the experimental resolution. Two lines between 150 and 220 MeV seemed to be most significant with a yield of about 10^{-3} per $\bar{p}$ annihilation. Similar experiments on ^4He did not show these structures. Therefore, if the significance of these lines - at present only about 3-4 standard deviations - could be secured they may well be identified with the $\bar{p}p$ system.

E. OUTLOOK

The progress experienced during the last decade was essentially achieved as a consequence of the greatly improved energy resolution originating from the use of solid state detectors. Hyperfine patterns in muonic atoms and strong interaction effects in hadronic atoms thus could be investigated in detail. In the near future a strong impetus may be expected in the muon and pion field from the new meson factories from which pion and muon beams of great intensity and high beam quality become available, which lead to stop densities increased by several orders of magnitude. Therefore, smaller amounts of material are needed thus facilitating the possibility of investigating isotopes and thin targets allowing an improved measurement of intensities particularly of low energetic lines.

On the other hand weak transitions such as 2s-2p transitions can be measured with high precision and also the very weak lines in hydrogen isotopes may be easier accessible. Coincidence measurements between various members of the X-ray cascade and between X-rays and nuclear γ-rays become possible which open a large field of experiments leading to a better understanding of the cascade but also to more information on strong interaction effects as well as to a more detailed study of the absorption mechanism of pions.

Compared with the possibilities in the muonic and pionic X-ray field the facilities with the heavier particles are rather modest. Also here improved beams should be possible if one is willing to invest in larger aperture beam elements. Recently suggestions have been made to decelerate antiprotons from the energy at which they are produced (the maximum of the production spectrum is about at 4 GeV for incoming protons of 20 GeV) to energies of a few hundreds of MeV where they easily could be stopped. In this deceleration process no increase of the phase space is expected and hence a beam of extraordinary quality leading to a $\bar{p}$ stop rate increased by many orders of magnitude as compared to conventional beams. It is clear that such a development would have a similar effect on the antiproton field as the meson factories on the pion and muon field.

REFERENCES

1. G.Backenstoss, H. Daniel, H. Koch, Ch.von der Malsburg, G.Poelz, H.P. Povel, H. Schmitt a. L. Tauscher, Phys. Lett. 43B (1973) 539.

2. L. Tauscher, Proc. 6th Int. Conf. High Energy Phys. and Nucl. Structure, Santa Fe 1975 (American Inst. of Phys. New York 1975) p.541.

3. V.N. Marushenko, A.F. Mezentsev, A.A. Petrunin, S.G. Skornyakov and A.I. Smirnov, JETP Lett. 23 (1976) 72.

4. D. Dugan, Y. Asano, M.Y. Chen, S.C. Cheng, E. Hu, L. Lidofsky, W. Patton, C.S. Wu, V. Hughes and D. Lu, Nucl. Phys. A254, (1975) 381, 396, 403.

5. B.L. Roberts, C.R. Cox, M. Eckhause, J.R. Kane, R.E. Welsh, D.A. Jenkins, W.C. Lam, P.D. Barnes, R.A. Eisenstein, J.Miller, R.B. Sutton, A.R.Kunselman, R.J. Powers and J.D. Fox, Phys. Rev. Lett. 33 (1974) 1181.

6. B.L. Roberts, C.R. Cox, M. Eckhause, J.R. Kane, R.E. Welsh, D.A. Jenkins, W.C. Lam, P.D. Barnes, R.A.Eisenstein, J.Miller, R.B. Sutton, A.R. Kunselman and R.J. Powers, Phys. Rev. Lett. 32 (1974) 1265.

7. F. Iachello and A. Landé, Phys. Lett. 35B (1971) 205. U.E. Schröder, Acta Phys. Austriaca 26 (1972) 248. T.E.O. Ericson and J.Hüfner, Nucl. Phys. 47B (1972) 205.

8. G. Backenstoss, A. Bamberger, I. Bergström, T. Bunaciu, J. Egger, R. Hagelberg, S. Hultberg, H. Koch, U. Lynen, H.G. Ritter, A. Schwitter and L. Tauscher, Phys. Lett. 43B (1973) 431.

9. L. Tauscher, G. Backenstoss, K. Fransson, H. Koch, A. Nilsson and J.De Raedt, Phys. Rev. Lett. 35 (1975) 410.

10. M.S. Dixit, A.L. Carter, E.P. Hincks, D.Kessler, J.S. Wadden, C.K. Hargrove, R.J. McKee, H. Mes and H.L. Anderson, Phys. Rev. Lett. 35 (1975) 1633.

11. R. Rafelski, B. Müller, G. Soff and W. Greiner, Ann. Phys. (N.Y.) 88 (1974) 419.
P.J.S. Watson and M.K. Sundaresan, Can. J. Phys. 52 (1974) 2037, V.W. Hughes, Proc. 6. Int. Conf. High Energy Phys. and Nucl. Structure, Santa Fe 1975 (American Inst. of Phys. New York 1975) p. 515.

12. F. Scheck, This proceedings.

13. R. Engfer, H. Schneuwly, J.L. Vuilleumier, H.K. Walter and A. Zehnder, Atomic and Nuclear Data Table 14 (1974) 509.

14. J. Blomqvist, Nucl. Phys. B48 (1972) 95.

15. L. Tauscher, CERN Preprint (1973).

16. A.Bertin, G.Carboni, J.Duclos, U.Gastaldi, G.Gorini, G. Neri, J. Picard, O. Pitzurra, A. Placci, E.Polacco, G.Torelli, A. Vitale and E. Zavattini, Phys. Lett.55B (1975) 411.

17. J.Bernabeu, T.E.O. Ericson and C.Jarlskog, Phys.Lett. 50B (1974) 467.

18. G. Feinberg and M.Y. Chen, Phys. Rev. D10 (1974) 190.

19. H. Schneuwly, L.Schellenberg, H. Backe, R. Engfer, W. Jahnke, E.Kankeleit, K.H. Lindenberger, R.M.Pierce, C. Petitjean, W.U. Schröder, H.K. Walter and A.Zehnder, Nucl. Phys. A196 (1972) 452.

20. L.M. Simons, Helv. Phys. Acta 48 (1975) 141.

21. G. Backenstoss, J. Egger, T. von Egidy, R.Hagelberg, C.J. Herrlander, H. Koch, H.P. Povel, A. Schwitter and L. Tauscher, Nucl. Phys. A232 (1974) 519.

22. G. Backenstoss, A. Bamberger, I. Bergström, P.Bounin, T. Bunaciu, J. Egger, S. Hultberg, H. Koch, M.Krell, U. Lynen, H. Ritter, A. Schwitter and R. Stearns, Phys. Lett. 38B (1972) 181.

23. M. Ericson and T.E.O. Ericson, Ann. Phys. N.Y. 36 (1966) 323.

24. G. Backenstoss, Ann. Rev. Nucl. Sci. 20 (1970) 467.

25. R. Abela, P. Blüm, R. Guigas, H. Koch, H. Poth, G. Backenstoss, M. Brandao, M. Izycki, L. Tauscher, P. Pavlopoulos and K. Zioutas, to be published Z. Physik A.

26. G. Backenstoss, J. Egger, H. Koch, H.P. Povel, A. Schwitter and L. Tauscher, Nucl. Phys. B73 (1973) 189.

27. W.A. Bardeen and E.W. Torigoe, Phys. Lett. 38B (1972) 135.

28. M. Alberg, E.M. Henley and L. Whilets, Phys. Rev. Lett. 30 (1973) 255.

29. S. Wycech, Nucl. Phys. B28 (1971) 541.

30. G. Backenstoss, T. Bunaciu, J. Egger, H. Koch, A. Schwitter and L. Tauscher, Z. Phys. A273 (1975) 131.

31. L.N. Bogdanova, H.D. Dalkarov and J.S. Shapiro, Ann. Phys. (N.Y.) 84 (1974) 261.

32. Basel-Karlsruhe-Stockholm group at CERN.

33. F. Myhrer and A.W. Thomas, Phys. Lett. 64B (1976) 59.

TABLE 1

Particle Properties and Bohr Radii of Atoms

		Hadrons			
	μ^-	π^-	K^-	$\bar{p}$	Σ^-
τ (ns)	2200	26	12	∞	0.15
m/m_e	207	273	966	1836	2343
$r_{B_{1s}} \cdot Z$ (fm)	256	194	54.7	28.8	22.6

TABLE 2

Topics connected with Exotic Atoms

Particle	μ^-	π^-	K^-	$\bar{p}$	Σ^-
Particle		m_{π^-}	m_{K^-}	$m_{\bar{p}}\mu_{\bar{p}}$	$m_\Sigma \mu_\Sigma$
Nucleus	Finite size of charge, magn. moment and el. quadr. moment	Nuclear Structure → nuclear surface Neutron distribution 2N Density Granular structure			
Interact.	El. magnetic Vacuum polarizat. Parity conservation	Strong Particle Nucleus- and Particle nucleon at zero energy (π,2N) s wave p wave			

TABLE 3

Determination of the π^--Mass

Transition energies and corrections applied (ref. 1)

	$_{53}I$ 5g→4f	$_{81}Tl$ 6h→5g
E_{exp}(keV)	237.136±0.017	301.733±0.015
Strong interaction (keV)	0.025±0.002	0.008±0.000
Vacuum polarization (keV)	1.036±0.010	1.303±0.010
Electron screening (keV)	-0.014±0.005	-0.075±0.005
Deduced m_{π^-}(MeV)	139.563±0.012	139.572±0.009
m_{π^-} (average) (MeV)	(139.569±0.008) MeV	

TABLE 4

Particle Masses (m) and magnetic moments (μ) obtained from atoms

Particle	m [MeV]		μ (nucl.magnetons)	
π^-	139.569 ±0.006	a)		
	139.5657±0.0017	b)		
K^-	493.688 ±0.030	a)		
	493.657 ±0.020	c)		
$\bar{p}$	938.179 ±0.058	d)	-2.791±0.021	d)
			-2.819±0.046	e)
Σ^-	1197.24 ±0.15	f)	$-1.40^{+0.41}_{-0.28}$ or $0.65^{+0.28}_{-0.40}$	f)
			-1.48 ±0.37	g)

a) CERN, Ge-detector, ref. 2)

b) Dubna, crystal spectrometer, ref. 3)

c) Columbia, Ge-detector, ref. 4) p. 381

d) Columbia, Ge-detector, ref. 4) p. 403

e) BNL, Ge-detector, ref. 5)

f) Columbia, Ge-detector, ref. 4) p. 396

g) BNL, Ge-detector, ref. 6)

TABLE 5

Test of vacuum polarization in muonic atoms[a)]

Transition / Energies (eV)	Ba $4f_{\frac{5}{2}}-3d_{\frac{3}{2}}$	Pb $5g_{\frac{7}{2}}-4f_{\frac{5}{2}}$	Ba $4f_{\frac{7}{2}}-3d_{\frac{5}{2}}$	Pb $5g_{\frac{9}{2}}-4f_{\frac{7}{2}}$
Measured transition energy E_m	441376±13	437754±12	433926±16	431363±14
Vacuum polarization order				
$\alpha(Z\alpha)$	2434.2	2188.6	2327.5	2105.0
$\alpha^2(Z\alpha)$	17	15	16	14
$\alpha(Z\alpha)^3$	-22	-45	-21	-43
$\alpha(Z\alpha)^5+\alpha(Z\alpha)^7$	- 1	- 6	- 1	- 6
Lamb shift + anomalous magnetic moment	9	8	- 8	- 7
Nucl. Pol.	15	7	12	6
Finite size	-145	-10	-56	- 4
Electron screening	-14	-79	-13	-79
Total calcul. energy $E_{calc.}$	441368	437754	433914	431340
$E_{calc}-E_m$	-8±13	-0±12	-12±16	-23±14

[a)] Data taken from ref.9, corrections from refs. 13-15

TABLE 6

Contributions to the energy difference between the 2 $P_{3/2}$ and 2 $S_{1/2}$ levels in $(\mu^- \, {}^4He)^+$. Energies in eV, rms-radius in fm = (1.650±0.025) fm (ref. 16)

Vacuum polarization						
1. order in α	2.order in α	Lamb shift	Nucl.finite size	Fine structure	Nucl.Pol.	Total
+1.6659	+0.0115	-0.0143	$-0.1053\langle r^2\rangle$	+0.1457	+0.0031	$+1.8199\langle r^2\rangle$ 0.1053
			-0.2867±0.0087			+1.5251±0.0087

TABLE 7: Examples for Strong Interaction Effects in Hadronic Atoms

Part.	Nucleus	Trans.	Energy (keV)	ε_{low}(eV)	Γ_{low}(eV)	Γ_{up} (eV)
π	^{3}He	2p-1s	10.692 ± 0.005	+ 37 ± 5	40 ± 15	
π	^{4}He	2p-1s	10.6987± 0.002	-75.7 ± 2.0	45 ± 3	$7.2 \pm 3.3 \cdot 10^{-4}$
π	Li	2p-1s	24.038 ± 0.004	- 568 ± 4	205 ± 15	
π	^{16}O	2p-1s	159.95 + 0.25	-15575± 250	7560 ± 500	11 ± 6
π	^{18}O	2p-1s	155.01 ± 0.25	-20510± 250	8670 ± 700	
π	Ne	2p-1s	238.35 ± 0.50	-33340± 500	14500 ± 3000	
π	A	3d-2p	168.88 ± 0.10	825 ± 100	1170 ± 170	
π	Co	3d-2p	384.74 ± 0.35	4570 ± 350	7370 ± 700	10.1± 2.1
π	Ba	4f-3d	582.99 ± 0.27	5440 ± 270	4300 ± 900	
π	Pb	5g-4f	575.56 ± 0.25	1730 ± 250	1100 ± 300	
K	C	3d-2p	62.73 ± 0.08	- 575 ± 80	1730 ± 150	0.98± 0.19
K	P	4f-3d	142.02 ± 0.08	- 315 ± 80	1440 ± 120	1.94± 0.33
K	S	4f-3d	161.56 ± 0.06	- 460 ± 50	2370 ± 120	3.25± 0.41
K	Ni	5g-4f	231.49 ± 0.07	- 180 ± 70	1020 ± 120	6.0 ± 2.3
$\bar{p}$	N	4f-3d	55.827 ± 0.050	3 ± 50	205 ± 70	0.13± 0.03
$\bar{p}$	^{16}O	4f-3d	73.438 ± 0.036	- 124 ± 36	320 ± 150	0.64± 0.11
$\bar{p}$	^{18}O	4f-3d	73.861 ± 0.042	- 189 ± 42	550 ± 240	0.80± 0.12
$\bar{p}$	S	5g-4f	140.440 ± 0.040	- 60 ± 40	650 ± 100	3.04± 0.70
Σ	C	4f-3d				0.031± 0.012
Σ	Ca	6h-5f				0.40± 0.22
Σ	Ba	91-8k				2.90± 3.5

ε_{low} and Γ_{l} shift and width of the lower level, Γ_{up} width of the upper level.

FIGURE CAPTIONS

Fig. 1: Lower part of the level scheme of exotic atoms. As an example a Ca atom is used to demonstrate the differences in the cascades where muons reach the 1s state whereas the hadronic cascades are terminated at excited levels by nuclear absorptions. Furthermore, the marking of names like "particle physics" in the scheme should indicate from which part of the cascade the relevant information is expected.

Fig. 2: Transitions between 2 fine structure doubletts.

Fig. 3: Lamb shift in ^{4}He.

Fig. 4: Scheme of levels which might be subject to parity mixing.

Fig. 5: Pionic (muonic) X-ray spectrum of ^{4}He (from ref. 21).

Fig. 6: Kaonic (Σ-hyperonic) X-ray spectrum of $^{31}_{15}$P (from ref. 22).

Fig. 7: Antiprotonic spectra of $D_2\,^{16}O$ and $D_2\,^{18}O$ which shows isotope effects due to strong interaction in the $4\rightarrow3$ transition, where the yield difference is most easily observed. The Ti lines originate from a Ti-container especially used to provide suitable calibration lines (from ref.2).

Fig. 8: Energy shift of the 2p level in pionic atoms vs Z (from ref. 25).

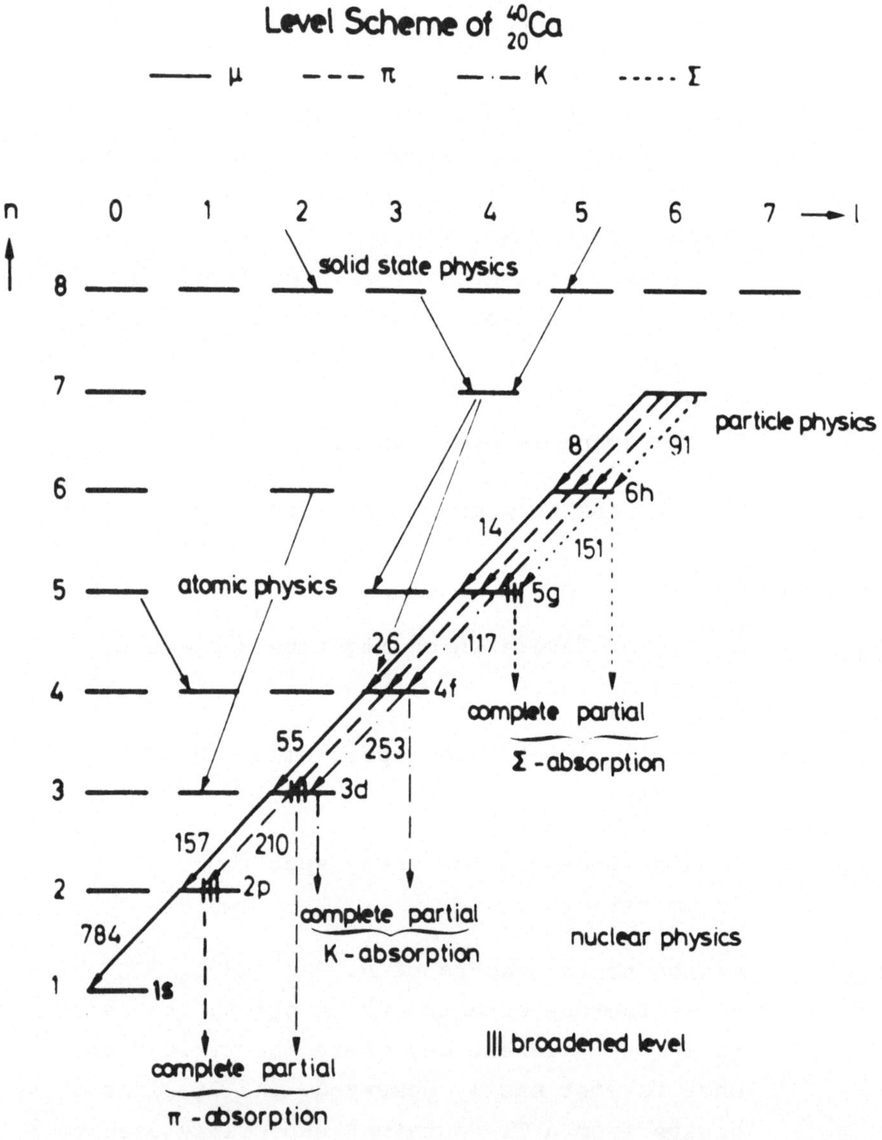
Level Scheme of $^{40}_{20}Ca$
μ
π
K
Σ
n
0
1
2
3
4
5
6
7
l
solid state physics
8
7
6
5
4
3
2
1
particle physics
atomic physics
8
91
6h
14
151
5g
26
117
4f
complete partial
Σ-absorption
55
253
3d
157
210
2p
complete partial
K-absorption
784
nuclear physics
1s
||| broadened level
complete partial
π-absorption

Fig.1

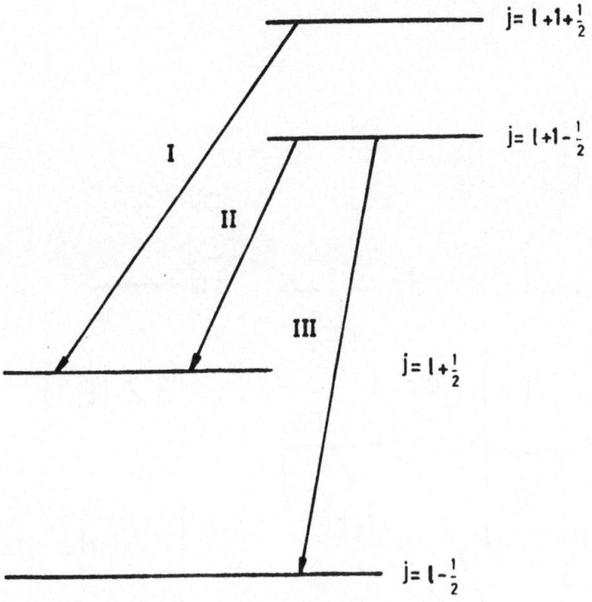

Fig. 2

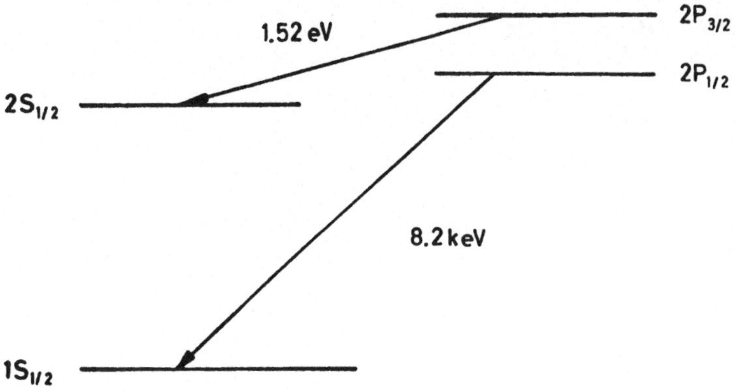

Fig. 3

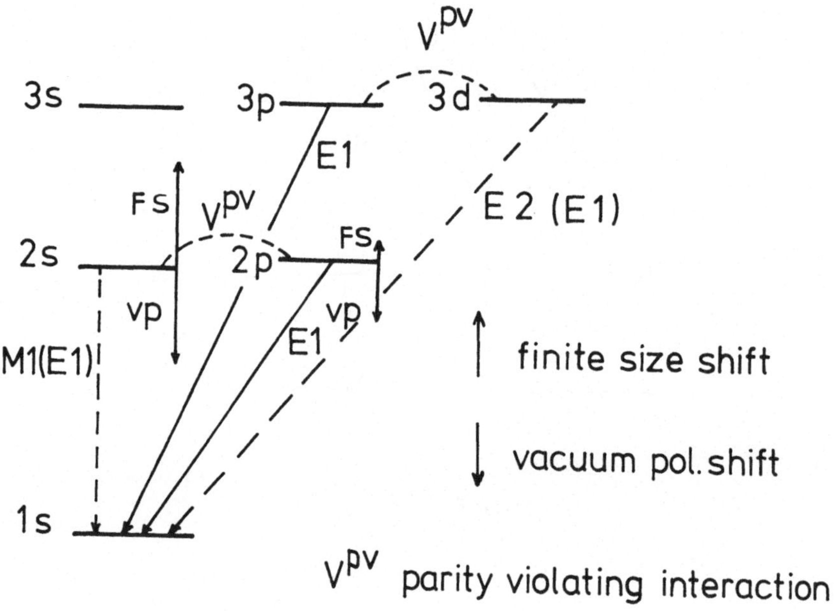

$$\Psi = \Psi_i + \eta \Psi_j$$

$$\eta = \frac{\langle \psi_j | V^{pv} | \psi_i \rangle}{|E_i - E_j|}$$

Fig. 4

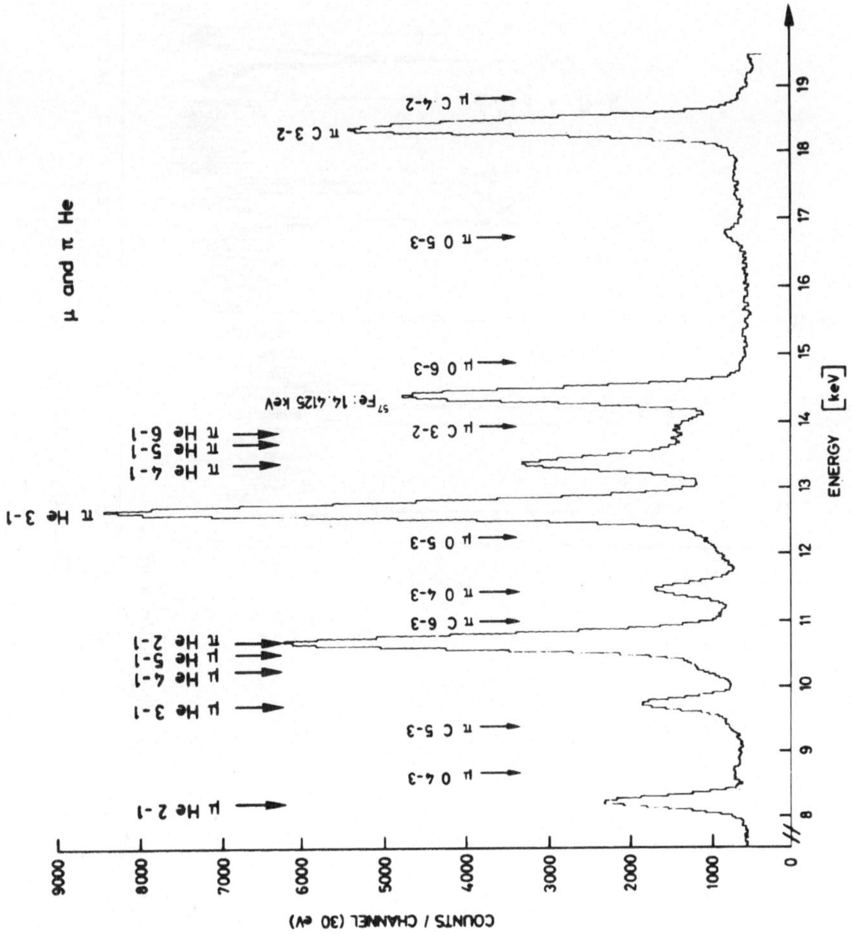
μ and π He
π He 3-1
μ He 2-1
μ He 3-1
μ He 4-1
μ He 5-1
π He 2-1
π He 4-1
π He 5-1
π He 6-1
μ O 4-3
π C 5-3
π C 6-3
π O 4-3
μ O 5-3
μ C 3-2
57Fe: 14.4125 keV
μ O 6-3
π O 5-3
π C 3-2
μ C 4-2
COUNTS / CHANNEL (30 eV)
ENERGY [keV]

Fig. 5

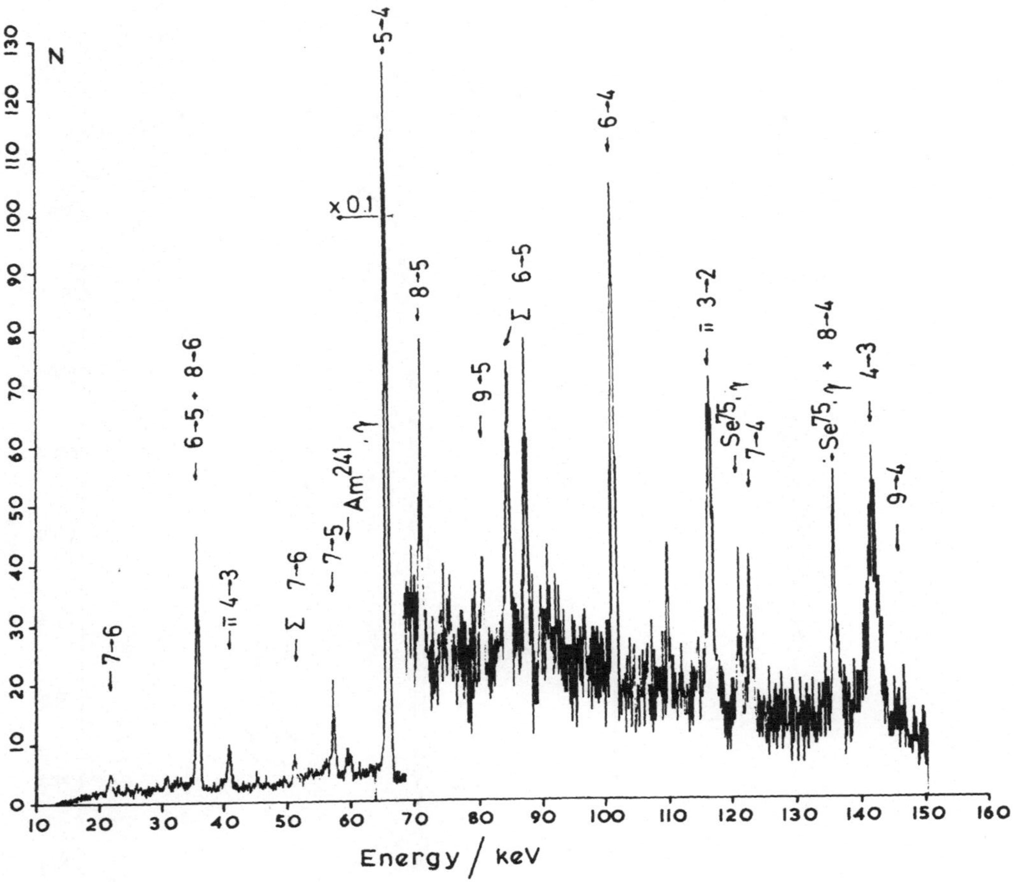
N
Energy / keV
x 0.1
7-6
6-5 + 8-6
π 4-3
Σ 7-6
7-5
Am241, γ
5-4
8-5
9-5
Σ 6-5
6-4
π 3-2
Se75, γ
7-4
Se75, γ + 8-4
4-3
9-4

Fig. 6

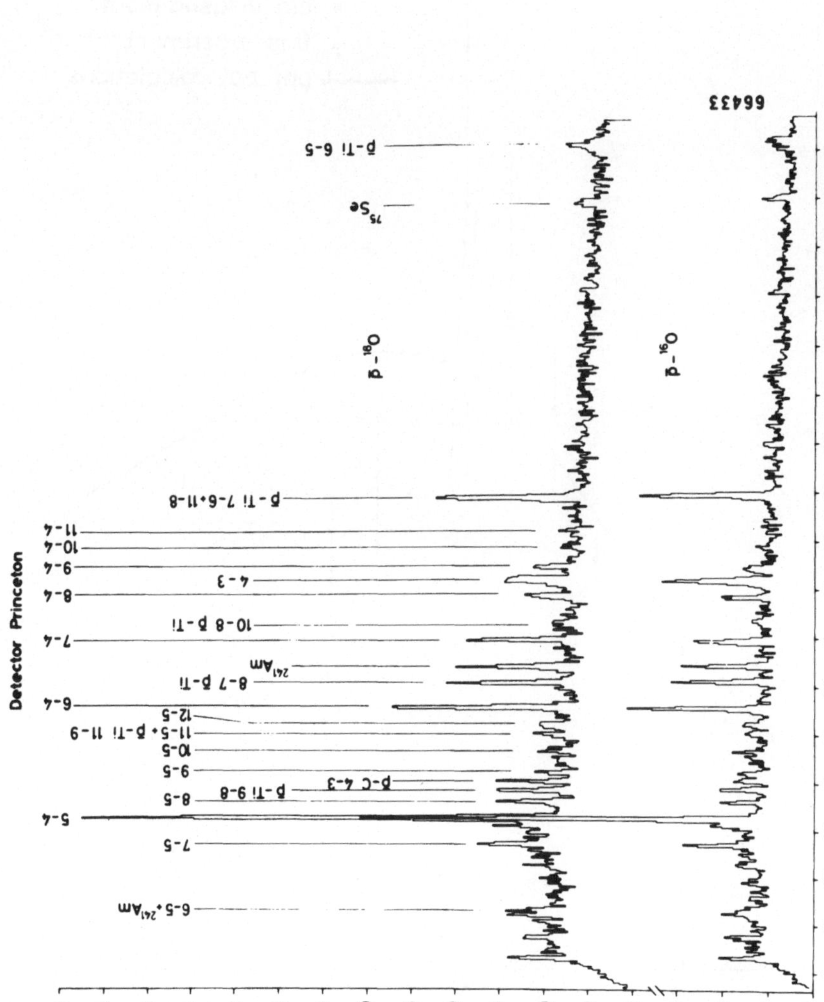
Detector Princeton
Counts / channel
keV
6-5+241Am
7-5
5-4
8-5
9-5
10-5
11-5+ β-Ti 11-9
12-5
6-4
β-Ti 9-8
β-C 4-3
8-7 β-Ti
241Am
7-4
10-8 β-Ti
8-4
4-3
9-4
10-4
11-4
β-Ti 7-6+11-8
β-18O
β-16O
75Se
β-Ti 6-5
66433

Fig. 7

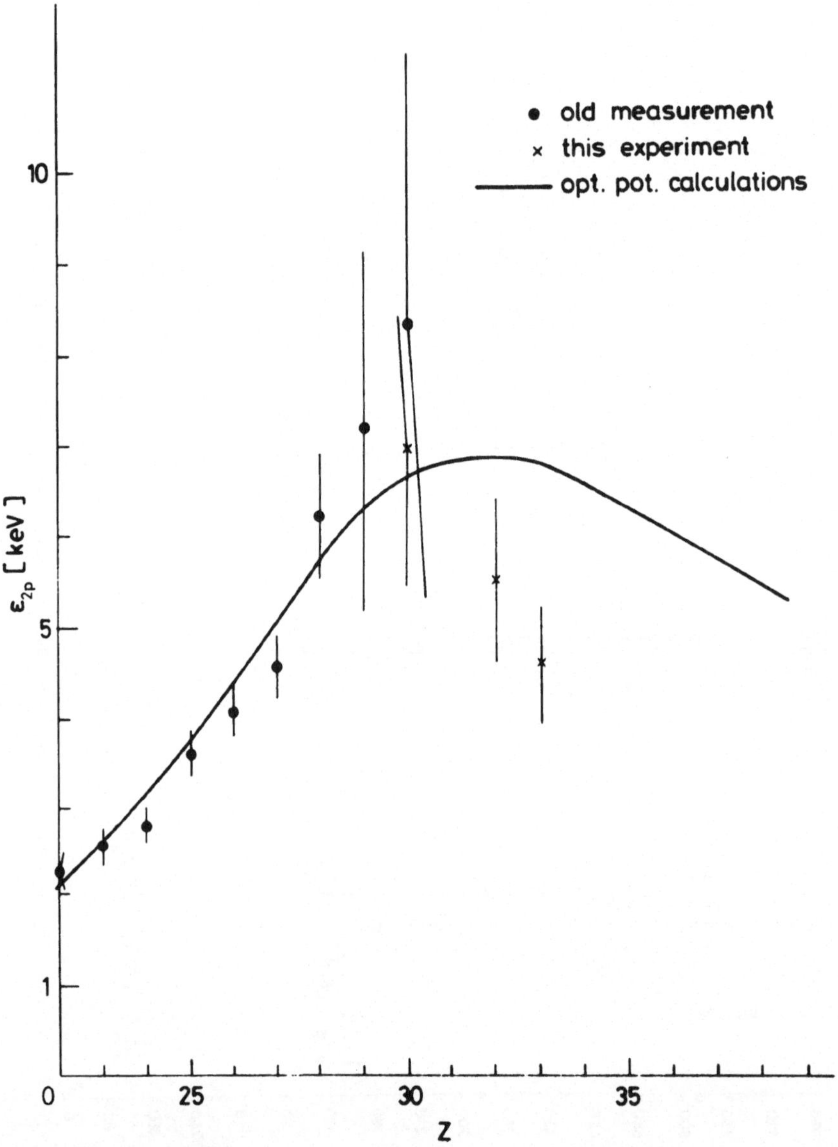

Fig. 8

Printed in Austria

Acta Physica Austriaca, Suppl. XVIII, 629–675 (1977)

PARTICLE PHYSICS AND EXOTIC ATOMS+

by

F. SCHECK
Johannes Gutenberg-Universität
Mainz, Germany

TOPICS

+Lecture given at XVI. Internationale Universitätswochen für Kernphysik, Schladming, Austria, February 24-March 5, 1977.

1. MUONIC ATOMS AND QUANTUM ELECTRODYNAMICS

There are essentially two reasons why muonic atoms are ideal systems for the purpose of testing predictions of Quantum Electrodynamics at low momentum transfers:

(1) In every muonic atom with nuclear charge Z one can choose orbits which are *hydrogen-like* to a very high degree of accuracy.

(2) Due to the particular *spatial structure* of muonic atoms, radiative corrections to the muonic energy levels are dominated by vacuum polarization due to virtual electron-positron pairs. This is in contrast to electronic atoms where this is a small effect as compared to the remainder of the Lamb shift (vertex corrections and anomalous magnetic moment).

Let us illustrate these two points. As to (1) consider the example of muonic lead. The r.m.s. radius of a circular Bohr orbit with main quantum number n is given by

$$\sqrt{\langle r^2\rangle_{n,\ell=n-1}} = na_B\sqrt{\frac{(n+1)(2n+1)}{2}}$$

with $a_B = \frac{1}{m_\ell Z_\alpha}$ and where m_ℓ is the mass of the bound negative lepton. The innermost electronic shell of the host atom is at about 1.12×10^3 fm from the nucleus, while the finite size of the nucleus extends to about 7 fm. Intermediate muonic orbits such as the 5 g or the 4 f states have r.m.s. radii of about 90 fm and 60 fm, respectively. For those orbits screening of the nuclear charge by the electronic shells must be small, while the nucleus still appears essentially as a point-like charge. As a conclusion, these intermediate orbits are very clean hydrogen-like orbits. All electromagnetic properties of such states are calculable to a high degree of precision.

As to point (2) we note that the dominant radiative correction is the vacuum polarization of order $\alpha(Z\alpha)$ due to virtual electron-positron pairs (see Fig. 1a). This contribution can be written in the form of an induced potential[+].

$$V_{VP}(\vec{r}) = -\frac{2\alpha Ze}{12\pi^2 r}\int_1^\infty e^{-2m_e xr}\left(1+\frac{1}{2x^2}\right)\frac{\sqrt{x^2-1}}{x^2}\,dx \quad . \qquad (1)$$

[+] If the finite size of the nucleus cannot be neglected then one must replace the factor $\frac{1}{r}$ in eq. (1) by $\int d^3r'\,\frac{\rho(\vec{r}\,')}{|\vec{r}\,'-\vec{r}|}$ as well as the r in the exponential by $|\vec{r}\,'-\vec{r}|$.

The potential V_{VP} causes a distortion of the pure Coulomb field $V_C = -\frac{Ze}{4\pi r}$ (viz. the corresponding expression that takes account of finite size) over a distance of the order of the Compton wave length of the electron $\bar{\lambda}_e \simeq 380$ fm. Indeed for large distances it drops off like

$$V_{VP} \underset{r \gg \frac{1}{m_e}}{\sim} e^{-2m_e r} / r^{5/2} \tag{2}$$

As a consequence, this distortion is felt very little in an electronic atom whose Bohr orbits are very much larger than $\bar{\lambda}_e$. (We recall that vacuum polarization contributes only about -17 Mc to the Lamb shift in hydrogen whose total magnitude is $\sim$1060 Mc.) Muonic orbits, however, lie well inside this range and therefore probe this additional potential where it is not asymptotic. In fact, it is found that vacuum polarization gives the bulk part of the muonic Lamb shift (i.e. about 98%, see below).

An expansion of eq. (1) suitable for the case of muonic orbits is [1]

$$V_{VP} \simeq -\frac{\alpha Ze}{4\pi^2} \left\{ \frac{2}{3\rho} (\ln\rho + \gamma) + \frac{5}{9\rho} - \frac{\pi}{2} + \rho - \frac{2}{9}\pi\rho^2 \right.$$

$$\left. + \frac{1}{6}\rho^3 (\ln\rho + \gamma) + \frac{7}{18}\rho^3 + O(\rho^5 \ln\rho) \right\} \tag{3}$$

where $\rho = m_e r = r/\bar{\lambda}_e$ and where $\gamma = 0.57721$ is Euler's constant. This expression is, of course, far from the asymptotic behaviour of eq. (2).

1.1 Radiative Corrections in Heavy and Medium-Weight μ-Atoms

The typical orders of magnitude can be read off Table 1, for the case of the $5g_{9/2}$ - $4f_{7/2}$ transition in muonic lead. Vacuum polarization of order $\alpha Z\alpha$ is always attractive. As it is more effective in the 4f-state than in the 5g state, due to the spatial size of these orbits, it increases the transition energy by about 2 keV. The higher corrections are small individually but furthermore have the tendency to cancel partly. As a consequence the $\alpha Z\alpha$ term is only off the total radiative correction by 1.6%.

Let us then turn to a brief discussion of the higher order radiative corrections. All of them can be written in the form of additional potentials which may then be inserted into the muon's Dirac equation or may be evaluated in perturbation theory.

(a) Order $\alpha^2 Z\alpha$

These terms are given by the diagrams of Figs.1b-1e. They have been evaluated by Blomqvist [1] starting from the complete vacuum polarization function to order α^2 as published earlier by Källén and Sabry [6]. Their total contribution to the 5g-4f transition in lead is 15 eV. Like the leading $\alpha Z\alpha$ term these contributions are attractive.

(b) Order $\alpha(Z\alpha)^{2n+1}$

The next important contribution to vacuum polarization in higher orders comes from the diagrams shown in Fig. 2. These diagrams represent the leading order effect of the distortion of the electron/positron propagators by the nuclear Coulomb field. Due to Furry's theorem the number of photon lines between the electron/positron loop and the nucleus (which is represented by an external potential) must be odd. The leading term is then of course the term $\alpha(Z\alpha)^3$. Such diagrams were considered first by Wichmann and Kroll [10] who showed that the corresponding polarization charge density is an analytic function of $Z\alpha$.

Exact expressions for as well as accurate power series expansions of the additional potential induced by these terms have been given by Blomqvist [1]. The resulting correction is found to be repulsive and in fact rather large. Even though $Z\alpha \simeq 0.6$ for lead, the series expansion in terms of powers of $(Z\alpha)$ converges well: Blomqvist finds the following relative contributions

$$\alpha(Z\alpha)^3 : \alpha(Z\alpha)^5 : \alpha(Z\alpha)^7 \simeq 1 : 0.12 : 0.02$$

for the 5g - 4f transition in $^{208}Pb_{\mu}$.
Calculations of these corrections to all orders $\alpha(Z\alpha)^{2n+1}$ and including the finite size of the nuclear charge density have been performed by a number of authors [11-14]. The finite size correction, in particular, has been found to reduce the value calculated for a point nuclear charge by about 10%.

(c) Remainder of Lamb shift

The remaining contributions from the vertex correction and the anomalous magnetic moment (which yield the bulk part of the Lamb shift in electronic hydrogen) can be evaluated from the approximate expression

$$V_{LS}(r) \simeq -\frac{\alpha Ze}{3\pi m_\mu^2}\{\ln(\frac{m_\mu}{2\Delta\varepsilon}) + \frac{11}{24} - \frac{1}{5}\}\rho(r) \tag{4}$$

where $\Delta\varepsilon$ is an average excitation energy (Bethe sum). The anomalous magnetic moment, finally, gives a contribution which can be approximated by

$$V_{AM}(r) \simeq \frac{a_\mu}{m_\mu^2}\frac{1}{r}\frac{dV_c(r)}{dr}\vec{\ell}\cdot\vec{s} + \frac{a_\mu}{4m_\mu^2}\nabla^2 V_c(r) \tag{5}$$

where V_c is the unperturbed Coulomb potential of the nucleus.

1.2 Other, non-radiative Corrections

We briefly mention a few other, non-radiative corrections that must be taken into account before comparing with experiment. Relativistic corrections to the non-relativistic reduced mass effect have been considered by Barrett et al. [3] and by Friar and Negele [4]. The resulting effect for the intermediate orbits that we consider here is small, about 2 eV for the 5g - 4f transition in lead. The finite size correction is small as expected. As the charge parameters of most nuclei are well known it can be calculated with sufficient accuracy.

The correction due to screening by the electronic shells of the host atom can be quite large (-83 eV for the example discussed above). It can be calculated rather accurately, however, by means of a Hartree-Fock treatment of the electronic shells [5]. The question as to which extent electronic shells are depleted by Auger effect and as to how quickly these shells are being refilled, before the muon makes the transition of interest, can be answered through independent experimental information (for a discussion see e.g. ref. 2).

Before we turn to other, higher-order radiative corrections let us have a look at the experimental situation as compared to the theoretical predictions.

1.3 Comparison with Experimental Data

Fig. 3 displays the differences between calculated and measured energies in muonic atoms from Barium to Lead, as a function of the transition energy. When the first data by Dixit et al. came out [7] (open circles in Fig. 3) the situation was very puzzling: There seemed to be a systematic discrepancy between theory and experiment, the theoretical values being too large by up to three standard deviations. Furthermore, the discrepancy was not so much dependent on Z but seemed rather to increase with increasing transition energy. One later experiment by Walter et al. [15] and by Vuilleumier et al. [2] partly confirmed these findings, although somewhat less pronounced (full circles in Fig. 3). It should be added that the discrepancy at first looked considerably more dramatic as the data were compared to an earlier calculation which was in error as to the sign

and the magnitude of the Kroll-Wichmann correction (order $\alpha(Z\alpha)^{2n+1}$). This error was corrected by Blomqvist [1]. The data in Fig. 3 are, of course, compared to his and other people's improved calculations.
This apparent discrepancy led to an intense theoretical activity in this field. The main lines of attack were

(i) speculations about possible anomalous muon-nucleon interactions - a question as old as muon physics itself -, stimulated among other things by the quest for new, possibly light particles in unified gauge theories (Higgs scalars) (for a summary see e.g. ref. 16);

(ii) search for breakdown or modifications of perturbative QED in the case of strong electric fields, i.e. for $Z\alpha$ not small as compared to 1;

(iii) evaluation of more intricate diagrams whose magnitude and sign is difficult to estimate such as the virtual Delbrück scattering diagram to which we turn below, in Sec. 1.4.

The situation has changed recently when two measurements by Tauscher et al. [8] and by Dixit et al. [9] (squares and triangles, respectively, in Fig. 3) were published. These new data disagree with the older data. Finally, the gold calibration standard has changed recently by about 10 eV (see discussion in ref. 2), thus decreasing the old discrepancy but disturbing a little the agreement with the new data [8].

1.4 Evaluation of the Virtual Delbrück Scattering Diagram

It is difficult, if not impossible, to estimate the sign and the magnitude of the diagrams of Fig. 4

which represent the order $\alpha^2(Z\alpha)^2$ correction due to virtual Delbrück scattering. A first estimate of this effect [17] gave a negative sign and a magnitude of about 30 eV for the case of lead - thus eliminating the previous apparent discrepancy between theory and experiment. However, another estimate by Rinker and Wilets [18] as well as very elaborate and detailed calculations by Borie [19] and by Fujimoto [20] found a contribution of the opposite sign and of the order of 1 eV only (muonic lead, as before).

1.5 Conclusion for the Case of Heavy Atoms

At present the theoretical predictions of radiative corrections in heavy muonic atoms are under good control and it is rather unlikely that any sizeable correction has been missed. The experimental situation, on the other hand, is not completely satisfactory and should definitely be checked in further detail. One step in this direction is already being done at present with the high-precision measurement of muonic X-rays in medium-weight μ-atoms by means of bent crystal-spectrometers to which we now turn.

1.6 Studies of Radiative Corrections in Medium-Weight Muonic Atoms

It seems difficult to push the accuracy of measurements with solid state detectors below the 30-50 ppm accuracy of the data mentioned above. If one is to use a bent crystal-spectrometer, however, it seems possible, with the higher resolution of this instrument and with increased statistics, to reach accuracies of the order of a few ppm. However, the resolution of the crystal-spectrometer is superior to the one of Ge-Li detectors only for X-ray energies below or equal to about 100 KeV. This limits the use of this technique to intermediate muonic transitions, such as 3d-2p transitions, in medium-weight nuclei (Z = 10-16). Such experiments are in progress at SIN [21], aiming at a precision determination of $3d_{5/2}-2p_{3/2}$ transition energies in systems like muonic ^{28}Si, ^{24}Mg, and ^{31}P. The accuracy in the determination of the transition energy has reached already the level of about 20 ppm - this means a test of the vacuum polarization to about 2% accuracy. It is hoped that this will be improved soon by another factor of ten. A preliminary result for the wavelength of the $(3d_{5/2} - 2p_{3/2})$ transition in ^{28}Si has been quoted in Ref. 22, which seems in agreement with the theoretical prediction within one standard deviation. Further results with the promised few ppm accuracy will be most welcome.

1.7 Light Muonic Atoms

The Lamb shift

$$\Delta \equiv E_{2p_{3/2}} - E_{2s_{1/2}} \tag{6}$$

in muonic Helium has been measured in a pioneering experiment by Zavattini and his collaborators [23]. The result of this experiment is listed in Table 2, together with the theoretical predictions [24],[25]. Here again vacuum polarization of order $\alpha(Z\alpha)$ is the dominant radiative correction as may be read off Table 2. Calculational details can be found in Ref. 24 and further references cited therein.

In the case of Helium there is a rather large correction from the nuclear finite size effect

$$\delta_{\text{finite size}} = -103.2 \cdot \langle r^2 \rangle_{He} \simeq -(281 \pm 9) \text{ meV} \qquad (7)$$

whose error stems from the uncertainty to which the r.m.s. radius of Helium is known from electron scattering[+]

$$\langle r^2 \rangle^{1/2}_{^4He} = (1.674 \pm 0.012) \text{ fm} . \qquad (8)$$

The error in this term is responsible for the bulk part of the total theoretical uncertainty as it is quoted in Table 2. In contrast to this the nuclear polarizability shift which lowers the 2s state by about 5 meV is not critical [26],[27], the estimated error being only about 2 meV. In fact, the experimental result may be used to determine the r.m.s. radius of Helium, assuming exact validity of the calculated radiative corrections. One finds

$$\langle r^2 \rangle^{1/2}_{^4He} = (1.6733 \pm 0.0030) \text{ fm} \qquad (9)$$

i.e. the error is reduced by a factor four as compared to the direct measurement (5). In conclusion even with the present uncertainties the radiative corrections are tested to ±0.25%.

[+] References are quoted in our Ref. 23).

These same corrections, finite size and polarizability, are very much smaller in muonic hydrogen. The finite size shift can be calculated from the unknown proton radius

$$\langle r^2 \rangle_p^{1/2} = (0.805 \pm 0.010) \text{ fm} \quad .$$

The polarizability shift of the 2s state may be estimated from the formula

$$\delta_{\text{Pol.}} = - \frac{1}{2} e^2 \alpha \langle \frac{1}{r^4} \rangle_{2s} \tag{10}$$

where α is the proton's electric dipole polarizability, $\alpha \simeq 10^{-3}$ fm^3 and where the expectation value of $1/r^4$ over the muonic 2s-state is readily calculated (cut off at small distances). Muonic-hydrogen, therefore, is an even cleaner test case then Helium - even though it is more difficult to measure. Two proposals at LAMPF and at SIN aim at a precision measurement of this Lamb shift which will provide us with another clean test of QED at low momentum transfers.

2. WEAK INTERACTIONS AND MUONIC ATOMS

2.1 Parity Violating Effects due to Weak Neutral Currents [28] - [32]

Gauge models unifying weak and electromagnetic interactions predict the existence of weak neutral currents. In the type of models as proposed by Weinberg and Salam, for instance, there is a neutral current interaction of the nucleons to the leptons (via the Z^0 intermediate boson) which contains both vector and axial vector pieces. It is believed that these interactions are the ones that are responsible for the $\Delta S = 0$, $\Delta Q = 0$ semileptonic processes which are now well established experimental facts. Here again, muonic (as well as electronic) atoms may yield important information that is complementary to the neutrino-induced reactions at high energies. Suppose the neutral vector boson couples to nucleons and leptons through both vector and axialvector currents of the form

$$J_\alpha^{(\Delta Q=0)} = \ell_\alpha^{(0)} + h_\alpha^{(\Delta S=0,\Delta Q=0)} \tag{1}$$

where the leptonic piece is given by

$$\ell_\alpha^{(0)} = \bar{e}\gamma_\alpha(a^{(e)}-b^{(e)}\gamma_5)e+\bar{\mu}\gamma_\alpha(a^{(\mu)}-b^{(\mu)}\gamma_5)\mu + \dots \tag{2}$$

and where the hadronic piece is of the form

$$h_\alpha^{(0,0)} = V_\alpha^{(0,0)} + A_\alpha^{(0,0)}. \tag{3}$$

As the mass of the neutral vector boson is probably large, of the order of 30 GeV, the resulting interaction between nucleons and leptons is of very short range and may be approximated by a contact interaction

$$H \propto J_\alpha^\dagger J^\alpha$$

If this is added to the standard electromagnetic interaction the interference between the vector-axial vector combination of the weak neutral hamiltonian with the vector-vector type electromagnetic interaction will lead to observable parity-violating effects. Specifically, there should occur spin-momentum correlations in atomic processes, i.e. in Auger and X-ray transitions.

Here again muonic atoms are good candidates for detection of such effects, for the following reasons:

(i) Given the short range of the weak neutral interaction all parity violating effects are proportional to the overlap of the atomic orbits with the nucleus. Due to its closeness to the nucleus a bound muon is preferred as compared to an electron.

(ii) The muon is produced polarized from pion decay. It is known that a good fraction of this initial polarization is preserved when the muon reaches the lower Bohr orbits. (Residual polarization for spin zero nuclei e.g. is about 17%.) This makes it possible to search for correlation effects between the muon spin and some momentum vector in the electromagnetic decay of the atom. An example is the angular asymmetry of a photon multipole transition with respect to the muon spin. Suppose we consider the electromagnetic transition from a specific atomic initial state i with given parity and angular momentum to another state f.

Let the transition amplitude be $T_{i\to f} \equiv T_0$. In presence of the weak neutral interaction parity is no longer a good quantum number. In particular if there is a state i' of opposite parity close to the initial state i, then these two states will be mixed. (The same statement applies to the final state f if it also has neighbour with odd relative parity.) Between the states i' and f there is also an electromagnetic transition amplitude $T_{i'\to f} \equiv T_1 \cdot T_1$ may have the same multipolarity as T_0 but has always opposite parity. The transition amplitude between the perturbed, parity-mixed states will have the form, written in a somewhat symbolical form

$$T \simeq T_0 \;(1 + \eta\; T_1/T_0) \tag{4}$$

where η is the parity-violating admixture coefficient of state i' into i,

$$\eta = \frac{\langle i' | H^{p.v.} | i \rangle}{E_i - E_i' - \frac{i}{2}(\Gamma_i + \Gamma_{i'})} \tag{5}$$

$H^{p.v.}$ is the parity odd hamiltonian due to the weak neutral interaction; E_i, $E_{i'}$ are the unperturbed energies of the two states; Γ_i, $\Gamma_{i'}$ are their radiative decay widths.

The observable parity-violating effects will be largest if

(i) the mixing parameter η is as large as possible. This will happen if the states i and i' are nearly degenerate and if they have a good overlap with the nuclear density.

(ii) the ratio $|T_1/T_o|$ is as large as possible. Therefore, T_o should be a strongly hindered but still measurable transition while T_1 should be an allowed and favoured transition. Two cases have been considered where these conditions can be optimized to some extent:

A) $|i\rangle \equiv |2s\rangle$; $|i'\rangle \equiv |2p\rangle$ going to $|f\rangle \equiv |1s\rangle$[28][29]

In this case T_o is a strongly hindered M1-transition between the 2s and 1s-states to which the highly favoured E1 transition between 2p and 1s-states is admixed (see Fig. 5(a)).

$$T_{2s \to 1s} \simeq T^{M1}_{2s \to 1s} + \eta \, T^{E1}_{2p \to 1s} \quad . \tag{6}$$

In light nuclei around Z = 3-4 the parameter η can become quite large, of the order of 10%, due to an almost complete degeneracy of the 2p and 1s-states in the nucleus+. In heavier nuclei where this degenerayy is less perfect, one finds $|\eta| \sim 10^{-7}$ and resulting forward-backward asymmetries of the emitted X-rays of the order of 10^{-5} to 10^{-6}.

B) $|i\rangle \equiv |3d\rangle$; $|i'\rangle \equiv |3p\rangle$ going to $|f\rangle \equiv |1s\rangle$[30]-[32]

The main transition here is an E2 transition between 3d and 1s, to which a small amount of E1 transition should be admixed (see Fig. 5(b)),

+ Around Z ≈ 3-4 the repulsive finite-size shift of the 2s-state is balanced by the attractive vacuum polarization.

$$T_{3d \to 1s} \simeq T^{E2}_{3d \to 1s} + \eta' \, T^{E1}_{3p \to 1s} \ . \qquad (7)$$

Model calculations predict a mixing parameter of the order of magnitude 10^{-6} (around Z = 30). When multiplied with T(E1)/T(E2), forward-backward asymmetries of the order of 10^{-5} are found.

Several experiments of type A or B have been proposed or are by now being carried out at SREL, CERN and SIN. First attempts at seeing an effect of type A have failed; the experiment along the lines of type B is still running at present [32].
Should such an experiment be successful, then the modest muonic atom will make an important contribution to a fundamental question in weak interactions, unravelling at once the nature of the weak $\Delta S = 0$, $\Delta Q = 0$ semileptonic interaction.

2.2 Exotic Muon Capture

Before leaving muonic atoms we would like to mention briefly one other topic which may have an important impact on our understanding of leptonic weak interactions: Neutrinoless muon capture, or muon-electron conversion. By this we mean one of the following exotic capture reactions from the 1s bound state of a muonic atom with the nucleus of charge number Z and mass number A

$$\mu^- + (Z,A) \to (Z-2,A)^* + e^+ \qquad (8)$$

$$\mu^- + (Z,A) \to (Z,A)^* + e^- \ . \qquad (9)$$

Here $(Z', A)^*$ denotes a discrete final nuclear state, either the ground state or a discrete excited state. As the muon is in a bound state, the positron/electron is monochromatic, $E_e \simeq m_\mu$. This gives a clean signature in the search for these processes. None of these processes have been seen so far, the experimental upper limit of the branching ratio being about 10^{-8} (as compared to ordinary muon capture)[+],

$$\frac{\Gamma(\mu \to e)}{\Gamma(\mu \to \nu_\mu)} < \sim 10^{-8} \quad . \tag{10}$$

The first of these processes (8) requires a double charge-exchange in a weak interaction i.e. the conversion of two protons into two neutrons. If there is a genuine isotensor weak interaction that allows $\mu^- \to e^+$ conversion[++] then this process has to involve either mesonic or N^* degrees of freedom in the nucleus [34]. Such diagrams are sketched in Fig. 6. The basic vertices are

$$(\mu^- e^+)(N^*N) \qquad \text{and} \qquad (\mu^- e^+)(N^*N^*)$$

while the vertex $(\mu^- e^+)(\pi^+\pi^-)$ of Fig. 6(a) does not contribute for antisymmetric nuclear states [34]. The capture rate is then given by the (unknown) coupling constants of the isotensor interaction times the probability of exciting N^* resonances in the appropriate intermediate steps.

[+] The experimental situation is summarized in Ref. 33.

[++] This is, of course, not possible within the ordinary lepton number schemes (additive or multiplicative). However, a scheme in which μ^- and e^+ have the same lepton number [35] would allow the process.

If there is no such new type of coupling, on the other hand, the capture rate will most likely be of order G_F^4 i.e. hopelessly small. For example, in gauge models without charged Higgs particles (or with the charged scalars "gauged away" appropriately), the process would go according to the diagramm of Fig. 7, where the intermediate leptonic state is thought to be a mixture of ν_μ and $\bar{\nu}_e$ (with non-vanishing mass difference) or of some other pair of neutral leptons. Even though gauge theories will give a convergent result for such processes these are likely to come out very small indeed.

The second process (9) is intimately connected with the decay of the muon into an electron and a photon

$$\mu^\pm \to e^\pm + \gamma \tag{11}$$

for which we know that $\Gamma(\mu\to e\gamma)/\Gamma(\mu\to e\nu\bar{\nu}) < 2.2\times 10^{-8}$. In either case we need to have a vertex $\langle e^-(p')|j^\alpha_{e.m.}|\mu^-(p)\rangle$ (with $j^\alpha_{e.m.}$ the electromagnetic current), connecting muon, electron and photon (real or virtual), see Fig. 8(a).

The processes (9) and (11) are forbidden if muonic lepton number is conserved and independently of whether the additive or multiplicative scheme applies. If, however, there is a small violation of muonic lepton number conservation then both processes are allowed and the rates are of order $Z\alpha^2G_F^2$ and αG_F^2, respectively. This could happen, for instance, through diagrams like the ones of Figs. 8(b)-(d) where (ν_μ, ν_e) or (N_μ, N_e) denotes a mixture of neutrinos or other neutral leptons [36] (with non-vanishing mass difference).

We do not want to go into detailed model considerations which are still controversial at this point. We just note a few general properties of the relevant matrix element $<e|j^{\alpha}|\mu>$. Covariance and gauge invariance require this vertex to have the form [37]

$$<e^-(p')|j^{\alpha}_{e.m.}(o)|\mu^-(p)> \; =$$

$$\frac{1}{(2\pi)^3}\overline{u_e(p')}\{G_1(k^2)[\gamma^{\alpha}-k^{\alpha}\frac{m_{\mu}-m_e}{k^2}]+G_2(k^2)[\gamma^{\alpha}+k^{\alpha}\frac{m_{\mu}+m_e}{k^2}]\gamma_5$$

$$+\frac{i}{m_{\mu}}[F_1(k^2) + F_2(k^2)\gamma_5]\sigma^{\alpha\beta}\;k_{\beta}\}\;u_{\mu}(p) \tag{12}$$

where $k = p-p'$ is the photon momentum.
This form still allows for parity violation in this decay. It is clear that $G_1(k^2)$ and $G_2(k^2)$ must be proportional to k^2, in order to have the rate for $\mu \to e\gamma$ finite. Therefore, only F_1 and F_2 contribute to $\mu \to e\gamma$, while all four form factors contribute to the process (9). If time reversal invariance holds true, (all existing models have this property), and with the electromagnetic current being hermitean, all four form factors are real.

The two exotic capture processes (8) and (9) as well as the decay (11) are being studied in new experiments at the meson factories. Some of these experiments are running at present and are expected to yield results soon. It is evident that these experiments could well turn out to be of utmost importance for our understanding of the leptons in general, and the nature of muonic lepton number in particular.

3. HADRONIC ATOMS

3.1 Line Widths, Energy Shifts and Low-Energy Hadron-Nucleon Scattering

The impact of hadronic atoms on particle physics is much less obvious than in the cases we have discussed above. For one, any theory of line widths and line shifts of the (Coulomb) bound states of hadronic atoms has to cope with the multiple scatterings of the hadron in the nucleus, under the effect of strong interactions. It is not enough to develop a multiple scattering theory for the case of nuclear matter. Indeed, the interaction of bound π^-, K^- or $\bar{p}$ etc., with the nucleus is very peripheral in the sense that the bound particle probes primarily the nuclear surface region where the nuclear matter density varies rapidly as a function of the distance from the center.

Furthermore, a description in terms of multiple scattering theory and of optical potentials is straightforward and simple only when the elementary system particle-nucleon behaves smoothly around and below threshold. If there are resonances or bound states of the elementary system nearby, then the construction of an equivalent optical potential becomes technically complicated and much less transparent than for the smooth case.

It is not the purpose of these lectures to go into the techniques of multiple scattering theories for hadronic atoms - these are spelled out in many original papers and reviews to which we refer the interested [38]. Instead we summarize here the general strategy in the construction of optical potentials for hadronic atoms in order to point out the contacts to the particle aspects which are involved.

The main idea is to construct an optical potential, within the framework of potential theory, whose main input are the on-shell hadron-nucleon scattering amplitudes at or below threshold, as well as bulk properties of the nucleus such as its ground state matter density, its geometrical shape and possibly some integral information on the two body densities (two body correlations). This is indeed feasible if

(i) the matter density ρ is low, in the sense that e.g. $|f_{HN}\rho^{1/3}|$ is smaller than unity, f_{HN} being the hadron-nucleon scattering amplitude at the relevant energy;

(ii) the elementary amplitudes f_{HN} behave smoothly around and below threshold;

(iii) the closure approximation can be applied to the nucleus. In this case only properties of its ground state are needed.

Also, if (ii) and (iii) hold then the "fixed scatterer approximation" can be used - an approximation that simplifies matters considerably. In particular, it can then be shown that only on-shell information on the hadron-nucleon scattering amplitude is needed.

Therefore, in the simplest case (low density, s-wave scattering only) the optical potential will be

$$V_{OPT} = - \frac{4\pi}{2m} \bar{a} \rho (\vec{r}) \qquad (1)$$

where $\bar{a}$ is the averaged hadron-nucleon s-wave scattering length, m being the hadron mass.

If the elementary system has inelastic channels open at threshold - this is the case for K^-N, $\bar{p}N$, Σ^-N but not for π^-N - the amplitude $\bar{a}$ is complex and the optical potential contains already an absorptive part. In the case of pions $\bar{a}$ is essentially real and absorption proceeds through a more complicated process involving two or more nucleons. Such an absorption process on two nucleons may be described roughly and in a rather phenomenological manner by adding an imaginary part to eq.(1) that is proportional to $\rho^2(\vec{r})$, the matter density squared.

Before we say more about such optical potentials, let us recall the experimental quantities that are accessible in the X-ray spectrum of hadronic atoms and that are sensitive to strong interactions. These are

(1) Line width and center-of-gravity energy shift of the last observable atomic state (called "critical state" or "lower state" in the literature),

$$\Gamma_{n_c} \text{ and } \varepsilon_{n_c} \; ; \tag{2}$$

(2) Line width of the last but one circular orbit

$$\Gamma_{n_c + 1} . \tag{3}$$

This quantity is determined by measuring the yield of the X-ray transition $(n_c + 1 \rightarrow n_c)$.

It is these quantities which are determined by the optical potential V_{OPT}. Anything that we may want to learn about the low-energy hadron nucleon scattering amplitudes has to be extracted from the experimental quantities (2) and (3) via the optical potential and the appropriate wave equation.

What helps in this analysis is the fact that in all directly observable atomic states, i.e. $n \geq n_c$, the Coulomb potential is still the dominant feature which determines the level structure, whilst the strong interaction acts more like a perturbation on this pattern. Indeed while observable widths and shifts never exceed $\sim$10 keV, the transition energies are several hundreds of keV+, i.e.

$$|\Gamma_{n_c}|, \; |\varepsilon_{n_c}| \; , \; |\Gamma_{n_{c+1}}| \ll |E_n - E_{n'}| \; . \qquad (4)$$

Nevertheless, the connection between real and imaginary parts of V_{OPT} with the widths and shifts is not simple and cannot be dealt with in perturbation theory. There can be important distortion effects which complicate the sensitivity of the experimental parameters to the input amplitudes. For instance, in K^- atoms the energy shifts are always negative (i.e. repulsive) even when the real part of V_{OPT} is made strongly attractive. The cause for this behaviour lies in the very strongly absorptive part of V_{OPT} and in the distortion of the kaonic state that follows from it. The optical potential and its connection to the hadron-nucleon scattering amplitudes at low energies has been studied extensively for pionic atoms, to some extent also for kaonic atoms. Much less

+ Such an ordering does not hold for $n < n_c$ and is actually reversed. But those states are observable only via a resonance admixture to higher states [39].

is known, however, in antiprotonic and Σ^- atoms. We therefore restrict our discussion to pions and kaons.

3.1.1 Optical potential for pionic atoms

Up to kinematical factors such as those which come from transforming from the pion-nucleon to the pion-nucleus centre-of-mass system, V_{OPT} is given by [40]

$$V_{OPT} = -\frac{4\pi}{2m_\pi}\{q(\vec{r}) - \vec{\nabla}\frac{\alpha(\vec{r})}{1-\frac{1}{3}\xi\alpha(\vec{r})}\vec{\nabla}\} \tag{5}$$

where

$$\begin{aligned} q(\vec{r}) &= (b_o-b_1)\rho_p(\vec{r}) + (b_o+b_1)\rho_n(\vec{r}) \\ &\quad + i\,B_{pp}\,\rho_p^2(\vec{r}) + i\,B_{pn}\,\rho_p(\vec{r})\,\rho_n(\vec{r}) \end{aligned} \tag{6}$$

and

$$\begin{aligned} \alpha(\vec{r}) &= (c_o-c_1)\rho_p(\vec{r}) + (c_o+c_1)\rho_n(\vec{r}) \\ &\quad + i\,C_{pp}\,\rho_p^2(\vec{r}) + i\,C_{pn}\,\rho_n(\vec{r})\rho_p(\vec{r}) \quad . \end{aligned} \tag{7}$$

Here b_o and b_1 are the isospin symmetric and antisymmetric combinations of pion-nucleon s-wave scattering amplitudes

$$b_o = \frac{1}{3}(a_1 + 2a_3) \tag{8}$$

$$b_1 = \frac{1}{3}(a_3 - a_1) \quad , \tag{9}$$

c_o and c_1 are the analogous (spin symmetric) p-wave scattering volumes, whilst the constants B_{pp}, B_{pn}, C_{pp} and C_{pn} are phenomenological parameters describing the absorption of the π^- on proton-proton or on proton-neutron pairs,

$$\pi^- + (pp) \rightarrow p + n; \; \pi^- + (pn) \rightarrow n + n \quad . \tag{10}$$

Typical values of these parameters are

$$\begin{aligned} b_o &= -0.029 \, m_\pi^{-1} & c_o &= 0.227 \, m_\pi^{-3} \\ b_1 &= -0.078 \, m_\pi^{-1} & c_1 &= 0.18 \, m_\pi^{-3} \end{aligned} \tag{11}$$

$$\frac{1}{A^2}(Z^2 B_{pp} + ZNB_{pn}) \simeq 0.043 m_\pi^{-4}; \; \frac{1}{A^2}(Z^2 C_{pp} + ZNC_{pn}) \simeq 0.076 m_\pi^{-6\,+} .$$

The salient features of this optical potential may be summarized as follows:

1. $b_o < 0$ implies that all 1s-shifts ε_{1s} must be negative (repulsive), while the sign of c_o is such that the shifts of all higher states $\ell \neq 0$ are positive,

   $$\varepsilon_{1s} < 0; \; \varepsilon_{2p}, \varepsilon_{3d}, \varepsilon_{4f}, \ldots > 0$$

   - this is indeed what is observed.

   Note however that also ε_{2p} will eventually become ne-

+ holds for homogeneous mixture of protons and neutrons, $\rho_p : \rho_n = Z:N$.

gative, as the nuclear charge is increased: As Z grows the pionic 2p-orbit is pulled towards the nucleus and feels more and more the relative s-wave interaction with nucleons in the nuclear periphery. The turnover should occur around $Z \gtrsim 35$ [41]. This has largely been confirmed by experiment [42].

2. Since $|b_o| << |b_1|$ there should be a strong dependence of ε_{1s} on $(\rho_p - \rho_n)$ i.e. on the nuclear isospin. This is indeed born out by experiment [43]. It would obviously be very interesting if muonic atom data could give us the value of b_o and settle the long debated sign of that quantity. However, an extraction of pion-nucleon scattering amplitudes from pionic atoms is obscured by

   (i) the uncertainties in the description of absorption (distortion effects).

   (ii) double scattering corrections which renormalize b_o. These corrections are important since they involve, among others, virtual charge exchange

$$\pi^- p \rightarrow \pi^o n \rightarrow \pi^- p$$

   and the correlations between pairs of nucleons.

   (iii) in the p-wave there is the additional uncertainty of the precise form of the Lorentz-Lorenz effect, which we do not discuss here [44], [46].

In summary, pionic atoms can at present do no more than to confirm the information from free pion nucleon scattering but cannot be used in trying to determine the pion-nucleon scattering lengths.

3.1.2 Kaonic atoms

At first sight things look much simpler for the case of kaons: Here there are absorption channels open right at threshold, such as

$$K^-N \to \pi\Lambda,\ \pi\pi\Lambda,\ \pi\Sigma,\ \ldots$$

so that absorption on one nucleon is possible and is in fact predominant. Therefore, in first approximation, the optical potential is proportional to the densities only, and to the complex s-wave scattering lengths [47]

$$V_{OPT} = -\frac{4\pi}{2m_K}\, q(\vec{r}) \tag{12}$$

with

$$q(\vec{r}) = \frac{1}{2}\,(A_o+A_1)\,\rho_p(\vec{r}) + A_1\,\rho_n(\vec{r})\ . \tag{13}$$

Here A_o and A_1 are the s-wave scattering lengths for isospin zero and one, respectively.
Here again, the extraction of scattering amplitudes from kaonic atom data is complicated by the following facts.

(i) For $Z = N = A/2$ and with $\rho_p(\vec{r}) = \rho_n(\vec{r})$ the potential (12), (13) is

$$V_{OPT} \simeq -\frac{4\pi}{2m_K}\,\frac{A_o+3A_1}{4}\,\rho_A(\vec{r}) \tag{14}$$

with

$$\frac{A_o+3A_1}{4} \simeq (-0.4 + i\ 1.6)\ \text{fm} \tag{15}$$

Thus V_{OPT} has a very large absorptive part (about 100 MeV in a square well of nuclear size) and there are enormous distortion effects.

(ii) In the isospin zero channel there is a resonance just below threshold which actually behaves almost like a bound state,

$$Y_o(1405) \text{ at } -27 \text{ MeV} .$$

It is technically difficult to incorporate this resonance into the optical potential which was derived under the assumption that the $\bar{K}N$ amplitudes are smooth below threshold. So long as this is not carried out properly kaonic level shifts and widths cannot be analyzed in terms of the elementary kaon-nucleon scattering amplitudes around threshold.

3.2 Magnetic Moments of Σ^- and of other Baryons

Intermediate orbits of antiprotonic and Σ^- atoms can be used for a measurement of the magnetic moments of these particles from the fine structure splitting. While the measurement on the antiproton yield the result

$$\mu_{\bar{p}} = -\mu_p \tag{16}$$

expected from CPT-invariance[+], the magnetic moment of the Σ^- is particularly interesting in the context of SU(3)-symmetry and of specific breaking mechanisms of this symmetry. The experimental situation is discussed in the lectures by G. Backenstoss at this meeting [48]. Therefore we omit the discussion of Σ^- atoms and of the method of extracting the magnetic moment from the X-ray spectrum. Instead we summarize here the symmetry predictions for the magnetic moments of baryons, and we refer the reader to ref. 48 for the complementary discussion of the relevant experiments.

(i) The magnetic moment operator is believed to consist of an isoscalar and an isovector part only. Then, from isospin symmetry alone one has the relationship

$$\mu_{\Sigma^+} + \mu_{\Sigma^-} = 2\mu_{\Sigma^o} . \tag{17}$$

(ii) From exact SU(3) symmetry, in which case the magnetic moment operator is the U-spin scalar member of an octet, one obtains the following relations [49]

$$\mu_{\Sigma^+} = \mu_p \tag{18}$$

$$\mu_\Lambda = \frac{1}{2}\mu_N = - \mu_{\Sigma^o} \tag{19}$$

$$\mu_{\Sigma^-} = \mu_{\Xi^-} \tag{2o}$$

Combining relations (18) and (19) with the isospin relation (17) one obtains

---

+ The result is quoted in ref. 48 where also the references to the original work are given.

$$\mu_{\Sigma^-} = - (\mu_P + \mu_N) = \mu_{\Xi^-} \tag{21}$$

(iii) In a broken SU(3) scheme some of these relations will be replaced by weaker sum rules. In particular, relations between baryons of different strangeness cannot be valid in this simple form any more. The intuitive physical reason for this is that the strange quark (s- or λ-quark) is heavier than the non-strange quarks (u,d- or p,n-quarks) and, therefore, will have a different intrinsic magneton from the one of the non-strange quarks.
Such a model for symmetry breaking within the framework of the quark model was proposed by Rubinstein, Socolow and the author in 1967 [50] - at a time where one could only dream of measuring the magnetic moment of Σ^-. In this model the relations (18) - (21) are replaced by the weaker sum rules

$$\mu_\Lambda + 3\mu_{\Sigma^+} = \frac{8}{3}\mu_P \tag{22}$$

$$\mu_{\Sigma^-} = - (\frac{4}{9}\mu_P + \frac{1}{3}\mu_\Lambda) \tag{23}$$

$$\mu_{\Xi^-} = \frac{4}{3}\mu_\Lambda + \frac{1}{9}\mu_P \quad . \tag{24}$$

These relations are such that the contributions of strange and non-strange quarks balance separately. This is unlike the SU(3)-relations which, in the language of the quark model, assume that λ-quark and n-quark have the same magneton.

(iv) In comparing these relations with experiment, one may want to keep track of the mass difference of the particles. We use the following admittedly arbitrary prescription. Let

$$M_i = \mu_i \frac{e}{2m_i} = \overline{\mu_i} \frac{e}{2m_N} \tag{25}$$

be the dimensioned magnetic moment of particle i. μ_i is the magnitude of the magnetic moment when expressed in its own magneton $\frac{e}{2m_i}$; whilst $\overline{\mu_i}$ is the magnitude of the magnetic moment when expressed in terms of the nuclear magneton $\frac{e}{2m_N}$. Note that experimental results are usually quoted in terms of $\overline{\mu_i}$.

We assume that the symmetry or quark model predictions hold for the quantities μ_i.

In Table 3 these relations are compared to the experimental values. The second column indicates the relation that is being tested; the third column gives the prediction without the mass correction factor, while the fourth contains that factor. The last column finally gives the experimental values[+].

The comparison with the data reveals a puzzling situation: While there is reasonable agreement for the cases of Λ and Σ^+, there are marked discrepancies for Σ^- and Ξ^-. The mass correction goes in the right direction for Λ; for Σ^+ the situation is not clear due to the large experimental error.

[+] Taken from ref. 51 where the references to original literature are found.

The differences between exact SU(3) and the quark model relations are appreciable for Σ^- and Ξ^- but both predictions disagree with the experimental values. In view of this situation it would therefore seem worthwhile to improve the accuracy on the magnetic moment of Σ^- from further measurements in Σ^- atoms.

REFERENCES

1. J. Blomqvist, Nucl. Phys. B48 (1972) 95;
   Similar calculations were also published by M.K. Sundaresan and P.J.S. Watson, Phys. Rev. Letters 29 (1972) 15, erratum in 29 (1972) 1122;and by T.L.Bell; Phys. Rev. A7 (1973) 148o.
2. J.L. Vuilleumier, W. Dey, R. Engfer, H. Schneuwly, H.K. Walter and A. Zehnder, Z. Physik A278 (1976) 1o9.
3. R.C. Barrett, D.A. Owen, J. Calmet, and H. Grotch; Phys. Letters 47B (1973) 297.
4. J.L. Friar and J.W. Negele; Phys. Letters 46B (1973) 5.
5. P. Vogel; Phys. Rev. A7 (1973) 63.
6. G. Källén and A. Sabry; Dan. Mat. Fys. Medd. 29 (1955) No. 17.
7. M.S. Dixit, H.L. Anderson, C.K. Hargrove, R.J. McKee, D. Kessler, H. Mes, and A.C. Thompson; Phys. Rev. Letters 27 (1971) 878.
8. L. Tauscher, G. Backenstoss, K. Fransson, H. Koch, A. Nilsson, and J. de Raedt; Phys. Rev. Letters 35 (1975) 41o.

9. M.S. Dixit, A.L. Carter, E.P. Hincks, D. Kessler, J.S. Wadden, C.K. Hargrove, R.J. McKee, H. Mes, and H.L. Anderson; Phys. Rev. Letters 35 (1975) 1633.

10. E.H. Wichmann and N.M. Kroll; Phys. Rev. 101 (1956) 843.

11. J. Arafune; Phys. Rev. Letters 32 (1974) 560.

12. L.S. Brown, R.N. Cahn, and L.D. McLerran; Phys. Rev. Letters 32 (1974) 562.

13. M. Gyulassy; Phys. Rev. Letters 32 (1974) 1393.

14. G.A. Rinker, and L. Wilets; Phys. Rev. A12 (1975) 748.

15. H.K. Walter et al.; Phys. Letters 40B (1972) 197.

16. L. Resnick, M.K. Sundaresan, and P.J.S. Watson; Phys. Rev. D8 (1973) 172.

17. M.Y. Chen; Phys. Rev. Letters 34 (1975) 341.

18. G.A. Rinker, and L. Wilets; Phys. Rev. Letters 34 (1975) 339.

19. E. Borie; Nucl. Phys. A267 (1976) 485.

20. D.H. Fujimoto; Phys. Rev. Letters 35 (1975) 341.

21. B. Aas et al.; proposal to SIN, No. R-71-01.

22. R. Eichler; thesis ETH Zürich 1976.

23. A. Bertin, G. Carboni, J. Duclos, U. Gastaldi, G.Gorini, G. Neri, J. Picard, O. Pitzurra, A. Placci, E.Polacco, G. Torelli, A. Vitale and E. Zavattini; Phys. Lett. 55B (1975) 411 and Nucl. Phys. A278 (1977) 381.

24. E. Borie; Z. Physik A275 (1975) 347.

25. E. Campani; Nuovo Cimento Letters 4 (1970) 982.

26. J.E. Bernabeu and C. Jarlskog; Nucl. Phys. B75 (1974) 59.

27. E.M. Henley, F.R. Krejs, L. Wilets; University of Washington, preprint 1975.

28. J. Bernabeu, T.E.O. Ericson, and C. Jarlskog; Phys. Letters 50B (1974) 467.

29. G. Feinberg, and M.Y. Chen; Phys. Rev. D10 (1974) 190.

30. L.M. Simons; Helv. Phys. Acta 48 (1975) 141.

31. Contributions by L.M. Simons and by E.M. Henley to the Proc. International Symposium on "Interaction Studies in Nuclei", Mainz 1975; H. Jochim and B. Ziegler eds., North Holland publishing.

32. L.M. Simons et al.; proposal to SIN No. R-75.03.

33. S. Fraenkel; "Rare and Ultra-rare Muon Decays" in Volume II of"Muon Physics", C.S. Wu and V.Hughes eds., Academic Press, New York, 1975.

34. L. Kisslinger; Phys. Rev. Letters 26 (1971) 998, erratum - 28 (1972) 869;
M. Rho, and M. Shuster; Phys. Letters 42B (1972) 45.

35. E.S. Konopinski, and H.M. Mahmoud; Phys. Rev.92 (1953) 1045.

36. T.P. Cheng, and Ling-Fong Li; preprint Dec. 1976.

37. S. Weinberg, and G. Feinberg; Phys. Rev. Letters 3 (1959) 111.

38. See for example, J. Hüfner, "Pions interact with Nuclei", Physics Reports 21C (1975) 1; F. Scheck; lectures at Vancouver Summer School 1975, Plenum Press, New York, 1976.

39. M. Leon; Nucl. Phys. A260 (1976) 461.

40. M. Ericson, and T.E.O. Ericson; Ann. Phys. (N.Y.) 36 (1966) 323;

M. Krell, and T.E.O. Ericson; Nucl. Phys. B11 (1969) 521.

41. M. Ericson, T.E.O. Ericson, and M. Krell; Phys. Rev. Letters 22 (1969) 1189.

42. L. Tauscher et al.; CERN preprint Jan. 1977; M. Leon et al.; Phys. Rev. Letters 37 (1976) 1135.

43. G. Backenstoß; Ann. Rev. Nucl. Science 20 (1970) 467.

44. F. Scheck and C. Wilkin; Nucl.Phys. B49 (1972) 541.

45. G. Fäldt; Nucl. Phys. A206 (1973) 176.

46. G. Baym, and G.E. Brown; Nucl. Phys. A247 (1975) 395.

47. T.E.O. Ericson, and F. Scheck; Nucl. Phys. B19 (1970) 45o.

48. G. Backenstoss; lectures at this meeting.

49. S. Coleman, and S.L. Glashow; Phys. Rev. Letters 6 (1961) 423.

50. H.R. Rubinstein, F. Scheck, and R.H. Socolow; Phys. Rev. 154 (1967) 16o8.

51. Review of Particle Properties; Rev. Mod. Phys. 48, Nr. 2.II (1976).

Table 1: Radiative and Other Corrections for the Example of the $5g_{9/2} - 4f_{7/2}$ transition in muonic ^{208}Pb

(Numbers taken from compilation in Ref. 2). The constants $m_\mu c^2 = 105.6595(3)$ MeV, $m_e c^2 = 0.5110041(16)$ MeV, $\alpha = 1/137.03602(21)$, and $\hbar c = 197.32891(66)$ MeV fm have been used in the calculations.

Effect	Contribut. (in eV)	Ref.
transition energy	429 343 (2)	2
reduced mass, relativistic corr.	2	3,4
finite size correction	-4	
electron screening	-83 (3)	5
nuclear polarization	6 (3)	
Radiative Corrections:		
order $\alpha Z\alpha$	2 106 (3)	
order $\alpha^2 Z\alpha$	15 (1)	
order $\alpha(Z\alpha)^{2n+1}$ (Kroll-Wichmann)	-43 (2)	
Lamb shift, remainder	-6 (1)	
total radiative correction	2 072	
total theoretical correction	1 993	
total theor. transition energy	431 336 (7)	
experimental transition energy	431 334 (10)	7,8,9

Table 2: Lamb Shift in Light Muonic Atoms
(Theoretical numbers are taken from Ref. 24)

	${}^{1}_{1}H_{\mu}$	${}^{4}_{2}He_{\mu}$
	(units: meV; 1 meV = 10^{-3} eV)	
fine structure	8.4	145.9
vacuum polarization of order $\alpha Z\alpha$	204.9	1 664.8
vac. pol., higher orders	1.5	11.6
Lamb shift, remainder	-0.6	-10.7
finite size effect	-3.4±0.1	-280.7±4
polarizability	2.10^{-2} *)	5±2
total theoretical	210.8±0.1	1 536±5
Experiment[23)]		1 527.5±0.3
theoretical value of Rinker et al. (quoted in ref.23)		1 527.3±4.2

*) our own estimate

Table 3: Predicted and Measured Magnetic Moments of Octet Baryons

Part.	Input	without mass correction	with mass correction	Experiment
Λ	rel. (19)	-0.96	-0.80	-0.67 ± 0.06
Σ^+	rel. (18)	2.79	2.20	2.62 ± 0.41
	rel. (22), μ_Λ^{exp}	2.71	2.16	
Σ^-	rel. (21)	-0.88	-0.69	-1.48 ± 0.37+)
	rel. (23), μ_Λ^{exp}	-1.02	-0.77	
Ξ^-	rel. (21)	-0.88	-0.63	-1.85 ± 0.75
	rel. (24), μ_Λ^{exp}	-0.58	-0.54	

+) The second, positive solution obtained by C.S. Wu et al. is omitted. See discussion in ref. 48.

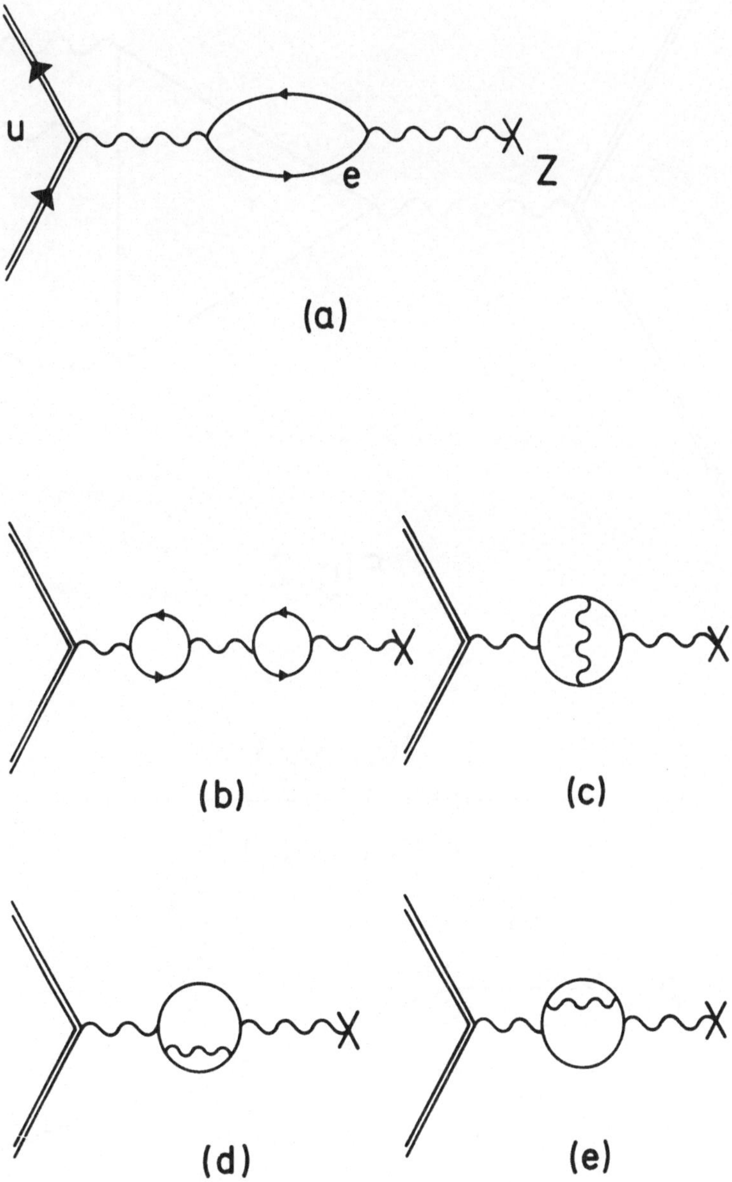

Fig. 1

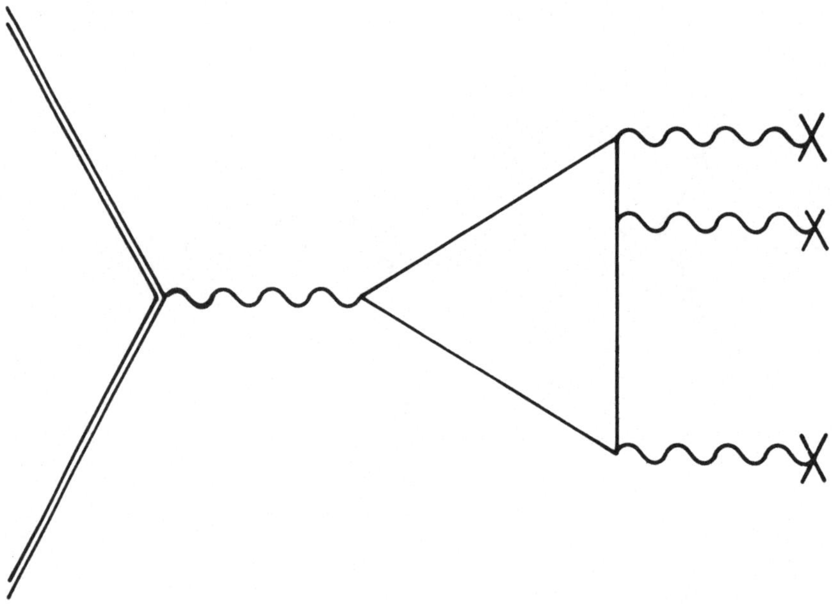

Fig. 2

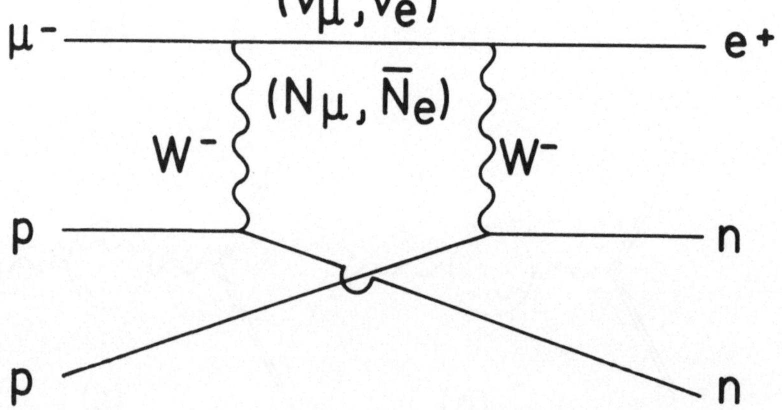

Fig. 7

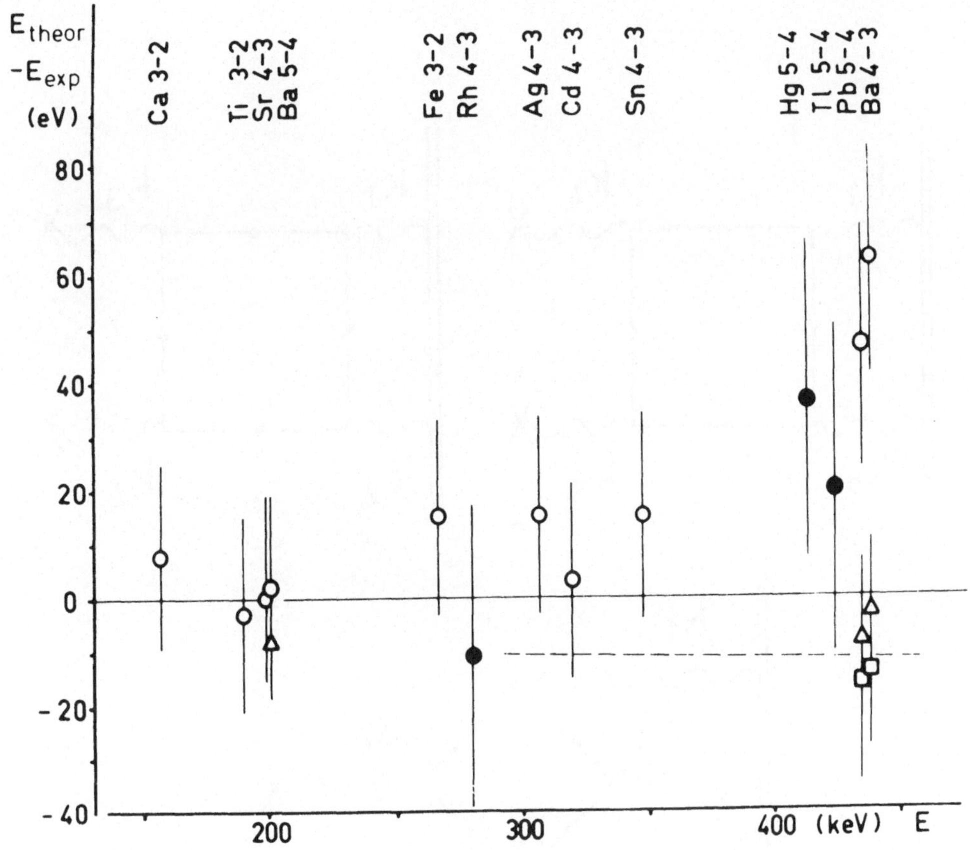

Fig. 3

Difference between predicted and measured transition energy in heavy muonic atoms as a function of transition energy. The comparison is made for the mean value of the resolved fine structure components. Open circles from Ref. 7; full circles from Ref. 2; triangles from Ref.9; squares from Ref. 8. The dashed line is the abscissa corresponding to the old value of the Gold calibration standard.

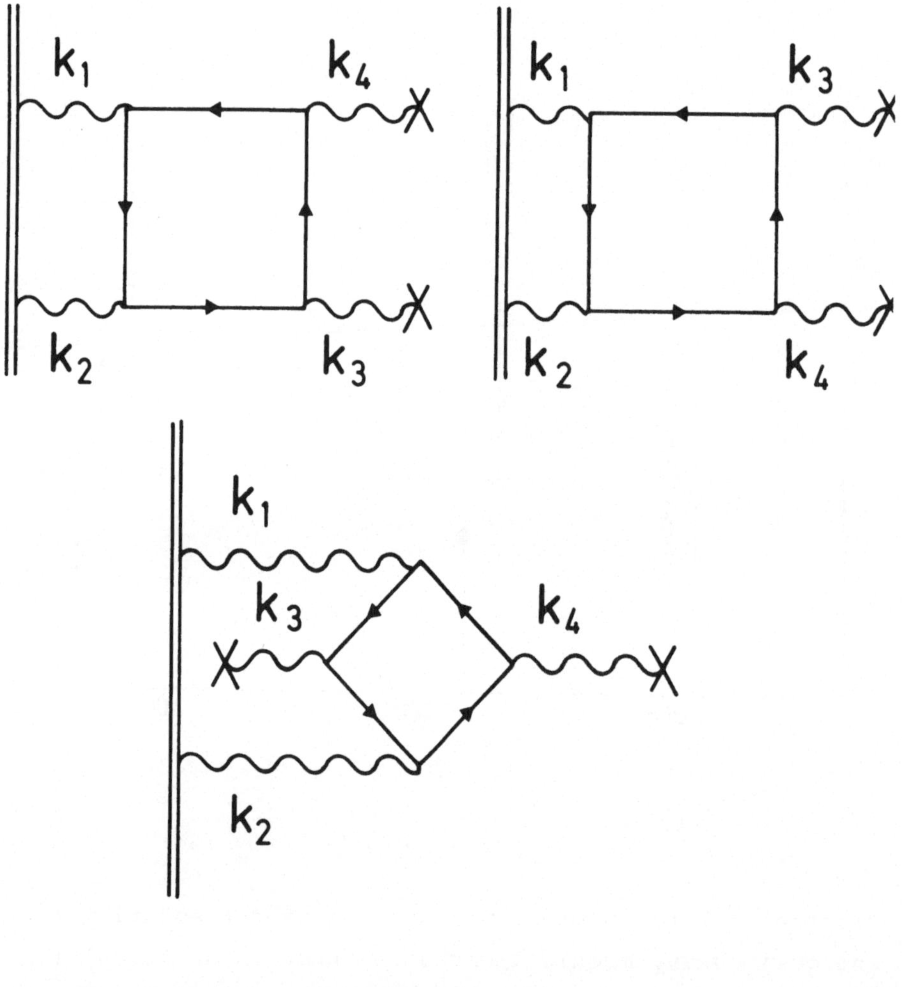

Fig. 4

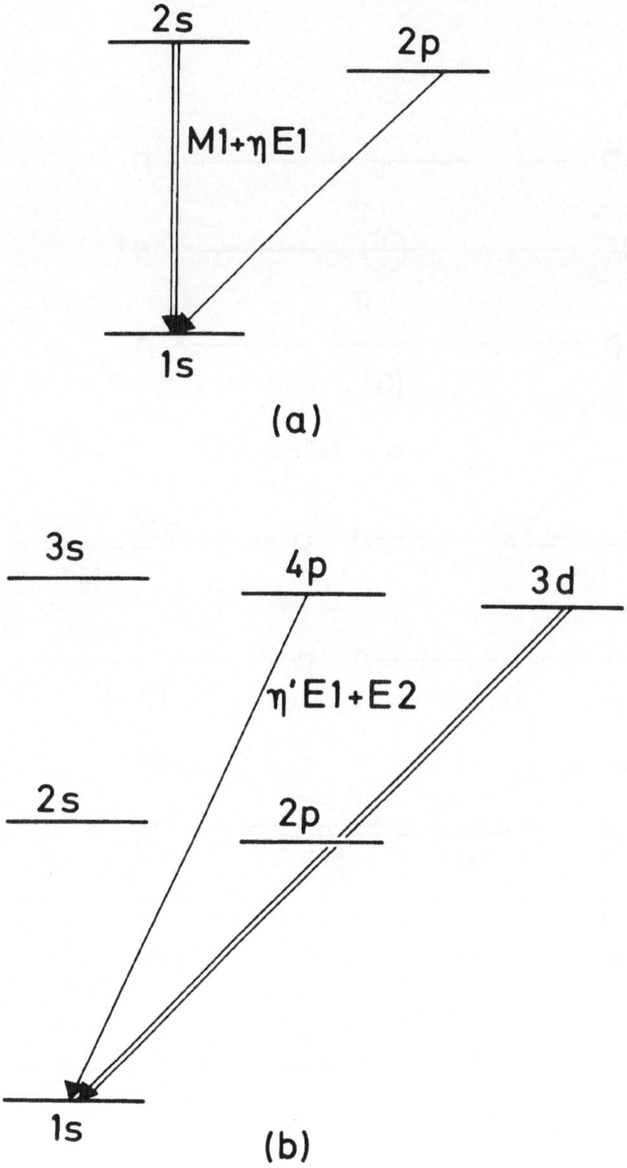

(a)

(b)

Fig. 5

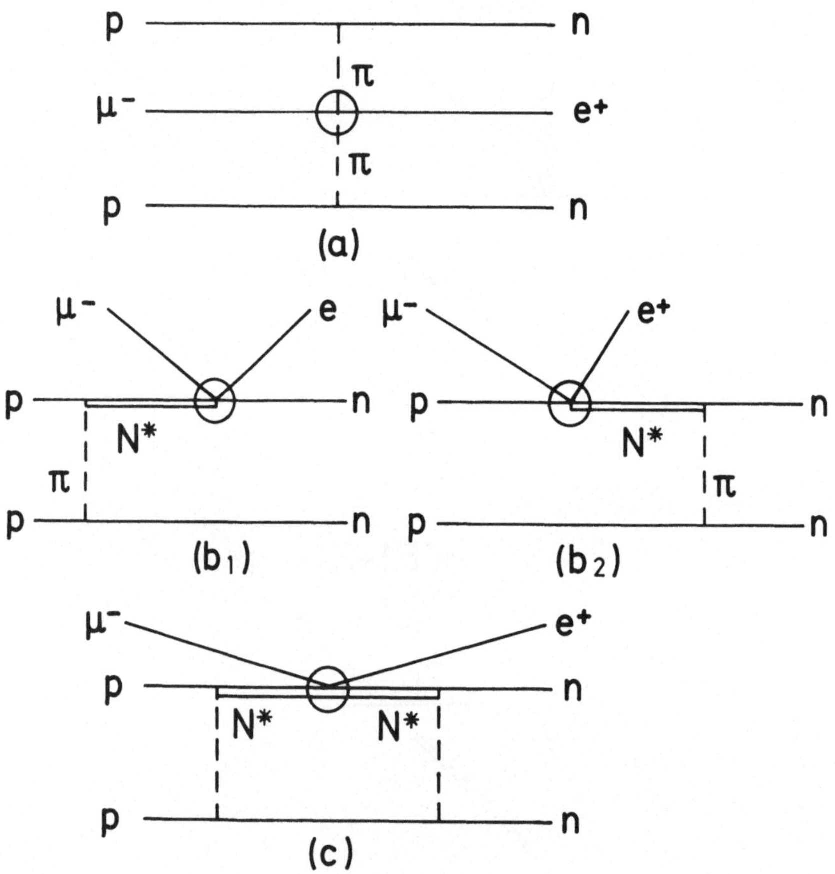

Fig. 6

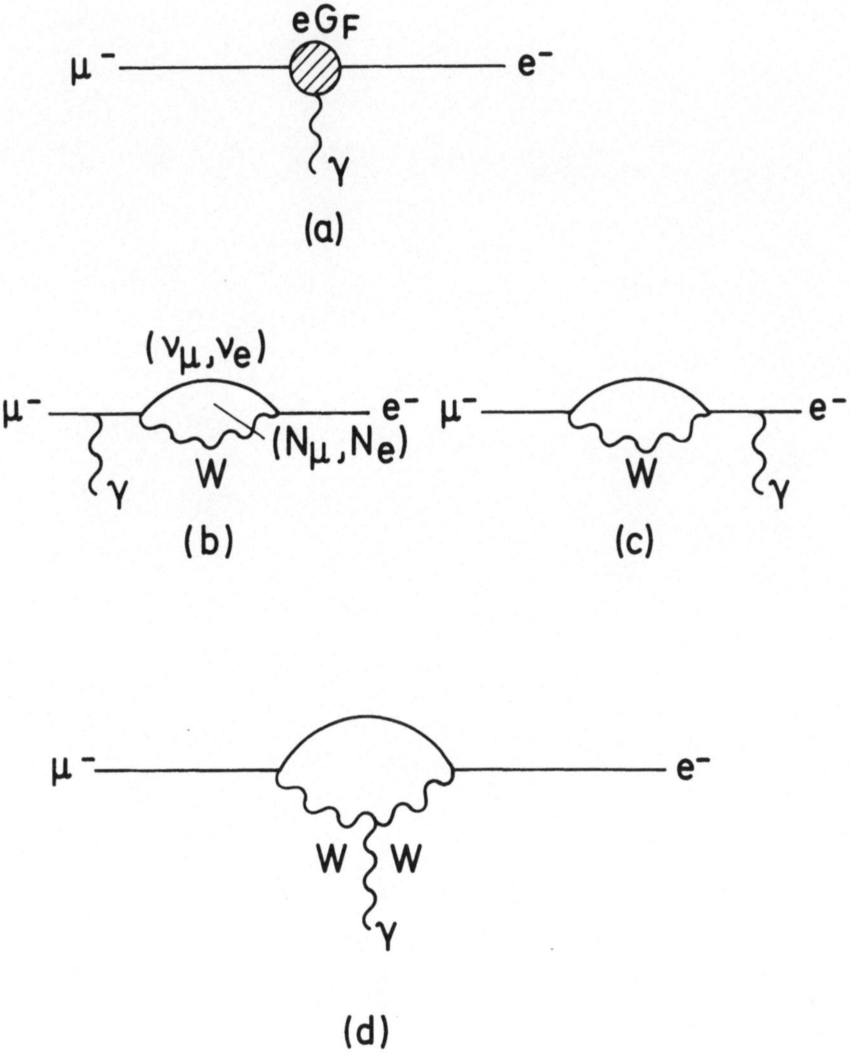

Fig. 8

Acta Physica Austriaca, Suppl. XVIII, 677–726 (1977)

MEDIUM ENERGY MESON PHYSICS[+]

by

W. WEISE
Institute of Theoretical Physics
University of Regensburg, W. Germany

1. INTRODUCTION

In these lectures I would like to discuss a number of topics related to the interaction of meson fields with the nuclear many-body system. I shall mostly concentrate on pions, since high quality pion beams are now becoming available at the meson factories, so that "pion optics", the scattering of pions from nuclear targets at various frequencies, is starting to become a subject of systematic investigations.

Inseparably connected with the pion-nuclear interaction is the dynamics of baryon resonances inside the nucleus. The $\Delta(1232)$ plays a particularly important role,

---

[+] Lecture given at XVI. Internationale Universitätswochen für Kernphysik, Schladming, Austria, February 24–March 5, 1977.

and our attempts to understand medium energy pion physics are in large part depending on a clarification of the way in which a Δ-isobar interacts with nucleons. (Similarly, kaons can be used to implant hyperons and hyperon resonances in nuclei. A few comments in this direction will follow in the last section of this survey.) The pion plays an interesting double role. Not only can it be employed as an external probing field, but it also generates the nuclear force. Once inside a nuclear environment, it does not make sense to distinguish between the pion as a probing particle and the pion as a generator of nuclear correlations.

Such aspects are best dealt with in a treatment of the pion-nuclear interaction as a true many-body response problem. Within such a framework, connections can be drawn between pion scattering (for pion frequencies $\omega \geq m_\pi = 140$ MeV) and situations where virtual pions (in particular, static pions with $\omega = 0$) are involved. In the latter case one studies the nuclear response function, in channels carrying pion quantum numbers, at low excitation energies. This is the domain of the Landau-Migdal (Fermi liquid) theory of nuclei; one of the Landau parameters in this theory, the one which essentially describes the static particle-hole interaction in isovector states of unnatural parity ($J^\pi = 0^-, 1^+, 2^- \ldots$) is strongly related to what one calls the Ericson-Ericson-Lorentz-Lorenz (EELL) correction in pion scattering. The same paramter is relevant to weak processes and plays an important role in the renormalisation of the axial vector coupling constant g_A in nuclei.

Another topic which has aroused considerable interest recently is pion condensation, i.e. the spontaneous formation of a real pion field in the ground state of dense nuclear or neutron matter. This is possibly a field of astrophysical interest. Although this leads somewhat outside the scope of laboratory medium energy physics, it is the opinion here that an understanding of meson dynamics in the nuclear many-body system will not be complete without extrapolations into regions with physically extreme conditions (like those present in neutron stars).

Before approaching the pion-nuclear many-body problem, it is essential to develop a model of the elementary pion-nucleon interaction both on- and off-shell. As a testing ground for off-shell properties, we shall use pion absorption in a simple system like the deuteron. The organisation of this paper will then be as follows:

a) Pion Absorption
b) Pion Scattering
c) Pion Condensation.

The main purpose is to outline a common dynamical basis for the description of these different subjects.

2. PION ABSORPTION

Pion absorption in a nucleus involves mainly more than one nucleon. This is so for simple kinematical reasons: at threshold, for example, 140 MeV are released without any momentum transfer. It needs (at least) two nucleons to add large momenta such that they sum up to zero. Pionic disintegration of the deuteron is particularly interesting because the two-nucleon wave functions are supposed to be quite well known so that one can hope to study in some detail mechanisms which take place in the close neighbourhood of the absorbed pion. Before that, we have to recall a few basic facts.

2.1 Pion-Nucleon Interaction

Medium energy pion-nucleon scattering is dominated by s- and p-wave pion interactions. S-wave scattering can be thought of as proceeding mainly through t-channel exchange of a ρ-meson, as shown in Fig. 1a. Ohter pieces, like the exchange of two pions coupled to isospin T = 0 (Fig. 1b), and pair ($N\bar{N}$)-exchange in the u-channel (Fig. 1c) tend to cancel (at least for on-shell pions at threshold) (see Ref.[1]).

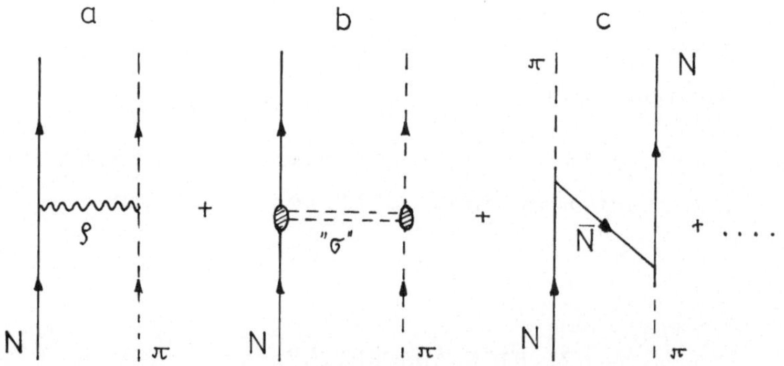

Fig. 1

Relevant pieces of s-wave pion-nucleon scattering.

P-wave pion-nucleon scattering involves predominantly nucleon and Δ(1232) intermediate states, as shown in Fig. 2.

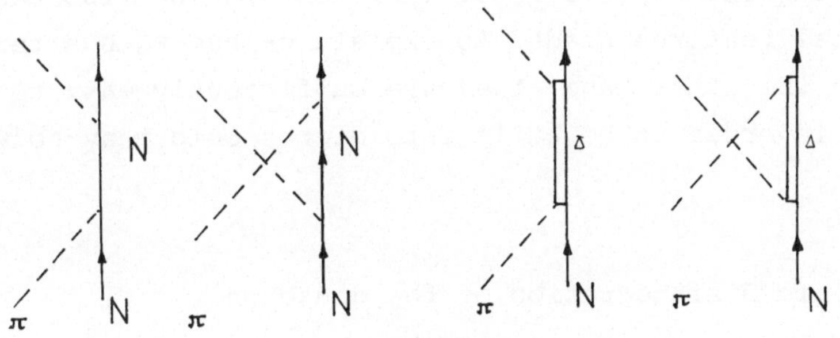

Fig. 2

The K-matrix for P-wave pion-nucleon scattering through nucleon and Δ-isobar intermediate states.

Such a description, which introduces a separate baryonic species (the Δ(1232)), can be shown to yield a scattering amplitude equivalent to Chew-Low theory [2],[3].

The basic ingredients are effective Lagrangians which couple pions to nucleons or Δ-isobars. In the non-relativistic limit, we write

$$\langle N'|L_{\pi NN}|N\pi^{\lambda}(q)\rangle = i\,\frac{f(q^2)}{m_\pi}\,\langle N'|\vec{\sigma}\cdot\vec{q}\,\tau_\lambda|N\rangle \,,$$

$$\langle \Delta|L_{\pi N\Delta}|N\pi^{\lambda}(q)\rangle = i\,\frac{f^{*}(q^2)}{m_\pi}\,\langle \Delta|\vec{S}\cdot\vec{q}\;T_\lambda|N\rangle \,, \qquad (1)$$

where the pion has charge λ and 4-momentum $q = (\omega,\vec{q})$. The quantities $f(q^2)$ and $f^{*}(q^2)$ are πNN and $\pi N\Delta$ form-factors, normalized such that $f^2(q^2 = m_\pi^2)/4\pi = 0.08$ and $f^{*}(m_\pi^2) = 2f(m_\pi^2)$ (according to Chew-Low theory). The $\vec{S}$ and $\underset{\sim}{T}$ are transition spin- and isospin operators connecting spin-isospin 1/2- and 3/2-states, respectively.

These effective Lagrangians will be the basis for all subsequent applications. They contain the important dynamical features needed to explain p-wave πN scattering. On the other hand, they are sufficiently easy to handle in order to be built into microscopic many-body schemes.

2.2 Pionic Disintegration of the deuteron

We would now like to investigate the reaction $\pi^+ d \to pp$, which is of fundamental importance for the understanding of pion absorption and production in more complex nuclei. The process is illustrated in Fig. 3.

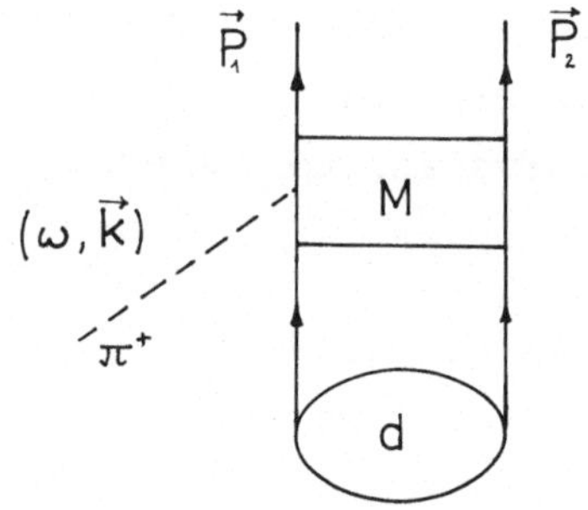

Fig. 3

Schematic picture of $\pi^+ d \to pp$. The M-box is supposed to contain mechanisms irreducibly linked to the pion absorption vertex. Final state correlations of the two outgoing protons are not shown here.

We note that the relative momentum of the two outgoing nucleons, $\vec{p} = \frac{1}{2}(\vec{p}_1 - \vec{p}_2)$, is approximately equal to $p \cong \sqrt{M\omega}$ (M is the nucleon mass) and reaches from 360 MeV/c at threshold to more than 500 MeV/c in the

Δ-resonance region. Therefore, this process is sensitive to short ranged mechanisms in the "M-box" of Fig. 3.

We can imagine what is contained in this "M-box" in lowest order (sse Fig.4).

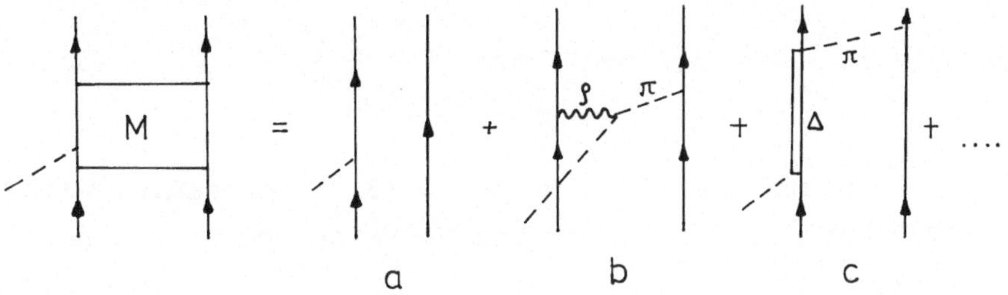

Fig. 4

Lowest order processes contributing to πd-absorption:

a) impulse approximation,

b) s-wave rescattering,

c) p-wave rescattering through Δ-isobar.

Let us first discuss the impulse approximation, Fig.4(a). The corresponding matrix element is simply

$$M_a = \langle pp | L_{\pi NN}(1) | \pi^+(k) d \rangle + (1 \leftrightarrow 2) \quad . \tag{2}$$

From Fig. 6 we observe that the $\pi^+ d \to pp$ cross section calculated with this one-body absorption mechanism has almost nothing in common with the experimental data+.

+ Relativistic corrections of order (ω/M), which are of the form $\langle N'(p') | \delta L_{\pi NN} | N(p) \pi^\lambda(k) \rangle = -i(\omega/4M) \langle N' | \vec{\sigma} \cdot (\vec{p}\,' + \vec{p}) | N \rangle$, have very little effect on this result.

These show a clear indication of the Δ-resonance, so that it is obvious that a rescattering process like Fig. 4(c) must be important. The corresponding matrix element is

$$M_c=\int\frac{d^4q}{(2\pi)^4}\langle pp|V_{12}^{\pi}(q)\frac{1}{\omega_{\Delta}-\omega-\frac{i}{2}\Gamma(\omega)}L_{\pi N\Delta}(1)|\pi^+(k)d\rangle+(1\leftrightarrow 2)\,. \quad (3)$$

Here $\omega_{\Delta} \simeq M_{\Delta} - M+\vec{k}^2/(2(M+\omega))$ and $\Gamma \simeq \frac{2}{3}\frac{f^*}{4\pi}\frac{k^3}{m_{\pi}^2}$, (3a)

the position and the width of the Δ-(3.3)-resonance, with M_{Δ} = 1232 MeV, and $V_{12}^{\pi}(q)$ is the NΔ-NN one-pion exchange interaction,

$$V_{12}^{\pi}(q) = \frac{f^*(q^2)f(q^2)}{m_{\pi}^2}\;\frac{(\vec{S}_1\cdot\vec{q})(\vec{\sigma}_2\cdot\vec{q})}{q^2 - m_{\pi}^2}\,\tau_1\cdot\tau_2 \;. \quad (4)$$

The transfered four-momentum squared,

$$q^2 = q_o^2 - \vec{q}^2 ,$$

is far away from the pion mass shell, large and negative. Consequently, the structure of the formfactors $f(q^2)$ and $f^*(q^2)$ becomes very important. One of the motivations to study this process has in fact been to extract information about these formfactors (see Refs.[4],[5]). We can think of them in terms of monopole forms,

$$f(q^2) = f(m_{\pi}^2)\,\frac{\Lambda_{\pi}^2 - m_{\pi}^2}{\Lambda_{\pi}^2 - q^2} , \quad (5)$$

and a similar expression for $f^*(q^2)$. Let us first take $\Lambda_\pi = \infty$ (zero range vertices) and calculate again the $\pi^+d \to pp$ cross section using both M_a and M_c. The cross section is now strongly overestimated. A finite cutoff Λ_π would bring it down, but at the expense of unreasonably small values of Λ_π (Ref. 5). It has been pointed out that ρ-meson rescattering according to Fig. 5 plays an important role in cutting down the cross section to the correct order of magnitude [6],[7].

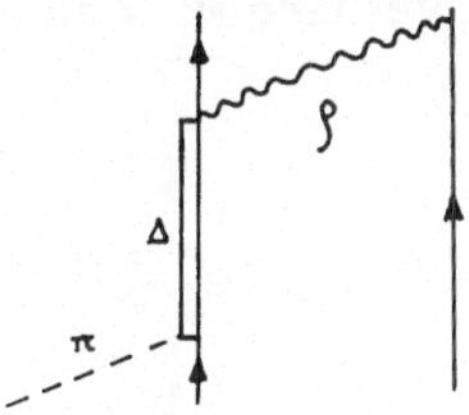

Fig. 5

Two-body absorption mechanism through rescattering of a ρ meson following Δ-isobar excitation.

To V_{12}^π of Eq. (3), we have to add the ρ meson exchange interaction in the $N\Delta \to NN$ transition channel,

$$V_{12}^\rho(q) = \frac{f_\rho^*(q^2)\,f_\rho(q^2)}{m_\rho^2}\,\frac{(\vec{S}_1\times\vec{q})\cdot(\vec{\sigma}_2\times\vec{q})}{q^2-m_\rho^2}\,(\underset{\sim}{T}_1\cdot\underset{\sim}{\tau}_2)\ . \qquad (6)$$

The tensor part of Eq. (6) has opposite sign as compared to the tensor part of one pion exchange, so as to act as a natural cutoff at short distance (the tensor force components are by far the dominant pieces at the large momentum transfers which we encounter here).

Parametrisations of the tensor coupling ρNN vertex,

$$f_\rho(q^2) = g_\rho \frac{m_\rho}{2M} (1+K) \frac{\Lambda_\rho^2 - m_\rho^2}{\Lambda_\rho^2 - q^2} , \qquad (7)$$

use $g_\rho^2/4\pi \simeq 0.5$ and K between 3.7 and 6.6, together with a probably large cutoff mass ($\Lambda_\rho \sim 2$ GeV). The most recent analysis [8] favours a strong coupling with K = 6.6. For the $\rho N\Delta$ vertex we can use the quark model relation,

$$f_\rho^* = \frac{6}{5} \sqrt{2}\, f_\rho$$

as a guide line.

The results [7] of Fig. 6 show the influence of the ρ exchange ΔN-interaction on the πd disintegration cross section, first in the limit of zero range π- and ρ-baryon vertices ($\Lambda_\pi = \Lambda_\rho = \infty$). Obviously, cutoffs of rather short range are now able to do the rest. In Fig. 7, such cutoffs have been added. The threshold behaviour is reproduced by including the s-wave pion rescattering process shown in Fig. 4a.

Introduction of a finite width mass distribution of the ρ-meson leads to only minor changes of these results [7]. Also, higher order rescattering effects, which can be treated to some extent in a coupled channels approach [8], do not seem to imply major modifications.

Since the pionic disintegration of the deuteron is obviously a sensitive probe of pion and Δ-isobar interactions, and since we are able to reproduce it with relatively simple effective Lagrangians, we are encouraged to use these also in more complicated situations. In fact, the parameters of the relevant $\pi N\Delta$ and $\rho N\Delta$ Lagrangians can all be determined (at least in principle) from other independent sources. The coupling constants and, in particular, the cutoff masses used here are in accordance with values appearing in the recent lieterature.

REFERENCES FOR SECTION 2:

1. J. Hamilton, Pion-Nucleon Interactions, in High Energy Physics, Vol. I (E.M.S. Burhop; ed.), Academic Press, N. Y. (1967), p. 193.
2. S. Barshay, G.E. Brown and M. Rho, Phys. Rev. Lett. 32 (1974) 787.
3. G. E. Brown and W. Weise, Phys. Reports 22 C (1975) 279.
4. C. Lazard, J.I. Ballot and F. Becker, Nuovo Cim. 65 B (1970) 117.
5. B. Goplen, W. Gibbs and E. Lomon, Phys. Rev. Lett. 32 (1974) 1012.
6. D. O. Riska, M. Brack and W. Weise, Phys. Lett. 61 B (1976) 41.
7. M. Brack, W. Weise and D. O. Riska, to be submitted to Nucl. Phys.
8. G. Höhler and E. Pietarinen, Nucl. Phys. B 95 (1975) 21o.
9. A. M. Green and J. A. Niskanen, Nucl. Phys. A 271 (1976) 503.

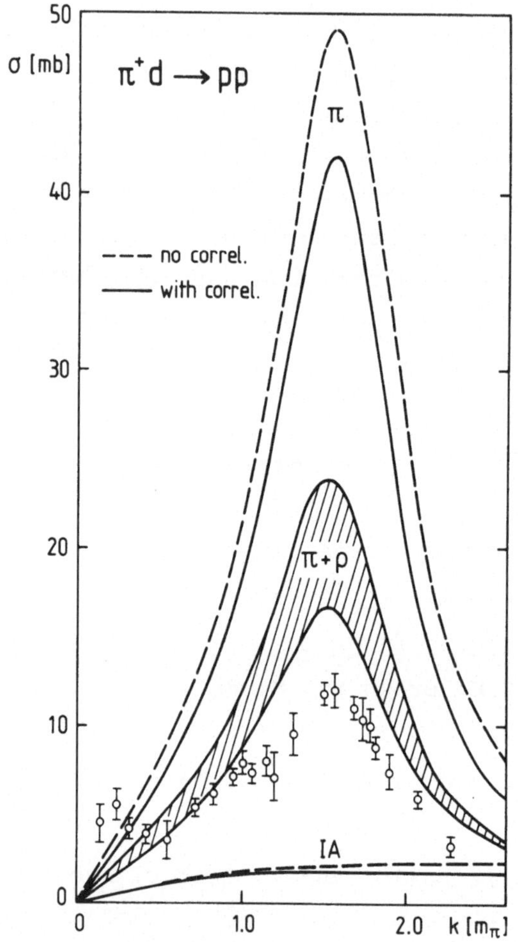

Fig. 6

Contributions of various one- and two-body absorption mechanisms to $\pi^+d \to pp$ as a function of incoming pion momentum k. (IA): Impulse approximation as shown in Fig.4a; (π): p-wave pion rescattering with IA added according to Fig. 4c, with zero range vertices ($\Lambda_\pi = \infty$). Dashed curves: plane outgoing waves; solid curves: final state wave function from Reid soft core potential.

$(\pi + \rho)$: ρ meson rescattering added as shown in Fig. 5. Upper and lower boundaries correspond to K = 3.7 and K = 6.6, respectively, together with $g_\rho^2/4\pi = 0.52$. ($\Lambda_\rho = \infty$ has been used here.) The deuteron wave function has been calculated with a Reid soft core potential (taken from Ref. [7]).

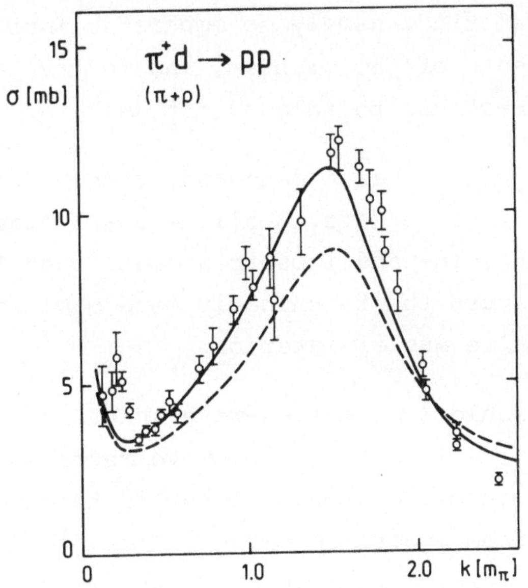

Fig. 7

"Best fit" to $\pi^+d \to pp$ cross section using s- and p-wave pion rescattering plus ρ meson exchange (also including impulse approximation); solid curve: K = 5.0, Λ_π = 1.2 GeV, Λ_ρ = 1.8 GeV. Dashed curve: K = 6.6, $\Lambda_\pi = \Lambda_\rho$ = 1.4 GeV. Otherwise same as Fig. 6.

3. PION SCATTERING

Pion elastic scattering from nuclei has been subject of intensive investigations in the recent literature. We refer to Refs.[1],[2] for reviews. We do not wish to reiterate here the various approaches which have been used to obtain more or less satisfactory fits to the existing data. Our purpose is more of a conceptual kind, namely to emphasise important many-body aspects of the problem, and to try to establish connections to related fields.

Pion scattering as a many-body response problem has been described in Refs.[3-5]; we shall start with pion optics in spin- and isospin symmetric nuclear matter and discuss the technically more complicated problem of finite nuclei later on.

At threshold ($\omega = m_\pi$), s-wave interactions should be important. The lowest order s-wave response of a nucleus to an incoming pion wave should receive its main contribution from a Hartree potential of the form shown in Fig. 8. We see immediately that, because of the isovector nature of the ρ-meson, such an interaction vanishes upon summation over an equal number of protons and neutrons. In fact, the s-wave part of the response in an N = Z nucleus comes from higher order processes where nuclear correlations are involved.

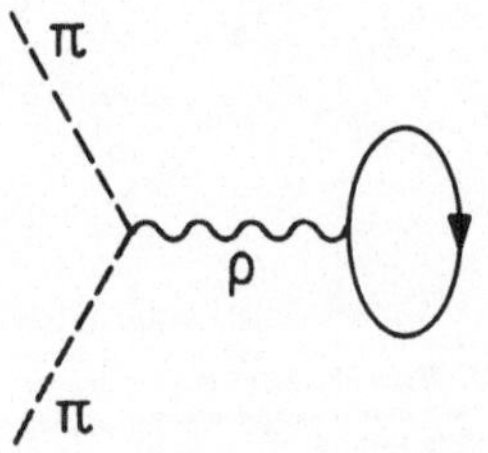

Fig. 8

Lowest order s-wave pion-nucleus interaction through ρ-exchange

We wish to focus on p-wave interactions, since these dominate as one moves away from threshold and approaches the region of the Δ_{33}-resonance.

3.1 P-wave Pion-Nucleus Interactions

The lowest order p-wave pion self-energy $\Pi^{(0)}(\omega,k)$ for a pion of four-momentum $q = (\omega,\vec{k})$ in nuclear matter can be represented as shown in Fig. 9, proceeding through nucleon-hole and Δ-isobar-hole intermediate states.

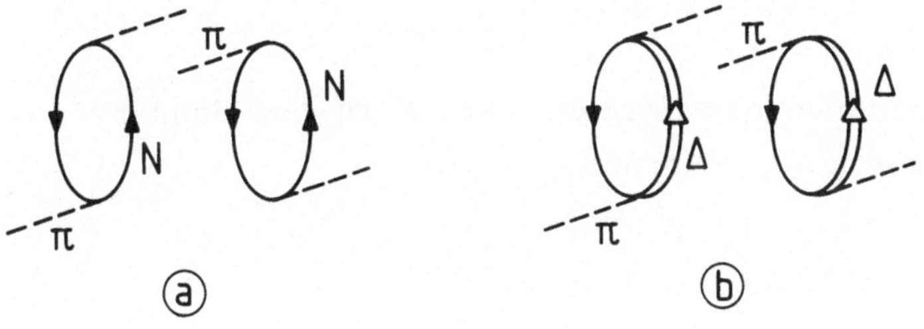

Fig. 9

Lowest order p-wave pion self-energy: (a) direct and crossed nucleon Born terms; (b) Δ-isobar-hole excitation, as obtained from Fig. 2 by "closing the nucleon loop".

Generally, the pion self-energy is related to the optical potential, V_{opt}, by

$$\Pi = 2\omega\, V_{opt} . \tag{8}$$

More precisely, we can write

$$\Pi^{(o)}(\omega,k) = -k^2\left[\frac{f^2(q^2)}{m_\pi^2} U_N^{(o)}(\omega,k) + \frac{f^{*2}(q^2)}{m_\pi^2} U_\Delta^{(o)}(\omega,k)\right], \tag{9}$$

where $U_N(o)$ and $U_\Delta^{(o)}$ are the relevant particle-hole Green's functions ("Lindhard functions"). For example, for pion frequencies much larger than the Fermi energy, $\omega >> k_F^2/2M$, we have

$$U_N^{(o)} \simeq \frac{\rho}{k^2/2M-\omega} + \frac{\rho}{k^2/2M+\omega} \tag{10}$$

for symmetric nuclear matter. Here ρ is the density and M the nucleon mass. Furthermore,

$$U_\Delta^{(o)} \simeq \frac{4}{9}\left[\frac{\rho}{\omega_\Delta - \omega} + \frac{\rho}{\omega_\Delta + \omega}\right] \tag{11}$$

where ω_Δ is given by eq. (3a). Note that in the static limit ($M \to \infty$), p-wave pion-nuclear scattering from a spin-saturated system with equal number of protons and neutrons goes completely through Δ-isobars.

The physical picture introduced in Fig. 9 is useful for at least three reasons:

a) it starts from microscopic effective π-baryon Lagrangians which can be used off - as well as on-shell.

b) it is possible to make contact with the Fermi-liquid theory of nuclei. In fact, for later purposes we note that at $\omega = 0$ and $k \to 0$ (the so-called Landau limit), we have

$$U_N^{(o)}(\omega = o,\ k \to 0) = N_o = \frac{2Mk_F}{\pi^2}\ , \tag{12}$$

$$U_\Delta^{(o)}(\omega = o,\ k \to o) = \frac{8}{9}\ \frac{\rho}{M_\Delta - M}\ . \tag{13}$$

(Here N_o is the density of states at the Fermi surface, k_F is the Fermi momentum. M is then usually interpreted as an effective mass.)

c) it provides a basis for a systematic incorporation of many-body corrections.

Such corrections appear in two ways in this scheme, as indicated in Fig. 10.

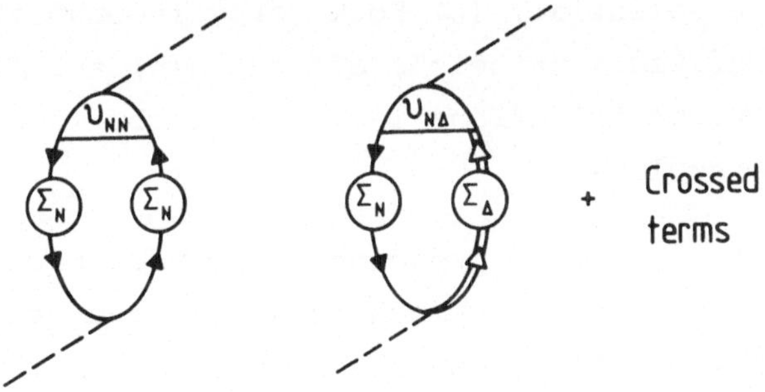

Fig. 10

Corrections to the lowest order pion optical potential: vertex corrections (V) and self-energy corrections (Σ) of particles and holes.

We shall first discuss self-energy interactions of nucleons and Δ-isobars, represented by Σ in Fig. 10. The nucleon-hole feels the binding potential of the surrounding nucleus (Σ_N). The Δ-isobar, once generated, can be thought of as moving in a complex optical potential (Σ_Δ). We illustrate this in more detail in Fig. 11.

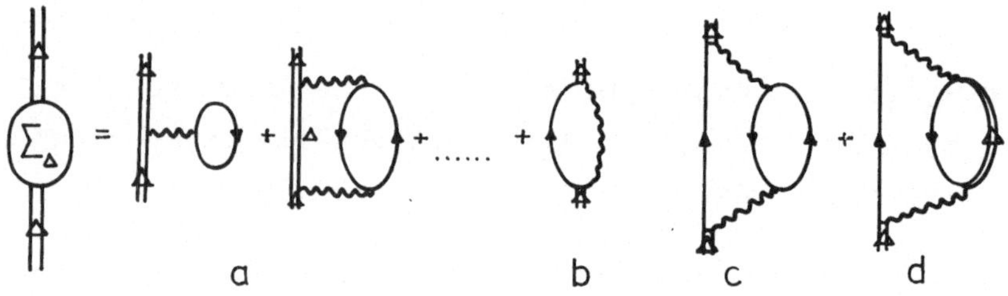

Fig. 11

Self-energy interaction of a Δ-isobar in a nucleus: (a) Hartree potential; (b) Fock term, incorporating lowest order Pauli principle effects; (c) and (d) higher order contributions involving isobar decay inside the nucleus.

The first terms (a) are responsible for isobar binding in the nucleus. Not very much is known about such binding effects, but on the basis of symmetry schemes, like the quark model, one would expect the isobar to move in a single particle potential similar to that for nucleons [5]. Term (b) includes the decay $\Delta \rightarrow N\pi$ inside the medium. For the nucleon, the phase space for decay is reduced by the Pauli principle; consequently, the decay width inside the nucleus is reduced by this agency. On the other hand, this reduction is overcompensated by the decay of the Δ into more complex many-body channels. For example, Fig. 11 c includes the decay channel $\Delta N \rightarrow NN$, the one which appeared to be most important in the pionic disintegration of the deuteron discussed in the last section.

Associated with these absorption channels are dispersive real parts which contribute to the shift of the Δ-isobar mass. Calculation of these shifts requires a careful treatment of pion and ρ-meson exchange ΔN interactions [5], [6]. The effect of self-energy interactions is a modification of, e.g., the isobar Lindhard function of eq. (11)

$$U_\Delta^{(o)} \rightarrow U_\Delta = \frac{4}{9} \frac{\rho}{\omega_\Delta + \Sigma_\Delta - \Sigma_N - \omega} + \text{crossed term}, \qquad (14)$$

where $\mathrm{Re}(\Sigma_\Delta - \Sigma_N)$ gives an energy dependent shift of the isobar mass. This shift is probably rather moderate, of the order of 10 MeV at $\omega \sim 2m_\pi$. The effective width of the isobar inside the nucleus is given by

$$\Gamma_\Delta = - 2 \,\mathrm{Im}\, \Sigma_\Delta$$

and is expected to be considerably larger than the free width, eq. (3a), as a consequence of the strong absorption channel $\Delta N \rightarrow NN$ and other collision broadening mechanisms.

A collection of many-body vertex corrections is shown in Fig. 12, including both Pauli and dynamical (short range) correlations.

Such vertex corrections have been studied in some detail in the literature. The most important pieces are those associated with particle-hole interactions other than one-pion exchange, shown in Fig. 12c. They are partly due to the fact that while a pion travels between two baryons, a short range repulsive core keeps

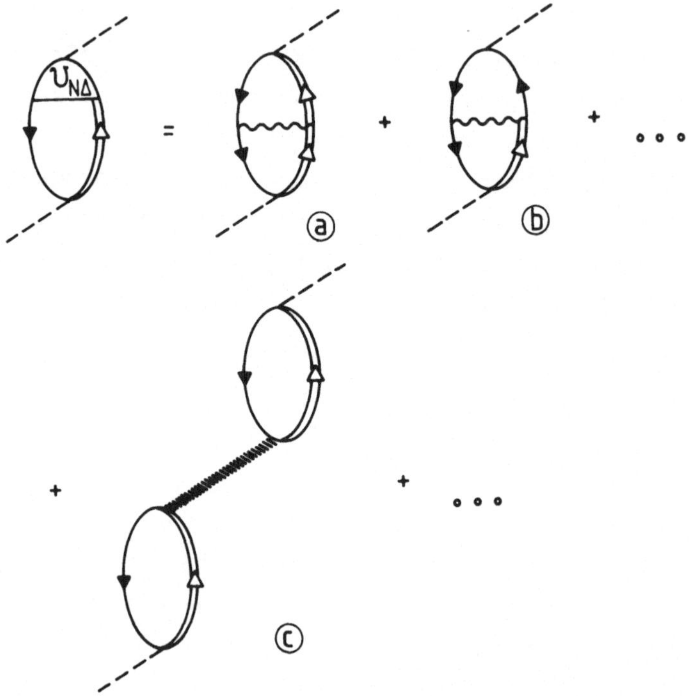

Fig. 12

Examples of vertex corrections to the pion self-energy. The shaded area in (c) is the full isobar-hole interaction in the relevant channel, minus the one-pion exchange, which is generated upon iteration of the self-energy in the wave equation.

them at a minimum distance. This is the source of the Ericson-Ericson-Lorentz-Lorenz (EELL) correction [7]. Another part comes from exchange of ρ mesons together with additional short range correlations [8]. Type (b) of Fig. 12 represents a class of Pauli corrections [9], which are important for static pions (ω = o), but enter less significantly in pion scattering problems. Finally, corrections of the type shown in Fig. 12 (a) are comparatively small [10].

We wish to speak about the combined effect in terms of a generalised EELL-correction. Since the underlying mechanisms are mainly short ranged (dominantly determined by the inverse mass of a vector meson), they can be approximated by a Fermi liquid parameter g', almost independent of frequency ω and momentum k. The fact that only one single g' governs both nucleon- and isobar-hole interactions is an assumption based on the validity of quark model schemes. In such a model short range repulsive correlations between all (non-strange) baryons will be the same.

Within that framework, the p-wave pion self-energy of eq.(9) now turns into[+]

$$\Pi(\omega,k)=2\omega V_{opt}(\omega,k)=-\frac{k^2}{m_\pi^2}\,\frac{f^2(q^2)U_N(\omega,k)+f^{*2}(q^2)U_\Delta(\omega,k)}{1+g'[(f^2/m_\pi^2)U_N+(f^{*2}/m_\pi^2)U_\Delta]}\,. \quad (15)$$

(Note that self-energy corrections are now supposed to be included in U_N and U_Δ). The magnitude of g' (the EELL correction) has been subject of recent activites. The "classical" value of Ref. [7] corresponds to g' = 1/3, which results from the action of a zero range hard core while a pion with zero range π-baryon interaction travels between two baryons. Recent developments have led to the conclusion that

---

[+] Here $f \equiv f(m_\pi^2)$, $f^* \equiv f^*(m_\pi^2)$.

1) finite range π-baryon vertices reduce g' to only a small fraction of 1/3, if only pions are considered [11];

2) most of g' comes from ρ-meson exchange combined with short range correlations generated mainly by ω-meson exchange [8],[10].

3) A lower limit g' $\gtrsim$ 0.5 is obtained both from a reaction matrix calculation using Reid's soft-core potential and from an investigation based on meson exchange mechanisms [12]. The precise value of g' depends sensitively on the ρ-baryon coupling constants. A g' twice as large as the "classical" EELL cannot be completely excluded [12].

3.2 Pion Elastic Scattering from Finite Nuclei

The adaption of the model discussed so far to finite nuclei is described Ref.[6]. The pion self-energy (or optical potential) now becomes non-diagonal in pion momenta, and the T-matrix is obtained from

$$(\vec{k}'|T(\omega)|\vec{k}) = \frac{1}{2\omega}(\vec{k}'|\Pi(\omega)|\vec{k}) + \int \frac{d^3q}{(2\pi)^3}(\vec{k}'|\Pi(\omega)|\vec{q})D_o(\omega,\vec{q})(\vec{q}|T(\omega)|\vec{k}), \tag{16}$$

where

$$D_o(\omega,\vec{q}) = (\omega^2 - q^2 - m_\pi^2 + i\delta)^{-1},$$

the free pion propagator. Clearly, if the full particle-hole structure of π is kept, solution of eq. (16) involves a diagonalisation in particle-hole (nucleon- and isobar hole) space [6],[13].

Lenz et al. [13] have carried out extensive calculations of π^{4}He scattering in terms of a coupling of the pion to Δ-nuclear states. From a fit to total cross section they deduce that there must be appreciable repulsion in the ΔN-interaction (besides one-pion exchange), together with a broadening of such Δ-nuclear states due to non-pionic decay mechanisms.

The source of such repulsive effects and, in particular, the role of isobar-hole interactions other than one-pion exchange is not discussed selectively in Ref. [13]. The significance of the EELL has been pointed out by Thies [14] for low energy π^{12}C scattering. We present his results in Fig. 13. Furthermore, in the (3.3) resonance region and beyond, Fig. 14 indicates that binding corrections are of importance in the neighbourhood of minima in $d\sigma/d\Omega$, and the EELL effect shows up especially at large angles.

The results Figs. 13 - 14 are both obtained within certain local approximations for the optical potential, whereas it has been pointed out [13], [6] that V_{opt} is basically a non-local quantity. We should take these results only as an indication of the relative importance of many-body effects usually not incorporated in standard "fixed scatterer" approximations.

As far as the sensitivity of pion elastic scattering to many-body effects is concerned, the following conclusions can be drawn: In the (3.3)-resonance, scattering in forward directions does not reveal anything more than scattering from a "black disk". Finer details become visible under backward angles, i.e. for large momentum transfer. For low energy scattering, the pion-nucleon

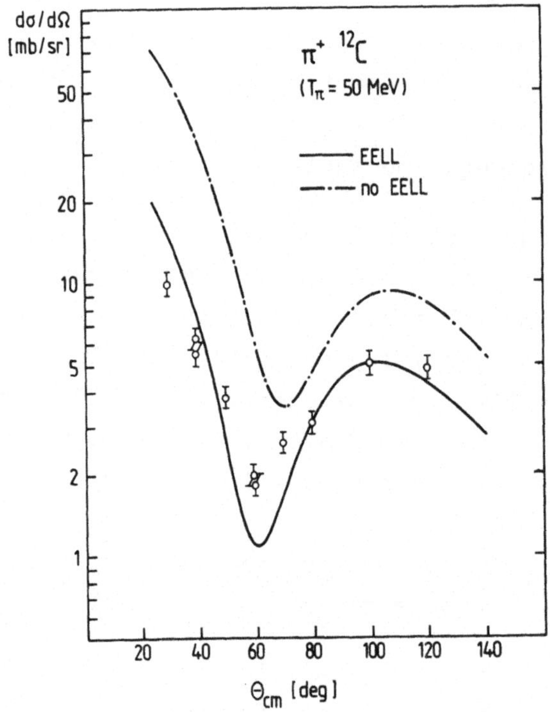

Fig. 13

Pion elastic scattering from ^{12}C at 50 MeV, calculated without (dot-dashed curve) and with inclusion of the EELL correction (g' = 0.4 has been used). Proper kinematical transformations and s-wave interactions are included in this calculation (Taken from Ref.[14]).

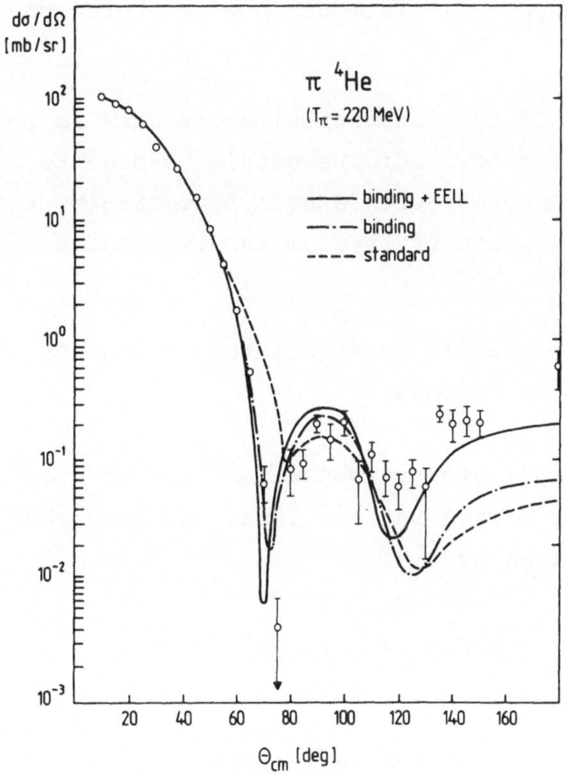

Fig. 14

Pion elastic scattering from ^4He at 220 MeV. Solid curve: EELL (g' = 2/3) and binding corrections included. Dotdashed curve: no EELL; dashed curve: neither binding corrections nor EELL taken into account. For details, see Ref.[6].

interaction is relatively weak, and the pion wave can penetrate into the interior sections of the nucleus, so that many body correlations show up there even in forward directions.

None of the theoretical approaches is truly satisfactory up to now. Efforts should be made to combine the scheme of Ref.[13] with a selective treatment of the many-body degrees of freedom as discussed above.

3.3 Relation to Landau-Migdal theory (the Landau limit of Pionic Response)

The full pion response function in nuclear matter (except for s-wave interactions, which we shall omit here) is given by

$$R(\omega,k) = \Pi(\omega,k)\ [1 + \Pi\ D_o(\omega,k)]^{-1}\ , \qquad (17)$$

in terms of the p-wave self-energy Π of eq. (15) and the free pion propagator D_o. This amplitude is illustrated in Fig. 15.

Eq. (17) can be rewritten

$$R(\omega,k) = -\frac{k^2}{m_\pi^2}[f^2(q^2)U_N(\omega,k)+f^{*2}(q^2)U_\Delta(\omega,k)](1+G'(\omega,k))^{-1}. \qquad (18)$$

Here G' is a generalisation of a Fermi liquid interaction in the sense of Landau theory. Suppose first that only

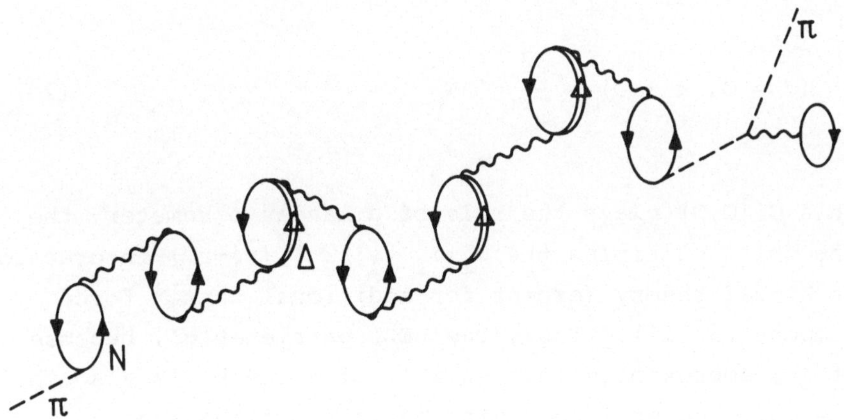

Fig. 15

Illustration of the π-nuclear response function. The wavy lines are the full (non-static) isobar-hole or nucleon-hole interactions. Not shown here are self-energy interactions of isobars and nucleons.

nucleons are present. Then, in our model

$$G'(\omega,k) = [\frac{f^2}{m_\pi^2} g' + \frac{f^2(q^2)}{m_\pi^2} g_{OPE}(\omega,k)]U_N(\omega,k) , \qquad (19)$$

where

$$g_{OPE} = \frac{k^2}{\omega^2 - k^2 - m_\pi^2} ,$$

the one-pion exchange interaction (up to a coupling constant). In the Landau limit, which corresponds to static pions (ω = o) at small momentum ($k \to 0$), we have, recalling eq. (12),

$$G' (\omega = 0, k \to 0) = \frac{f^2}{m_\pi^2} g' N_o . \quad (20)$$

This G'(0,0) plays the role of a Landau parameter, the one which multiples the $\vec{\sigma}_1 \cdot \vec{\sigma}_2 \, \tau_1 \cdot \tau_2 \, \delta^3(\vec{r}_1 - \vec{r}_2)$-interaction in Migdal theory (except for additional tensor force components [12]). Thus, the EELL parameter g', because of its approximate independence of ω and k, is also the one which governs low lying nuclear excited states carrying pion quantum numbers, e.g. isovector magnetic resonances. The influence of Δ-isobar degrees of freedom, less pronounced at $\omega = 0$, gives corrections of the order of 10% to these conclusions [12].

It is interesting to note that phenomenological determinations [15] from low energy magnetic nuclear properties give values around $g' \simeq 0.6$ or even larger (tensor force components are, however, omitted in these evaluations).

This last discussion, as far as Δ-isobar corrections at ω = o are concerned, is of importance for the renormalisation of g_A, the axial vector coupling constant in nuclei [16],[17].

REFERENCES FOR SECTION 3

1. J. Hüfner, Phys. Reports 21 C (1975) 1.

2. M.M. Sternheim and R. R. Silbar, Ann. Rev. Nucl. Sci. 24 (1974) 249.

3. S. Barshay, V. Rostokín and G. Vagradov, Nucl. Phys. B 59 (1973) 189.

4. C.B. Dover and R. H. Lemmer, Phys. Rev. C 7 (1973) 2312.

5. G.E. Brown and W. Weise, Phys. Reports 22 C (1975) 279.

6. W. Weise, Nucl. Phys. (in print).

7. M. Ericson and T.E.O. Ericson, Ann. of Phys. 36 (1966) 323.

8. G. Baym and G.E. Brown, Nucl. Phys. A 247 (1975) 395.

9. J. Delorme and M. Ericson, Phys. Lett. 60 B (1976) 451.

10. E. Oset and W. Weise, Phys. Lett. 60 B (1976) 141.

11. J. Hüfner and F. Iachello, Nucl. Phys. A 247 (1975) 441.

12. G. E. Brown, S. O. Bäckman, E. Oset and W. Weise, to be published in Nucl. Phys.

13. M. Hirata, F. Lenz and K. Yazaki, Ann. of Physics. (To be published);
    F. Lenz, Proc. Pittsburgh Conference on Meson-Nuclear Physics (1976).

14. M. Thies, Phys. Lett. 63 B (1976) 43.

15. P. Ring and J. Speth, Nucl. Phys. A 235 (1974) 315.

16. M. Rho, Nucl. Phys. A 231 (1974) 493.

17. J. Delorme, M. Ericson, A. Figureau and C. Thevenet, Ann. of Phys. 1o2 (1976).

4. PION CONDENSATION

A pion condensate is a system of interacting nucleonic matter whose ground state contains a strong pionic component. The question under which conditions (density ρ, temperature T) such a boson condensate will exist as an equilibrium configuration of strongly interacting matter, has aroused considerable interest and theoretical activity in recent times, largely stimulated by the proposals of Migdal [1] and independently by Sawyer and Scalapino [2]. In fact, Migdal has claimed that a pion condensate should already be present in ordinary finite nuclei, and this should have consequences on a number of nuclear phenomena; Saywer and collaborators revived the astrophysical interest in this question, since the equation of state for dense matter, and therefore the properties of neutron stars, should be influenced by the presence of a pion condensate.

Although there is hardly any disagreement about the existence of a pion condensate at sufficiently high baryon density and low temperature, the precise location of the threshold for pion condensation is still a point of controversy, especially in isospin-symmetric nuclear matter. For neutron matter, some progress has been made, and quantitative estimates indicate that the critical density for the phase transition is presumably not smaller than about 0.35 fm^{-3} (twice the density of normal nuclear matter).

Here I intend to give a brief survey of the main lines of thought which have developed so far [3]. I would like to describe how the critical conditions for pion condensation are arrived at, and how this continues, for the example of cold neutron matter, into an equation of state at high densities, with possible astrophysical implications. For nuclear matter and finite nuclei, the discussion still remains qualitative and is confined to simple estimates.

4.1 Threshold conditions

Consider a normal Fermi sea of nucleons (N) at low density. Such a system will essentially behave like a non-interacting Fermi gas. If this system is adiabatically compressed, then at some density it might be energetically favourable to create real pions spontaneously in the ground state, through an equilibrium reaction

$$N \leftrightarrow N' + \pi . \tag{21}$$

This will happen if there is a sufficiently large net attraction in the pion-nucleon interaction such that the total energy of the system is lowered by having a non-zero expectation value for the pion field in the ground state.

We would first like to approach the condensation threshold from low densities (assuming zero temperature, unless stated otherwise). In order to do so, we have to specify what we mean by a real pion in the nuclear medium, and we have to clarify the equilibrium condition for the process, Eq. (21). A combination of both will enable us to give a criterion for the onset of pion condensation. This has been worked out by Baym [4], and we shall give a simple illustration of his arguments. Later on, we shall consider an alternative approach through minimization of a field theoretically founded model Hamiltonian, as proposed by Baym et al. [5] and by Campbell et al. [6].

We recall that a real pion of frequency ω and momentum $\vec{k}$ in free space satisfies the energy-momentum relation

$$D_o^{-1}(\omega,\vec{k}) = \omega^2 - \vec{k}^2 - m_\pi^2 = 0 \tag{22}$$

i.e. the spectrum has two branches,

$$\omega = \pm\sqrt{\vec{k}^2 + m_\pi^2} \quad ;$$

the $D_o(\omega,k)$ is the free pion Green's function. If $D_o^{-1} \neq 0$, we call this pion virtual.

For a real pion in an extended nuclear environment, we have instead of Eq. (22):

$$D^{-1}(\omega,\vec{k};\rho) = \omega^2 - \vec{k}^2 - \Pi(\omega,\vec{k};\rho) = 0 \ , \tag{23}$$

where $\Pi(\omega,\vec{k};\rho)$ is the pion self energy[+].

By virtue of the frequency dependence of Π, the spectrum $\omega(k)$ obtained by solving eq. (23) will in general have more than just two branches. At low density,

---

[+] Equivalently, we can speak of there being a pole in the pion response function $R(\omega,\vec{k};\rho)$ of eq. (17).

two of these branches will be pion-like, i.e. they approach $\omega = \pm\sqrt{\vec{k}^2 + m_\pi^2}$ as $\rho \to 0$. Other branches will correspond to collective excitations of the system carrying the quantum numbers of a pion. (Such states are well known to exist in finite nuclei; for example, the $J^\pi = 0^-$, $T = 1$ state in ^{16}O at 13 MeV above the ground state is one of them.)

In pion scattering, a real pion is simply added from external sources. The question of pion condensation is instead whether at some density the ground state of a system of many nucleons is capable of spontaneously creating real pions. If this is going to take place as an equilibrium process as in eq. (21), then the pion frequency must obey

$$\omega = \mu_N - \mu_{N'} , \qquad (24)$$

where μ_N and $\mu_{N'}$ are nucleon chemical potentials, i.e. the energies required to remove a single nucleon from the Fermi surface. (For a non-interacting Fermi gas, $\mu_N = \varepsilon_F^{(N)}$, the Fermi energy). Thus the threshold for pion condensation is marked by the condition that eq. (23) be fulfilled at a pion frequency given by eq. (24). The lowest density for which this occurs is the critical density for the phase transition.

Note that, in normal nucleonic matter at low density, pions are not completely absent, since the system interacts through one-pion exchange. But these pions are generally virtual, i.e. they do not satisfy eq.(23).

As far as the condensate frequency ω of Eq. (24) is concerned, we observe a principal difference between a system with N = Z (symmetric nuclear matter), and N >> Z (e.g. neutron star matter). In the latter case, the favourable process to start with is

$$n \leftrightarrow p + \pi^-$$

so that

$$\omega = \mu_n - \mu_p .$$

Note that the creation of a π^- is necessarily accompanied by a particle-hole excitation carrying the quantum numbers of a π^+. For neutron matter with a small fraction (2.5 %) of protons, the difference between n and p chemical potential approaches m_π at about $\rho \simeq 2\rho_o$ [7] ($\rho_o = 0.17 fm^{-3}$, the density of ordinary nuclear matter), so that at this point the pion rest mass in eq. (23) is compensated by ω.

Conversely, in symmetric (N = Z) nuclear matter, because of isospin symmetry, equal amounts of π^o, π^+ and π^- participate, and we clearly have $\omega = 0$ (i.e. the pion field is static).

We now have to specify once again the pion-self energy, the s-wave part of which we shall simply write as

$$\Pi_s = -4\pi \{b_o(\omega,k)\rho + b_1(\omega,k)(\rho_n - \rho_p)\} \qquad (25)$$

for the example of a π^- in a nuclear medium. Here the part proportional to $\rho_n - \rho_p$ is supposed to be generated from the Hartree potential of Fig. 8, which gives $b_1 \stackrel{\sim}{=} -0.09\ m_\pi^{-1}$, independent of ω and k, whereas b_o is a small background (at least at threshold, $\omega = m_\pi$ and k = 0), where $b_o \stackrel{\sim}{=} -0.005\ m_\pi^{-1}$. We shall neglect this term. Note that Π_s is repulsive for neutron matter.

The p-wave part of the pion self-energy is given by eq. (15). For symmetric (N = Z) nuclear matter, where condensation appears at $\omega = 0$, we choose for the Lindhard functions U_N and U_Δ the forms of eqs. (12-13); for a π^- in neutron matter, where the condensate frequency is large (of the order of the pion mass), we have

$$U_N \simeq \frac{2\rho}{k^2/2M+\omega}\,, \qquad U_\Delta \simeq \frac{2}{3}\left\{\frac{1}{\omega_\Delta-\omega} + \frac{1}{3}\,\frac{1}{\omega_\Delta+\omega}\right\}\rho\ . \tag{26}$$

Note that condensation will appear at finite momentum k, so as to take advantage of the attractive p-wave interaction.

4.2 Threshold for pion condensation in neutron matter

The threshold density ρ_c for neutron matter has been calculated in Ref.[8].In table 1, the threshold parameters are listed for various situations. First, if Δ isobars are disregarded, but short range correlations (g' = 0.45 was used in ref.[8]) between nucleons are included, the critical density is far above three times nuclear matter density ρ_o. If the Δ isobar is introduced, but its short range repulsive interaction is left out, then the p-wave self energy receives sufficient attraction to pull ρ_c down to 0.5 ρ_o! But as soon as a repulsive core in the ΔN interaction is taken into account, ρ_c moves up to twice ρ_o again.

TABLE 1

Threshold parameters for π^- condensation in neutron matter. The chemical potentials of Sjöberg [7] (for a 2.5% supplement of protons to the neutron sea) have been used throughout. First row: contributions of nucleon-hole excitations only, including the full particle-hole interaction g_{NN}. Second row: isobar-hole excitations have been added, but only one-pion exchange isobar-nucleon interactions are included. Third row also takes into account isobar-nucleon correlations due to rho meson exchange and a short distance cutoff (d = 0.4 fm), as described in the text. Numbers without parentheses are obtained including s-wave pion-nucleon interactions. For comparison, numbers in parentheses show results if s-wave interaction is omitted. Critical densities are given in units of $\rho_o = 0.5\ m_\pi^3$ (nuclear matter density).

	ρ_c/ρ_o	$\omega_c[m_\pi]$	$k_c[m_\pi]$
nucleons only	>3 (1.9)	- (1.0)	- (2.5)
incl. isobars OPE only	0.5 (0.45)	0.5 (0.46)	2.4 (2.2)
incl. isobars + correlations	2.0 (1.35)	1.05 (0.87)	2.9 (2.2)

This clearly demonstrates the importance of isobars and the short range core in the baryon-baryon force. The calculation of Ref.[8] has left out a number of small effects; most of them tend to push ρ_c even higher. We shall not go into details of this, but refer to Refs. [8] and [3] for a discussion.

4.3 Equation of state for dense neutron matter

The threshold conditions eq. (23) together with (24) signalize the onset of an instability, but do not allow a description of the system in the condensed phase. Such an effort is necessary, however, to derive an equation of state for dense neutron matter, which serves as an input in discussions of neutron star models.

We shall now preceed to give a brief description of how to construct the energy density of neutron matter in the presence of a pion condensate, following a method proposed by Campbell et al. [6]. We shall not go into formal details of their model which is based on rather deep field theoretical considerations, but we hope to synthesize the main physical ideas behind this.

To start with, we note that the condensed pion field is described by three parameters: frequency ω, momentum k and, in addition, the expectation value ϕ of the field, which vanishes in the normal phase and acquires a finite value in the condensed state. We therefore wish to construct the energy density of the system in the following form:

$$E(\rho;\omega,k,\phi) = E_0(\rho) + E_\pi(\omega,k;\phi) + E_{\pi N}(\rho;\omega,k,\phi) . \quad (27)$$

Here

$$E_0(\rho) = \frac{3}{5}\frac{k_F^2}{2M}\rho + v(\rho) \quad (28)$$

is the energy density of the normal system ($v(\rho)$ is the potential energy density). In practical applications, we shall use the result of Pandharipande [9] for E_0.

The term E_π contains the energy of the condensed pion field, including contributions from $\pi\pi$ interactions. The term $E_{\pi N}$ is the contribution from interactions of the condensed pion field with the surrounding nucleons. The ground state energy density is then obtained as a function of baryon number density ρ by minimizing eq. (27) with respect to ω, k and ϕ.

To derive E_π and $E_{\pi N}$, Campbell et al. [6] have introduced a model Hamiltonian which combines P-waves, S-wave πN and $\pi\pi$ interactions on the basis of chiral symmetry (the σ model)$^+$. In the simplest version of this model, the field expectation value ϕ is measured in terms of an angle θ of a chiral rotation,

$$\phi = F_\pi\, e^{-iqx} \sin\theta \quad (29)$$

and is accompanied by a scalar field

$$\sigma = F_\pi \cos\theta \quad (30)$$

$^+$ Au and Baym [5] have shown how to relate this Hamiltonian to Weinberg's non-linear Lagrangian.

such that $\sigma^2 + \phi^2 = F_\pi^2$ is a chiral invariant. Here $F_\pi = 94.5$ MeV is the pion decay constant which in turn determines the strength of s-wave and $\pi\pi$ interactions.

For a situation in which only non-relativistic nucleons couple to the condensed pion field (no Δ isobars, no short range correlations), Baym et al. [10] succeeded to diagonalize the σ model Hamiltonian and give a closed form for the energy density, $E(\rho,\theta)$, after minimization with respect to ω and k:

$$E(\rho,\theta) = E_o(\rho) - \frac{1}{8}\,\frac{\rho^2}{F_\pi^2}\,\frac{g_A^2(g_A^2-1)\sin^2\theta}{1+(g_A^2-1)\sin^2\theta} + 2F_\pi^2 m_\pi^2 \sin^2\frac{\theta}{2} \quad ; \qquad (31)$$

here

$$g_A = 2\,F_\pi\,f/m_\pi = 1.36 \ (\text{for } f^2/4\pi = 0.08).$$

The physical origin of these terms is intuitively clear: the attractive piece is mainly governed by f^2, the p-wave πN coupling strength, whereas the third term, proportional to F_π^2, includes the net repulsion from s-wave and $\pi\pi$ interactions. The question of pion condensation is now whether the system will reduce its total energy (as compared to E_o) by acquiring a finite expectation value for the pion field (i.e. a finite angle θ in the language of the σ model). This will obviously happen only beyond a certain critical density ρ_c, which can be obtained upon minimization of E with respect to θ. The result is

$$\rho_c = \frac{2\,F_\pi^2\,m_\pi}{g_A\,\sqrt{g_A^2-1}}\,, \qquad (32)$$

and shows clearly the competition of repulsive and attractive interactions (in the numerator and denominator of eq. (32), respectively) in establishing ρ_c. In this oversimplified model, we would end up with $\rho_c \simeq 1.5\,\rho_o$.

Isobars have to be incorporated, of course, as well as short range correlations. The way to do this is described in Refs. [10],[11]. In fact, the short range repulsion simply leads to an effective reduction of g_A in eqs. (31,32)

$$g_A \rightarrow (1-g')g_A,$$

where g' is the same (approximately constant) parameter that has already appeared in the previous section. Secondly, the inclusion of isobars leads to an effective enhancement of the p-wave coupling strength, and it turns out in Ref.[11] that (to an extremely good approximation) this can be incorporated by a parameter ξ such that finally

$$g_A^* = (1 + \xi)\ (1 - g')\ g_A \tag{33}$$

is the effective p-wave interaction strength, to be inserted in eq. (31-32) instead of g_A. This g_A^* is the key parameter of the model. Determination of ξ, which depends on the $\pi N\Delta$ coupling strength, leads to $\xi \simeq 0.8 - 0.9$, almost independent of density [11].

We are now in a position to study the equation of state (at zero temperature), i.e. the pressure P as a function of baryon number density ρ:

$$P(\rho) = [\rho \frac{\partial}{\partial\rho} - 1]\ E(\rho) \ . \qquad (34)$$

A typical result is given in Fig. 16 and shows a pronounced first order phase transition around $\rho_c \sim 2\rho_o$. (Of course, everything depends crucially on g_A^*. We have chosen g' = 0.5 in this example, so that $g_A^* \cong 1.3$ results). Obviously, the system is able to get rid of a large amount of pressure by the transition into a pion condensed phase.

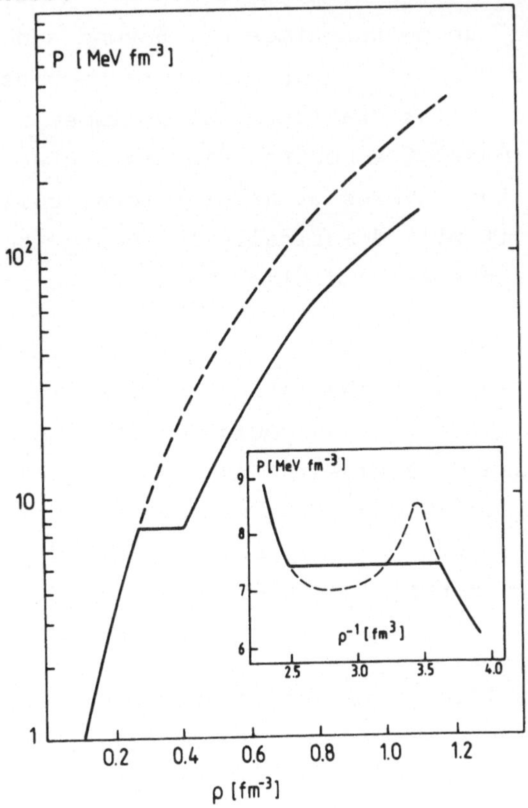

Fig. 16

Pressure P as a function of density for pion condensed neutron matter (solid line) as compared to normal neutron matter (dashed line). Insertion shows the phase transition region in more detail. (Taken from Ref.[11]).

4.4 Some astrophysical implications

The stability of dense stars is mainly a question of balance between the pressure and the gravitational force. Given the equation of state, the properties of such objects (e.g. mass and radius as a function of central density) can be obtained by solving the TOV (Tolman-Oppenheimer-Volkoff) equation of hydrostatic balance including effects of general relativity. Neutron stars are the densest possible stable configurations of matter, extending from central densities ρ of about $\rho_o = 0.17\ \mathrm{fm}^{-3}$ up to ten times ρ_o. Beyond a certain critical star mass M_{crit}, the pressure becomes insufficient to resist gravitational collapse [12],[13]. The critical mass of a neutron star is a crucial quantity for the physics of gravitational collapse. For example, it will drastically influence the ratio between neutron stars and black holes to be found in nature [13].

We can ask what the influence of a pion condensate on this critical mass would be. Since the presence of the condensate reduces the pressure in the star, M_{crit} should be lowered. This effect is shown in Fig.17. It indicates a reduction of about 10 % of M_{crit} with respect to the normal (uncondensed) star. Much stronger effects appear at intermediate densities.

Results like these are of course subject to considerable uncertainties, both in the equation of state for normal neutron matter as well as in the model describing condensation, because of the impossibility to check their validity in such extreme situations.

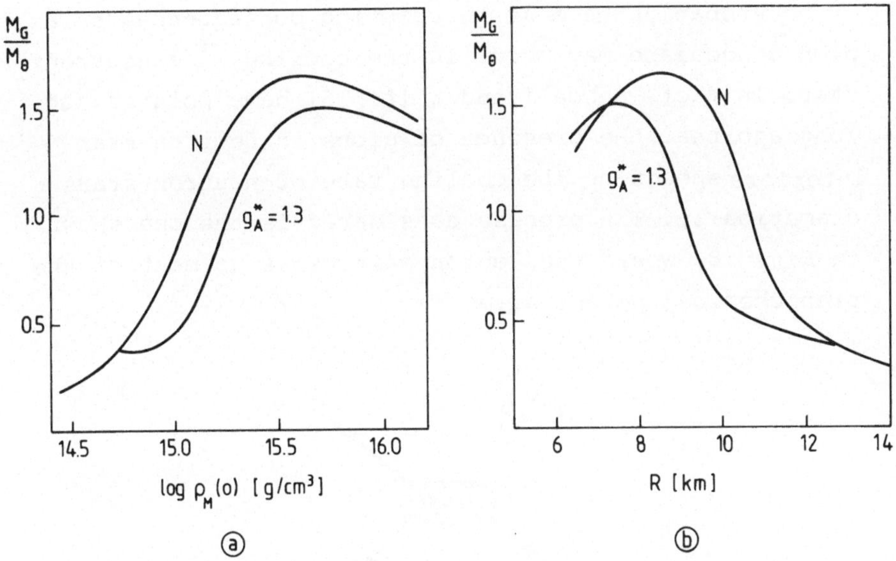

Fig. 17

(a) Total gravitational mass of neutron star (in solar units) as a function of mass density in the center of the star, calculated with the equation of state of Fig. 16, for normal (N) and pion condensed neutron matter; (b) same as function of star radius.

From the empirical point of view, neutron star data are not (yet) conclusive to rule out certain equations of state. Most reliable estimates of neutron star masses are based on the study of binary X-ray sources like Centaurus X-3 and Herculis X-1. For the neutron star in Herculis X-1, a mass of M ≃ 1.2 solar masses seems to be within the range of possibilities [13], which would not be too far away from the critical mass indicated in Fig. 17.

Probably the most interesting consequences of a pion condensate may occur in the cooling of a neutron star. In fact, Bahcall and Wolf [15] have pointed out long ago that the presence of pions in neutron star interiors speeds up the cooling rate of neutron stars dramatically. The process considered is the one shown in Fig. 18, where the lepton pair picks up most of the pion chemical potential.

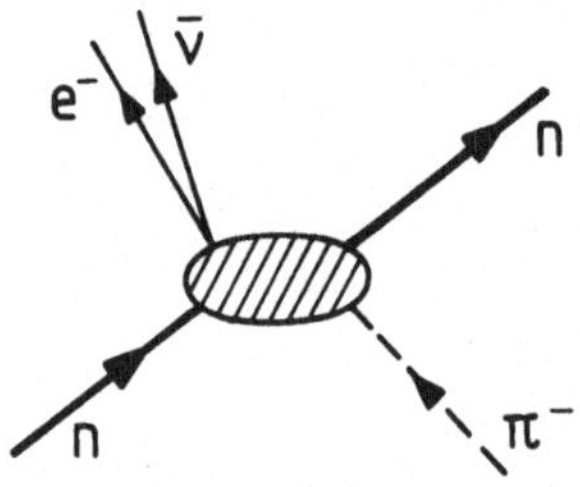

Fig. 18

Beta-decay of condensed pions as cooling mechanism for neutron stars.

Maxwell et al. [16] point out that the neutrino luminosity obtained in this way is such that a pion condensate formed at a temperature of 10^{11}K would radiate a sizeable fraction of a solar mass within a millisecond in terms of an intense burst of anti-neutrinos. This would help to explain why the Grab pulsar has cooled down to about 10^6 in a relatively short amount of time. According to Bahcall and Wolf, cooling from 10^{11}K to 10^6K takes place in a matter of days in the presence of pions, whereas the usual Urca process would need no less than 10^5 years.

4.5 Pion condensation in nuclear matter and finite nuclei

As we have mentioned, pion condensation in isospin-symmetric nuclear matter should occur at frequency $\omega = 0$, unlike neutron matter, where a finite ω could be used to cancel the rest mass. At first sight, this might seem to disfavour condensation, but on the other hand, repulsive πN s-wave interactions are absent in this case.

The determination of the critical density ρ_c depends crucially on assumptions about the coupling of virtual Δ isobars to nucleons and about the nature of short range baryon-baryon correlations, represented by the parameter g' in eq. (15). This is shown in Fig.19, where a model estimate of ρ_c as a function of g' is presented, using the threshold condition of eq. (3) in the form

$$k^2 + m_\pi^2 + \Pi(\omega = 0, k; \rho) = 0 , \tag{35}$$

together with eq. (10). The critical wave number in this model comes out to be $k_c \cong 2\, m_\pi$.

Migdal et al. [17] have estimated $\rho_c \sim 0.5\, \rho_o$, but it should be noted that the short range repulsion they use in their $N\Delta$-interaction is extremely weak, thus favouring condensation.

In a situation like this, one might ask whether pion condensation in finite nuclei would have experimentally detectable consequences. Migdal [18] discusses a number of possibilities.

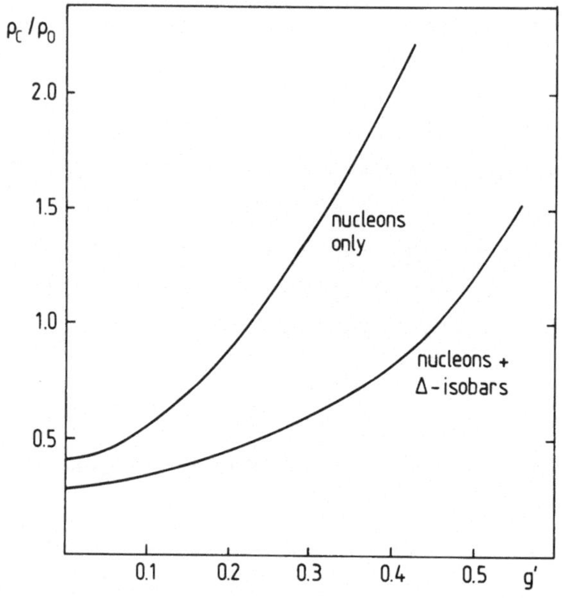

Fig. 19

Critical density (in Units if normal nuclear matter density), as a function of correlation paramter g', for symmetric nuclear matter, obtained by solving eq. (35). Included are NN and NΔ vertex factors of the form $[(\Lambda_\pi^2 - m_\pi^2)/(\Lambda_\pi^2 + k^2)]^2$ with $\Lambda_\pi = 1$ GeV; an effective nucleon mass $M^* = 0.8$ M has been used. Upper curve is calculated without isobars.

For example, he suggests that the condensate field should induce a modulation of the nuclear density of the form

$$\frac{\Delta\rho(r)}{\rho_0(r)} = \zeta^2 \frac{\sin 2k_c r}{2k_c r} ,$$

where $k_c \cong 2\, m_\pi$ is the critical wave number and ζ depends on the condensate field strength. Such a modulation would add a band of Fourier components with $q = 2\,k_c$ to the nuclear charge form factor $F(q^2)$. Experiments do not seem to give a conclusive indication about the presence of such an effect.

Brown [3] points out that, since a π condensate in a finite nucleus should lead to a collective state of pion quantum numbers degenerate with the ground state, such a phenomenon should simply be seen in nuclear spectra. At least there should be a tendency of a downward movement of e.g. 0^-, T = 1 state by residual interactions, which is not the case.

Finally, one might argue that if ordinary nuclei are not sufficiently dense to establish a strong condensate, one could look for it in medium energy heavy ion collisions where nuclear matter might be compressed up to $\rho \sim 4\,\rho_o$ and beyond in the front of a shock wave [19]. A primary question is of course whether such a compression is slow enough in order to meet the equilibrium conditions for a condensate. Furthermore, dense matter in such a situation is highly excited, with temperatures probably not much smaller than the pion mass itself. Nevertheless, the claim has been made [20] that a pion condensate might play an important role in supporting the formation of a shock wave. Such investigations would require a detailed study of the equation of state for dense nuclear matter at finite temperature.

REFERENCES FOR SECTION 4

1. A.B. Migdal ZhETF 61 (1971) 2210; JETP (Sov. Phys.) 34 (1972) 1184;

A.B. Migdal et al., ZhETF 66 (1974) 443.

2. R.F. Sawyer, Phys. Rev. Lett. 29 (1972) 382;
D.J. Scalapino, Phys. Rev. Lett. 29 (1972) 386;
R.F. Sawyer et al., Phys. Rev. D7 (1972) 953;
Rev. D7 (1973) 1580.
3. G.E. Brown and W. Weise, Phys. Reports 27 C (1976) 1.
4. G. Baym, Phys. Rev. Lett. 30 (1973) 1340.
5. C.K. Au and G. Baym, Nucl. Phys. A 236 (1974) 500.
6. D. Campbell, R. Dashen and J.R. Manassah, Phys. Rev. D 12 (1975) 979.
7. O. Sjöberg, Nucl. Phys. A 222 (1974) 161.
8. S.O. Bäckman and W. Weise, Phys. Lett. 55 B (1975) 1.
9. V. Pandharipande, Nucl. Phys. A 174 (1971) 641.
10. G. Baym et al., Phys. Lett; 58 B (1975) 304.
11. W. Weise and G. E. Brown, Phys. Lett. 58 B (1975) 300.
12. G. Baym and C. Pethick, "Neutron Stars", Ann. Rev. Nucl. Sci. (1975).
13. R. Ruffini, "The Physics of Gravitationally Collapsed Objects", in: Neutron Stars, Black Holes and Binary X-ray Sources, Astroph. and Spa. Sci. Lib., Vol.48 (1975).
14. O. Maxwell and W. Weise, Phys. Lett. 62 B (1976) 159.
15. J. N. Bahcall and R.A. Wolf, Phys. Rev. 140 B (1965) 1452.
16. O. Maxwell, G.E. Brown, D.K. Campbell, R.F. Dashen and J.T. Manassah, preprint (1976), to be published.
17. A.B. Migdal, O.Markin and I. Mishustin, ZhETF 66 (1974) 443.

18. A.B. Migdal, N.Kirichenko and G. Sorokin, Phys. Lett. 50 B (1974) 411.

19. M.I. Sobel, P.J. Siemens, J.P. Bòndorf and H.A. Bethe, Nucl. Phys. A 251 (1975) 502.

20. V. Ruck, M. Gyulassi and W. Greiner, Z. Phys. A 277 (1976) 391.

5. SHORT DIGRESSION ON KAON-NUCLEAR INTERACTIONS

Just as the pion is an agency to generate nucleon resonances, in particular the Δ-isobar, inside a nucleus, kaons can be used to inject hyperons and hyperon resonances, so that, at least in principle, the interaction of these objects with the nuclear environment can be studied. Experimental information to date includes mainly kaonic atoms [1] and (K,π) strangeness exchange reactions [2].

In kaonic atoms, the kaon-nuclear dynamics [3] is essentially determined by the Λ(1405) which appears as an isospin-zero resonance (sometimes called the Y_0^*-resonance) in the $\bar{K}p$ amplitude below threshold. The Λ(1405) couples strongly to the πΣ system, and all of these objects will interact with surrounding nucleons. This leads to modifications in the properties of the Λ(1405) inside the nucleus [4]. From a careful analysis of kaonic shifts and widths, one can hope to obtain information about such modifications [5].

Strangeness exchange (K,π) reactions are used to investigate excited states of Λ-hypernuclei.

Recent studies [6] indicate that

a) the interaction of a $\Lambda(1115)$ with nucleons in the nucleus seems to be much weaker than the corresponding nucleon-nucleus interaction.

b) In particular, spin-orbit splitting in hypernuclei seems to be strongly reduced as compared to the spin-orbit interaction in nuclei.

Especially the magnitude of the spin-orbit coupling is a valuable source of information about the detailed nature of scalar and vector boson exchange mechanisms in the Λ-nucleon interaction [7].

REFERENCES FOR SECTION 5

1. G. Backenstoß et al., Nucl. Phys. B 73 (1974) 189.
2. B. Povh, Rep. Progr. Phys. 39 (1976) 823.
3. M. Alberg, E. Henley and L. Wilets, Ann. of Phys. 96 (1976) 43.
4. J. M. Eisenberg, Phys. Rev. C 14 (1976) 2343.
5. W. Weise and L. Tauscher, Phys. Lett. 64 B (1976) 424.
6. J. Hüfner and A. Bouyssy, Phys. Lett. 64 B (1976) 276.
7. R. Brockmann and W. Weise, to be published.

Acta Physica Austriaca, Suppl. XVIII, 727-788 (1977)

INTERACTIONS OF HADRONS WITH NUCLEI AT HIGH ENERGY+

by

G. ALBERI
Istituto di Fisica Teorica
Miramare, Trieste

1. INTRODUCTION

Strong interactions at high energy look already so complicated, when we consider two-body scattering, that it seems useless to study the interaction of hadrons with nuclei, where we are faced also by the complication of the many body problem. We will see, by going through, many examples, that it is not so: indeed, at high energy, we can describe in most cases the nucleus just as a matter distribution. This is true for instance in the case of elastic scattering and coherent production at small momentum transfers.

Usually the matter distribution is assumed to be equal to the charge distribution, known from electron scattering

+Lecture given at XVI. Internationale Universitätswochen für Kernphysik, Schladming, Austria, February 24-March 5, 1977.

data: actually from theoretical considerations we expect the neutron distribution to be different from the proton distribution. This effect, however, is too small to be important at the present stage in coherent production experiments, or multiparticle production in nuclei. The idea in these experiments is to learn about the diffractive states of the hadrons and their interactions with the nuclear environment or about the dynamics of multiparticle production; but for very precise proton elastic scattering experiments the difference between proton and neutron distribution could be relevant, to explain discrepancies between theory and experiments. Some time ago two groups of nuclear experimenters, one in Paris and the other in Leningrad, got interested in high energy physics and performed various experiments of elastic and inelastic scattering of protons on nuclei. As we shall see, the first analysis of the experimental data, successively confirmed, has shown that a simple picture of spherical matter distribution, is failing dramatically for high momentum transfers in inelastic scattering. These discrepancies have been already qualitatively explained with more sophisticated nuclear models, which were invented before to explain classic spectroscopic data; but in the long run these data may become a new and independent source of nuclear information. In this analysis we are clearly faced with the complication of the many body problem, but the use of Glauber-eikonal approach carries tremendous simplifications and makes possible to extract microscopic nuclear structure information from the data.

The guide line in the first half of this report is Glauber theory: as applications, we will consider elastic and inelastic scattering of proton, coherent and semi-coherent production, and regeneration of K^0.

In the second part we will consider the multi-particle production data and a few phenomenological models, which explain experimental data making hypothesis on the time evolution of the multiple production process.

2. GLAUBER THEORY AS TOOL FOR NUCLEAR AND PARTICLE PHYSICS

2.a Elastic and inelastic scattering of protons

The most remarkable fact about Glauber theory [1] is that it can be derived from Watson multiple scattering theory [2]

$$T = \sum_{i=1}^{A} t_i + \sum_{ij}{}' t_i G t_j + \sum_{ijk}{}' t_i G t_j G t_k + \dots \tag{2.1}$$

in the approximation of neglecting in the amplitudes and Green functions the Fermi motion and the binding of the nucleons. In this approximation and at high energies the Green function is in coordinate space,

$$G(\vec{b}-\vec{b}', z-z') = -\frac{i}{v}\,\delta^{(2)}(\vec{b}-\vec{b}')\,\theta(z-z')\,e^{ik(z-z')} \tag{2.2}$$

where $\vec{b}$ is the transverse coordinate of the projectile (impact parameter). This is the eikonal form of the Green function and describes the forward propagation of the incident particle. The θ-function can be written as Fourier transform of the eikonal propagator,

$$\theta(z) = \frac{1}{2\pi i}\int_{-\infty}^{+\infty} \frac{e^{-i(k-k'_{\parallel})z}}{k-k'_{\parallel} + i\varepsilon}\, dk'_{\parallel} = -\frac{1}{2} + \frac{1}{2\pi i}P\int_{-\infty}^{+\infty} \frac{e^{-iq_{\parallel}z}}{q_{\parallel} + i\varepsilon}\, dq_{\parallel}$$

and it can be separated in an off-shell part and on-shell part. It was shown [2] that the sum of the first A terms of the expansion (2.1) calculated with the on-shell part of the Green function, is exactly equivalent to the sum of the whole series. The reason is that there is a cancellation between the first A terms, with the off-shell part of the Green function, and the rest of the series+.

Using this formal result one can write the well known Glauber formula (Fig. 1)

$$F(\vec{q}_{\perp},\{\vec{r}_j\}) = \frac{ik}{2\pi}\int d^2b\, e^{i\vec{q}_{\perp}\cdot\vec{b}}\{1 - \prod_{j=1}^{A} (1-\gamma(\vec{b}-\vec{s}_j))\}^{++}$$

$$\vec{r}_j \equiv (\vec{s}_j, z_j) \tag{2.3}$$

for the scattering operator in the coordinate space of the A-nucleons.

$\gamma(b)$ is the profile function of the elementary scattering, and it is related to the scattering amplitude by a Fourier-Bessel transform:

+This cancellation was first shown to occur by Harrington[2b] and is called Harrington cancellation.

++The equivalent formula for the total cross-section can be derived from a probabilistic model for multiple production using unitarity [2c].

$$f(q_\perp) = \frac{ik}{2\pi} \int e^{i\vec{q}_\perp.\vec{b}} \, \vec{\gamma}(b) d^2b \quad . \tag{2.4}$$

For a product wave function

$$<\vec{r}_1, \ldots \vec{r}_A|0> = \prod_{i=1}^{A} \psi(r_j)$$

one writes for elastic scattering

$$<0| \prod_{j=1}^{A} (1 - \gamma \, (\vec{b}-\vec{s}_j))|0> = (1 + \tfrac{i}{A}\chi(b))^A \tag{2.5}$$

where

$$\chi(\vec{b}) = A \int d^2s \; \rho(\vec{s},z)\gamma(\vec{b}-\vec{s}) \; .$$

$A\rho(r)$ is the nuclear density distribution,

$$\rho(r) = \psi^2(r) \; .$$

Now

$$(1 + \frac{i}{A} \chi(\vec{b}))^A = \sum_{k=1}^{A} \frac{1}{k!}(i\chi)^k \; a_k \tag{2.6}$$

$$a_k = \frac{1}{A^k} A(A-1)\ldots(A-k+1) \sim 1 - \frac{1}{2A} k(k-1) \quad .$$

So we can "exponentiate", up to terms $O(1/A)$,

$$(1 + \frac{i}{A}\chi(b))^A \sim e^{i\chi(b)} \tag{2.7}$$

and obtain for the elastic scattering

$$<0|F|0> = \frac{ik}{2\pi} \int d^2b \; e^{i\vec{q}_\perp \cdot \vec{b}} \; (1 - e^{i\chi(b)}) \; . \tag{2.8}$$

This formula is called also optical limit of Glauber theory or optical model: one appreciates its simplicity already for light nuclei like ^{12}C, but it becomes the only possibility for large nuclei like ^{208}Pb. One might have some doubt on the expansion for a_k for k close to A: but Franco [3] has shown in a careful numerical analysis that for a large nucleus A = 180, only the first 20 terms in the sum are important due to the term 1/k! which is rapidly decreasing with k.

Ahmad [4] performed an optical model analysis of the recent data [5] of elastic scattering of proton on heavy nuclei. His results for ^{39}K (Fig. 2) and for ^{40}Ca (Fig. 3) are in beautiful agreement with experiment, assuming a neutron distribution equal to the proton distribution. The proton distribution is taken from electron scattering data. The same calculation for ^{48}Ca (Fig. 4), fails, but one can take $R_n = R_p + 0{,}2$ f, and the theory is adjusted to the experiment. Again there is failure for ^{208}Pb (Fig.5a) but adjusting the diffuseness parameter for the neutron distribution we get again good agreement with experiments. It appears from this analysis that the neutron distribution is more diffused, than the proton distribution ($t_n = 0{,}65$ f, $t_p = 0{,}54$ f).

It is clear from these results that the diffraction pattern with minima and maxima is sensitive to the shape of the nuclear surface: the reason is that for high momentum transfer, the oscillating term $e^{i\vec{q}_\perp \cdot \vec{b}}$ in (2.8), cancels the contribution of the internal region of the nucleus, where $\chi(b)$ is almost constant with b (see Fig. 5b).

Formula (2.3) can be fully exploited, using a microscopic model for the nucleus, like harmonic oscillator model. For one-particle one-hole excitation in the final state one can write the following formula for the profile function [6]

$$\langle \tilde{\psi}_f | \prod_{\ell=1}^{A} (1-\gamma_\ell) | 0 \rangle = -\frac{2}{ik} \int d^2q' \, e^{-i\vec{q}' \cdot \vec{b}} \, f(\vec{q}') \times$$

$$\times Y^*_{J_f M_f}(\hat{q}') \, \frac{\hat{p}}{\hat{h}\,\hat{J}_f} \langle p \| \hat{O}_{J_f} \| h \rangle \sum_{m_h} D^{h,m_h}_{h,m_h} \tag{2.9}$$

where

$$\langle p \| \hat{O}_\lambda \| h \rangle = (-1)^{2h+\lambda-1/2-p} \frac{\hat{h}}{\sqrt{4\pi}} \times$$

$$\times \langle h\ 1/2;\ p-1/2 | \lambda\, 0 \rangle \langle R_p | j_\lambda | R_h \rangle \ ;$$

$\hat{J} = (2j+1)^{1/2}$ (p, h are the angular momentum of the particle and hole;

D^{h,m_h}_{h,m_h} is the distortion factor in the unexcited part of the nucleus.

These formulas put a link between the old work in the shell model and high energy scattering. For the inelastic scattering of protons from ^{12}C, one can use the old work of Gillet and Vinh Mau [7] and calculate the proton excitation of the 2^+ and 3^- levels, explored experimentally by the Saclay group [8]. The results show disagreement with the data at high momentum transfers (Fig. 6)$^+$. It was tried [9],[10] also to exploit (2.3) with the α-model for ^{12}C and the results (Fig. 7,8) look definitely better than with the shell model. These results cannot be taken as definite proof for the cluster structure, but they show that, being the proton strongly absorbed in the nucleus the shape of the nucleus is much more important here than for instance electron scattering. There is clearly a difference if the proton is absorbed by a cigar shape or a disc shape or a spherical shape: the cluster model for ^{12}C carries necessarily a deformation.

2.b Coherent regeneration of neutral K-mesons

We word coherent, often used in this context, means simply that the nucleus remains in the ground state: the word comes actually from atomic physics. For X-rays scattering from atom, the cross section at small angles

$$\frac{d\sigma}{d\Omega} = |\langle 0| \sum_j e^{i\vec{q}.\vec{r}_j} |0\rangle|^2 \frac{d\sigma}{d\Omega} \quad \text{(Thompson)} \tag{2.10}$$

$^+$This agreement is present also for a simple DWIA calculation [6b].

where $\vec{q}$ is the momentum transfer and $\vec{r}_j$ the coordinates of the electrons. For q a << 1 where a is the atomic radius

$$\frac{d\sigma}{d\Omega} \sim Z^2 \qquad \text{because} \qquad e^{i\vec{q}.\vec{r}_j} \sim 1$$

i.e. there is complete coherence between the electrons. At large angles one can excite all the atomic states so that

$$\frac{d\sigma}{d\Omega} = \sum_f <0|\sum_j e^{-i\vec{q}.\vec{r}_j}|f><f|\sum_i e^{i\vec{q}.\vec{r}_i}|0>$$

$$= <0|\sum_{ij} e^{i\vec{q}(\vec{r}_i-\vec{r}_j)}|0>$$

which gives

$$\frac{d\sigma}{d\Omega} \sim Z$$

because the interference term vanish being q >> 1/R i.e. the electrons interact with the incident photon in a completely incoherent way.

So coherent regeneration of K^o means simply elastic scattering of K^o from nuclei: the word regeneration refers to the unique feature of the K^o beams, to be not pure, but superposition of so called K_S and K_L, eigenstates of CP

$$|K^o> = \frac{1}{\sqrt{2}} (|K_S> + |K_L>) \quad . \tag{2.11}$$

This is the initial state, but even in the vacuum the relative percentage of K_S and K_L change because they decay with different life time. This change is called regeneration. If we wait long enough, all the K_S have decayed and we are left with a pure beam of K_L: being

$$|K_L\rangle = \frac{1}{\sqrt{2}}\left(|K^o\rangle - |\bar{K}^o\rangle\right) \tag{2.12}$$

we can regenerate K_S, through a slab of heavy material, because K^o and $\bar{K}^o$ are transmitted in a different way. The final state is

$$\frac{1}{\sqrt{2}}\left(f_+|K^o\rangle - f_-|\bar{K}^o\rangle\right) = \frac{(f_+-f_-)}{\sqrt{2}}|K_S\rangle + $$

$$+ \frac{(f_++f_-)}{\sqrt{2}}|K_L\rangle$$

where

$$f_\pm = \frac{ik}{2\pi}\int(1 - e^{i\chi_\pm(b)})\, e^{i\vec{q}.\vec{b}}\, d^2b \quad .$$

Since both K_S and K_L decay weakly in 2π due to CP violation, one can study the decay law in the interference region between K_S and K_L. In this region the decay law is sensitive to the phase of the regeneration amplitude

$$I_{2\pi}(t) \sim e^{-\Gamma_S t}\, |\rho(p)|^2 \,\{1 + \frac{|\eta_{+-}|^2}{|\rho(p)|^2}\, e^{\Gamma_S t}$$

$$+ 2 \frac{|\eta_{+-}|}{|\rho(p)|} e^{\frac{1}{2}\Gamma_S t} \quad \cos(\Delta m t - \phi)\} \tag{2.13}$$

where ρ (p) is the transmission regeneration amplitude

$$\rho = 2\pi i \Lambda_S N \quad \frac{f_+ - f_-}{k} \quad \frac{1 - e^{i(\frac{\Delta m}{\Gamma} - \frac{1}{2})\ell}}{1 - 2i \frac{\Delta m}{\Gamma_S}} ,$$

$\Lambda_S = \beta\gamma \tau_S$ mean decay length of K_S,

$N =$ density of the scattering centers,

$\ell = L/\Lambda_S$ length of the transmitter,

$$\eta_{+-} = \frac{\langle\pi^+\pi^-| S|K_L\rangle}{\langle\pi^+\pi^-| S|K_S\rangle} = |\eta_{+-}| e^{i\phi_{+-}} ,$$

$$\phi = \phi_{+-} - \arg [i(f_+ - f_-)]$$

$$- \arg \{\frac{1 - e^{(i\frac{\Delta m}{\Gamma_S} - \frac{1}{2})\ell}}{1 - 2i\frac{\Delta m}{\Gamma_S}}\} .$$

Since we know from the regeneration in vacuum ϕ_{+-}, we can determine here the argument of the nuclear regeneration phase. It turns out to be consistent with the Regge pole picture of K^0N scattering [11], which gives for the regeneration amplitude:

$$f_+ - f_- = 3\rho - \omega ;$$

for an I = 0 target, like ^{12}C only ω can be exchanged and therefore we have clean information on the ω Regge trajectory. The results are consistent with the energy behaviour of the regeneration amplitude, as shown in Fig. 9. This information can be extracted from the forward regeneration amplitude. The shape of the differential regeneration cross section is sensitive to the neutron distribution. The reason is that the regeneration amplitude on nucleon is twice as large for the neutron than for the proton as easily explained by the quark model (Fig. 10), furthermore the regeneration amplitude is sensitive to the shape of the nuclear surface: this can be seen from the form of the profile in impact parameter space of the regeneration amplitude (Fig. 11). There are actually some data [12] which would indicate that R_n = 7,65 f for lead, but one should await for better experimental data, before drawing any conclusions (Fig. 12).

2.c Coherent and semicoherent production

As we have already seen for the case of isospin, the coherent and semicoherent[+] reactions satisfy to selection rules. The s-channel selection rules are based on the following relation between the helicity amplitudes [13]

$$\langle\lambda_f|T|\lambda_i\rangle = \nu_f\nu_i(-)^{\lambda_i-\lambda_f} \langle-\lambda_f|T|-\lambda_i\rangle \qquad (2.14)$$

[+]Semicoherent means that the nucleus is left in an excited state.

following from the invariance of T under parity transformation: ν_f, ν_i are the natural parities of the initial and final state. For projectiles of spin 0 the relation becomes

$$\langle 0|T|0\rangle = \nu_f \nu_i \langle 0|T|0\rangle \tag{2.15}$$

which means that $\nu_f = \nu_i$. This helicity amplitude is the only one, different from zero, exactly in the forward direction due to angular momentum conservation. So the natural parity of the produced object, has to be the same of the projectile in a coherent reaction. P.E. for a 0^+ target, only the unnatural parity series can be produced by an incident pseudoscalar $0^- \to 0^-, 1^+, 2^-, 3^+$ .....

$(\pi \to \eta, A_1, A_3, \ldots \quad K \to Q)$.

For a semicoherent reaction, where the natural parity of the nucleus changes, we produce objects with different natural parity from the projectile.

P.E. for $0^+ \to 0^-$, only the natural parity series can be generated from a pseudoscalar $0^- \to 0^+, 1^-, 2^+$

$(\pi \to \varepsilon, \rho, A_2, \ldots \quad K \to K^* \ldots)$.

These rules hold only at zero angle and for zero spin projectiles.

The t-channel selection rules are based on a Regge picture of the hadron nucleus interaction. For spin 0 nucleus, in the forward direction, the natural parity is conserved at the nuclear vertex. For coherent scattering $0^+ \to 0^+$ only the natural series of Regge poles can be exchanged $[\nu=(-)^J]$ and for the semicoherent $0^+ \to 0^-$, the unnatural series $[\nu = (-)^{J+1}]$.

The isotopic spin conservation in the t-channel allows the exchange of isoscalar only (ω,η,f,ϕ) for coherent interaction with a I = 0 target. For a semi-coherent interaction I = 0 → I = 1, only isovectors can be exchanged $(\pi, A_1, A_2, \rho \ldots)$.

The dynamics of coherent production is less firmly established theoretically than the elastic scattering: while for elastic scattering the ordering of the scatterers was not important, because of the Harrington cancellation, here we have to maintain it for physical reasons. Indeed the elastic scattering of the produced object has to follow in time the production and the elastic scattering of the projectile has to precede the production. Actually the Harrington cancellation was shown only for commuting amplitudes. Harrington has shown that the ordered form of Glauber theory holds for the deuteron in the case of isospin dependent amplitudes [14], but he warned also that it is true only if the isospin dependent potentials do not overlap.

So for the moment, we take the ordered form of Glauber theory as an ansatz, with physical foundations. Since we have to allow for production, we assume here that the scattering amplitudes are operators in the space of several particles or resonances (π, A_1, A_3) and give Hamiltonians between different sectors of the space. In a quark model picture of the elementary particle these operators would be defined in terms of the quark physical observables (spin, isospin...).

Being α the projectile and β the final product, the coherent production amplitude [15] is

$$<\beta|\hat{F}|\alpha> = \frac{ik}{2\pi} \int d^2b \; e^{i\vec{q}_{\perp}.\vec{b}} (\delta_{\alpha\beta} -$$

$$-<\beta| \int d^3r_1 ... d^3r_A |\psi(\vec{r}_1, ... \vec{r}_A)|^2 \times$$

$$\times Z \prod_{j=1}^{A} [1-\gamma(\vec{b}-\vec{s}_j, z_j)] |\alpha>) \qquad (2.16)$$

where Z is the ordering operator $\theta(z_1-z_2)$, $\theta(z_2-z_3)$. ....$\theta(z_{A-1}-z_A)$; $\gamma(b,z)$ may depend also on z because in production, the nucleon recoils longitudinally to conserve energy

$$\gamma(\vec{b},z) = e^{izq_{\parallel}} \gamma(\vec{b})$$

and

$$q_{\parallel} \sim \frac{m_\beta^2 - m_\alpha^2}{2k} \quad .$$

Defining

$$\phi(\vec{b},z) = \int d^3r_1 ... d^3r_A |\psi(\vec{r}_1 ... \vec{r}_A)|^2 Z \prod_{i=1}^{A} (1-\gamma(\vec{b}-\vec{s}_j)\theta(z-z_j))$$

we find that it satisfies the differential equation

$$\frac{\partial}{\partial z}\phi(\vec{b},z) = - \sum_{j=1}^{A} \int d^3r_1 ... d^3r_A |\psi(\vec{r}_1 ... \vec{r}_A)|^2 \times$$

$$\times\ Z\{\delta(z-z_j)\gamma(\vec{b}-\vec{b}_j,z_j)\ \prod_{i\neq j}(1-\gamma(\vec{b}-\vec{s}_i)\theta(z-z_i))\}$$

which becomes for a factorized wave function

$$|\psi(\vec{r}_1,\ldots\vec{r}_A)|^2 = \prod_{i=1}^{A}\rho(\vec{r}_i)$$

$$\frac{\partial}{\partial z}\Phi^A(\vec{b},z) = -A\int d^2s\ \rho(\vec{s},z)\gamma(\vec{b}-\vec{s},z)\Phi^{A-1}(\vec{b},z). \qquad (2.17)$$

For large A, $\Phi^A \sim \Phi^{A-1}$. Projecting on the initial and final state and inserting all possible intermediate states, we get a coupled channel optical equation

$$\frac{\partial}{\partial z}\psi_\beta(\vec{b},z) = \sum_\delta \frac{U_{\beta\delta}(\vec{b},z)}{2ik}\ e^{izq_{\parallel}(\beta,\delta)}\ \psi_\delta(\vec{b},z) \qquad (2.18)$$

for

$$\psi_\beta(\vec{b},z) = \langle\beta|\ \Phi(\vec{b},z)|\alpha\rangle$$

and

$$U_{\beta\delta}(\vec{b},z) = -\ 2ik\ A\int d^2s\ \rho(\vec{s},z)\ \gamma_{\beta\delta}(\vec{b}-\vec{s},z)$$

which becomes in the zero range approximation for the production amplitude the usual form for the optical potential

$$U_{\beta\delta}(\vec{b},z) = -\ 4\pi\ A\ \ f_{\beta\delta}(0)\ \rho(\vec{b},z)\ . \qquad (2.19)$$

Once we solve the optical coupled channel equation, we have the coherent production amplitude

$$F_{\beta\alpha}(\vec{q}_\perp) = -\frac{ik}{2\pi}\int d^2b\ e^{i\vec{q}_\perp\cdot\vec{b}} \langle\beta|(\Phi(\vec{b},\infty)-1)|\alpha\rangle \quad . \qquad (2.20)$$

For incident π and only A_1 as excited state of the pion, there are only two coupled equations ($\alpha \equiv \pi$, $\beta \equiv A_1$)

$$\frac{\partial}{\partial z}\psi_\alpha(\vec{b},z) = \frac{U_{\alpha\alpha}}{2ik}\psi_\alpha(\vec{b},z)$$

$$\frac{\partial}{\partial z}\psi_\beta(\vec{b},z) = \frac{U_{\beta\alpha}}{2ik}\psi_\alpha(\vec{b},z) + \frac{U_{\beta\beta}}{2ik}\psi_\beta(\vec{b},z) \quad .$$

In the first equation there is no coupling term, because we neglect the possibility for the A_1 to convert back to the pion (weak-coupling). The first equation is easily solved

$$\psi_\alpha(\vec{b},z) = \exp\left[-\frac{2\pi}{ik} f_{\alpha\alpha}(0) A \int_{-\infty}^{z} \rho(\vec{b},z')dz'\right];$$

ψ_α describes the absorption of π inside the nucleus. Even the second equation is easily solved, neglecting $q_\parallel$.

$$\psi_\beta(\vec{b},z) = \frac{f_{\beta\alpha}(0)}{f_{\beta\beta}(0)-f_{\alpha\alpha}(0)}\left[\exp\left(-\frac{2\pi}{ik} f_{\alpha\alpha}(0)A\int_{-\infty}^{z}\rho(\vec{b},z')dz'\right)\right.$$

$$\left. - \exp\left(-\frac{2\pi}{ik} f_{\beta\beta}(0)A\int_{-\infty}^{z}\rho(\vec{b},z')dz'\right)\right] \quad . \qquad (2.21)$$

So the coherent production amplitude is

$$F_{\beta\alpha}(\vec{q}) = \frac{ik}{2\pi} \frac{f_{\beta\alpha}(0)}{f_{\beta\beta}(0)-f_{\alpha\alpha}(0)} \int d^2b[\exp(-\frac{2\pi i}{k}AT(b)f_{\alpha\alpha}(0))$$

$$-\exp(-\frac{2\pi i}{k}AT(b)\ f_{\beta\beta}(0))] \qquad (2.22)$$

which shows to be sensitive to the β-N amplitude. This information was extracted from experimental data (Fig.13) of 3π and 5π production on several nuclear targets [16].

The results were obtained, with a simultaneous fit, of the integrated coherent cross section on several nuclei, with a formula of the type (2.22) including the longitudinal momentum transfer (see Fig. 13 e 14). The 3π-N cross section was about equal to the π-N cross section on the A_1 peak (1.2 GeV) and it was slightly increasing with the mass. The same was done for the 5π case and the result for the 5π-N cross section was 1/3 of the π-N (Fig. 15). The first result was easily digested by the theorist, who could explain it thinking the A_1 as a quark excitation of the π; but the second one, regarding the 5π, was suggesting a very peculiar dynamical mechanism.

The first proposal by Van Hove and Wilkin [17], was based on the possibility of virtual mass transition, inside the nucleus, which by constructive interference would enhance the coherent cross-section, decreasing the absorption.

An other possibility, would be a two step process $\pi \to A_1 \to A_5(5\pi)$, which again would reduce the absorption. The two step mechanism was studied by Von Bochmann and Margolis [18] for the case $\pi \to A_1 \to A_3$: using (2.16) as starting point they obtained the formula

$$F_{\pi A_1}(q) = f_{\pi A_1}(0) \int d^2b\, e^{i\vec{q}_\perp\cdot\vec{b}}\, e^{-\frac{1}{2}\sigma' T(b)}\, T(b) \times$$

$$\times\ (1 - (1/2)\sigma'\ R\ T(b))$$

where

$$R = \frac{f_{\pi A_1}(0) f_{A_1 A_3}(0)}{f_{\pi A_3}(0) f_{\pi\pi}(0)}$$

and

$$T(b) = \int_{-\infty}^{+\infty} \rho(\vec{b},z)\,dz\ .$$

In the rough approximation [18b] that $(1- \frac{1}{2}\sigma' RT(b)) \sim e^{-(1/2)\sigma' RT(b)}$ the two step mechanism reduces or enhances the absorption according to the sign of R.
It is clear that also in the case of 5π, one could get the desidered result with ad hoc values for R: this parameter should be calculated in a dynamical theory or confirmed by other experimental data, to make sure that this interpretation is correct.

From the fact that $\sigma' \sim \sigma_{\pi N}$ and not $\sim 3\sigma_{\pi N}$, one could draw the conclusion that there is not time for the diffractive state of the pion to develop into three pion inside the nucleus. Also considering A_1 as $\pi\rho$-system and taking in account of the mutual shadowing [16], one cannot fit the data. If the Deck model [19] is used to describe the dynamics of diffractive excitation of the π, one obtains more absorption than acutally observed. We see that in these experiments we do not determine only the interaction of unstable systems with nucleon, but we get also information on the time development of diffractive excitation. More precise information should be given by experiments of the same type at SUPERCERN; indeed the energy dependence of the phenomenon is connected to the time dependence because of the Lorentz time dilation.

In a coupled channel model (2.22) the energy dependence of the phenomenon would come only from the energy variation of the elementary amplitudes $f_{\alpha\alpha}(0)$ and $f_{\beta\beta}(0)$. A possible picture of diffraction, consistent with the coupled channel model [20], is the analogous of vector dominance for hadron. The vector dominance is founded on the hypothesis that the photon is a superposition of the bare photon and all the diffractive states with the same quantum number of the photons, like ρ, ϕ, ω (vector mesons). For photons, the optical model approach was used to analyze data of ρ, ϕ, ω coherent photoproduction on several nuclei [21] and the results for the ρ-N, ϕ-N... cross sections were perfectly consistent with the quark model. However, we have to say that for the photon case, all the resonances are well established and not overlapping in mass; therefore the situation is much cleaner than in the π case, where P.E. it was found a copious production of 0^- state in the 3π system [22], and there is no established resonance having those quantum numbers. Even the $A_1(1^+)$, it is not clear yet if it is a resonance or not. The experimental results for the interaction cross section would indicate that the 0^- is an expanded $\pi\rho$ system [22] already in the nucleus, (Fig. 16) being the $\sigma \sim 50$ mb.

But a resonable two step model $\pi \rightarrow A_1 \rightarrow 0^-$ [23] can explain the large value for the cross section and the mass dependence, assuming as usual that both 0^- and A_1 travel inside the nucleus with contracted size, expanding only outside the nucleus.

Here we have discussed only the diffractive production, which correspond to no exchange of quantum numbers in the t-channel, but in principle one could consider excited states of the nucleus, which have different isospin and different spin-parity from the ground state. This type of reaction, which we call semicoherent, being the final nuclear state only slightly different from the ground state, was studied experimentally by Piccioni and Koester [24]. Unfortunately only the states with the same natural parity of the ground state were produced, due to limited energy resolution. Some recent experiment of deuteron breakup [25], have shown that the unnatural parity transition $1^+ \rightarrow 2^+$ is possible for the deuteron and it is due to a ΔN resonant state [26] at 300 MeV excitation energy of the deuteron. The peak (Fig. 17) is 50 MeV large and can easily be detected, in high energy physics experiments. This transition occurs also for mediumlight nuclei and could be considered as semicoherent reaction.

3. MULTIPARTICLE PRODUCTION

If it is possible to detect the target nucleus, in the final state, one can perform experiments of coherent production of multiparticle systems

$$h + A \rightarrow A + X \tag{3.1}$$

and use in this way the selection rules in the t-channel for coherent reactions. As shown in Fig. 18, in coherent reactions only the so called "pomeron" exchange survives, with small correction from other I = 0 trajectories, like ω for instance. The "pomeron" exchange means that mechanism, not yet well understood, which describes elastic scattering and diffraction dissociation at asymptotic energies and predicts constant values for the total hadron-nucleon cross sections. This exchange has been defined in the past as a Regge exchange with intercept $\alpha(0) = 1$ and $\alpha'(0)$ small, $\sim 2\ (\mathrm{GeV}/c)^{-2}$, in such a way to give constant cross section $\sigma \sim s^{\alpha(0)-1}$ and slow shrinking of the forward elastic peak

$$\frac{d\sigma}{dt} \sim e^{(b + \alpha' \log s)t} .$$

This type of experiments was performed already, but only with the deuteron as a target [27]: in this way it is possible to fit the inclusive cross section with a triple Regge formula including only the Pomeron and the other I = 0 trajectories (ω, ...)

$$\frac{d^2\sigma}{dt d\nu} = \frac{1}{s^2} \sum_{ijk} G_{ijk}(t) \left(\frac{s}{\nu}\right)^{\alpha_i(t)+\alpha_j(t)} (\nu)^{\alpha_k(0)} \qquad (3.2)$$

where $\nu = M_X^2 - m_p^2 - t$ and G_{ijk} is the triple Regge coupling (Fig. 18).

Especially at large masses, where we expect the typical $1/M^2$ dependence of the triple-pomeron, the situation should be much less ambiguous, than for $pp \to p + X$ (see Fig. 19, 20). In the above study, one uses only the t-channel selection rules: in the following we will study the incoherent multiparticle production, and exploit the unique possibility given by nuclear targets, to study the process at a non-asymptotic stage: the idea was already in the treatment of coherent diffraction, but here the information is more detailed because we detect particles with different energies.

Before proceeding some definitions are in order. The reaction studied here is of the inclusive type

$$h\,A \to \pi + X\ . \tag{3.3}$$

Not always the π is identified, but in the multiparticle production at high energies, pions are most abundantly produced. Since there is a universal strong cut off in transverse momentum $e^{-(p_\perp/4)^2}$, the data for the inclusive cross section are presented as function of a longitudinal variable, which is called rapidity,

$$y = \frac{1}{2}\ln\frac{E + p_\parallel}{E - p_\parallel}\ . \tag{3.4}$$

The rapidity is just the boost parameter, to transform from the target frame, to the moving particle frame. In most experiments only the angle is measured and one measures actually the so called pseudorapidity,

$$\eta = -\ln\tan\frac{\theta_{lab}}{2} \sim y^{lab}\ . \tag{3.5}$$

This approximation is excellent provided $p_\perp^2 \gg \mu^2$, being μ the pion mass. From conservation of particle number

$$\int_{o}^{y_{max}} dy \frac{d\sigma}{dy} \sim \sum_{n} n \sigma_n \qquad (3.6)$$

and the average multiplicity is

$$<n> = \frac{\sum n\sigma_n}{\sigma_{inel}} = \frac{1}{\sigma_{inel}} \int_{o}^{y_{max}} dy \frac{d\sigma}{dy} = \int_{o}^{y_{max}} dy \frac{dn}{dy} \qquad (3.7)$$

which shows the intimate connection between the average multiplicity and the rapidity distribution.

A large portion of the data on multiparticle production were collected with the emulsion technique. This technique is based on the activation of grains by ionization of the particles. Ionization depends strongly on the velocity and on the charge of particles and for light relativistic particles the ionization becomes extended because of the electromagnetic cascade: so the tracks are divided in black, grey and shower particles (relativistic). The grey tracks are commonly identified with protons and the black ones are heavier particles: both types of tracks correspond to nuclear fragments and are called heavy tracks. The shower tracks are pions and their multiplicity n_s can be compared on the multiplicity in Hydrogen: we exclude in the count of produced particles the heavy tracks, because they are not connected with the elementary production process.

Measuring the angle of the shower track, one get the pseudo-rapidity distribution. Comparing it with the same distribution for p-p scattering (Fig. 21), we realize that for $y > y_c$ the two rapidity distributions are indistinguishables while for $y < y_c$, there is a considerable excess of particles produced in nuclei. Integrating on rapidity one get the average multiplicity, which can be compared with the average multiplicity on hydrogen. The ratio

$$R_A = \frac{\langle n \rangle_{PA}}{\langle n \rangle_{PP}}$$

is plotted in Fig. 22: since $\langle A \rangle$ in emulsions ~ 65, it is clear from the figure that the multiplicity does not depend strongly on A. An intra-nuclear cascade model gives a much stronger dependence on A, as shown by Montecarlo calculations [28]. Even a simple model [28], where only the leading particle is allowed to produce other particles in successive interactions gives too large multiplicity. Since the leading particle has a fraction F of the incident energy s

$$F = 0.48 + \frac{2.3}{s}$$

the leading particle after the first scattering has energy Fs, after the second F^2s, and so on. So the average multiplicity is

$$\langle n \rangle_A = \langle n(s) \rangle_P + \langle n(Fs) \rangle_P + \ldots \langle n(F^{\bar{\nu}}_s) \rangle_P$$

where $\bar{\nu}$ is the average number of collisions. Using the law

$$\langle n \rangle = a + b \ln s$$

one get

$$<n>_A = \bar{\nu}<n>_p + \frac{1}{2}\bar{\nu}(\bar{\nu}-1)\ln F \quad .$$

The growth with $\bar{\nu}$ of the $<n>_A$ predicted by this formula is too strong, to be comparable with cosmic rays data (see Fig. 22). This can be easily checked with more recent data for the A dependence of $<n>_A$. If one includes in the calculation also the possibility for the secondaries to produce, the result go obviously in the wrong direction, giving more multiplicity.

To explain the deviation of experimental data from the predictions of the leading particle cascade, Gottfried developed a model [28],[29] which is based on the two following assumptions:

1) In the interaction of the incident hadron with the target nucleon, a distribution of expanding hadronic matter is created. This hadronic matter is still in a preliminary stage of evolution and behaves in a rather singular way. At the time of the impact, is completely contracted and it is expanding with time. Its rapidity distribution is just a box in Gottfried's model and is divided in two parts: the front part which is fast and the back part, which correspond to slow secondaries (see Fig. 23),

$$H \rightarrow H_1 + H_2 \quad .$$

The reason why it is expanding with time is that different points in the y-distribution have different velocities.

2) In successive interactions <u>the front part of the spectrum behaves like a hadron</u> and produces again a y-distribution of hadronic matter, which again is separated

in a front part and in a back part. The back part of the incident y-distribution has <u>not enough energy</u> to have inelastic interaction (see Fig. 24).

When the resulting hadronic matter distribution is well outside the nucleus, the front part develop in a certain number of pions n_{H_1} while the second part in n_{H_2}, in such a way that $n_{H_1} + n_{H_2} = \langle n \rangle_P$ is the average multiplicity at the incident energy. The remaining $(\bar{\nu}-1)$ back part of the spectrums, develop in $(\bar{\nu}-1)n_{H_2}$ pions, where n_{H_2} is the average multiplicity on proton for the average energy of the slice. Therefore the total multiplicity is

$$\langle n \rangle_A = \langle n(E) \rangle_P + (\bar{\nu}-1)\langle n(E') \rangle_P \ . \tag{3.8}$$

Since the probability that one target nucleon suffers an inelastic collision is $\sigma_{in}(P)/\sigma_{in}(A)$, the average number of collisions is

$$\bar{\nu} = \frac{A\ \sigma_{in}(P)}{\sigma_{in}(A)} \sim A^{1/3} \ . \tag{3.9}$$

Since we want the front part of the y-distribution to behave like a hadron, after a certain time interval, time t between two collisions, it must have the size of the hadron. The time evolution of the limits of the y-distribution is classical and we require that

$$(\beta_1-\beta_2)t = \tau = \frac{\tau_o}{\gamma_1} = \frac{2\tau_o}{e^{\bar{y}_1}} \tag{3.10}$$

where $\bar{y}_1 = (y_1 + y_2)/2$ and τ_o is the hadron size in its rest frame. Being $\gamma = \cosh y$ and $\eta = \beta\gamma = \sinh y$,

$$\beta = \frac{\sinh y}{\cosh y} = \frac{1 - e^{-2y}}{1 + e^{-2y}} \underset{y>>1}{\sim} 1 - 2e^{-2y} \quad . \tag{3.11}$$

Substituting this expression of β, in (3.10) we get the following expression of y_2,

$$y_2 = \frac{y_1}{3} + \frac{2}{3} \ln \left(\frac{t}{\tau_o} \left(1 - e^{-2(y_2-y_1)}\right)\right) . \tag{3.12}$$

This gives the head of the spectrum of H_2, which correspond to the incident energy in the corresponding p-p interaction. From a numerical calculation of y_2 as function of the time interval t (Fig. 25), we realize that the time evolution of the slice is very slow within the nuclear matter, whose size is of the order of τ_o. Therefore, we take in account only the constant term. Being.

$$y_2 = \ln \frac{E'}{m} \tag{3.13}$$

this means that

$$E' = E^{1/3} \quad .$$

Furthermore being the average multiplicity on hydrogen

$$\langle n \rangle \sim b \ln E$$

this gives

$$R_A = 1 + \frac{1}{3} (\bar{\nu} - 1) \ . \tag{3.14}$$

This simple law cannot be verified in detail with data from emulsion, because of many different elements present in the emulsion.

A novel technique, based on a Cerenkov counter, was used by Busza et al. [30] to count the multiplicity of relativistic particles, produced by 100 and 175 GeV/c protons on several nuclear targets. The results of this experiment are shown in Fig. 26, where the average multiplicity is given against the average number of collisions. The linear dependence of (3.14) is perfectly reproduced in the experimental data, however the parameters, originally given by Gottfried, are not adequate: other models, of similar types, seem to be in better shape [30]. Even if there is some truth in the Gottfried model, its predictive power is limited to the average multiplicity or at most the multiplicity distribution [31]: indeed it is not possible to reproduce a realistic rapidity distribution.

A qualitative understanding of the data comes also from some space time considerations on the multipheripheral model [32]. This model is based on the assumption that the multiparticle production is described by the Feynman graph of Fig. 27a): there is an additional constraint, that the momentum transfers across the vertical lines have to be small, which means that on vertices there is a form factor, also necessary for peripheral exchange reactions. A proper treatment of this diagram in the old fashioned perturbation theory [33],[34] , has shown that in the target frame, it takes a time $\tau = \frac{E}{\mu^2}$, before all the particle exchanges complete: therefore in space or time the diagram looks as in Fig. 27b). So for a fast particle produced in the first collision it is impossible to interact again with a nucleon of the target, which it is possible for the slow particles. Being small the momentum transfers across the vertical exchanges, the fast particles in the target frame are upwards and the slow ones are downwards. The critical energy is $E_c = R\mu^2$ and the critical rapidity is $y_c = \ell n(2R\mu^2)$ (see Fig. 28a). This simple argument is however in apparent contradiction with unitarity because in the total cross section this would mean that the exchange of one pomeron is the dominant term, i.e. $\sigma_A \sim A$, while experimentally $\sigma_A \sim A^{2/3}$. Also the elastic double scattering in the deuteron would vanish at high energies because there is no time for the hadron to reform and rescatter again (see Fig. 29).

Using this simple ideas, Nikolaiev [35] has written down transport equation, for the partons, emitted by the ladder in the first collision. The main idea of the paper is to assume that parton do not interact with nuclear matter, but after a time of the order of $\tau_o e^y$ they convert to hadrons and then they interact. The function which describes this transition of the parton created in the point x is

$$P(z, x|y) = \theta(z-x + \tau_o e^y) \phi(z|y) \qquad (3.15)$$

where $\phi(z|y)$ is the parton distribution at the point z (only the forward dimension is taken in account). The parton distribution is given by the equation

$$\phi(z|y) = \delta(z)P(Y,y) + \frac{1}{\lambda(z)} \int_y^Y dy_1 \, P(y_1,y)H(z|y) \qquad (3.16)$$

where $P(Y,y)$ is the inclusive spectrum of partons produced by a hadron of rapidity Y and $H(z|y)$ is the hadron spectrum at the point z: $\lambda^{-1}(z) = \sigma_{in}\,\rho(z)$ is the inverse mean free path; while the first term comes from the primary hadron, the second refers to the secondary hadrons. The hadron distribution is given by the transport equation

$$\frac{\partial}{\partial x}H(x|y) = -\lambda(x)H(x|y) - \int_0^x dz\left[\frac{\partial}{\partial x}P(z,x|y)\right]. \tag{3.17}$$

Inserting (3.16) in (3.15) and substituting the result in (3.17), we get an integro-differential equation, which can be solved by usual methods and becomes an integral equation

$$H(x|y) = \theta(x-\tau_o e^Y)P(Y,y)\exp\left[-\int_{\tau_o e^Y}^{x}\lambda(t)dt\right]$$

$$+\int_0^{x-\tau_o e^Y} d\tau\,\lambda(\tau)\,e^{-\int_{\tau+\tau_o e^Y}^{x}\lambda(t)dt}\int_y^Y dy_1\,P(y_1,y)H(\tau|y_1). \tag{3.18}$$

The result is easily connected with experiments

$$\frac{1}{\sigma_{in}}\frac{d\,\sigma_{in}}{dy} = H(x \gtrsim R + \tau_o\,e^Y|y)\;.$$

The numerical results for

$$R_A(y) = \left(\frac{1}{\sigma_{in}}\frac{d\,\sigma_{in}}{dy}\right)_A \Big/ \left(\frac{1}{\sigma_{in}}\frac{d\,\sigma_{in}}{dy}\right)_P$$

are shown in Fig. 31: there is a remarkable agreement with experiment.

The only free parameter in the theory is the time life of the partons in their rest frame τ_o; the value of the parameter required by the data is much smaller, than expected from early scaling in deep inelastic electron scattering or from multiperipheral model (τ_o = .07/m_N = .074 GeV^{-1}). More recently, Bialkowski et al. [36] considered a cluster model for production in nuclei, with a freezing time as parameter: the model is very similar to the previous ones and the result for the freezing time is also similar (τ_o = .046 GeV^{-1}). In the same paper, however, the major emphasis is given to a different version of this model, where the dressing up with gluons of the cluster is accelerated by the presence of hadronic matter. This actually ammounts to a multiplicative factor in the so called "immature" probability $\theta(g) = e^{-tv/\tau_o\gamma v}$, which determines the suppression of multiplicative processes through the factor $1 - \theta(y)$: this could be interpreted as probability for the cluster to dress-up with gluons and become a hadron. Although the results of the two models are relatively similar, one should prefer the latter because the conversion from "parton" to hadrons does not occur in a deterministic way, as in the former one. Although the parton model [35] and the immature cluster model [36], are giving results in agreement with experiment, the situation is very unsatisfactory because of the small value of the freezing time in the first case and the somewhat arbitrary assumption about the acceleration of dressing in the second case.

The bremsstrahlung model for multiple particle production is giving analogous predictions for the inclusive distribution in the central region, but it gives different results for the leading particle spectrum [37].

4. CONCLUSIONS

Summarizing, we have considered here some informations one can obtain from the interaction of hadrons from nuclei. The nuclear structure information is still not definite, because due to the complexity of many body problem, we want to have the same information from other sources, to make sure that what we have learned from high energy proton interaction is correct. Also the information on the cross section of the unstable particles, like 3π or 5π systems is still very incomplete, because we have still a very primitive understanding of dynamics of diffraction dissociation; we have to say that for the case of incident photons, the situation is much clearer because the unstable particles are established resonances and the dynamics is understood in terms of vector dominance. For the ρ-N case, the numbers extracted from heavy nuclei were confirmed on the deuteron. If we sum together the experimental evidence in coherent production of 3π and in multiparticle production, we realize that we have learned something definite, that is at high energies the fast particles or partons, need a certain time or longitudinal space to interact with nucleons or in other words parton need a certain time to convert to strong interacting hadrons. However, it is clear that much theoretical work is needed to formulate the statement in a more rigorous fashion. There are also more experimental data, like for instance correlation data [38], or rapidity distribution of π^+ and π^- in emulsions [39], which show very interesting features. These data can be very useful, in guiding gently the theorists towards the right answer to the problems of strong interaction dynamics.

REFERENCES

1. R.J. Glauber, in "Lectures in Theoretical Physics", Vol.I° ed. by W.E. Brittin and L.G. Dunham, Interscience Publ. N.Y. 1959.

2. J.M. Eisemberg, Ann. of Phys., 71 (1972) 542.

2b. D.Harrington, Phys. Rev., 184 (1969) 1745.

2c. A. Bialas, M.Blezinski and W. Czyz, Krakow preprint, Report N. 951/PH.

3. V. Franco, Phys. Rev., 6C (1972) 748.

4. I. Ahmad, Nucl. Phys., A247 (1975) 418.

5. G.D. Alkhazov et al., Phys. Lett., 42B (1972) 121 and JETP Lett., 10 (1973) 101.

6. G. Alberi, M. Gmitro and L.H. Hambro, Ref. TH 2178 CERN to be published in Nuovo Cimento.

6b. J. Saudinos and C. Wilkin, Annual Review of Nuclear Science, 1974.

7. V. Gillet and N. Vinh Mau, Nucl. Phys., 54 (1964) 321.

8. R. Bertini et al., Phys. Lett., 45B (1973) 119.

8b. G.D. Alkhazov et al., JETP Lett., 10 (1973) 101.

9. Y. Abgrall, J. Labarsouque and B. Morand, Nucl. Phys., A232 (1974) 235.

10. Z.A. Kan and I. Ahmad,tobe published in Pramana.

11. V.L. Telegdi in "High Energy Physics and Nucl. Structure", 1975 (Santa Fè and Los Alamos), A.I.P.

12. H. Foeth et al. in "High Energy Physics and Nucl. Structure", S. Devons Ed., N.Y. 1970 (Plenum Press).

13. A.D. Martin and T.D. Spearman, "Elementary Particles", N. Holl. 1970.

14. D.R. Harrinton, Nucl. Phys., B59 (1973) 305.

15. G.V. Bochmann, Phys. Rev., D6 (1972) 1938.

16. C. Bemporad et al., Nucl. Phys., B33 (1971) 397.

17. L. Van Hove, Nucl. Phys., B46 (1972) 75.
Rogers and C. Wilkin, Nucl. Phys., B.

18. G. Von Bochmann and B. Margolis, Nucl. Phys., B14 (1969) 609.

18b. G. Fäldt, to be published in the Proceeding of the Conf. on Multiple Production in Nuclear, Trieste, 1976.

19. R.T. Cutler, Phys. Rev., D10 (1974) 824.

20. A.S. Goldhaber, Phys. Rev., D7 (1973) 765.

21. H. Alvensleben et al., Phys. Rev. Lett., 24 (1970) 786.

22. G. Bellini in "High Energy Collisions Involving Nuclei" ed. by G. Bellini, L. Bertocchi and P.G. Raincota (Bologna, 1975) p. 271.

23. L. Bertocchi and D. Treleani, Nuovo Cimento, 34A (1976) 193.

24. W. Mehlop et al. in "High Energy Collisions Involving Nuclei" ed. by G. Bellini et al. (Bologna, 1975) p.271; L.J. Köster, ibidem.

25. B.S. Aladashvili et al., Nucl. Phys., A274 (1976) 486.

26. G. Alberi and F. Baldracchini, INFN/AE-76/8 to be published in the Proceeding of the Conference on Multiple Production on Nuclei, June 1976.

27. Y. Akimov et al., FERMILAB-Pub-76/36-EXP and Phys. Rev. Lett., 35 (1975) 766.

28. K. Gottfried in "High Energy Physics and Nuclear Structure", 1973, ed. by G. Tibell, North Holland-Amsterdam;
K. Gottfried, Phys. Rev. Lett., 32 (1974) 957.

28b. B. Andersson, Nucl. Phys., B95 (1975) 237.

29. L. Bertocchi, Lecture Notes of Ecole d'Etè de Gif sur Yvette, 1975.

30. W. Busza et al., Phys. Rev. Lett., 34 (1975) 839;
W. Busza in "High Energy Physics and Nuclear Structure", 1975, Santa Fè, ed. A.I.P.

31. G. Calucci in "High Energy Collisions Involving Nuclei", Trieste, 1974, Editrice Compositori-Bologna.

32. D. Amati, S. Fubini and A. Stanghellini, Nuovo Cim. 26 (1961) 896.

33. J. Koplik and A.H. Mueller, Phys. Rev., D12 (1975) 3638.

34. J.H. Weiss, Acta Phys. Pol., B7 (1976) 851.

35. N.N. Nikolaiev, Landau Institute preprint to be published in"Multiparticle Production on Nuclei at Very High Energy",Trieste, 1976;
G.V. Davidenko and N.N. Nikolaiev, Yadernaia Fisika, 24 (1976) 772.

36. G. Bialkowski, C.B. Chiu and D.M. Tow, University of Texas, ORO 294, December 1976.

37. A. Bialas and L. Stodolsky, Acta Phys. Pol., B7 (1976) 845.

38. G.Baroni et al., Univ. of Roma, preprint, Jan. 1977.

39. S.A. Azimov et al., to be published in Proceeding of the Conference on Multiparticle Production in Nuclei, Trieste, June 1976.

FIGURE CAPTIONS

Fig. 1: Notation in Glauber formula (2.3).

Fig. 2: Elastic differential cross sections for $p^{39}K$ scattering at 1 GeV. The experimental points are from the Leningrad group (Ref. 8b).

Fig. 3: Same as for Fig. 2, but for ^{40}Ca.

Fig. 4: Elastic differential cross section for $p^{48}Ca$ scattering at 1 GeV. Dahsed curve: neutron distribution the same as the proton distribution. Solid curve: neutron distribution different from the proton distribution. Dotted curve: same as above but without the Coulomb scattering. The experimental points are from the Leningrad group (Ref. 8b).

Fig.5a: Elastic differential cross section for $p^{208}Pb$ scattering at 1 GeV. Dotted curve: neutron distribution the same as the proton distribution. Solid curve: neutron distribution different from the proton distribution. The experimental points are from the Saclay group (Ref. 8).

Fig.5b: For large q the oscillating term $e^{i\vec{q}_{\perp}\cdot\vec{b}}$ cancels the contribution of the inner part of the nucleus in the integral (2.8).

Fig. 6: Test of shell-model wave function. Cross-section of the inelastic scattering of 1 GeV protons (Ref. 8). Figures 6 and 6a correspond to the excitation of the 3^- and 2^+ states of ^{12}C respectively.

Fig. 7: Calculation $p^{12}C \rightarrow p^{12}C^*(2^+)$, with the 3 model for ^{12}C [10].

Fig. 8: Calculation of $p^{12}C \rightarrow p^{12}C^*(3^-)$ with the 3 model for ^{12}C (Ref. 9).

Fig. 9: Status of regeneration amplitude in carbon.

Fig.10: Quark picture for K^o-N and K^o-P interaction.

Fig.11: Nuclear thickness and regeneration amplitude as function of impact parameter.

Fig.12: Experimental results of Foeth et al. on Pb compared to optical model predictions.

Fig.13: Example of differential cross section $d^2\sigma/dM_3\,dt'$ for Ta target. Low quality events are excluded from the experimental data. The curve is calculated from the optical model expression (11) and includes an incoherent contribution (dashed line) of the form (14) neglecting the Pauli principle.

Fig.14: Integrated coherent cross-section of 3π as function of A, for different mass intervals (Ref. 16).

Fig.15: The same as 14 for 5π.

Fig.16: Separation of the 0^- in 3π coherent production (Ref. 22).

Fig.17: Mass distribution of the two protons in the charge exchange reaction $pD \rightarrow n(pp)$ at 1.65 GeV/c incident proton momentum. The experimental data are from Ref. 25.

Fig.18: Selection of exchanges in coherent production.

Fig.19: Differential cross-section for $pp \rightarrow Xp$ vs. M_x^2 for s - 500 GeV^2 and $|t| = 0.025\ (GeV/c)^2$, obtained from the listed $|t|$-values using slopes given in the references. For the extrapolation of the CNLM and Rutgers - Imperial College data a slope of 6 $(GeV/c)^{-2}$ was used (Ref. 27).

Fig. 20: Inclusive distribution of proton in p.p.collisions.

Fig. 21: Comparison between pseudorapidity distribution for p-p and p-A collisions (Ref. 30).

Fig. 22: R_A as function of E_{LAB} and predictions of leading particle cascade model (taken from Ref. 29).

Fig. 23: The two slices in the rapid.distr.of hadr.matter.

Fig. 24: Gottfried model in picture (taken from Ref. 30).

Fig. 25: Behaviour with $\xi = t/\tau_o$ of the slice separation in Gottfried hadron.

Fig. 26: Comparison between Busza data [30] and: a) the Gottfried model (Ref. 28); b) the theory of Bialkowski (Ref. 36).

Fig. 27: Multiperipheral model in the traditional Feynman graph formulation (a) and in the old-fashioned perturbation theory (b).

Fig. 28: Interaction with another target nucleon is possible only if the energy of the particle is sufficiently slow $y < y_{crit} = \ell n\left(\frac{R}{\mu^2}\right)$.

Fig. 29: In the old-fashioned perturbation theory there is no elastic rescattering on the deuteron at high energy.

Fig. 30: Parton model description, of multiple production in nuclei.

Fig. 31: $\dfrac{\dfrac{1}{\sigma_{in}} \left.\dfrac{d\sigma}{dy}\right|_A}{\dfrac{1}{\sigma_{in}} \left.\dfrac{d\sigma}{dy}\right|_{p-p}} = Ry$ as function of y, in emulsion.

Comparison with Davidenko and Nikolaiev theory (Ref. 35).

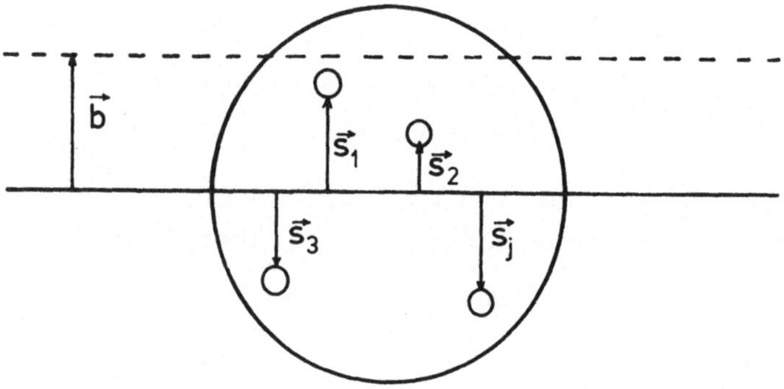

Fig.1

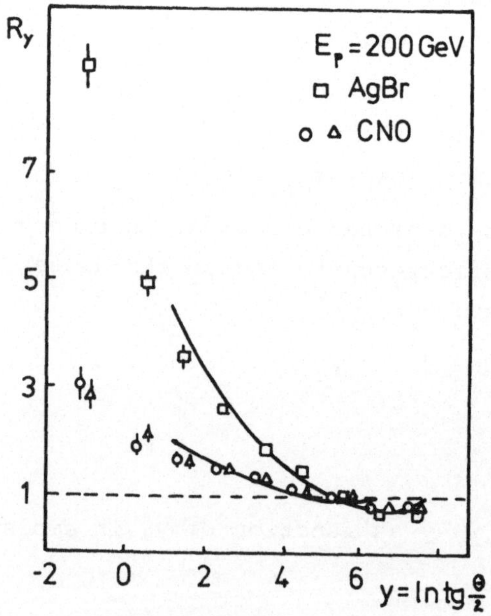

Fig. 31

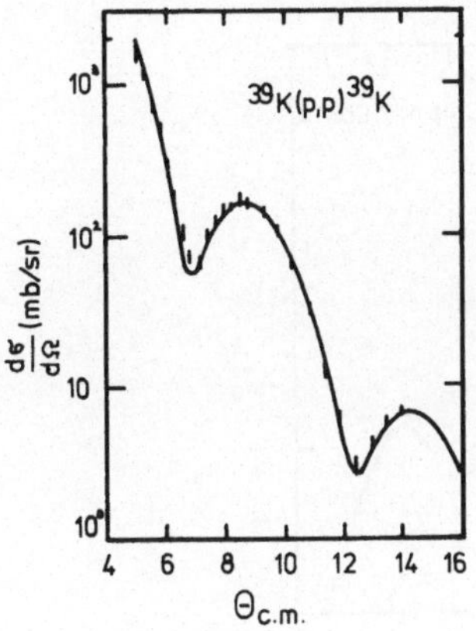
$^{39}K(p,p)^{39}K$
$\frac{d\sigma}{d\Omega}$ (mb/sr)
10^3
10^2
10
10^0
4 6 8 10 12 14 16
$\Theta_{c.m.}$

Fig.2

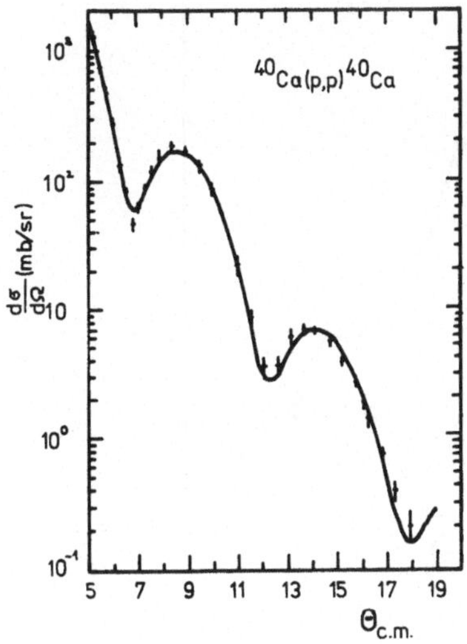
$^{40}Ca(p,p)^{40}Ca$
$\frac{d\sigma}{d\Omega}$ (mb/sr)
10^3
10^2
10
10^0
10^{-1}
5 7 9 11 13 15 17 19
$\Theta_{c.m.}$

Fig. 3

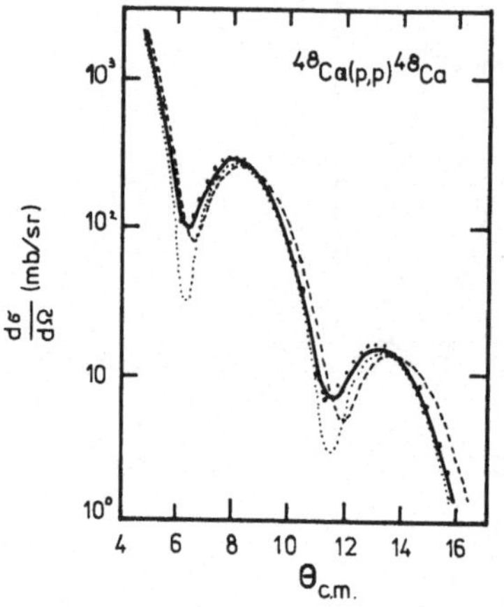

48Ca(p,p)48Ca
10³
10²
10
10⁰
dσ/dΩ (mb/sr)
4
6
8
10
12
14
16
θc.m.

Fig. 4

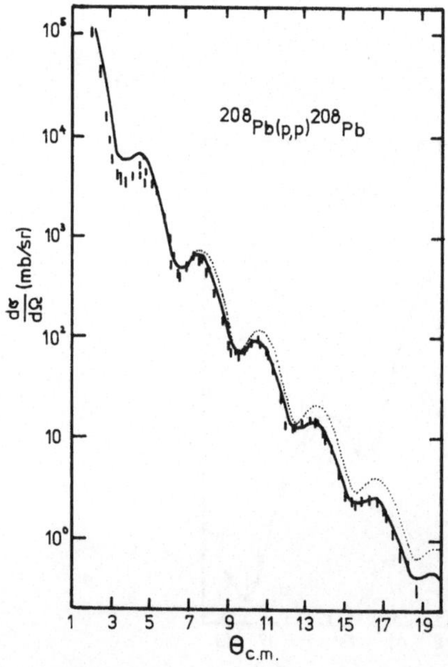

208Pb(p,p)208Pb
10⁵
10⁴
10³
10²
10
10⁰
dσ/dΩ (mb/sr)
1
3
5
7
9
11
13
15
17
19
θc.m.

Fig. 5a

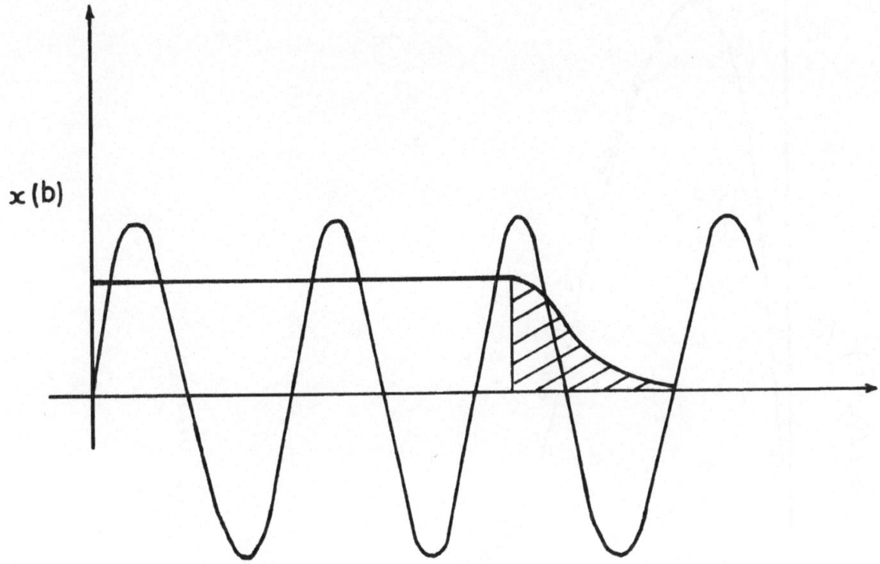

Fig. 5b

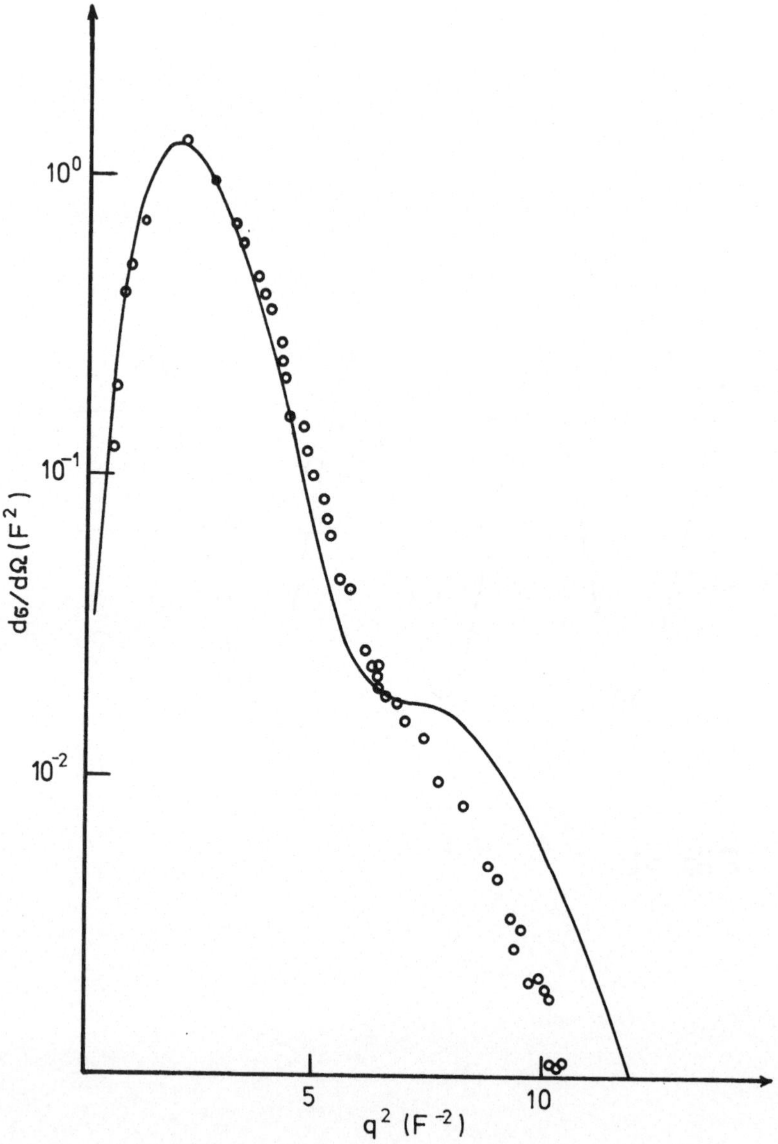

Fig. 6

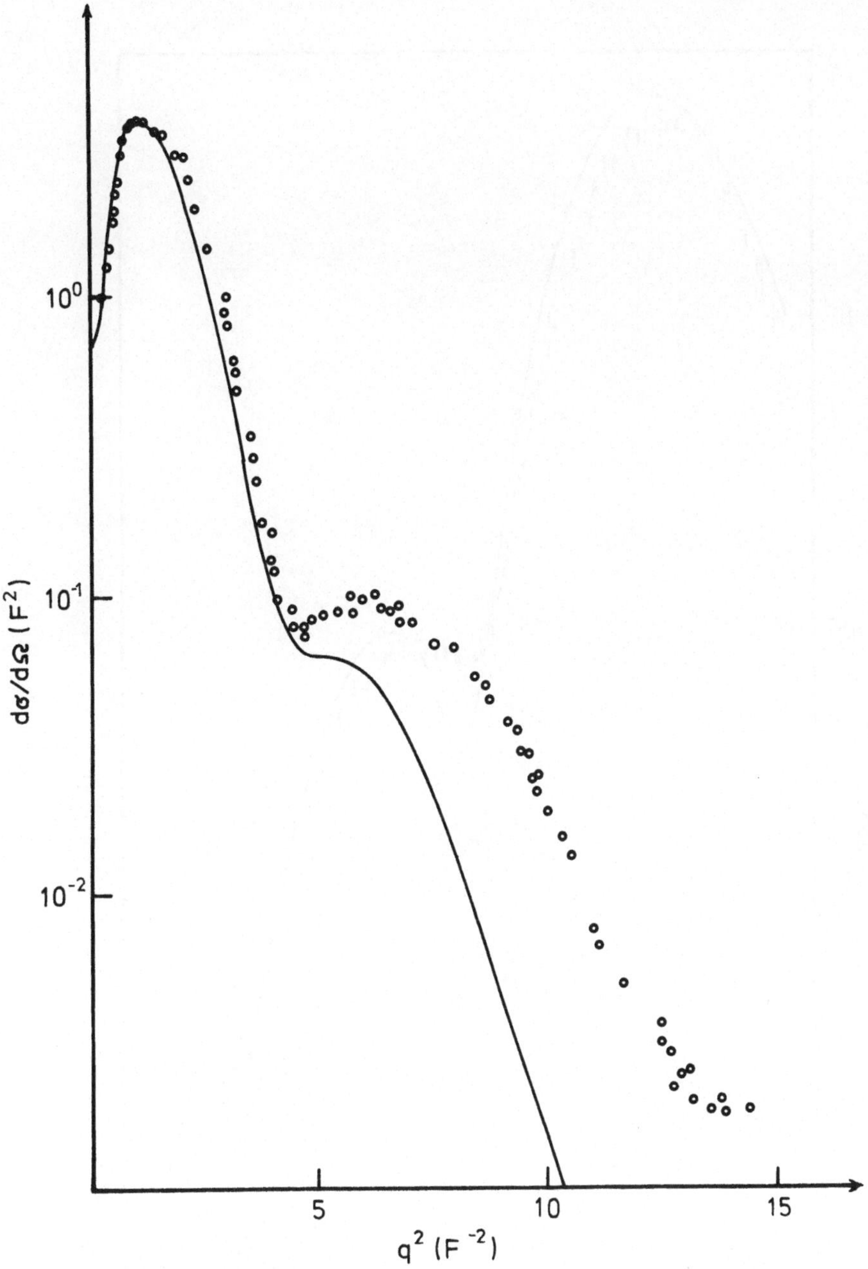

Fig. 6a

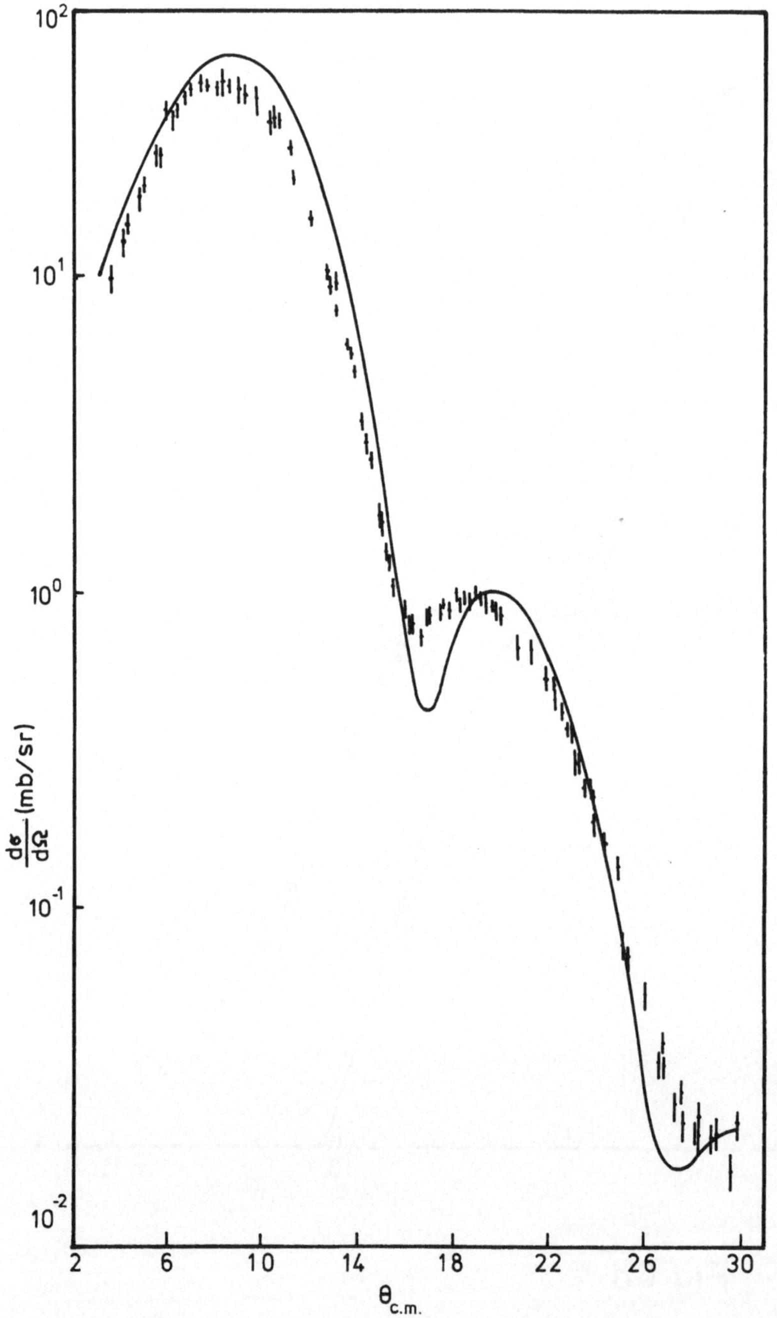

Fig. 7

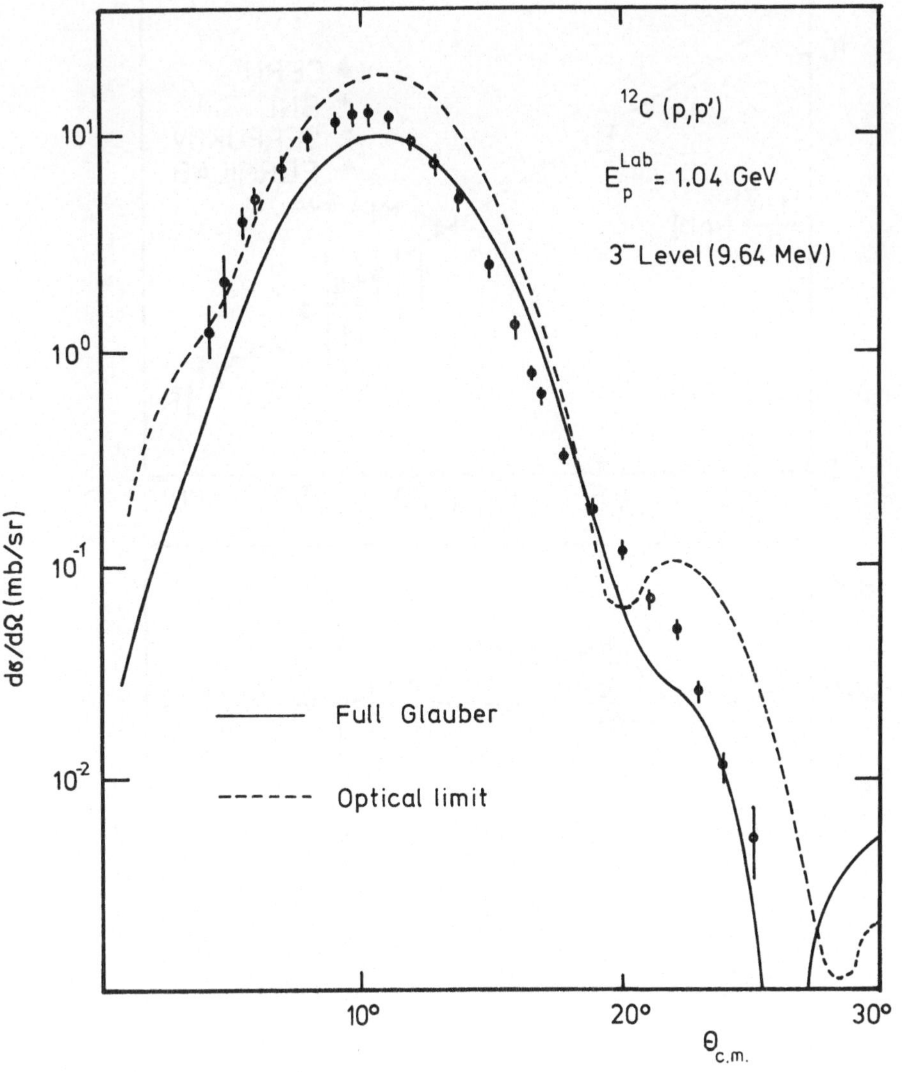

$^{12}C\,(p,p')$
$E_p^{Lab} = 1.04$ GeV
3^- Level (9.64 MeV)
10^1
10^0
10^{-1}
10^{-2}
dσ/dΩ (mb/sr)
Full Glauber
Optical limit
10°
20°
30°
$\Theta_{c.m.}$

Fig. 8

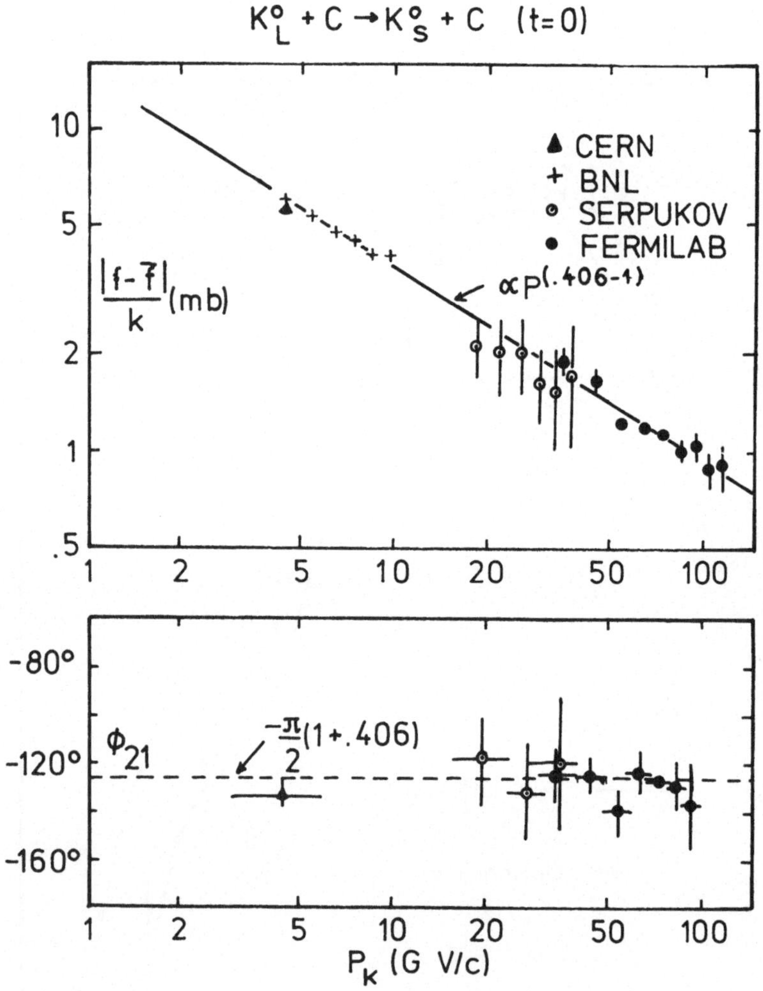
$K^o_L + C \rightarrow K^o_S + C \quad (t=0)$
10
5
2
1
.5
$\frac{|f-\bar{f}|}{k}$ (mb)
CERN
BNL
SERPUKOV
FERMILAB
$\propto P^{(.406-1)}$
1
2
5
10
20
50
100
-80°
-120°
-160°
ϕ_{21}
$\frac{-\pi}{2}(1+.406)$
1
2
5
10
20
50
100
P_K (G V/c)

Fig.9

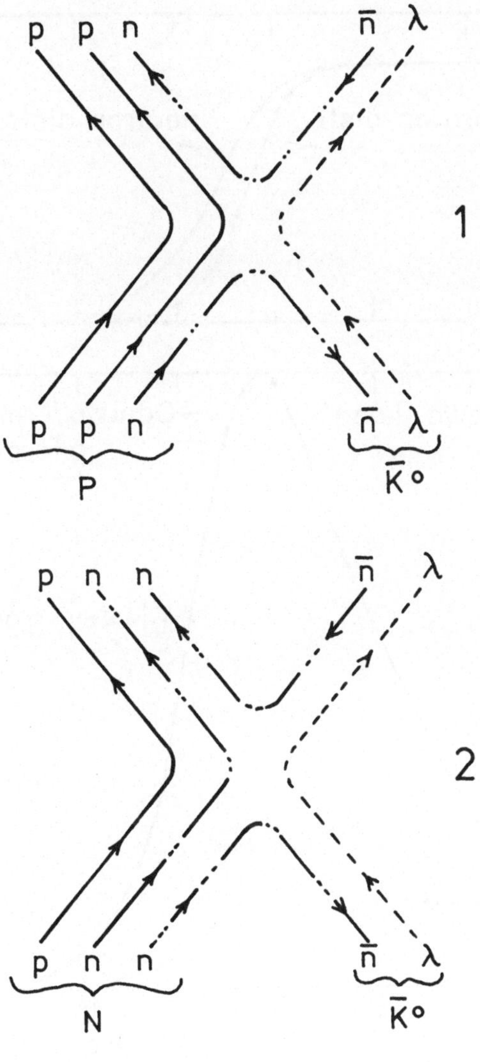

Fig. 10

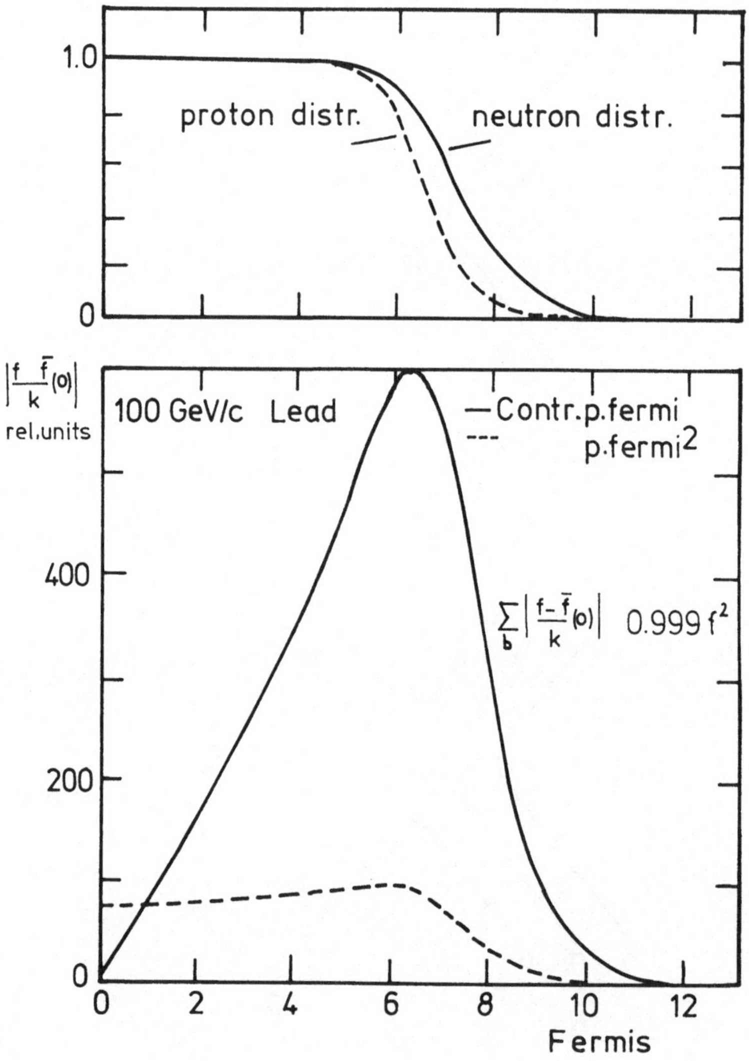
1.0
0
proton distr.
neutron distr.
$\left|\frac{f-\bar{f}}{k}(0)\right|$
rel.units
100 GeV/c Lead
— Contr.p.fermi
--- p.fermi^2
400
200
0
$\sum_b \left|\frac{f-\bar{f}}{k}(0)\right|$ 0.999 f^2
0
2
4
6
8
10
12
Fermis

Fig. 11

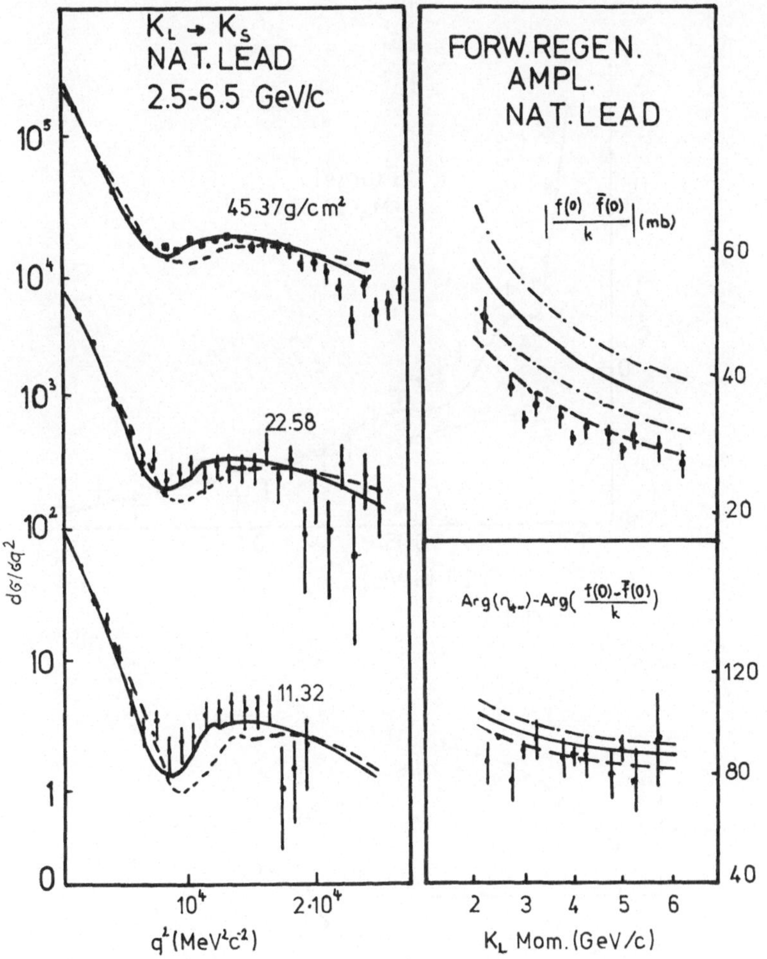

$K_L \to K_S$
NAT. LEAD
2.5-6.5 GeV/c
45.37g/cm²
22.58
11.32
$d\sigma/dq^2$
q^2 (MeV²c⁻²)
FORW. REGEN. AMPL. NAT. LEAD
$\left|\frac{f(0)-\bar{f}(0)}{k}\right|$ (mb)
$Arg(\eta_{+-}) - Arg(\frac{f(0)-\bar{f}(0)}{k})$
K_L Mom. (GeV/c)

Fig.12

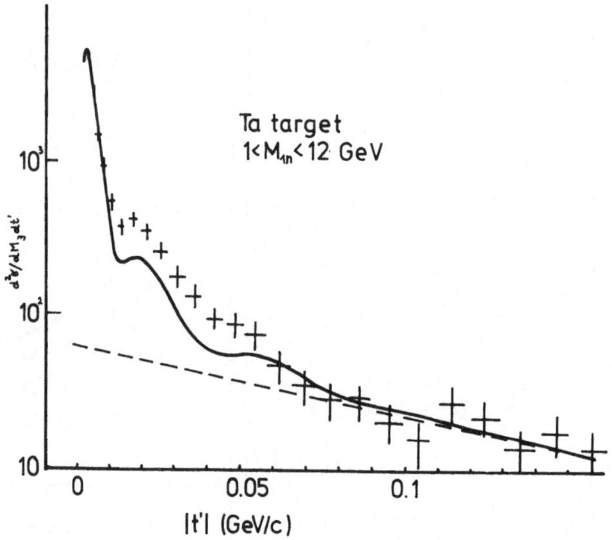
Ta target
1<M4n<12 GeV
10³
10²
10
0
0.05
0.1
|t'| (GeV/c)

Fig.13

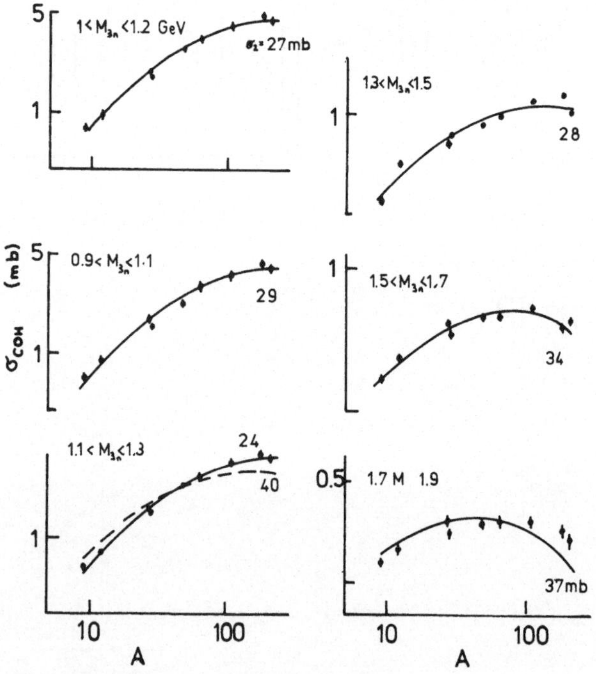
1<M3n<1.2 GeV
σ1=27mb
1.3<M3n<1.5
28
0.9<M3n<1.1
29
1.5<M3n<1.7
34
σCOH (mb)
1.1<M3n<1.3
24
40
0.5
37mb
10
100
A

Fig.14

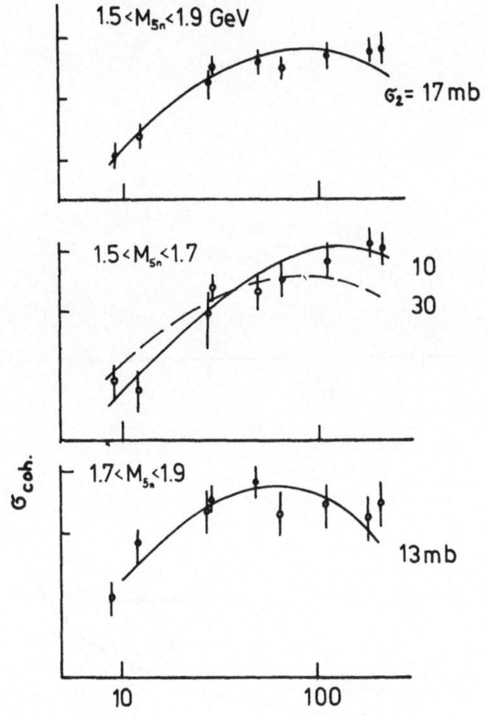
1.5<M5n<1.9 GeV
σ2= 17mb
1.5<M5n<1.7
10
30
σcoh.
1.7<M5n<1.9
13mb
10
100

Fig.15

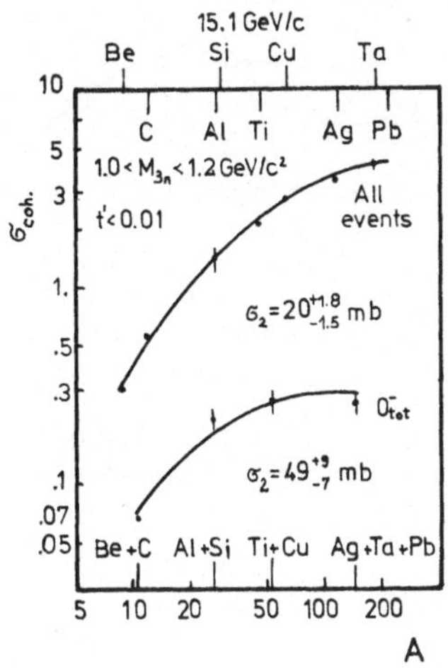
15.1 GeV/c
Be
Si
Cu
Ta
C
Al
Ti
Ag
Pb
1.0<M3n<1.2 GeV/c²
t'<0.01
σcoh.
All events
σ2=20+1.8 −1.5 mb
σtot
σ2=49+9 −7 mb
Be+C
Al+Si
Ti+Cu
Ag+Ta+Pb
10
5
3
1.
.5
.3
.1
.07
.05
5
10
20
50
100
200
A

Fig.16

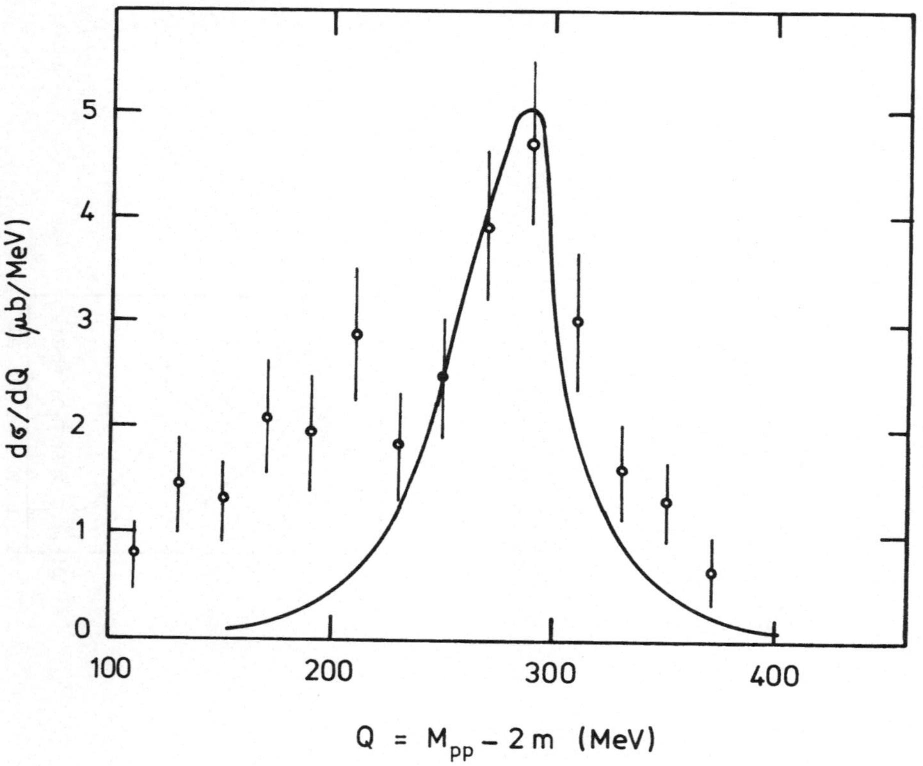

Fig.17

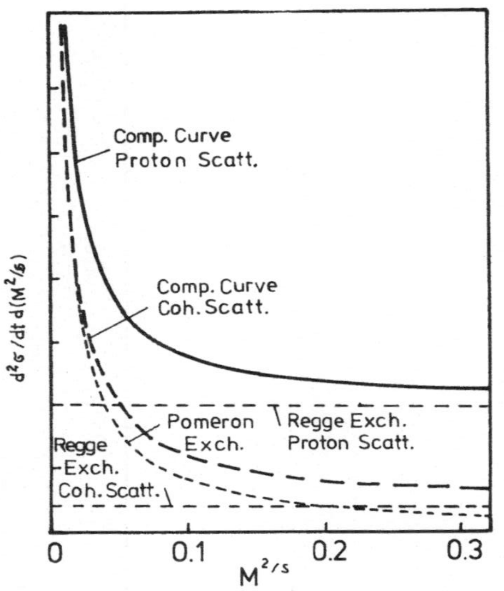

Comp. Curve
Proton Scatt.
Comp. Curve
Coh. Scatt.
Pomeron
Exch.
Regge Exch.
Proton Scatt.
Regge
Exch.
Coh. Scatt.
d²σ / dt d(M²/s)
0
0.1
0.2
0.3
M²/s

Fig. 18

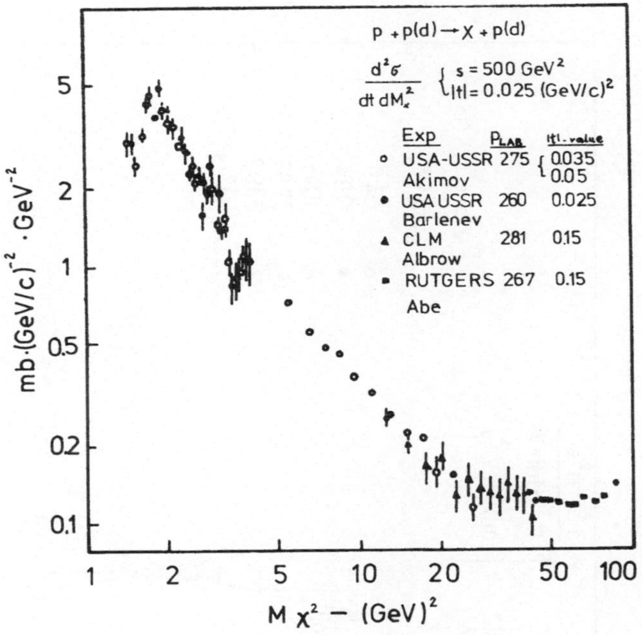

p + p(d) → X + p(d)
d²σ / dt dM²ₓ { s = 500 GeV² ; |t| = 0.025 (GeV/c)²
Exp
P_LAB
|t|-value
USA-USSR 275 { 0.035 ; 0.05
Akimov
USA USSR 260 0.025
Barlenev
CLM 281 0.15
Albrow
RUTGERS 267 0.15
Abe
mb·(GeV/c)⁻²·GeV⁻²
5
2
1
0.5
0.2
0.1
1
2
5
10
20
50
100
M X² — (GeV)²

Fig. 19

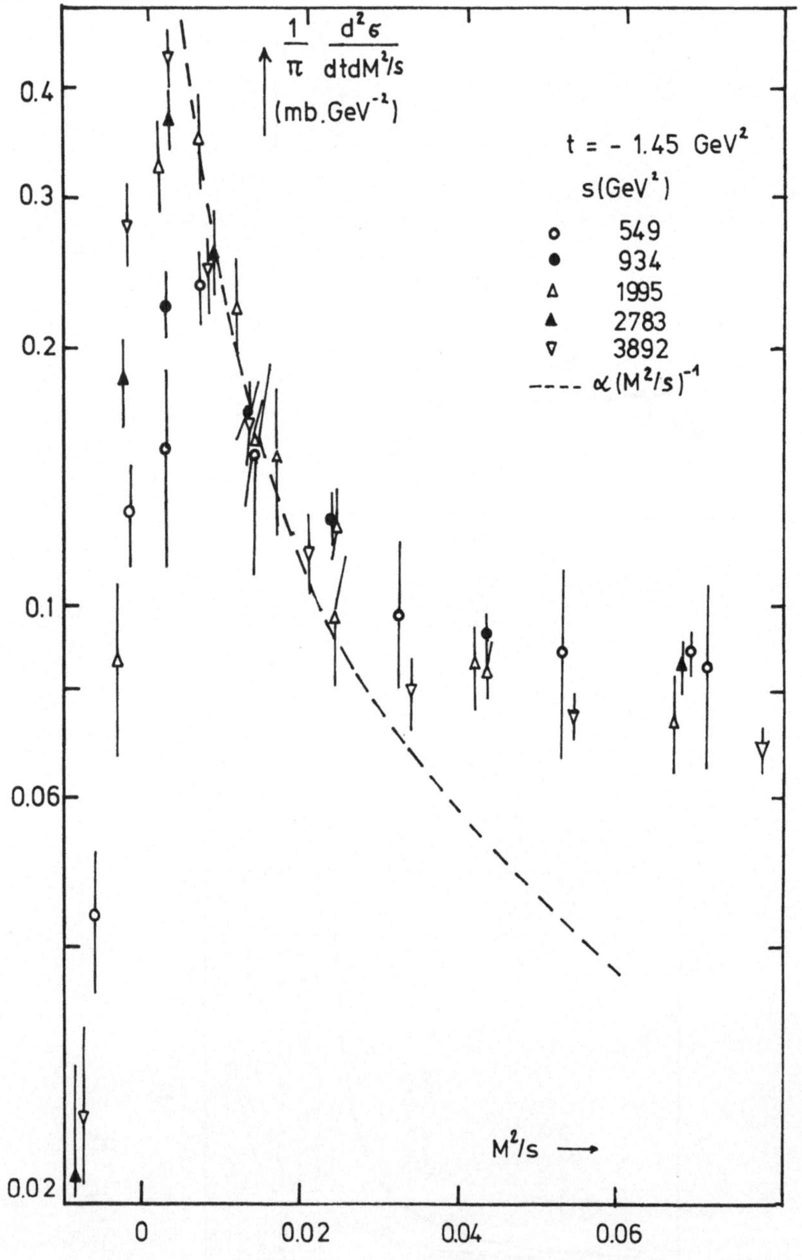

$\frac{1}{\pi} \frac{d^2\sigma}{dt dM^2/s}$
(mb.GeV^{-2})
t = - 1.45 GeV^2
s(GeV^2)
549
934
1995
2783
3892
---- $\propto (M^2/s)^{-1}$
0.4
0.3
0.2
0.1
0.06
0.02
0
0.02
0.04
0.06
M^2/s

Fig. 20

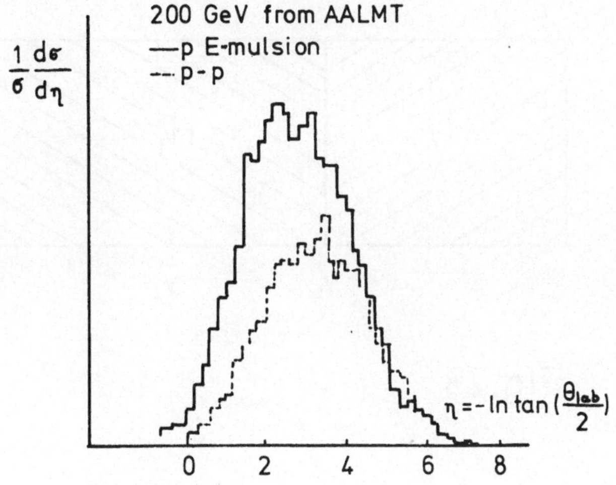
200 GeV from AALMT
—p E-mulsion
--- p-p
$\frac{1}{\sigma}\frac{d\sigma}{d\eta}$
$\eta = -\ln\tan(\frac{\theta_{lab}}{2})$
0
2
4
6
8

Fig. 21

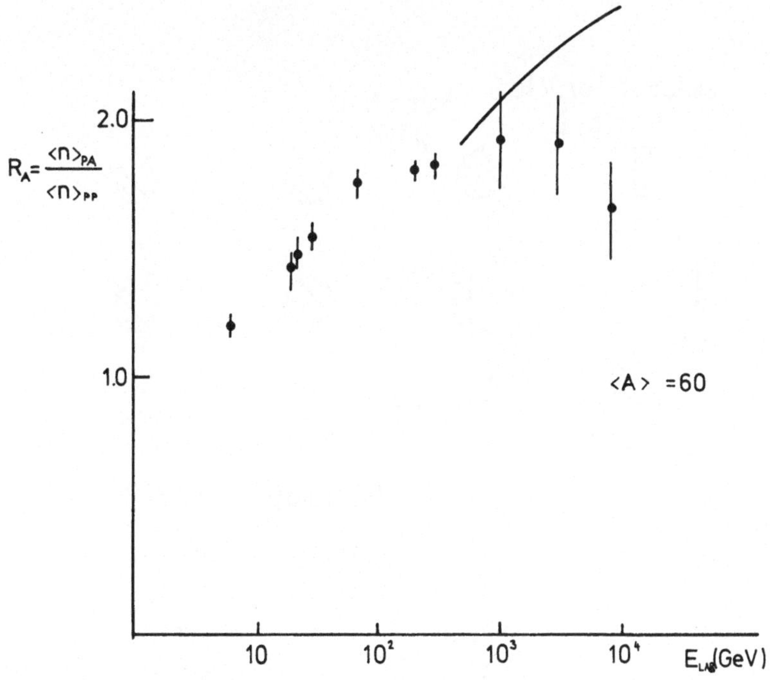
$R_A = \frac{\langle n \rangle_{PA}}{\langle n \rangle_{PP}}$
2.0
1.0
⟨A⟩ =60
10
10^2
10^3
10^4
E_{LAB}(GeV)

Fig. 22

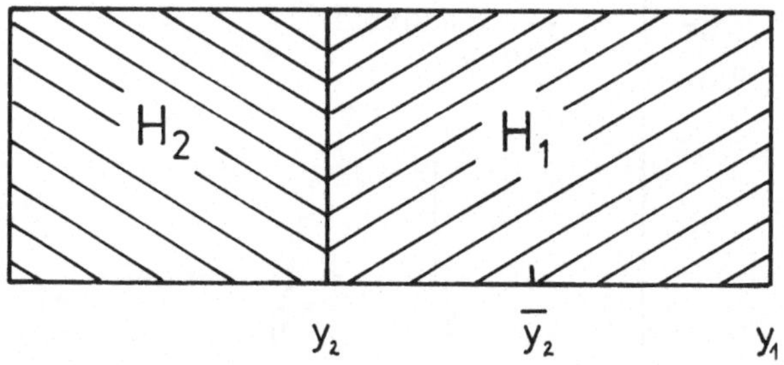
H2
H1
y2
ȳ2
y1

Fig. 23

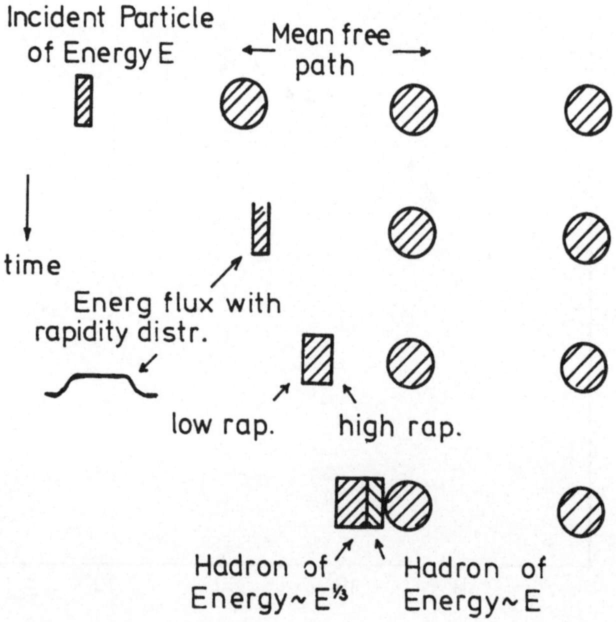
Incident Particle
of Energy E
Mean free
path
time
Energ flux with
rapidity distr.
low rap.
high rap.
Hadron of
Energy ~ E⅓
Hadron of
Energy ~ E

Fig. 24

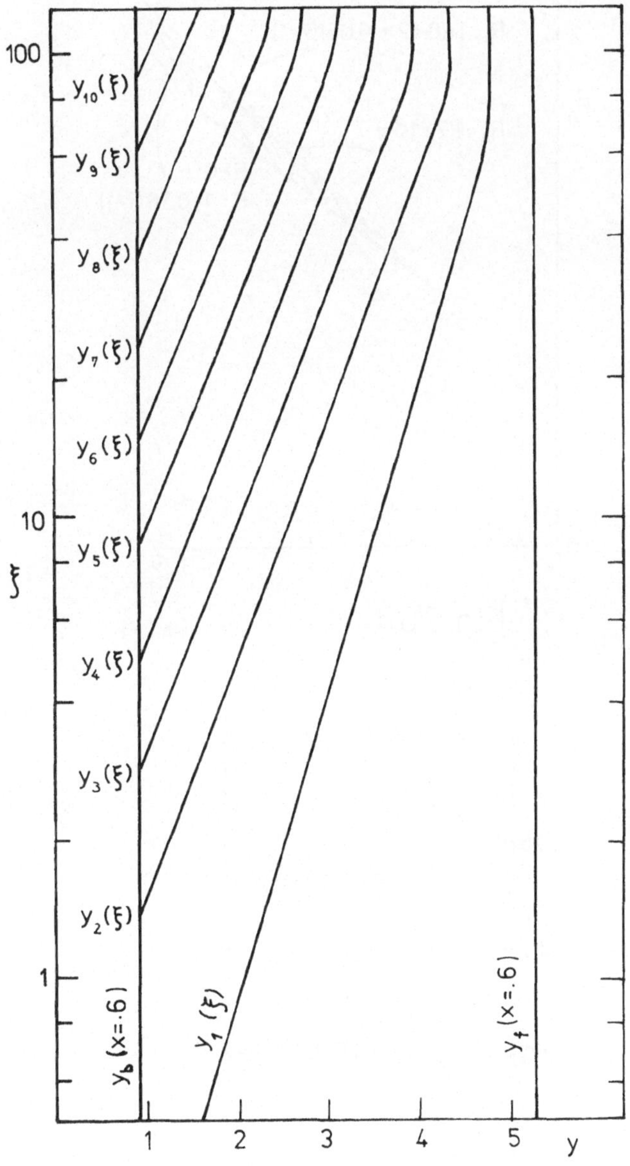
100
10
1
ξ
$y_{10}(\xi)$
$y_9(\xi)$
$y_8(\xi)$
$y_7(\xi)$
$y_6(\xi)$
$y_5(\xi)$
$y_4(\xi)$
$y_3(\xi)$
$y_2(\xi)$
$y_1(\xi)$
$y_b(x=.6)$
$y_t(x=.6)$
1
2
3
4
5
y

Fig. 25

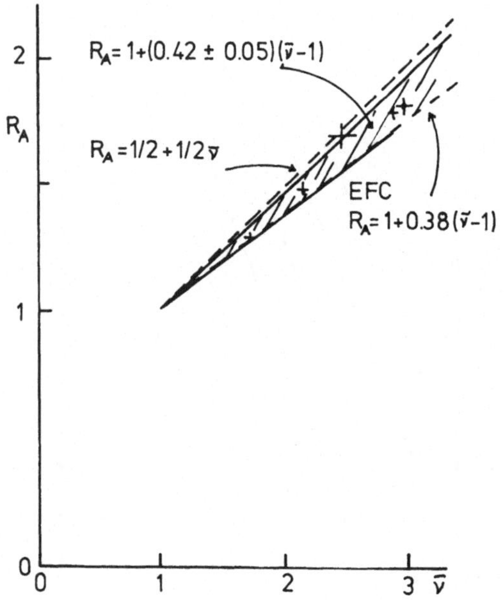

Fig. 26a

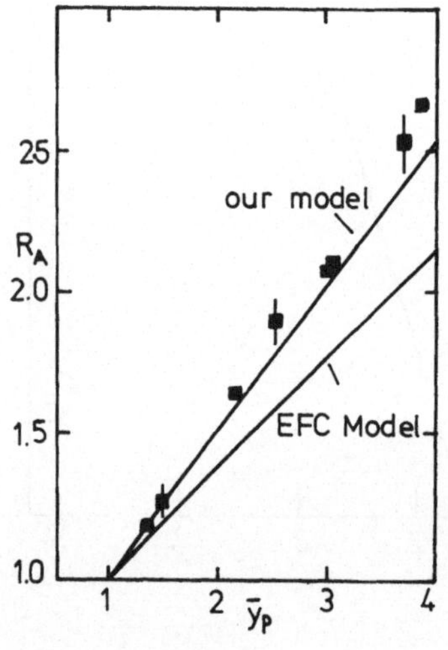

Fig. 26 b

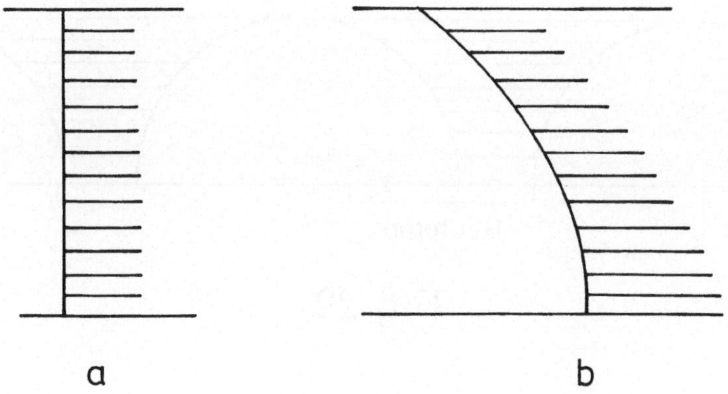

Fig.27

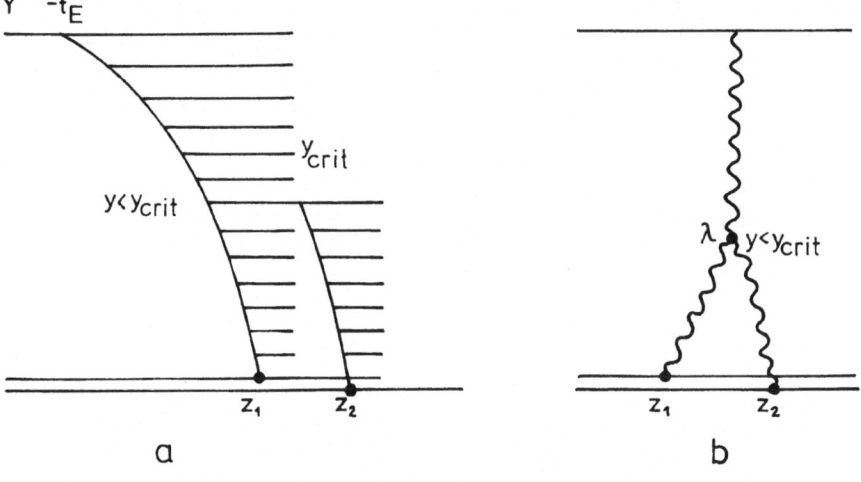

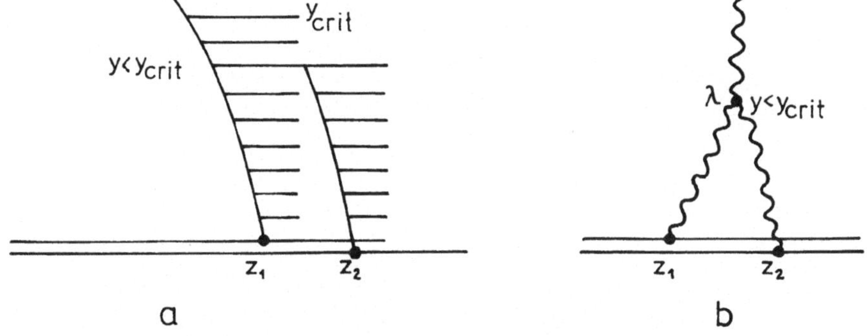

Fig.28

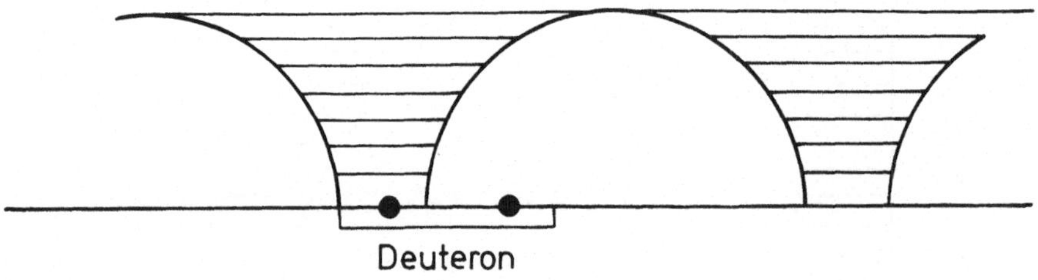

Fig. 29

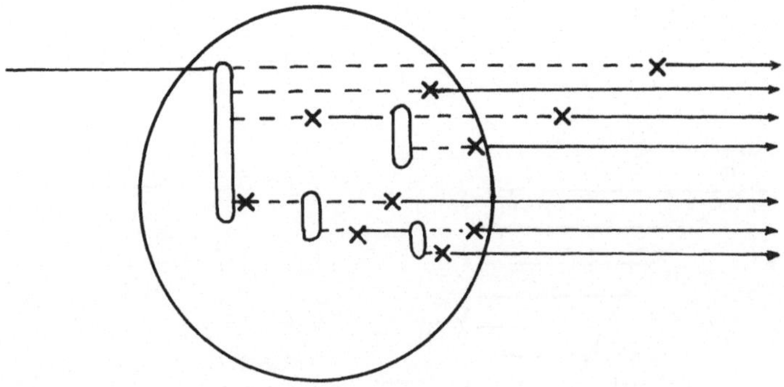

Fig. 30

Acta Physica Austriaca, Suppl. XVIII, 789–834 (1977)

COSMIC GAMMA RADIATION+

by

G. BÖRNER
Max-Planck Institut für Physik und Astrophysik
München, Germany

I. INTRODUCTION

The energy spectrum of the cosmic gamma radiation extends over 9 orders of magnitude from 10^5 eV to 10^{14} eV (or equivalently covers the frequency range from 10^{19} sec^{-1} to 10^{26} sec^{-1}). Above 10^{14} eV cosmic gamma rays are absorbed very efficiently through pair production on the 2.7^{o}K black body background radiation (cf. Lüst & Pinkau 1967). These gamma rays give direct information on the interaction of highly relativistic electrons with photons, matter, and magnetic fields, and on the interaction of energetic nuclei with matter. Because of their low absorption they can traverse not only our own galaxy, but can reach us from extremely early epochs of the universe. One thus hopes through gamma ray observation to learn about the sites of nucleosynthesis, the location

+ Lecture given at XVI. Internationale Universitätswochen für Kernphysik, Schladming, Austria, February 24-March 5, 1977.

of discrete sources of cosmic rays, the conditions in the universe at an early cosmological epoch, and the properties of particle acceleration mechanisms in specific localized sources such as Quasars, Radio galaxies, or pulsars. Although gamma ray astronomy holds these potentially great rewards, the experimental progress has been slow and beset by many difficulties.

There are extremely few extraterrestrial quanta, although the energy flux is $\frac{dE}{E}$ from 1 keV to 100 MeV, i.e. there are equal amounts of energy in successive decades of the energy. But whereas the number of photons with energy above 1 keV in the γ-ray range is $\approx$ 15 cm^{-2} sec^{-1} sr^{-1} (which is about 100 times the charged cosmic ray flux at medium latitudes), the number of gamma photons above 100 MeV is $2x10^{-4}$ cm^{-2} sec^{-1} sr^{-1} (about 10^{-3} of the c.c.r.f.). Since charged particles generate secondary gamma rays locally with high efficiency, one has a high instrument background just because of cosmic rays hitting the detection equipment. On the other hand the actual detection rate of extraterrestrial gamma rays is very low: The detector aboard OSO-3-(Clark et al. 1968) had an effective aperture of about 0.5 cm^{-2} sr, and registers cosmic gamma ray photons at the rate of 1 per 3 hours.

Balloon experiments have additional difficulties with gamma radiation produced by nuclear collision of cosmic ray particles in the air (production of π^o and the decay $\pi^o \rightarrow 2\gamma$), and bremsstrahlung of relativistic particles in the atmosphere.

Only in recent years have vastly improved detectors flown on satellites (Kraushaar et al. 1972; SAS-2

(NASA experiment) and COS-B (European collaboration)) yielded conclusive and interesting results in the range around 100 MeV. In the range of a few MeV the Compton-telescope (Schönfelder (1975)and references therein) has similarly improved the experimental situation.

The detection of low energy gamma rays $\leq$10 MeV occurs through the photoeffect or the Comptoneffect. Omnidirectional scintillationcounters are widely used in this energy range: A NaJ or CsJ crystal is surrounded by a thin plastic scintillator in anticoincidence arrangement to reject charged particles. With these counters one has big problems in estimating instrumental background effects, especially below 3.5 MeV, where atoms in the NaJ crystal activated by cosmic ray particles radiate (see Schönfelder (1975)). A real step forward, and still the best instrument to date in the range <10 MeV is the Compton-telescope developed by V. Schönfelder and his colleagues at the Max-Planck-Institute für Extraterrestrische Physik in Garching/München (cf. Schönfelder (1975)). The instrument (see fig. 1) consists of 2 detectors which are 120 cm apart, surrounded by active anticoincidence counters which reject charged particles. A gamma ray is registered by two subsequent Compton collisions, first in the upper detector (15x15x15 cm^3) then in the lower one (60x60x15 cm^3). A time of flight measurement guarantees that gamma rays are registered only if the pulse in the upper detector is a few nanoseconds earlier than in the lower. For each event the pulse heights in the two detectors are measured; they are a measure of the energy losses in both detectors, and from these values the energy and the direction of the incoming gamma ray can be determined (Schönfelder (1975)).

Above 10 MeV the gamma rays are registered by the pair production effect. The presently used detectors consist of spark chambers which give a 3 dimensional picture of the V-shaped tracks of the electron-positron pair. Charged particles are rejected by anticoincidence counters. One obtains information in the direction of the incident γ-ray, with an angular resolution of 2° at 100 MeV and $\propto E^{-2/3}$. Information on the energy can be derived from the opening angle of the pair $\bar{\alpha} \propto E_{\gamma}^{-1}$, and from Coulomb-scattering of the electrons.

Observations from the ground are possible for the very high energy (10^{11} eV) gamma rays. A high energy gamma ray hitting the atmosphere induces a cascade shower (pair production) and the Cerenkov emission of these energetic electrons can be detected. The radius of the Cerenkov light cone at the ground is about 100 m providing a large detective area (10^{4} m^{2}) and hence rather high counting rates. The background in this approach is extremely high, however, being a sum of cascade Cerenkov light from high energy charged cosmic ray particles, starlight and other light in the night sky. Because of these formidable background problems claims of detection of high energy gamma rays (from the Crab, from Cent) have to be treated with caution. Grindlay (1976) gives an account of the present situation.

II. THE DIFFUSE COMPONENT

(II.1) Extragalactic Diffuse Radiation

The experimental and theoretical situation has been reviewed recently by Schönfelder (1975), and I will just pick out a few salient points from the discussion given by him.

In the X-ray region one has found besides various discrete sources (galactic and extragalactic) an isotropic, diffuse component. Below 100 keV the anisotropy of this diffuse component has an upper limit ≤ 4 % in angular intervals of 10° (Schwartz (1970); Fabian & Sanford (1971)). Such an extragalactic radiation is also indicated by the OSO-III data around 100 MeV (Kraushaar et al. (1972)) since the radiation comes from high galactic latitudes and is isotropic within the statistics (only 266 quanta, therefore $\pm$ 17 % statistical fluctuation in each angle of 1 sterad). The SAS-B satellite observations (Fichtel et al. (1975)) also find a flux of photons >50 MeV from high galactic latitudes, which must therefore be of extragalactic origin.

Similarly the ballon-flown Compton-telescope experiments of Schönfelder et al. (1976) (see also Schönfelder (1975)) did not find a difference in flux between regions with high and low galactic latitudes. The experimental results are shown in fig. 2 (taken from Schönfelder et al. (1976)). You can see that a straight line extrapolation of the X-ray measurements below 100 keV with a spectral shape of

$11x10^{-3} E^{-2.3}$ Photons cm^{-2} sec^{-1} sr^{-1} MeV^{-1}

is consistent with the measurements around 100 MeV. The experiments between 0.5 MeV and 1.5 MeV (e.g. Apollo XV) have uncertain corrections, and the Compton-telescope did not measure in this range. This introduces an uncertainty in the specific straight line extrapolation, but the excess above 1 MeV seems to be real feature of the diffuse component. Above 20 MeV the spectrum steepens somewhat and meets the straight line extrapolation value at 100 MeV. The isotropy of the diffuse flux has not been established so far, since most experiments looked only at particular regions of the sky, Kraushaar et al. (1972) being the only overall survey.

Many speculations as to the origin of the diffuse component have been advanced, but no convincing generally accepted models exist. Further refinement of the observations may give some clarification, e.g. to the possible contribution of localized extragalactic sources. In the X-ray regime already 10-20 % of the diffuse flux can be attributed to the superposition of discrete sources such as Quasars, galaxies, clusters of galaxies. No such estimate is possible as yet in the gamma energy range. One can only point out that the sum of the radiation of all normal galaxies above 100 MeV (radiating $1.8x10^{42}$ sec^{-1} as our own galaxy) would contribute only 3 % of the total flux. But there may be far more energetic sources, perhaps a higher density of quasars or radio galaxies at earlier epochs in the universe ($n_1 \approx$ 1000 n_o (present density), were n_1 corresponds to a redshift of z = 2), which might contribute a large part of the observed diffuse flux. As long as one does not have observations of extragalactic sources, this question cannot be discussed reasonably, and therefore one should also consider the two models which want to explain the diffuse gamma radiation over the whole spectrum by one specific physical process.

The first model attributes the production of gamma rays to intergalactic electrons which undergo collisions with the photons of the 2.7°K universal blackbody background (inverse Compton effect). The isotropically distributed quanta of the 2.7°K background with an average energy of $\varepsilon_o = 6 \times 10^{-4}$ eV have after the collision an energy

$$E_\gamma = 4/3\ \varepsilon_o\ \gamma^2 \tag{2.1}$$

where $\gamma m_o c^2$ is the electron energy, and the condition $\gamma\varepsilon_o \ll m_o c^2$ is assumed to be satisfied. The total cross section in this case is just the classical Thomson cross section

$$\sigma_T = 0.665 \times 10^{-24}\ \text{cm}^2 \tag{2.2}$$

and the differential cross section for a single electron of energy E scattering on a photon of energy ε_o to give a photon of energy E_γ is (Ginzburg & Syrovatskii (1964)):

$$\sigma(E_\gamma, \varepsilon_o, E) = \frac{3}{32}\varepsilon_o^{-2}\gamma^{-4}\left(2E_\gamma - E_\gamma^2/\gamma^2\varepsilon_o + 4E_\gamma \ln\frac{E_\gamma}{4\varepsilon_o\gamma^2} + 8\varepsilon_o\gamma^2\right) . \tag{2.3}$$

For an electron intensity spectrum $I_e(E)$ (cm^{-2} sec^{-1} sr^{-2} MeV^{-1}) isotropically scattering on an isotropic photon field, the resulting γ-ray spectrum due to scattering along a path length L is

$$I_\gamma(E_\gamma) = L \int_0^\infty n_{ph}(\varepsilon_o)d\varepsilon_o \int_{E_{min}}^\infty \sigma(E_\gamma,\varepsilon_o,E)I_e(E)dE \qquad (2.4)$$

$$E_{min} = m_o c^2 \; (E_\gamma/4\varepsilon_o)^{1/2} \quad .$$

For $I_e(E) \propto E^{-\alpha}$, and an isotropic black body radiation field

$$n_{ph}(\varepsilon_o) = \frac{8\pi}{h^3c^3} \frac{\varepsilon_o^2}{\exp(\varepsilon_o/kT)-1} \qquad (2.5)$$

one finds

$$I_\gamma(E_\gamma) \propto E_\gamma^{-(\alpha+1)/2} \quad . \qquad (2.6)$$

In a model of this type an equilibrium spectrum of the electrons in intergalactic space has to be determined. Brecher & Morrison (1969) find $I_e(E) \propto E^{-3.6}$ for $E > 100$ MeV (below 100 MeV energy loss by cosmological expansion dominates and $\alpha = 2.6$), where they take the radiogalaxies as sources for the intergalactic electrons. Then the photon spectrum above 30 eV (eq. (2.1)) has a slope of $E^{-2.3}$. This fits well the X-ray observations above 40 keV and the gamma ray spectrum around 100 MeV. But the diffuse X-ray spectrum seems to change its slope below 10 keV (e.g. Pal (1973)), which requires somewhat artificial assumptions of this model. In addition the excess above 1 MeV cannot be explained, and the required electron intensities are very high (normal galaxies should emit 10 times as many electrons as our own galaxy). Evolutionary models of the universe have, of course, enough free parameters to fit any desired shape of the present electron intensity spectrum.

The second model is the mechanism of bremsstrahlung by nonthermal electrons in the intergalactic gas. This process can be thought of as the transition of an electron from a state (E_o, p_o) to a state $(E, \underline{p})$ in the field of a nucleus whose recoil energy is neglected. In this transition a photon of energy $E_\gamma = E_o - E$ is emitted. This process is extremely inefficient, because the radiation length in intergalactic matter is X_o = 65 g/cm^2, whereas the present column density in a closed universe is 0.1 g/cm^2. Thus even strong evolution models which place the photon production at early epochs (z=10; Arons et al. (1971)) require unreasonably high electron intensities ($n_{e,o}$ = 0.5 eV/cm^3 compared to 10^{-2} eV/cm^3 for the energy density of electrons in our galaxy).

The inverse Compton model requires an additional component to explain the excess in the intensity of the diffuse gamma radiation between 1 and 100 MeV. Most of the proposals for these features rely on production mechanisms which took place at early stages in the history of the universe (z = 100). The models are very speculative, but of great cosmological significance, since gamma rays between 1 MeV and 10 GeV can reach us from early times with redshifts $z \gtrsim 70$ (epoch $t(z) = t_o/(1+z)^{3/2}$ for an Einstein-de Sitter universe) for a present mean density of $n_o = 10^{-5}$ cm^{-3} ($z \gtrsim 300$ for

$n_o = 10^{-7}$ cm^{-3}). These gamma rays contain information about much earlier stages in the history of the universe than e.g. the quasars ($z \lesssim 3.5$). Stecker (1969; 1971) has proposed models which ascribe the excess to redshifted quanta from the decay

$$\pi^o \rightarrow 2\gamma .$$

In the rest frame of the π^o meson both gamma quanta are emitted isotropically (i.e. with equal probability in all directions) with energy $m_\pi c^2/2 \approx 68$ MeV. Transformation to the laboratory system, where π^o has the kinetic energy E_π, shows that the quanta are preferentially emitted in the direction of motion of the meson. The intensity distribution after integration over all directions of emission is constant around 68 MeV and zero otherwise with the boundary given by

$$E_{\gamma \max \atop \min} = \frac{1}{2} (E_\pi + m_\pi c^2) \pm \frac{1}{2} (E_\pi^2 + 2E_\pi m_\pi c^2)^{1/2} . \qquad (2.7)$$

In fig. 3 a few simple decay spectra are shown.

Gamma emission from early epochs in the universe could be observed now (because of the low absorption cross section, mean free path ≈ 65 g/cm^2), but the maximum would be redshifted from 68 MeV to $(1/(1+z)) \cdot$ 68 MeV. For redshifts between $z = 0$ and $z = 70$ this radiation would now be observed in the energy range 1 MeV to 70 MeV, i.e. could correspond to the observed excess in the diffuse background. In his first model, Stecker (1969) has suggested that the π^o mesons are produced by interactions of cosmic rays with the intergalactic gas. A good fit to the observations can be obtained with the assumption of an Einstein-deSitter model ($q_o = 1/2$, $n_o = 10^{-5}$ cm^{-3}), where the injection of the cosmic rays happens burst-like around z = 100 (E.deS. age: 10^7 years). Each one of 20 % of all present galaxies had to inject 10^{62} ergs in cosmic rays within 10^7 to 10^8 years. The present intensity of intergalactic cosmic rays would still be 10^{-4} of galactic intensity. Instead of an Einstein-deSitter model present-day cosmological evidence rather favours an open model, with $q_o \approx 0$ and e.g. $n_o \approx 10^{-7}$ cm^{-3}. This would destroy the good fit, because the absorption would be less and radiation with high z would be relatively more intense, such that a normalization at 100 MeV would give a 1 MeV excess too large by an order of magnitude. Many other parameters can, of course, be changed accordingly to restore the fit (e.g. more constant injection of cosmic rays etc.).

Stecker's (1971) second model invokes π^o production by matter-antimatter annihilation in a baryonsymmetric universe. Such a universe consists of equal amounts of matter and antimatter, which became separated immediately after the big bang origin. Now the separate regions of matter and antimatter are of the size of clusters of galaxies, and annihilation takes place at their boundaries. The reaction

$p + \bar{p} \rightarrow \pi^+ + \pi^- + \pi^o$ at thermal energies

(E_p < 286 MeV) is considered. The energy of the π^o mesons lies in the range 5 MeV $\leq E_\pi \leq$ 923 MeV. Similarly the gamma ray spectrum $f_A(E_\gamma)$ has a flat maximum around 68 MeV, drops sharply towards the sides and is zero below 5 MeV and above 923 MeV. The spectrum now is obtained by integrating over all the history

$$I_A(E_\gamma) = \int_{z=o}^{\infty} \frac{(1+z)^{2+x}}{(1+2q_o z)^{1/2}} f_A\,(E_\gamma(1+z))e^{-\tau(E_\gamma,z)}\,dz \;.(2.8)$$

The factor $(1+z)^x$ takes into account the dependence of the annihilation cross-section on temperature. Stecker (1971) takes x = 0.36 for z < 150, x = -1/2 for z > 150. This model gives a good fit to the observations if again one uses the Einstein-deSitter model without absorption ($\tau = 0$) ($I_A(E_\gamma) \propto E^{-2.5}$). An open cosmological model ($q_o = 0$) would not give a good fit, the spectrum would become too steep ($I_A(E_\gamma) \propto E^{-3}$).

We see that further experimental and theoretical work is necessary, before one can judge whether Stecker's models are more than extremely interesting speculations.

It has also been proposed that the excess is due to inverse Compton scattering of extragalactic protons with energy $>10^{15}$ eV on the 2.7^oK background and on starlight (Strong et al. (1973)). This sounds somewhat artificial, because the most natural explanation for the change in slope of the differential cosmic ray spectrum around 3×10^{15} eV ($\propto E^{-2.6}$ below, $\propto E^{-3.2}$ above) is that the high energy protons cannot be contained by the magnetic field of our galaxy, and this intensity loss appears as a steepening of the spectrum (Peters (1959)).

(II.2) Galactic Component

High energy gamma rays from the galactic disk have been seen by the SAS-2 satellite (>35 MeV) and the CosB satellite (>70 MeV). Although the contribution of point sources is difficult to exclude, a diffuse component clearly seems to be present. In figure 4 (Thompson et al. (1976)) the SAS-2 results for the high energy gamma ray emission (>100 MeV) as a function of galactic longitude are shown, summed between $-10^{\circ} <$ $b'' < 10^{\circ}$. One clearly sees the dominance of the emission from the plane over the diffuse flux (1×10^{-4} cm^{-2} sec^{-1} sr^{-1}: a factor 3 above the isotropic diffuse flux), and the high intensity of the galactic center region. Thompson et al. (1976) further claim that distinct peaks in their data at $l'' = 315^{\circ}$, 330°, 345°, 0°, 35° are correlated with the position of spiral arms in our galaxy. This may well be the case, but the presently available data are not yet conclusive. One should perhaps rely on further observations of CosB which will give a more detailed sky map. A survey of the galactic disk by Cos-B (Caravane collab. 1976) in the regions $244^{\circ} < l'' < 2840^{\circ}$and $350^{\circ} < l'' < 20^{\circ}$ indicates the existence of a wide component and a narrow one ($<4^{\circ}$ in galactic latitude above 300 MeV) (see fig. 5). The sensitivity of the Cos-B experiments permits the investigation of these features in more detail (fig.6 from Caravane Coll. (1976)). These measurements suggest that most of the high energy galactic gamma rays originate

i) from close-by regions whose latitude-profile is compatible with a distance $\leq$ 1 kpc (these are from $-10^{\circ} < b'' < -2^{\circ}$, $2^{\circ} < b'' < 10^{\circ}$);

ii) from distant regions, compatible with a thin line source ($-2^{\circ} < b'' < 2^{\circ}$) indicating a distance > 3 kpc (2.5×10^{-4} cm^{-2} sec^{-1} sr^{-1}).

At l" = 270° such a line source is not observed, implying that the contribution from the outer galaxy is very weak. In contrast to the SAS-2 data a feature at l" = 0° is not found. It cannot be excluded that the enhancement toward the galactic centre is due to a localized source.

Models for a production mechanism of these galactic gamma rays have to consider the interaction of cosmic rays, interstellar matter, and magnetic fields. Since the gamma rays are not absorbed at all, they bring direct information on the physical conditions at the sites where they have been produced.

According to Thompson et al. (1976) the data suggest that above 50 MeV the gamma rays come from π^{o} decay, whereas below 50 MeV inverse Compton scattering on starlight or electron bremsstrahlung could be the relevant mechanism. In principle the production mechanism could be decided by observations, since the Compton gamma ray spectrum (as well as the bremsstrahlung spectrum) is a power law which is produced by a power law distribution of electrons whereas the π^{o} decay gives a spectrum with a broad peak around 68 MeV. Good energy resolution ($\leq$20 %) will be necessary to determine the spectral shape.

The π^{o} are produced by collisions of high energy cosmic rays with interstellar matter, and therefore the gamma ray flux is proportional to the product of the density of interstellar matter with the density of cosmic rays at a specific location in the galaxy. Since the matter distribution (deduced from 21 cm line observation HI, 2.6 mm CO line-H_2) shows less contrast between the galactic center and the outer regions of the disk than the gamma ray intensity profile, it has to be concluded that the cosmic ray density is not uniform in the galaxy. The correlation with the 150 MHz synchrotron map (Landecker & Wielebinski (1970)), which shows a strong center to anticenter contrast, lends support to the idea that in the galactic disk the sum of kinetic gas pressure, cosmic ray pressure and magnetic field pressure is in equilibrium with the gravitational attraction of the gas. Thus the cosmic ray density is coupled to the matter density and the magnetic field distribution in the galaxy.

Strong peaks in the gamma ray intensity distribution offer additional clues: The 35° peak (SAS-2), e.g. would correspond to a line of sight tangent to a spiral arm at a radius of 6 kpc. This spiral feature has also been observed in H_2 measurements indicating the molecular hydrogen as a principal source of the galactic gamma readiation.

The observations require a strong galactic variation of the cosmic rays (within 7 kpc of the galactic center), which excludes a purely extragalactic origin of the nuclei of the cosmic radiation. Supernovae and Pulsars may well be able to produce the required cosmic ray intensity within the galaxy. Detailed models are, however, not yet available.

III. GAMMA RAY LINES

Chupp et al. (1973) have established gamma spectroscopy as an observational science by detecting gamma ray lines in the large (3B) solar flares of August 4 and 7, 1972. Hard photons from these flares ranged in energy up to almost 10^7 eV, and consisted of both emission lines and continuum. Fig. 7 (from Chupp et al. (1973)) shows the observed lines at 0.51, 2.2, 4.4, and 6.2 MeV, as well as two Co lines, which were produced in the detector for calibration purposes. Detailed theoretical work before and after the observations (Ramaty & Lingenfelter 1967, 1972) supports the conclusion that these lines are due to positron annihilation, neutron capture on hydrogen, deexcitation of C^{12} and O^{16} nuclei, and that the continuum is most likely bremsstrahlung of relativistic electrons.

The basic mechanisms in solar flares are interactions between accelerated charged particles and the ambient solar atmosphere. Gamma ray lines from excited nuclei are prompt emissions, because the excited levels decay in time intervals which are much shorter than any of the characteristic times of the flare. The neutrons result mainly from the disintegration of He^4 nuclei in proton-alphaparticle collisions. Downward moving neutrons are thermalized and undergo capture in the photosphere at a column density of 10^{17} protons/cm^2. Since neutrons can also be captured by He^3 to form tritium without emission of gamma rays, the observed line intensity together with other flare parameters can be used to place an upper limit on the photospheric abundance of He^3, which is important for theories of element synthesis. Positrons in solar flares result from the decay of π^+ mesons and of various radioactive nuclei produced by the nuclear reactions. The finite half lives of these positron emitters and the finite capture time of the neutrons in the photosphere result in a considerable time delay with respect to the prompt nuclear deexcitation lines. This has been verified by data from the flare of

August 7, 1972 (Chupp et al. (1973)), when at a time when all prompt emissions were very small, the 2.2 MeV and 0.51 MeV line were still observable.

Extraterrestrial and extrasolar gamma ray lines may be expected from decay processes during nucleosynthesis. The suggestion that the heavier elements are synthesized in supernova explosions leads to the expectation of line structure either from close by supernovae or in the diffuse background. Theoretical estimates indicate that with slightly improved energy resolutions in present detectors such structures might be observed. Clayton & Silk (1969) have investigated the silicon burning processes in a supernova explosion which synthesize the elements up to Fe^{56}. β-capture transforms Ni^{56} to Co^{56} (6.1 days halflife), and Co^{56} to Fe^{56} (77 days halflife), and with these processes gamma lines are omitted: $Ni^{56} \rightarrow Co^{56}$ gives lines at 0.163, 0.276, 0.472, 0.745, 0.812, 1.56 MeV with an average emission of 2.1 quanta per decay. $Co^{56} \rightarrow Fe^{56}$ gives decays at 0.511, 0.847, 1.03, 1.24, 1.76, 2.02, 2.60, 3.26 MeV with an average emission of 2.8 quanta per decay reaction. Because of the short halflives of 6 resp. 77 days one has to be so lucky as to find a supernova in our galaxy which is right now undergoing explosion. If, however, all the Fe^{56} in the universe has been produced that way, the integral effect should be observable. For a matter density of $\rho_o = 1.7\text{x}10^{-31}$ g/cm^3 in the present universe one has $n_{Fe} = 2.3\text{x}10^{-12}$ g/cm^3, and then (neglecting absorption) one finds a total flux of all $Ni^{56} \rightarrow Co^{56} \rightarrow Fe^{56}$ lines of

$$2.7\text{x}10^{-2}\ \text{cm}^{-2}\ \text{sec}^{-1}\ \text{sr}^{-1}$$

comparable to the total diffuse flux above 500 keV.

To determine the spectral distribution, one has to take into account the redshift of the lines which have been emitted at earlier stages in the universe. In fig. 8 (after Clayton & Silk (1969)) you can see the spectral distribution calculated for an Einstein-deSitter cosmos ($t_o = 11.8 \times 10^9$ years, start of nucleosynthesis at $t = 2 \times 10^9$ years). Exponentially decreasing rate of nucleosynthesis/galaxy (e^{-2} of initial rate today - solid line) and constant rate (dashed line) give the same qualitative behaviour. The steps in the spectrum occur at the energy of each line, and their size is proportional to the average rate of nucleosynthesis today. Measurements of these steps (if they are not smeared out by absorption) seem possible in the near future. According to Schönfelder (1975) flux measurements of 10^{-5} cm^{-2} sec^{-1} and an energy resolution of a few keV are necessary and will soon be achieved. Then important ideas on element formation can be put to test. In his book Chupp (1976) extensively discusses many relevant experimental and theoretical aspects.

IV. GAMMA RAY PULSARS

The Crab pulsar (PSR 0532, $l'' = 185^o$, $b'' = -6^o$), has a period $T = 0.033$ sec, a derivative $dT/dt = 420 \times 10^{-5}$ sec/sec, and is at a distance of 2 kpc in the center of the supernova remnant of the Crab supernova of 1054 A.D. This pulsar has been observed in gamma rays by the SAS-2 satellite (Fichtel et al. 1976) and by the Cos-B satellite (Caravane Coll. (1976)). It emits pulses in the energy range between 50 MeV and 2000 MeV with the precise period known already from radio, optical, and X-ray observations of this source. The flux above 100 MeV is (Fichtel et al. 1976)

$$(3.2 \pm 0.9) \times 10^{-6} \text{ cm}^{-2} \text{ sec}^{-1} ,$$

and at least a fraction of 35 % of the flux above 50 MeV is pulsed (according to Cos-B data), as compared to a pulsed fraction of 10 % for the X-ray emission between 2-12 keV. At 100 keV the pulsed fraction increases to 40 %. Above 100 MeV the pulsed intensity according to SAS-2 data is $(2.2 \pm 0.7)\times 10^{-6}$ cm^{-2} sec^{-1}, so that at higher energies the total gamma ray enhancement is due to pulsed emission. A superposition of the Cos-B data for the X-ray lightcurve (2-12 keV) and the gamma ray light curve (50 MeV) shows a very good coincidence in phase (fig. 9) with the double pulse structure characteristic for the Crab. The spectrum of the Crab gamma rays is consistent with a power-law extrapolation from the X-ray energies for both the pulsed and the total intensities.

While this gamma ray flux from the Crab had more or less been expected, and had already been indicated by balloon observations, a big surprise was in store when the pulsar in the Vela supernova remnant (PSR 0833, T = 0.089 sec, dT/dt = 120×10^{-15}) was observed. This pulsar had long been known only as a radio source and has only recently been detected in optical light (magnitude $25^{m}.6$). No pulsed emission in X-rays has been detected so far. (Rappaport et al. (1974) give an upper limit of 3×10^{-11} erg cm^{-2} sec^{-1} for 1.5-10 keV). The surprising result (first found by Thompson et al. (1975), and confirmed and improved with better statistics by the COS-B group (1976) was that the Vela pulsar emits gamma radiation pulsed at the radio period, with a lower limit of the pulsed fraction of the intensity of 85 % (100 MeV). The lightcurve for the gamma pulses is shown in fig. 10, taken from the above mentioned publication by the COS-B group (1976). It is interesting to note that there is a clear double pulse, whereas the radio pulse is single. Neither of the two peaks (38 msec separation) is in phase with the radio pulse. This is in contrast to the Crab pulsar, where the pulses from radio through gamma ray energies are in phase. A comparison of the Crab and Vela pulsar in two gamma ray energy ranges, however, shows a strong similarity: For both pulsars the pulse shape is determined by two narrow peaks separated by 0.42 of the period. This evidence may indicate that the gamma ray emission is a common property of the pulse formation mechanism, while at lower energies the mechanism is obscured by other complicated processes.

The pulsed flux of the Vela pulsar above 50 MeV is $1.3x10^{-5}$ cm^{-2} sec^{-1} and above 100 MeV is $1.0x10^{-5}$ cm^{-2} sec^{-1}. The lower values of SAS-2 (a factor 2 lower - Thompson et al. (1975)) have been due to calibration problems and have been corrected meanwhile to the COS-B value (addendum to Thompson et al.).

To convey some feeling for the importance of these observational results I want to review briefly some of the theoretical ideas on pulsars (Ruderman (1972)): It is now generally accepted that pulsars are rotating neutron stars. A neutron star is a massive, extremely dense, and small object. Typically it has a mass of 1 M_o and a radius of 10 km. The physics of cold, dense matter-nuclear physics at 10 times nuclear density - tells us that such massive bodies can exist in equilibrium after they have exhausted their nuclear fuel, because the repulsive core of the nucleon-nucleon interactions balances gravity. The interior of such stars would also be a field of close contacts of astrophysics with high energy physics, but here we have to leave it for a future Schladming school.

The formation of such a small and dense object in a supernova eventwill also lead to a strong magnetic field. Because of the high electron number to be expected in such a star, the electrical conductivity will be very large, and the magnetic flux will be frozen in. Thus if we start with a normal star of radius $R = 10^{11}$ cm, $M = 1\ M_o$, $\rho = 1\ g/cm^2$ and $B = 100$ Gauss, we will end up with a neutron star of $R = 10^6$ cm, $\rho = 10^{15}\ g/cm^3$, $M = 1\ M_o$, and a magnetic field $B = 10^{12}$ Gauss. These strong magnetic fields provide the link of communication between the rotating neutron star and the observer. All that is observable is the electromagnetic radiation, produced by charged particles accelerated in the strong magnetic field of the pulsar and reaching us as continuous radiation or in pulsed form. The source of energy for this radiation is the loss of rotational energy of the neutron star.

$$\dot{E} = I\ \Omega\ \dot{\Omega} \tag{4.1}$$

(I: moment of inertia, $\Omega = \frac{2\pi}{T}$).

It haus been shown by Goldreich & Julian (1969) that despite the strong gravitational attraction and the assumption $T = 0$ in a neutron star, there cannot be a vacuum outside of the star. Because of the high conductivity in the interior, the electric field is given by

$$\underline{E} - (\underline{\Omega} \times \underline{r})/c \times \underline{B} = 0 \quad . \tag{4.2}$$

If it is assumed that outside is a vacuum, the matching of the interior solution to an outside solution consistent with a dipolar $\underline{B}$-field gives an electric field component parallel to $\underline{B}$ outside of the star, which will accelerate the surface charge layer of the star. Thus a rotating magnetic neutron star must possess a magnetosphere, composed of charged particles travelling outwards along the magnetic field lines. But then very quicly (4.1) becomes valid in the magnetosphere also, as long as inertial effects of centrifugal acceleration of the plasma are ignored, and therefore

$$\underline{E} \cdot \underline{B} = 0 \tag{4.3}$$

everywhere in the magnetosphere. What then is the stationary state? Because magnetosphere corotation cannot hold exactly at the "light cylinder"

$$r_L = c/\Omega \tag{4.4}$$

$\underline{E} \cdot \underline{B} = 0$ is only an approximation within too. But there is no consensus among pulsar model builders where the condition fails.

Only the component of $\underline{E}$ parallel to $\underline{B}$ can give a charged particle a relativistic energy (if $|E| \ll |B|$). For the vacuum fields $\underline{E}$ is quadrupolar

$$(\underline{E} \cdot \underline{B})_{vac} \simeq \frac{\Omega r}{c} B_o^2 \left(\frac{R}{r}\right)^4 \tag{4.5}$$

(B_o: surface field, canonical value 10^{12} G; $R = 10^6$ cm).

Even for $(\underline{E} \cdot \underline{B})_{magn.} = (\underline{E} \cdot \underline{B})_{vac} \times 10^{-6}$ a charged particle near the Crab pulsar could acquire 10^{12} eV after it has travelled 10^6 cm from the stellar surface. The sign of $\underline{E}$ changes between pole and equator, so that the outward current flux also has one sign in polar cap and an opposite one in a surrounding sheath. Assumptions about $\underline{E} \cdot \underline{B}$, and therefore about the region of particle acceleration, exist in great variety (Ruderman 1972): From

i) acceleration well beyond the light cylinder in a dipole wave radiation field, over

ii) acceleration just within r_L, where corotation can be maintained only by relativistic azimuthal velocities, to

iii) acceleration near the star by $\underline{E} \cdot \underline{B} \neq 0$ (Ruderman & Sutherland (1975), Sturrock (1971).

In Sturrock's (1971) model particle acceleration near the star produces extreme relativistic electrons moving along curved magnetic field lines. Thereby they emit photons ("curvature radiation") of maximal energy

$$E_\gamma = 3/2\ \gamma^3\ \hbar c/\rho \qquad (4.6)$$

ρ: radius of curvature, $E_e = \gamma m_o c^2$.

Radiation reaction limits the electron energy to about 10^{14} eV near the pulsar. The radiated photons have energies of the order 10^9 eV, sufficient to produce

$e^- - e^+$ pairs in the strong magnetic field. The mean free path ℓ of $\hbar\omega > 2\,mc^2$ photons (Erber 1966) is

$$\ell = \frac{4.4}{e^2/\hbar c}\;\frac{\hbar}{mc}\;\frac{B_q}{B_\perp}\,\exp\left(\frac{4}{3\chi}\right)$$

$$\chi = \frac{\hbar\omega}{2mc^2}\,\frac{B_\perp}{B_q} \qquad B_q = \frac{m^2c^3}{e\hbar} = 4.4\times10^{13}\,G \qquad (4.7)$$

$B_\perp$: component perpendicular to photon path.

The photons are produced almost parallel to B ($B_\perp = 0$ means ℓ infinite) and therefore must go a distance $\approx$ radius of curvature to meet B large enough for significant pair productions. But an electric field acts like a $B_\perp$ component (Daugherty & Lerche 1975), and the outcome of all this is a very short mean free path of γ-rays for pair production. The electron-positron pairs will again emit γ-rays, these will produce new pairs, and so on. Eventually some γ-radiation will escape from this inner magnetosphere plasma of relativistic electrons and positrons. The ratio of the gamma radiation to the outgoing particle flux has not been computed reliably so far. A drawback of this model seems to be the difficulty to efficiently produce 10^{14} eV electrons leaving the magnetosphere, which the observations suggest as a main energy loss process for the Crab pulsar: Since the X-ray flux from the Crab nebula has a power law spectrum and shows polarization (Weisskopf et al. (1976)), it must be the synchrotron radiation of relativistic electrons moving in the nebula's magnetic field. It appears likely that the unpulsed part of the gamma ray flux is due to the same mechanism. The detection of a constant flux from the Crab nebula at energies above 35 MeV implies the existence of electrons of 10^{13} to 10^{14} eV for a magnetic field of 10^{-3} to 10^{-4} Gauss in the nebula. (For synchrotron radiation the critical frequency is given by (Ginzburg & Syrovatskii (1964))

$$\nu_{CT} = \frac{3}{4\pi} \frac{e}{mc} \gamma^2 B_{\perp} \quad sec^{-1} \quad . \tag{4.8}$$

Such electrons must be supplied by the pulsar and radiate close to the pulsar. It is then not surprising that a shift from a predominantly unpulsed flux at X-ray energies to a predominantly pulsed flux at gamma-ray energies occurs.

It is also interesting to note that the pulsed gamma luminosity of the Vela pulsar is only 1/4 that of the Crab pulsar, whereas in X-rays the upper limits on the pulsed flux from the Vela pulsar (3×10^{-11} erg cm^{-2} sec^{-1} given by Rappaport et al. (1974)) give less than 10^{-3} for the luminosity ratio of the two pulsars. This discrepancy between gamma rays and X-rays is difficult to understand, if Vela and Crab are only different in their rotation period, because then the incoherent emission of X-rays should not scale drastically different from the gamma-ray emission.

A search for other pulsars has been carried out by SAS-2, and two more candidates have been found (Thompson et al. (1976)): PSR 1818 (T = 0.6 sec) with a pulsed flux above 35 MeV of 2×10^{-6} cm^{-2} sec^{-1} and

PSR 1717 (T = 0.742 sec) with the same pulsed flux. Apparently the X-ray source Cyg X-3 has also been seen at gamma-ray energies, as well as another strong point source γ195+5 with an intensity above 100 MeV of $5x10^{-6}$ cm^{-2} sec^{-1}.

We can be sure of many more exciting results to come.

V. COSMIC GAMMA BURSTS

To discovery of the phenomena to be described here was an outcome of the Limited Test Ban Treaty of 1963. From 1964 six pairs of "Vela" satellites have been launched into large circular orbits around the earth - about 240 000 km in diameter. The satellites carry gamma-ray detectors to register the gamma-rays from nuclear weapon detonations on the earth. But each satellite was also equipped with 6 small detectors, designed to respond to the gamma radiation emitted by radioactive debris as it expanded from behind the moon, after a bomb test on the backside which would otherwise be shielded from the Vela satellites. The detectors have a fairly steady counting rate due to the flux of charged cosmic ray and solar wind particles, and this is used to indicate that the instruments are working. A sudden rise in the counting rate, however, leads to registration of the following 15 minutes of data, which are stored and later transmitted to earth, toegether with the exact time at which the rate increase occurred. Very frequently one satellite is triggered this way by bursts of charged particles from the sun. But it has been found (Klebesadel et al. (1973)) that every now and then two or more satellites were triggering at nearly the same time, and at a rate several hundred times too high to explain it by accident. It has now firmly been established that one is seeing here short bursts of gamma radiation in space. Since the satellite orbit is 0.8 light seconds in diameter, any such blast of radiation would never need more than 0.8 sec to trigger one detector system and then cross the orbit to trigger another satellite. Exact measurements of the time delay ≤ 0.8 sec between the triggering of two satellites can give information on the source direction. Klebesadel et al. (1973) in utilizing this have been able to show that the gamma ray bursts must be of extrasolar origin. Between 1969 and the end of 1973 about 33 events had been registered by the Vela, Imp-6 and OSO-7 satellites (Strong (1975)), that is an average of 9 to 10 per year.

Figure 11 gives a gamma burst (72-4 in Strong's (1975) classification) detected by Vela 5A and 6A. The time duration of about 6 to 7 sec with much shorter spikes during the burst is typical. Figure 12 shows the recording of event 72-1 by the Soviet satellite Kosmos 461, which does not trigger, but records everything all the time (Mazets et al. (1974)). There have also been bursts with durations of 1 sec, and up to 80 sec. The energy flux in a burst ranges from 3×10^{-6} erg cm^{-2} up to 10^{-4} erg cm^{-2}. The characteristic photon energies are of the order of several 100 keV, with similar spectra for all bursts. In fig. 13 a measured spectrum from Apollo XV is compared with a spectrum calculated for thermal bremsstrahlung of electrons (Anzer & Börner (1976)). Assuming a source for these bursts radiating isotropically at a distance of d parsec, we require luminosities between 3×10^{32} d^{2} erg and 10^{35} d^2 erg. At distances of 1 kpc therefore, these short pulses of radiation require energies of 10^{39} ergs/sec, i.e. they are for a short time among the most energetic objects in our galaxy.

These mysterious observations have naturally led to a flood of speculative theoretical papers: From the crashing of comets on the surface of old neutron stars, to the very ingenious model of accelerated dust grains with $\gamma \gtrsim 10^3$, which scatter solar photons via a fluorescence process into gamma energies when they break up near the orbit of Jupiter, virtually no imaginable astrophysical scenario has been left out. Anzer & Börner (1974) and Ruderman (1975) have given a critical review of many of the proposed models. Most of these models have a weak point, which makes them practically unacceptable. Apparently the requirement to liberate a large amount of energy (10^{40} ergs) in a short time (<10 sec), exclusively in gamma rays of $\gtrsim$ 300 keV is a very severe restriction. A very serious attempt to meet these requirements has been made by Anzer et al. (1976), who want to attribute a certain class of very similar bursts (4 bursts from same direction, 6×10^{-5} erg/cm^2, 6 sec duration) to the binary X-ray source Cyg X-1. Their model therefore invokes spherical accretion on a compact object (black hole) from the stellar wind of a companion star. The gamma bursts arise, when occasional flares in the companion drive shocks in the wind, which then produce a short increase in the accretion rate on the compact object. In the interval between flares the steady accretion from the wind leads to an X-ray source behaviour compatible with observations. The model can reproduce the required gamma burst chracteristics, it needs however, a rather large explosion energy on the companion star of 10^{42} erg.

REFERENCES

U. Anzer, G. Börner, 1974, Astron. & Astrophys. 40, 123.

U. Anzer, G. Börner, 1976, MPI-PAE/Astro 77.

U. Anzer, G. Börner, P. Mészáros, 1976, Astron. & Astrophys. 50, 305.

J. Arons, R. McCray, J. Silk, 1971, Ap. J. 170, 431.

K. Brecher, P. Morrison, 1969, Phys. Rev. Letters 23, 802.

The Caravane Collaboration for COS-B, K. Benneth, G.F. Gignami, G. Boella, R. Buccheri, J.J. Burger, A. Cuccia, W. Hermsen, J. Higdon, G. Kanback, L. Koch, G.G. Lichti, J. Masnov, H.A. Mayer-Hsselwander, J.A. Paul, L. Scarsi, P.G. Shukla, B.N. Swanenburg, B.G. Taylor and R.D. Wills, 1976, Proc. NASA Symp. on "Structure and Content of the Galaxy and Cosmic Gamma Rays".

E.L. Chupp, D.J. Forrest, C. Reppin, A.N. Suri, 1973, NASA SP-339, 165.

E.L. Chupp, 1976, Gamma-Ray Astronomy, D. Reidel Holland/ USA.

D.D. Clayton, J. Silk, 1969, Ap.J. 158, L43.

J.K. Daugherty, I. Lerche, 1975, Astrophys. Space Sc. 38, 437.

T. Erber, 1966, Rev. Mod. Phys. 38, 626.

A.C. Fabian, P.W. Sanford, 1971, Nat. Phys. Sci. 231, 52.

C.E. Fichtel et al., 1975, Ap.J. 198, 163.

V.L. Ginzburg, S.I. Syrovatskii, 1964, The Origin of Cosmic Rays (Macmillan Comp. N.Y.).

P. Goldreich, W.H. Julian, 1969, Ap.J. 157, 869.

J.E. Grindlay, 1976, Proc. NASA-Symp. on "Structure and Content of the Galaxy and Cosmic Gamma Rays".

R.W. Klebesadel, I.B. Strong, R.A. Olson, 1973, Ap.J. 182, L85.

W.L. Kraushaar et al., 1972, Ap.J. 177, 341.

T.L. Landecker, R. Wielebinski, 1970, Aust. J. Phys. Astrophys., Suppl. 16.

R.E. Lingenfelter, R. Ramaty, 1967, High Energy Nuclear Reactions in Astrophysics (ed. B.S.P. Stern) Benjamin N.Y.

R. Lüst, K. Pinkau, 1967, Electromagnetic Radiation in Space (Dr. Reidel, Dordrecht).

E.P. Mazets, S.V. Golenetskii, V.N. Il'inski, 1974, Zh. E.T.F. Pis'ma 19, 126.

A.E. Metzger et al., 1974, Ap.J. 130, 693.

Y. Pal, 1973, Proc. IAU Symp. No. 55, X- and Gamma Ray Astronomy, p. 279.

B. Peters, 1959, Proc. Cosmic Ray Conf., Moscow, 3, 157.

R. Ramaty, R.E. Lingenfelter, 1972, NASA preprint "High Energy Phenomena on the Sun" (X-693-73-193), 301.

S. Rappaport, H. Bradt, R. Doxsey, A. Levine, G. Spada, 1974, Nature 251, 471.

M.A. Ruderman, P.G. Sutherland, 1975, Ap.J. 196, 51.

M.A. Ruderman, 1975, VII. Texas Symp. on Relativ. Astrophys. (N.Y. Acad. Sci.), p. 164.

M.A. Ruderman, 1972, Ann. Rev. Astron. Ap. 10, 427.

V. Schönfelder, 1975, Fortschritte d. Physik 23, 1.

V. Schönfelder, U. Graser, J. Daugherty, 1976, Ap. J. (to be published).

D.A. Schwartz, 1970, Ap. J. 162, 439.

F.W. Stecker, 1969, Ap. J. 157, 507.

F.W. Stecker, 1971, NASA SP-249.

A.W. Strong, J. Wdowczyk, A.W. Wolfendale, 1973, 13th Intern. Conf. on Cosmic Rays - Denver 1, p. 27.

I.B. Strong, 1975, VII. Texas Symposium on Relativistic Astrophysics (N.Y. Acad. Sci., 1975), p. 145.

P.A. Sturrock, 1971, Ap. J. 164, 529.

D.J. Thompson, E.C., Fichtel, D.A. Kniffen, H.B. Ögelmann, 1975, Ap. J. 200, L79.

D.J. Thompson, C.E. Fichtel, R.C. Hartman, D.A. Kniffen, R.C. Lamb, 1976 Proc. NASA Symp. on "Structure and Content of the Galaxy and Cosmic Gamma Rays".

M.C. Weisskopf et al., 1976, Columbia Univ. preprint.

FIGURE CAPTIONS

Fig. 1: Schematic picture of a Comptontelescope and illustration of a double Compton collision.

Fig. 2: Various experimental results on the photon energy.

Fig. 3: Some ideal γ-ray spectra emitted from the decay of ideal spectra of neutral pions. (From F.W.Stecker:1971, NASA Sp-249).

Fig. 4: SAS-2 data of (>100 MeV) gamma rays. Distribution along the galactic plane, summed from $b'' = -10^{\circ}$ to $b'' = +10^{\circ}$.

Fig. 5: Latitude profile of the gamma ray emission summed from $l^{II} = 350^{\circ}$ to $l^{II} = 20^{\circ}$.

Fig. 6: Longitude profile of the gamma ray emission in the energy range 300-2000 MeV summed from $-2^{\circ} < b^{II} < 2^{\circ}$ (curve a) and from $-10^{\circ} < b^{II} < -2^{\circ}$ and $2^{\circ} < b^{II} < 10^{\circ}$ (curve b).

Fig. 7: Pulse-height spectra recorded in the time interval 0623. 0632 UT on August 4, 1972 (from Chupp (1976), p. 155).

Fig. 8: The composite $^{56}Ni \rightarrow {}^{56}Co \rightarrow {}^{56}Fe$ γ-ray spectrum in a specific Einstein-de Sitter universe. The solid line is an exponentially decreasing rate of nucleosynthesis/galaxy $=e^{-2}$ of the initial rate today. The dashed line is a constant rate of nucleosynthesis/galaxy. For comparison, the dotted line is the background spectrum observed on the Apollo-15 spacecraft by Trombka et al., (1973), (heavy solid dots are data points).

Fig. 9: Gamma-ray light curve of NP 0532, obtained by the synchronization method, compared with the X-ray light curve. The pulsar period is divided into bins of ∿2 ms.

Fig. 10: Gamma-ray light curve of PSR 0833, obtained through solar system barycentric analysis, for the period 20 October to 8 November 1975. The pulsar period of 89.2 ms is divided into bins of ∿1 ms.

Fig. 11: Event 72-4, on 1972 May 14, beginning 13591 s UT.

Fig. 12: Intense burst of gamma-radiation in the range 0.05-0.3 MeV. Kosmos 461 data on event 72-1. All three prominent peaks appear in the Vela data together with a fourth peak at ∿63565 s UT. In the Vela data the count rate for the two later peaks is reduced relative to the Kosmos 461 data, indicating a softening of the spectrum (after Mazets et al.).

Fig. 13: Comparison between the observed spectrum (Metzger et al., 1974) and calculated spectra of thermal bremsstrahlung for temperatures of 2.2×10^9 K (dashed line) and 4.5×10^9 K (full line).

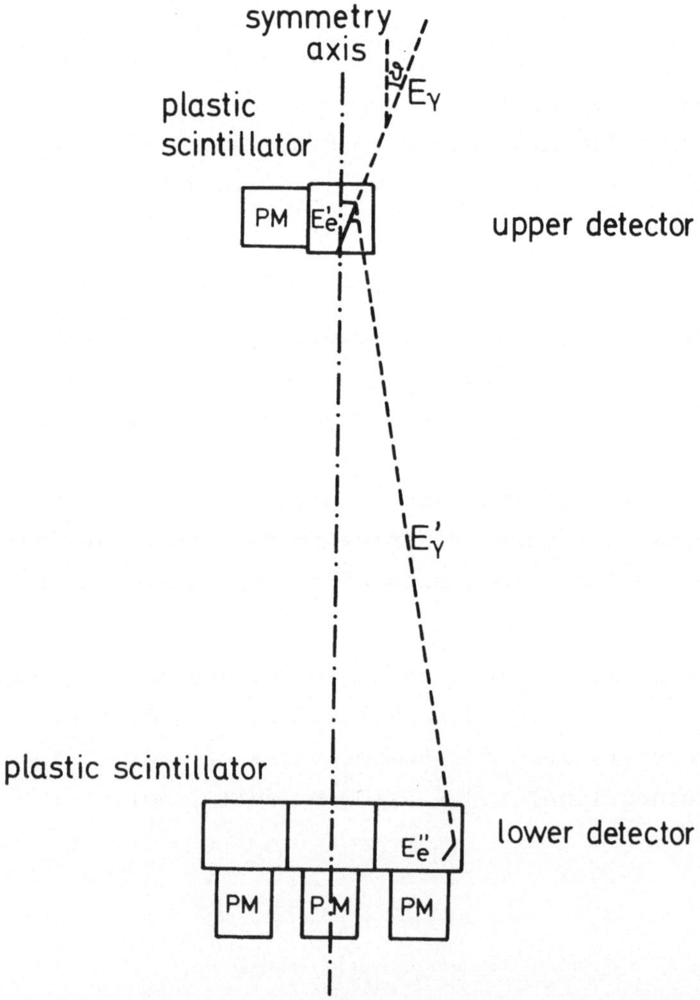
symmetry
axis
plastic
scintillator
ϑ
E_γ
PM
E_e'
upper detector
E_γ'
plastic scintillator
E_e''
lower detector
PM
PM
PM

Fig.1

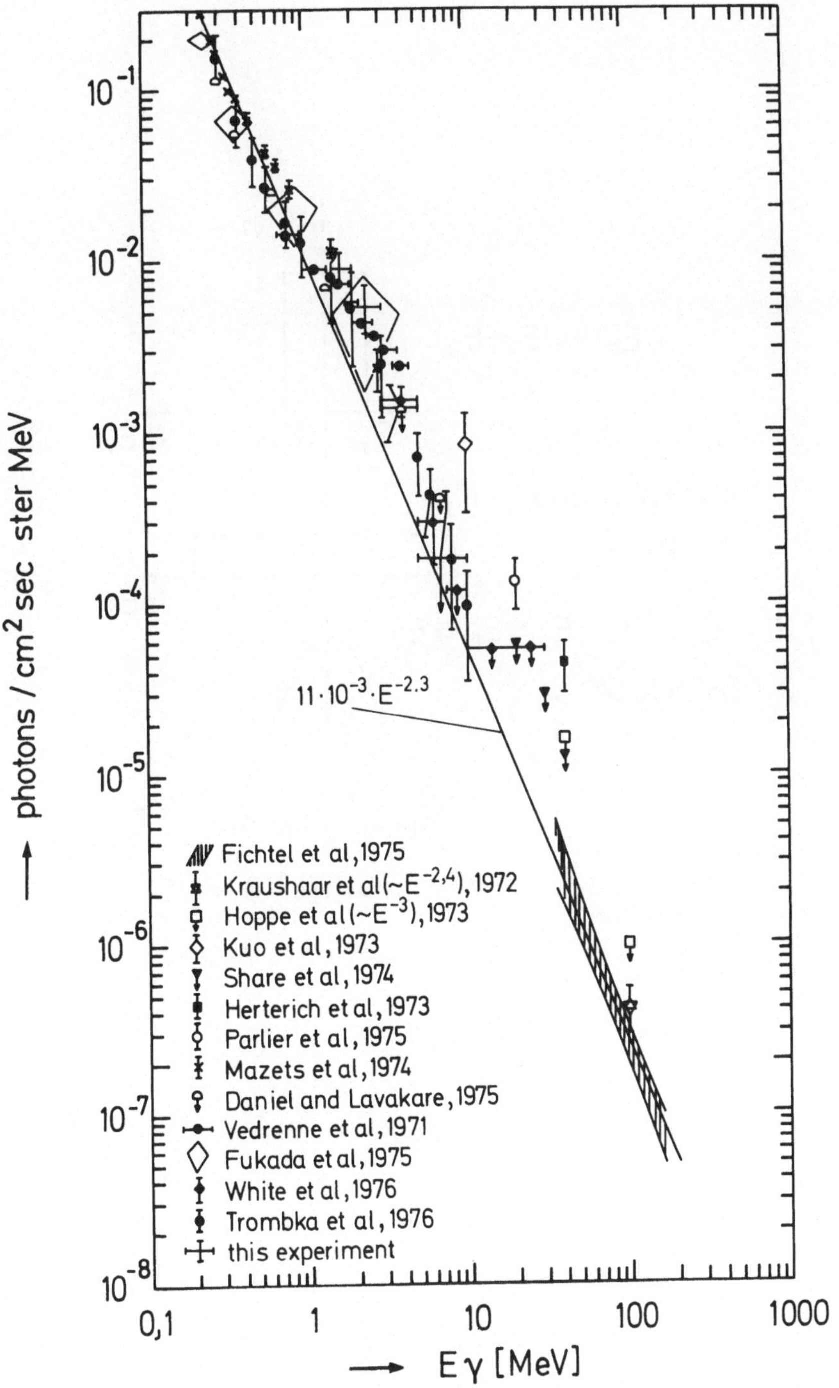
→ photons / cm² sec ster MeV
10^-1
10^-2
10^-3
10^-4
10^-5
10^-6
10^-7
10^-8
0,1
1
10
100
1000
→ Eγ [MeV]
11·10^-3·E^-2.3
Fichtel et al, 1975
Kraushaar et al (~E^-2,4), 1972
Hoppe et al (~E^-3), 1973
Kuo et al, 1973
Share et al, 1974
Herterich et al, 1973
Parlier et al, 1975
Mazets et al, 1974
Daniel and Lavakare, 1975
Vedrenne et al, 1971
Fukada et al, 1975
White et al, 1976
Trombka et al, 1976
this experiment

Fig. 2

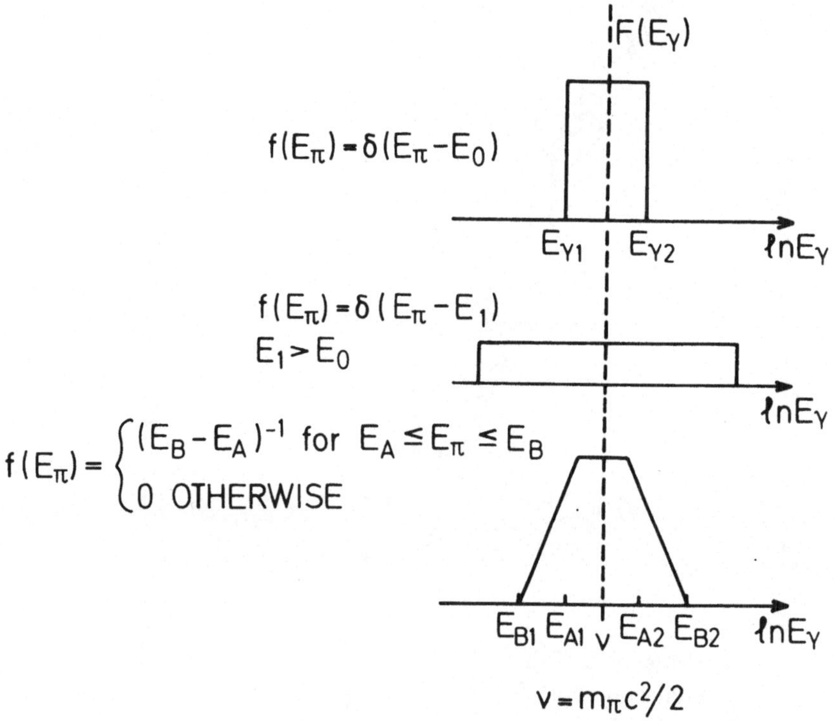
F(Eγ)
f(Eπ)=δ(Eπ−E0)
Eγ1
Eγ2
ℓnEγ
f(Eπ)=δ(Eπ−E1)
E1>E0
ℓnEγ
f(Eπ) = (EB−EA)^-1 for EA ≤ Eπ ≤ EB
0 OTHERWISE
EB1
EA1
v
EA2
EB2
ℓnEγ
v=mπc²/2

Fig. 3

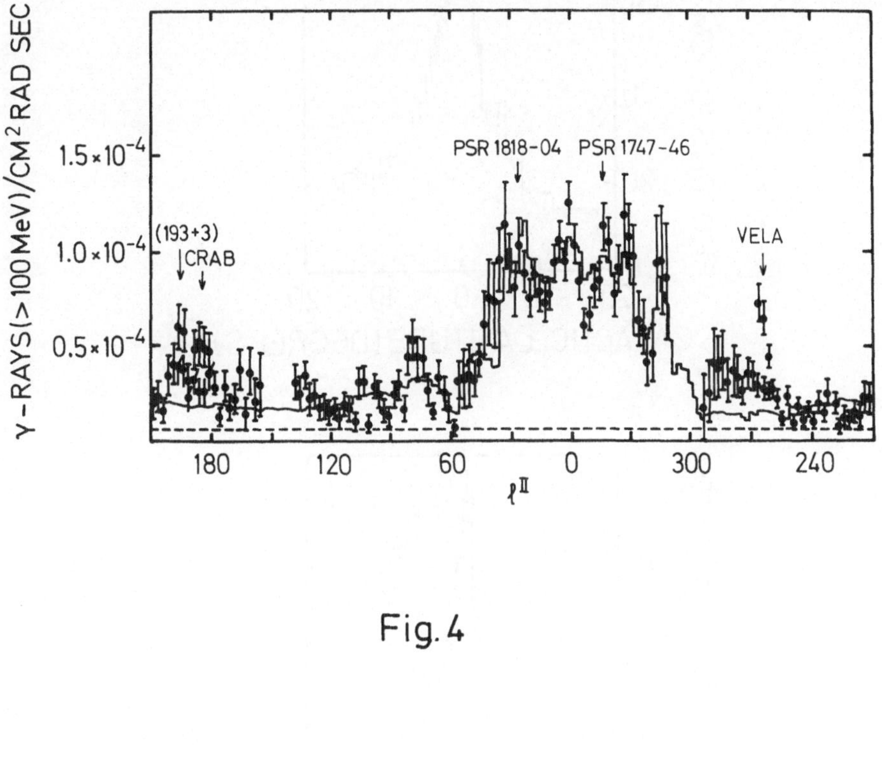
γ-RAYS(>100 MeV)/CM² RAD SEC
1.5×10⁻⁴
1.0×10⁻⁴
0.5×10⁻⁴
(193+3)
CRAB
PSR 1818-04
PSR 1747-46
VELA
180
120
60
0
300
240
ℓII

Fig. 4

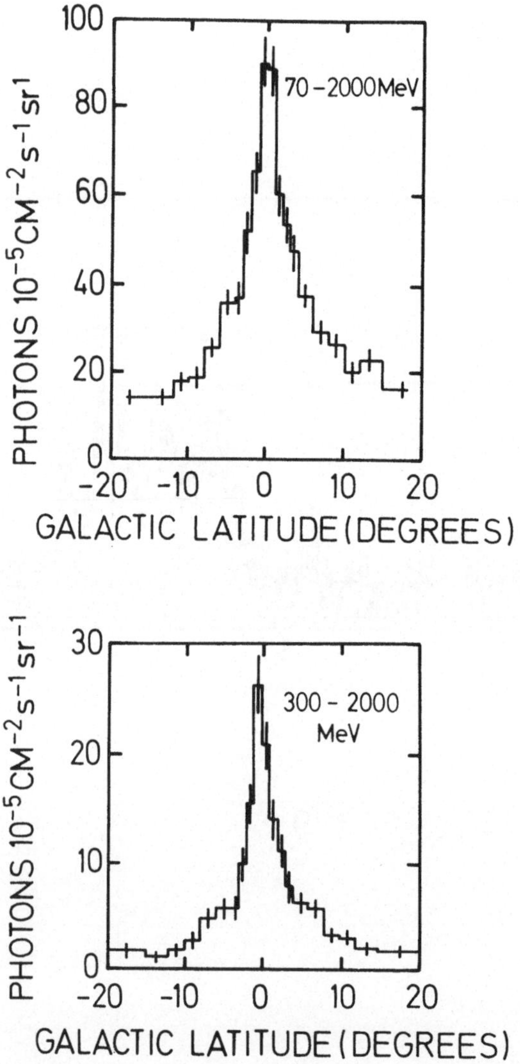
100
80
60
40
20
0
PHOTONS 10^{-5}CM^{-2}s^{-1}sr^{-1}
70 - 2000MeV
-20 -10 0 10 20
GALACTIC LATITUDE (DEGREES)
30
20
10
0
PHOTONS 10^{-5}CM^{-2}s^{-1}sr^{-1}
300 - 2000
MeV
-20 -10 0 10 20
GALACTIC LATITUDE (DEGREES)

Fig.5

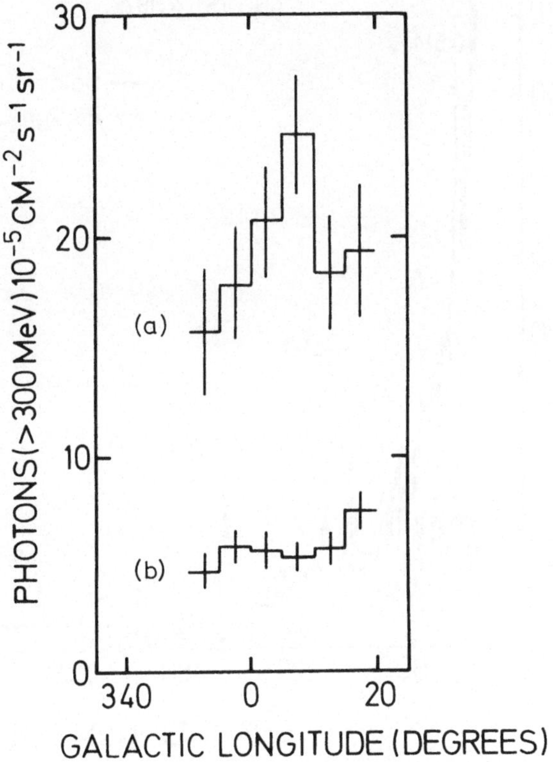

30
20
10
0
340
0
20
(a)
(b)
PHOTONS(>300 MeV) 10^{-5} CM^{-2} s^{-1} sr^{-1}
GALACTIC LONGITUDE (DEGREES)

Fig.6

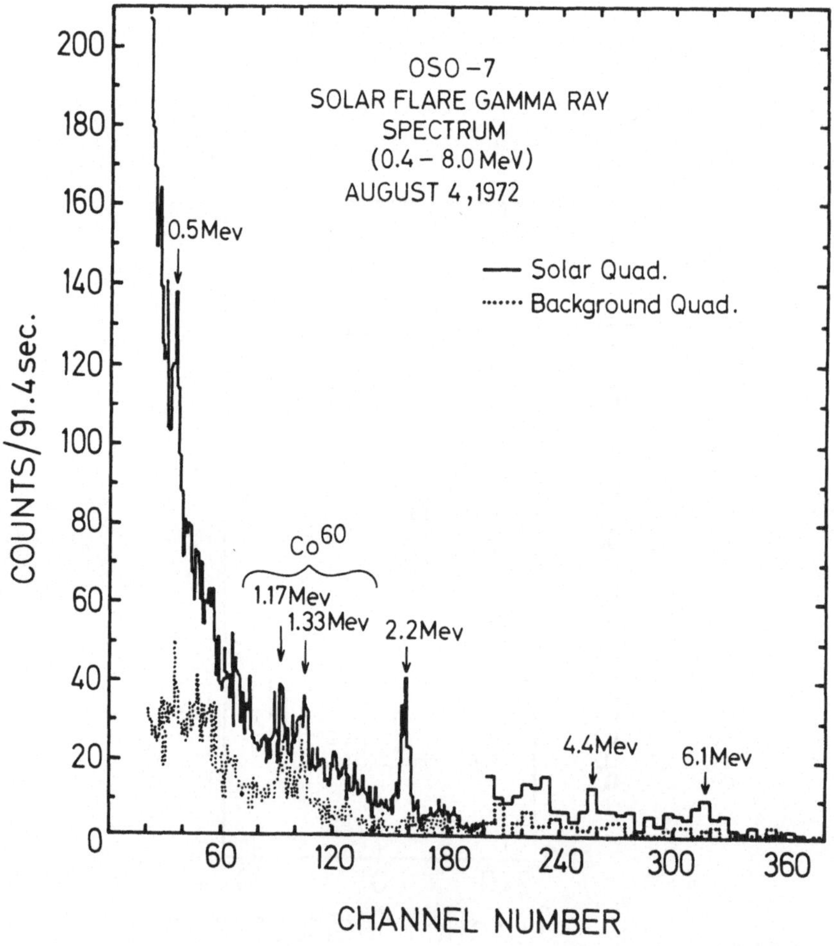
OSO-7
SOLAR FLARE GAMMA RAY
SPECTRUM
(0.4 - 8.0 MeV)
AUGUST 4, 1972
Solar Quad.
Background Quad.
0.5Mev
Co60
1.17Mev
1.33Mev
2.2Mev
4.4Mev
6.1Mev
COUNTS/91.4 sec.
CHANNEL NUMBER
0
20
40
60
80
100
120
140
160
180
200
60
120
180
240
300
360

Fig. 7

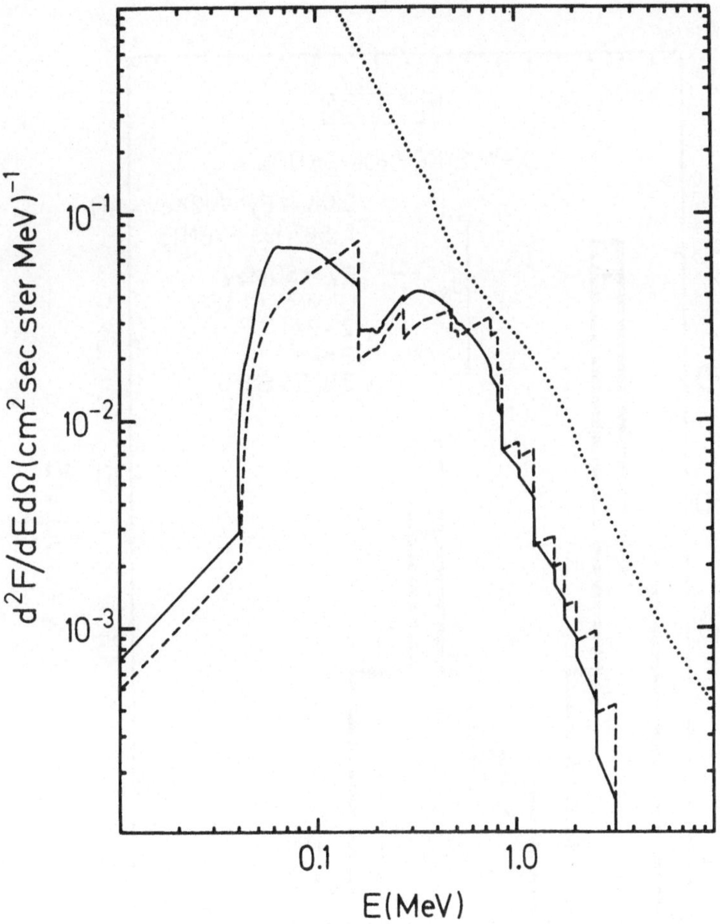

Fig. 8

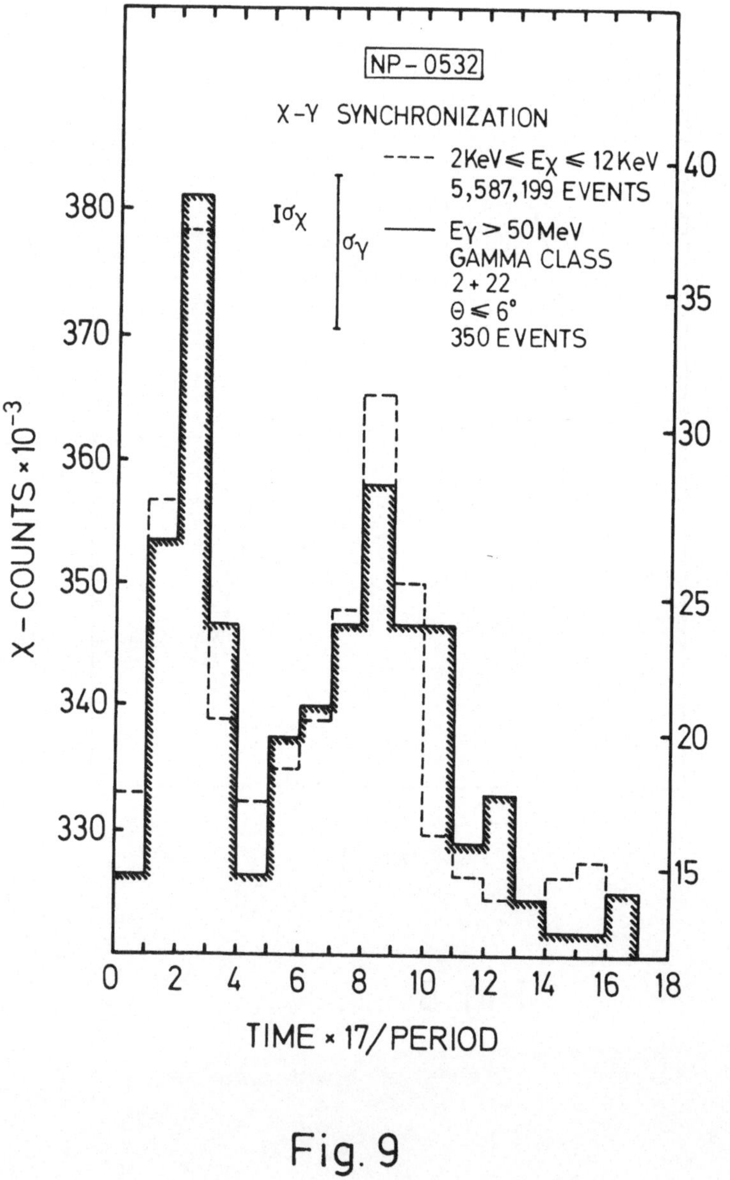
NP-0532
X-Y SYNCHRONIZATION
---- 2KeV ≤ Eχ ≤ 12KeV
5,587,199 EVENTS
Iσχ
σγ
— Eγ > 50MeV
GAMMA CLASS
2+22
Θ ≤ 6°
350 EVENTS
X-COUNTS × 10⁻³
γ-COUNTS
380
370
360
350
340
330
40
35
30
25
20
15
0
2
4
6
8
10
12
14
16
18
TIME × 17/PERIOD

Fig. 9

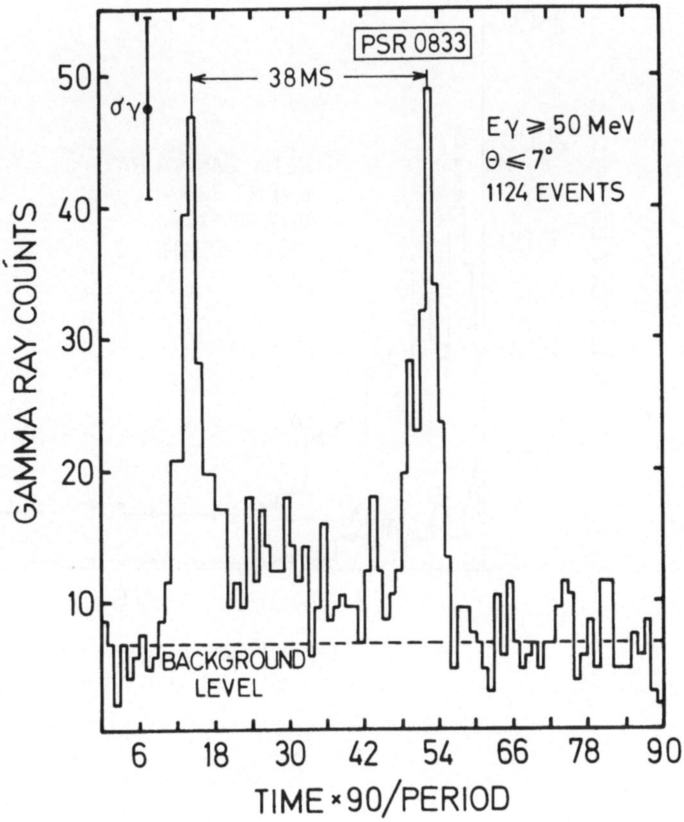
PSR 0833
38MS
σγ
Eγ ≥ 50 MeV
Θ ≤ 7°
1124 EVENTS
GAMMA RAY COUNTS
50
40
30
20
10
BACKGROUND
LEVEL
6
18
30
42
54
66
78
90
TIME × 90/PERIOD

Fig.10

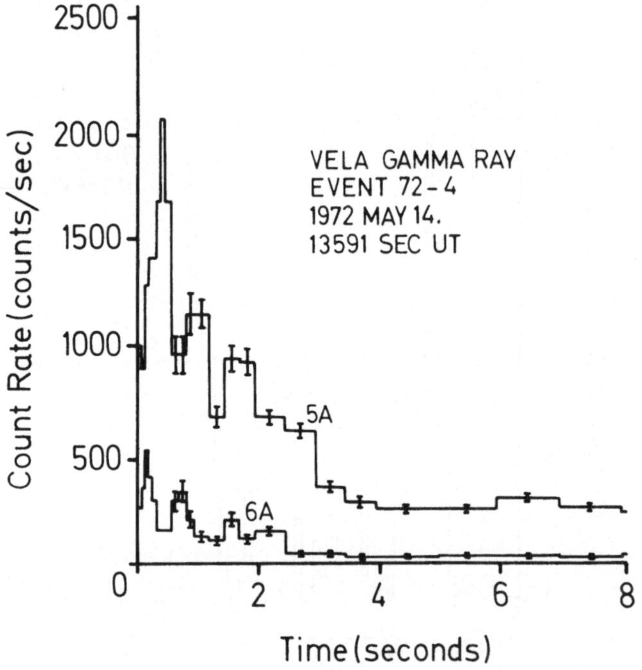
2500
2000
1500
1000
500
0
Count Rate (counts/sec)
VELA GAMMA RAY
EVENT 72-4
1972 MAY 14.
13591 SEC UT
5A
6A
2
4
6
8
Time (seconds)

Fig.11

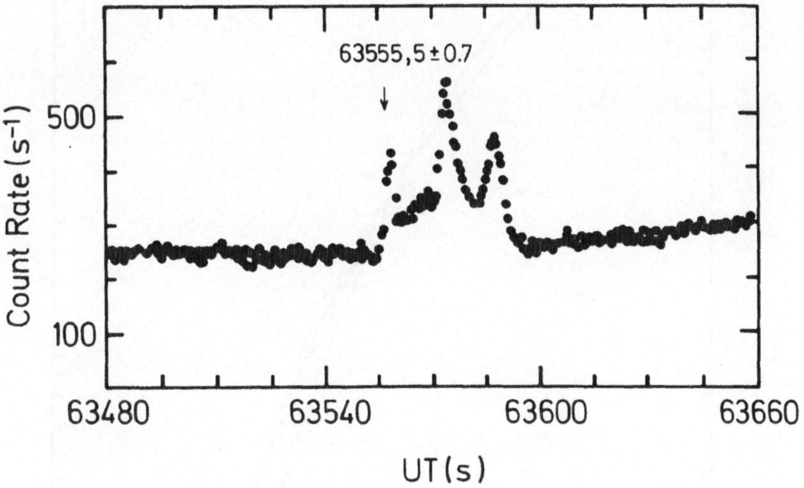

Fig.12

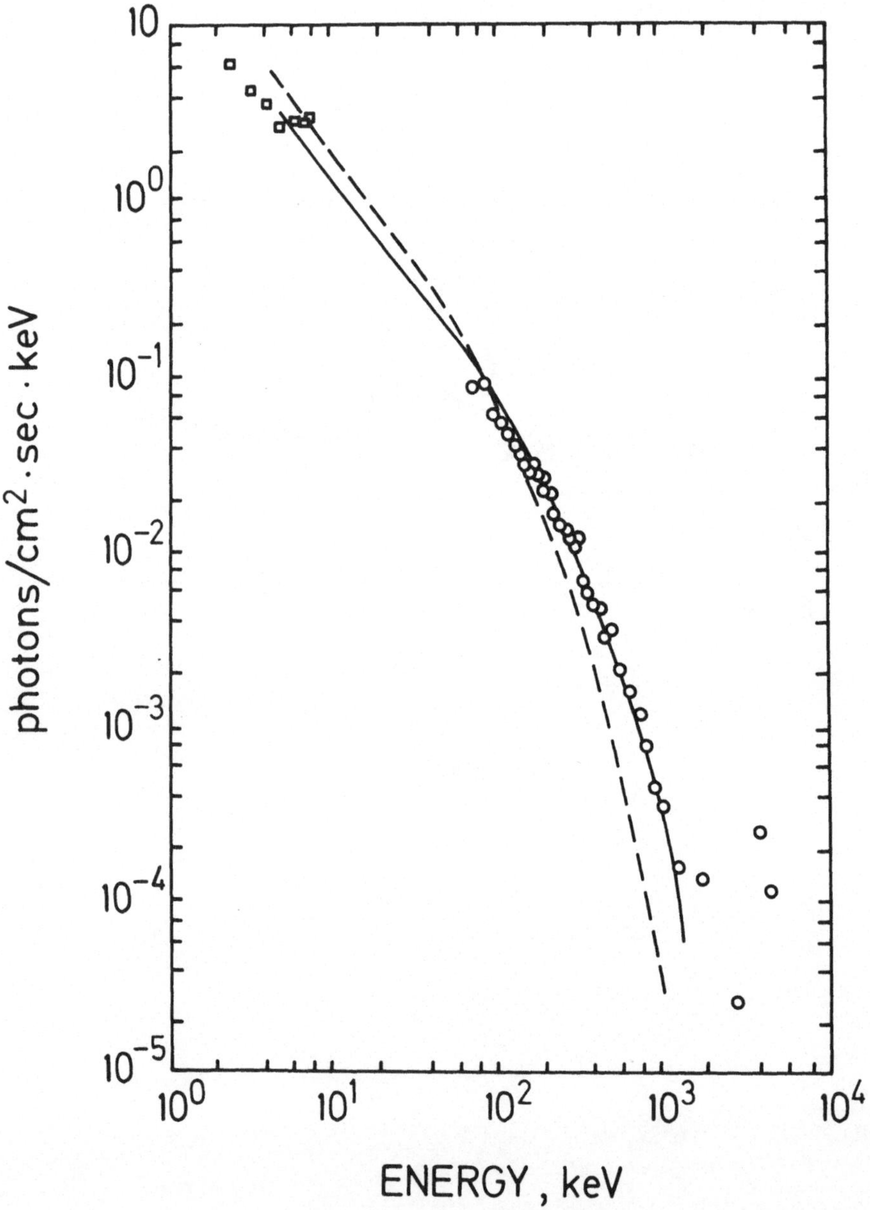
10
10^0
10^{-1}
10^{-2}
10^{-3}
10^{-4}
10^{-5}
photons/cm^2 · sec · keV
10^0
10^1
10^2
10^3
10^4
ENERGY , keV

Fig.13

Acta Physica Austriaca, Suppl. XVIII, 835–872 (1977)

"SPLENDEURS ET MISÈRES" OF HAWKING'S EFFECT+

by

P. HÁJIČEK
Institute for Theoretical Physics
Berne, Switzerland

1. INTRODUCTION

At the beginning of 1974, Hawking obtained the following result: a Schwarzschild black hole will emit particles as if it were a hot body with a temperature given by

$$T_H = \frac{\hbar c^3}{16\pi^2 Gk} \cdot \left(\frac{1}{M_H}\right) \; {}^{o}K = 10^{-6} \left(\frac{M_\theta}{M}\right) \; {}^{o}K$$

where M_H (and M_θ) is the mass of the hole (and the Sun) in kg, ħ, G, k the Planck, Newton, Boltzmann constant, respectively, and c the light velocity [1]. (In fact,

+Lecture given at XVI. Internationale Universitätswochen für Kernphysik, Schladming, Austria, February 24-March 5, 1977.

the result was a bit more general including black holes with non-zero angular momentum and charge, but we restrict ourselves to the Schwarzschild spacetime, in order to suppress the geometrical, purely general relativistic aspects as far as possible; for recent calculations with rotating charged hole, see [2]).

A popular explanation of this effect is analogous to, but, say, one step more difficult than, that of particle creation by strong Coulomb fields as discussed here by Prof. Müller. Roughly speaking, a particle can have a negative binding energy only beyond the horizon in Schwarzschild space-time. Thus, one needs a sort of tunnel-effect: outside the horizon, a virtual pair is created. One particle of the pair has negative energy, it crosses the horizon and becomes real. The second particle can, then, bring its positive energy to infinity.

Remark first that T_H is proportional to the <u>inverse</u> of mass. There are, therefore, only two ways of coexistence of such a hole with an infinite thermal bath of a constant temperature T_B ($T_B = T_H$ is instable):

1) $\underline{T_H > T_B}$: the hole radiates in an escalating way. If $T_B = 0$ (vacuum), then the mass decrease of the hole is governed by [3]:

$$\frac{dM_H}{dt} = - \frac{\hbar c^4}{G^2} \frac{\alpha}{M_H^2}$$

(because $T_H \sim M_H^{-1}$, radiating area $\sim M_H^2$, and the power $\sim AT_H^4 \sim M_H^{-2}$), where α is a numerical coefficient depending on which particle species can be emitted at a significant rate. A typical lifetime is [3]:

$$\tau \sim 10^{66} \left(\frac{M_H}{M_\theta}\right)^3 \text{ yr} ,$$

and some typical powers emitted [3]:

for $M_H >> 10^{14}$ kg:

$$P = 3.5 \times 10^{33} \left(\frac{kg}{M_H}\right)^2 \text{ Js}^{-1}$$

$$\alpha = 2 \times 10^{-4}$$

81.4 % ...... neutrinos

16.7 % ...... photons

1.9 % ...... gravitons

for 5×10^{11} kg $<< M_H << 10^{14}$ kg:

$$P = 6.3 \times 10^{33} \left(\frac{kg}{M_H}\right)^2 \text{ Js}^{-1} , \quad \alpha = 3.6 \times 10^{-4} ,$$

45 % .... electrons and positrons,

45 % .... neutrinos

9 % .... photons

1 % .... gravitons .

As M_H decreases, kT_H increases so that more and more massive particles can be emitted. The end (even with constant α) looks like an explosion. The exact form of it is uncertain, because it depends on strong interaction theory and the theory of quantum fields in very strongly ensued spacetimes (for some estimates, see [4]). In any case, a sort of γ-ray barst should then, result.

2) $\underline{T_H < T_B}$: the hole graws through accreation of the surrounding material of the bath and it cooles to get practically O-temperature, everything swallowing object, as it is usually believed. This is the fate of all holes resulting in collapse: $M_H \gtrsim 3\ M_\theta$, and they are definitely cooler than the background radiation.

The behaviour can be modified, if the bath is finite so that it gets more warmed than the hole by the hole radiation [5], or if it is not of constant temperature (the Universe).

The astrophysical importance of the effect depends on how many small black holes can exist [6] (the holes of mass larger than $\sim 10^{12}$ kg live longer than the Universe).

From the theoretical point of view, the effect seems extremely interesting.

1) It limits the predictability of the dynamical development of physical systems more than the Heisenberg's uncertainty principle, because it leads to a maximally mixed state (all states with the same energy having the same probability) evolving from a pure one. As Hawking puts it: "God not only plays dices, he sometimes throws them where they cannot be seen" [7].

2) The process consisting of a collapse with a subsequent evaporation of the black hole violates the conservation

laws of the baryon and lepton charges. In particular, the baryons are practically all converted to entropy, because black holes begin to emit baryons first when $kT_H \sim m_{proton} c^2$, that is

$$\frac{M_H}{M_\theta} \sim 5.10^{-20} \; !$$

There are even proposals of direct quantum jumps of this sort for microscopic systems [8].

3) It tends strongly to unifying the theoretical physics. Notice that all main physical constants -h, c, G, k- appear in Hawking's formula. There are hints, how to quantize gravity, how to modify quantum field theory in the presence of gravity and how to generalize statistical physics [9].

4) There are also suggestions from other directions that black holes are thermodynamic systems. Let us give a very short sketch of it.

Consider a neutral stationary black hole with total energy E and total angular momentum J. These two parameters describe the states of such a hole completely ("no-hair-theorems" by Carter and Robinson). Consider now a classical test particle moving along a geodesic, having the four-momentum p_i and falling into the hole (Christodoulou). In a suitable coordinate system, the particle energy and angular momentum with respect to infinity are cp_o and $-p_3$, respectively. We have the conservation law

$$dE = cp_o, \quad dJ = -p_3 .$$

The condition that p_i be non-spacelike and future oriented at the point where the particle crosses the horizon is

$$cp_o \geq -c\Omega p_3 ,$$

where

$$\Omega = \frac{j}{2m(m^2 + \sqrt{m^4 - j^2})}$$

$$j = \frac{GJ}{c^3} , \quad m = \frac{GE}{c^4} .$$

(Ω is independent of where the particle crosses the horizon.) We have, therefore, an inequality for dE and dJ in such a process:

$$dE - c\Omega dJ \geq 0 .$$

The surface area A of the black hole is

$$A = 8\pi(m^2 + \sqrt{m^4 - j^2}) ,$$

and its differential in terms of dE and dJ reads

$$dA = \frac{8\pi}{\kappa} (dE - c\Omega dJ) ,$$

where

$$\kappa = \frac{\sqrt{m^4 - j^2}}{2m(m^2 + \sqrt{m^4 - j^2})}$$

is the so-called surface gravity of the hole. Hence, any of our processes satisfy

$$dA \geq 0 .$$

We also have

$$dE = \frac{\kappa}{8\pi} dA + c\Omega dJ .$$

The last two equations are much like the second and the first law of thermodynamics, if one conjectures

$$S = \lambda . A , \quad T = \frac{1}{\lambda} \cdot \frac{\kappa}{8\pi} ,$$

where S and T would be entropy and temperature of the black hole and λ some constant.

An attempt to explain the equation $S = \lambda \cdot A$ and to find the value of λ is due to Beckenstein. He estimates the number N of all (microscopic, quantum) star configurations which can collapse to the same black hole state as described by E and J.

He finds

$$S = k \lg N \doteq \frac{k}{\hbar c} A ,$$

with $\lambda \doteq \frac{k}{\hbar c}$.

The explanation of the equation $T = \frac{\kappa}{8\pi}$ is just Hawking's effect (giving $\lambda = \frac{k}{4\hbar c}$).

2. QUANTUM FIELD THEORY IN CURVED SPACETIMES

For the sake of simplicity, we consider a scalar, hermitian, massless field ϕ which is minimally coupled to gravity

$$g^{ij} \nabla_i \nabla_j \phi = 0 , \tag{1}$$

where g^{ij} is the contravariant metric tensor and ∇_i the corresponding covariant derivative. Three comments:

1) The calculation for existing particles (neutrinos, photons, gravitons), which are not scalars, is analogous to the scalar one ([1],[3]).

2) The holes with $M_H >> 10^{14}$ kg radiate only <u>massless</u> particles in considerable rates; on the other hand, if $kT_H >> mc^2$ for some particle mass m, then these particles will be radiated mostly in ultra-relativistic modes and the approximation $m = 0$ should be good.

3) The gravitation comes in (1) through g^{ij} and ∇_i and remains classical, i.e. not quantized. This is an approximation analogous, e.g. to the WKB-approximation in the quantum theory of solitary waves. (There is a deep relation between solitary waves and black holes.) An attempt to justify it is as follows (there is no good justification at the time being, because a full-fledged quantum theory of gravity which is necessary for that is not known). The occupation numbers are roughly estimated by

$$N_{min} \sim \frac{E_{tot}}{h\nu_{max}} \quad .$$

In a spacetime with collapse resulting in a black hole of mass M_H, the total energy is

$$E \sim M_H c^2 ,$$

and the "highest frequency"

$$\nu_{max} \sim \frac{c}{R_g}$$

where

$$R_g = \frac{2G}{c^2} M_H$$

is the curvature radius at the horizon. Then

$$N_{min} \sim \left(\frac{R_g}{R_P}\right)^2 ,$$

where

$$R_P = \sqrt{\frac{Gh}{c^3}} \sim 5 \times 10^{-35} \text{ m}$$

is the Planck radius. Thus, N's are really very high for $M_H >> M_P \sim 10^{-8}$ kg.

In the classical description of General Relativity Theory, the gravitational field is the spacetime geometry (M,g) (M stands for manifold structure - e.g. $\mathbb{R}^4$ for Minkowski spacetime - , g for the metric). In this section, we do not require that (M,g) be a black hole spacetime, but only the following two points:

1) The curvature radius at any point of M is much larger than R_P.

2) (M,g) is globally hyperbolic.

1) justifies the use of semiclassical approximation, and 2) means that M can be covered by a family of Cauchy hypersurfaces similarly to Minkowski spacetime, covered by t = const.-hyperplanes.

We choose one of the Cauchy hypersurfaces, S_i, say, as initial, and a second, S_f, as final in M (in analogy to the usual choice $S_i = (t\to-\infty)$, $S_f = (t\to+\infty)$ in Minkowski spacetime).

The field ϕ in (1) is an operator-valued distribution; its algebraical properties (commutation rules) on M can be obtained by a straightforward and natural generalization from flat background (see, e.g., [10]). As the next step, it is usual to construct the Fock space of incoming fields (given at S_i) and to describe the evolution of ϕ from S_i to S_f as a unitary transformation in this space. First, one looks for the corresponding automorphism of the field algebra; this can be reduced to a c-number problem exactly as in the theory of linear quantum fields in external potentials [11],[12]. Let us collect all well-known results we shall need:

<u>Theorem:</u> Let F_n, G_r be complex functions on M, $n \in I_1$, $r \in I_2$, I_1, I_2 some index sets, satisfying:

a) $\{F_n, F_m^* | n,m \in I_1\}$, $\{G_r, G_s^* | r,s \in I_2\}$ are two bases in the Banach space of all complex solutions of (1) which are square integrable together with their first partial derivatives along a fixed Cauchy hypersurface in M.

b) $(F_n, F_m) = \delta_{nm}$, $(F_n, F_m^*) = 0$, $n, m \in I_1$,

$(G_r, G_s) = \delta_{rs}$, $(G_r, G_s^*) = 0$, $r, s \in I_2$,

where

$$(F,G) = \frac{i}{2}\int_S (F^* \frac{\partial G}{\partial x^a} - \frac{\partial F^*}{\partial x^a} G)\sqrt{-g}\,\cdot$$

$$g^{ab}\,\varepsilon_{cdeb}\,\frac{\partial x^c}{\partial u^1}\,\frac{\partial x^d}{\partial u^2}\,\frac{\partial x^e}{\partial u^3}\,du^1\,du^2\,du^3 \qquad (2)$$

is the Klein-Gordon scalar product, S an arbitrary Cauchy hypersurface in M, given parametrically by

$$x^i = x^i\,(u^1,\,u^2,\,u^3)$$

(the value of the product is independent of S chosen), g is the determinant of the metric tensor g_{ij} and ε_{abcd} is the Levi-Civita symbol.

c) $\{F_n | n \in I_1\}$ define the positive frequency near S_i;

$\{G_r | r \in I_2\}$ define the positive frequency near S_f.

c*) Let there be an isometry Ψ of a neighbourhood of S_i onto a neighbourhood of S_f and a map ψ of I_1 onto I_2 such that

Ψ_* (Cauchy data of F_n on S_i) = Cauchy data of $G_{\psi(n)}$ $\forall n$, where Ψ_* is the map of tensor fields induced by Ψ (for scalar fields simply $\Psi_*^{-1}(f) = f \circ \Psi$).

Let the transformation between the two bases be

$$G_r = \sum_n (\alpha_{rn} F_n + \beta_{rn} F_n^*) \ . \tag{3}$$

Let a_n be the annihilation operator of a particle in the mode F_n:

$$\phi = \sum_n (F_n a_n + F_n^* a_n^+) \ ,$$

and similarly b_r for G_r.

Then:

1) The inverse of (3) is

$$F_n = \sum_r (\alpha_{rn}^* G_r - \beta_{rn} G_r^*) \ .$$

2) The a's and b's are related by the Bogoljubov transformation

$$b_r = \sum_n (\alpha_{rn}^* a_n - \beta_{rn}^* a_n^+) \ . \tag{4}$$

2^*) The Bogoljubov transformation (4) and the maps Ψ and ψ define a field algebra automorphism α by

$$\alpha(a_n) = b_{\psi(n)} \qquad \forall n \in I_1 \ .$$

α describes the development of the field from S_i to S_f in the conventional quantum field theory [11], [12].

The Fock space $F([F_n])$ of incoming fields is, as usual, the symmetric tensor algebra over the vector space $[F_n]$ ($[F_n]$ is the linear span of the set $\{F_n\}$ determined by the Klein-Gordon norm):

$$F([F_n]) = C \oplus_{\perp} [F_n] \oplus_{\perp} ([F_n] \otimes_s [F_n]) \oplus_{\perp} \dots$$

and the operator a_n^+ is defined by

$$a_n^+ |\xi\rangle = a_n^+ |\xi_o, \xi_1, \dots, \xi_m, \dots\rangle =$$

$$= |0, \xi_o F_n, \dots, \sqrt{m}\, \xi_{m-1} \otimes_s F_n, \dots\rangle ,$$

where ξ_m is the component of the vector $|\xi\rangle$ in $\underbrace{[F_n] \otimes_s \dots \otimes_s [F_n]}_{m \text{ times}}$.

The eq. (4) can be regarded as the definition of b's on $F([F_n])$. In case that c*) is satisfied, one looks, as a next step, for the unitary operator S on $F([F_n])$ which implements α:

$$b_{\psi(n)} = S^{-1} a_n S \quad \forall n \in I_1 .$$

There are many α's, for which S does not exist.

We shall need only the expectation values of some polynomials in b's and b^+'s in the incoming vacuum.

These can be calculated, at least formally, without c^*) being satisfied, if one uses (4) and the standard action of a's on the incoming vacuum. They are clearly invariant under all transformations of the basis $\{F_n, F_n^*\}$ which do not "mix frequencies", i.e. of the form

$$F'_n = \sum_m \alpha_{nm} F_m , \qquad F'^*_n = \sum_m \alpha^*_{nm} F^*_m . \tag{5}$$

Of course, neither α, nor S were invariant under (5), if they had a sense (c^*) satisfied). By a transformation of the type (5), we shall be able to bring (3) to a very special form: both bases can be split into two parts

$$\{F_n\} = \{X_n, Y_n\} , \qquad \{G_n\} = \{U_n, V_n\}$$

such that (3) becomes

$$\begin{aligned} U_n &= \text{ch}\,\chi_n \quad X_n - \text{sh}\,\chi_n - Y^*_n , \\ V_n &= -\text{sh}\,\chi_n \cdot X^*_n + \text{ch}\,\chi_n \cdot Y_n , \end{aligned} \tag{6}$$

and $\{\chi_n\}$ is a positive sequence. The creation operators corresponding to these modes are denoted as follows,

$$X_n \ldots a^+_n , \qquad U_n \ldots b^+_n ,$$

$$Y_n \ldots c^+_n , \qquad V_n \ldots d^+_n ,$$

and satisfy

$$a_n = \operatorname{ch}\chi_n \cdot b_n - \operatorname{sh}\chi_n \cdot d_n^+ ,$$
$$c_n = -\operatorname{sh}\chi_n \cdot b_n^+ + \operatorname{ch}\chi_n \cdot d_n . \tag{7}$$

The initial vacuum $|0i\rangle$ and the final vacuum $|0f\rangle$ are defined by

$$a_n|0i\rangle = c_n|0i\rangle = 0 \qquad \forall n ,$$
$$b_n|0f\rangle = d_n|0f\rangle = 0 \qquad \forall n . \tag{8}$$

The normalized state with k_n particles in the mode U_n and ℓ_n particles in the mode V_n is

$$|K,L,f\rangle = \prod_n \frac{1}{\sqrt{k_n}} \frac{1}{\sqrt{\ell_n}} (b_n^+)^{k_n} (d_n^+)^{\ell_n} |0f\rangle ,$$

where $K = \{k_n\}$, $L = \{\ell_n\}$ are sets of occupation numbers of the U-particles and V-particles, respectively.

The expansion of $|0i\rangle$ into the states $|K,L,f\rangle$ is easily calculated, using (7) and (8):

$$|0i\rangle = \frac{\prod_n \left[\sum_k \frac{(\operatorname{tgh}\chi_n)^k}{k!} (b_n^+)^k (d_n^+)^k\right] |0f\rangle}{\prod_n \operatorname{ch}\chi_n} .$$

The normalization factor $(\prod_n \operatorname{ch}^2 \chi_n)^{-1/2}$ converges, if and only if

$$\bar{N} = \sum_n sh^2 \chi_n = \langle 0i| \sum_n b_n^+ b_n |0i\rangle < \infty \tag{9}$$

(if c*) is satisfied, then (9) is the necessary and sufficient condition for S to exist [11]). Now, remark that our expansion of $|0i\rangle$ can be used to calculate the expectation values of polynomials in b's and b^+'s even in case that (9) is not satisfied. We have to write it as follows

$$|0i\rangle = \prod_n \frac{\exp(tgh\, \chi_n \cdot b_n^+ d_n^+)}{ch\, \chi_n} |0f\rangle$$

and notice that

$$\langle 0f| \left[\frac{\exp(tgh\chi_n \cdot b_n^+ d_n^+)}{ch\, \chi_n}\right]^+ \left[\frac{\exp(tgh\chi_n \cdot b_n^+ d_n^+)}{ch\, \chi_n}\right] |0f\rangle = 1$$

for any n.

Consider the case that the V-particles are inobservable, and so a part of information contained in $|0i\rangle$ is superfluous. Then, it is sufficient to know the density matrix ρ, which is obtained by taking the trace of $|0i\rangle\langle i0|$ over the orthonormal basis $|L,f\rangle$ [12], [13]. An easy calculation yields [14]: ρ is diagonal in the basis $|K,f\rangle$ with

$$\langle K,f|\rho|K,f\rangle = \prod_n \frac{(tgh^2 \chi_n)^{k_n}}{ch^2 \chi_n} \quad . \tag{10}$$

Observe that ρ becomes a Gibbs' distribution, if

$$\frac{\lg(\mathrm{tgh}^2 \chi_n)}{E_n} = \text{constant independent of } n, \tag{11}$$

where E_n is the energy of the U_n-particle. The corresponding temperature is

$$T = -\frac{1}{k}\frac{E_n}{\lg(\mathrm{tgh}^2\,\chi_n)} . \tag{12}$$

3. SPHERICALLY SYMMETRICAL COLLAPSE

In this section, we describe the gravitational field, choose proper bases and calculate the transformation coefficients.

The spacetime

First, the question should be answered, why we choose to work in a collapse space-time and not in a static black hole one. The reason is that attempting to extend the static black hole solution analytically into the past one comes to a boundary, beyond which the metric is not more static (the so-called past horizon), and then to a bad singularity outside the black hole. This is a strong suggestion that such a space-time has not much physical meaning, and that all black holes originate in a truly dynamical process of collapse.

Our simplifying assumptions are: the collapsing star is spherically symmetrical and everywhere outside of it pure vacuum (the star does not rotate, the energy density of the star radiation is neglected and the star, toegether with its gravitational field, forms an isolated system). The absence of rotation is a strong idealization, but the result is not crucially dependent on it ([1],[2]). The star should further be static during $-\infty < t < t_o$ (corresponding to, say, the long time spent on the main sequence), then collapsing so that a black hole is left over, which is again static from some time on (Picture 1).

The metric outside the star has the form

$$ds^2 = (1-\frac{R_g}{r})c^2dv^2-2cdvdr-r^2(d\theta^2 + \sin^2\theta d\phi^2) \quad , \qquad (13)$$

and the coordinates v, r, θ, ϕ have the following meaning: v is the advanced time, that is v = const are spherically symmetrical null hypersurfaces whose null geodesic generators converge towards future; their parametrization v is chosen such that the proper time of very far away ($r \gg R_g$), static (r = const, θ = const, ϕ = const) observers coincide, up to an additive constant, with v. r = const-sections of these null hypersurfaces are two-dimensional spheres with total surface $4\pi r^2$. This is the definition of r, rather than e.g. the distance from some center. θ and ϕ are the usual spherical coordinates on these spheres, and they are constant along the null geodesic generators of v = const-hypersurf. v,r,θ,ϕ are called advanced Eddington-Finkelstein coordinates and (13) is nothing but the well-known Eddington-Finkelstein extension of Schwarzschild solution to vacuum Einstein's eqs..

The coordinates v,r,θ,ϕ can be extended to the inside of the star: v,θ an ϕ remain constant along the extension of the null geodesics generating $v = const$ outside, r is constant along spherically symmetrical sections of $v = const$ and equal to $r = \sqrt{\frac{P}{4\pi}}$, where P is their surface. Because of the spherical symmetry, the $v = const$ hypersurfaces form the past half-light- coni of the central points (the center of the star at different times). The metric inside the star will not have the form (13). It had to solve the inhomogeneous Einstein's equations, where the source term, energy-momentum tensor, would contain the details of the star structure. We shall not need the exact form of this metric, but only the fact that the central points are completely regular points of spacetime, and the spherical symmetry.

As to the future half-light-coni of the central points, we can extend their null geodesics generators up to the surface of the star and further, qualitatively, the following thing will happen. As the advanced time increases, the star becomes smaller and denser and the gravitational fields inside and in its neighbourhood stronger. The light rays will be bent more and more as the coni start at larger and larger values of v, until a point at $v = v_o$, say, comes so that its future half-light-cone emerges at the surface of the star with just non-diverging generators. We can choose v in such a way that $v_o = 0$. It should be emphasized that this effect is not due to some interaction of the light with the material of the star, but purely to the gravitational pull towards center similarly to the bending of light rays at the limb of the Sun (instead of light, we should rather imagine neutrino). The light coni of the central points with $v > 0$ emerge even converging at the surface of the star.

The next development of these null hypersurfaces can already be described quantitatively (because we know the exact form of the metric). The cone of the central point with $v = 0$ comes smoothly over to the hypersurface $r = R_g$ of the metric (13): $r = R_g$ is namely the maximal r in the Eddington-Finkelstein spacetime where there is an outgoing nondiverging spherically symmetric null hypersurface (for example, Minkowski spacetime has no such hypersurface at all). This hypersurface has the equation $r = R_g$, so the signal from the central point with $v = 0$ can never reach infinity. We call this hypersurface up to its vertex at $v = 0$ horizon. This is, because the far away static observer can see, if they wait sufficiently long time, up to horizon, but never beyond it.

The future half-light-coni of the central points with $v < 0$ satisfy the following relation outside the star:

$$v = u + 2c^{-1}r + 2c^{-1} R_g \lg (r-R_g) \ , \tag{14}$$

where u is constant along each such cone. u is chosen such as to coincide, up to an additive constant, with v along the trajectory of any far away static observer. u is, therefore, the retarded time normalized to be the proper time of far away observers. u can be extended to the inside of the star, if one takes it constant along the light cone.

The equation (14) shows that u is singular at the horizon. The meaning of the singularity is as follows. Consider the null hypersurfaces H_n defined by $u = n$, where n is an integer. These H_n cut the trajectory of a far away static observer regularly in intervals of 1 sec. On the other hand, consider a radially falling observer which crosses the horizon at the moment $\hat{v}$ of the advanced time v. His trajectory in a neighbourhood of the crossing point is given by

$$r = r(v) = R_g + A(v-\hat{v}) + \frac{1}{2} B (v-\hat{v})^2 + \dots$$

and crosses, pro his unit advanced time v the number $\nu = \frac{du}{dv}$ of such hypersurfaces. The function u(v) is calculated from (14):

$$u = v - 2c^{-1} r(v) - 2c^{-1} R_g \lg [r(v) - R_g] ,$$

thus

$$\nu \sim 2c^{-1} R_g \frac{r'(v)}{r(v)-R_g} \sim \frac{2c^{-1}R_g}{v-\hat{v}} , \qquad (15)$$

where only the leading term has been left. Hence, the v-frequency of cutting H_n's diverges at the horizon. This is the well-known effect of the frequency shift between an observer near the horizon and a far away one. The leading term in (15) is independent of A, B, etc., that is, of the velocity of the radial trajectory. This holds even for trajectories that would cross the horizon inside the star: propagation of light rays a finite distance through regular spacetimes can only finitely deform their geometry. Applying this to the central trajectory, we obtain

$$\nu = 2c^{-1} R_g \frac{1}{v} \quad . \tag{16}$$

Further extension of the null geodesic generators of H_n through the vertex at $r = 0$ yields the null hypersurface $v = v_n$, where v_n is the v-coordinate of the intersection of H_n with $r = 0$. This means that a far away static observer will see the frequency ν given by (16), if his advanced time is near $v = 0$.

A deep past, far away observer ($t \to -\infty$, $r \to \infty$) will see the signals along H_n to come almost regularly first, then with higher and higher frequency that will diverge near $v = 0$ according to (16); at $v > 0$, he will see no signal at all. In this way, the propagation through the spacetime deformes the signal considerably (Picture 2). The cause of this deformation clearly is the great qualitative change of the space as a whole, which comes about during the collapse: initially - almost Euclidian, finally - a black hole. This deformation leads to the mixing of positive and negative frequencies and yields, in this manner, a particle creation from vacuum. The leading term (16), which does not depend on the details of the collapse, determines the spectrum of the created particles.

The Choice of Bases

In our spacetime, there are two regions, where it is clear what a positive frequency means:

T^- : v, θ, ϕ finite, $\quad r \to \infty$,

T^+ : u, θ, ϕ finite , $r \to \infty$.

As one sees from (13), the metric is flat at T^-, v, r, θ and ϕ being the usual Minkowski spacetime advanced time and spherical coordinates, respectively, and similarly for u, r, θ, ϕ near T^+. Asymptotic solutions to (1) at T^- and T^+ are, therefore, identical with those for Minkowski spacetime, so we have the boundary conditions for the initial and final bases:

$$F_{\omega\ell m}(v,r,\theta,\phi) \to \frac{e^{-i\omega v}}{\sqrt{2\pi\omega}} \frac{\hat{F}_{\omega\ell}(r)}{r} Y_{\ell m}(\theta,\phi) \ , \tag{17}$$

$$U_{\omega\ell m}(u,r,\theta,\phi) \to \frac{e^{-i\omega u}}{\sqrt{2\pi\omega}} \frac{\hat{U}_{\omega\ell}(r)}{r} Y_{\ell m}(\theta,\phi) \ , \tag{18}$$

where $\omega > 0$, ℓ is an arbitrary non-negative integer and m an integer satisfying $-\ell \le m \le \ell$; $\hat{F}_{\omega\ell}(r)$, $\hat{U}_{\omega\ell}(r)$ are radial functions obliged to satisfy

$$\lim_{r\to\infty} \hat{F}_{\omega\ell}(r) = \lim_{r\to\infty} \hat{U}_{\omega\ell}(r) = 1 \ ,$$

and $Y_{\ell m}(\theta,\phi)$ are the orthonormal spherical harmonics [15].

Do the natural boundary conditions determine the bases? We observe:

1) The dynamical equation (1) for ϕ includes only the interaction of ϕ with the gravitation; the interaction of ϕ with the matter of the star is switched off. Thus,

even if (17) together with (1) determines the bases uniquely, the resulting bases will only be an approximation to the proper ones.

2) In Minkowski spacetime, the boundary condition (17) at $\mathcal{I}^-$ together with the regularity at $r = 0$ determine uniquely the incoming basis [15]. That is to say, for massless fields, the points ($t \to -\infty$, r = const) cannot influence the fields. The same holds for static Schwarzschild spacetime.

On the other hand, for U's, the regularity at the center is of no use; it must be replaced by some boundary condition at the horizon. The asymptotical behaviour of the solutions of (1) in the limit $u \to \infty$ ($r \to R_g$) outside the star can be found easily [16]:

$$U_{\omega \ell m} \to \sum_{\omega'} [K_{\omega\omega'\ell}\, e^{-i\omega' u} + L_{\omega\omega'\ell}\, e^{-i\omega' v}]\, Y_{\ell m}(\theta,\phi), \quad (19)$$

where $K_{\omega\omega'\ell}$, $L_{\omega\omega'\ell}$ are some constants. The eq. (1), the boundary conditions (18) and all L's determine K's uniquely. Let us choose, e.g.

$$L_{\omega\omega'\ell} = 0 \qquad \forall\ \omega,\omega',\ell \ . \quad (20)$$

Then, U's are uniquely determined, in particular

$$K_{\omega\omega'\ell} = K_{\omega\ell}\ \delta(\omega - \omega') \ .$$

However, U's and U^*'s do not form a complete system. One sees easily that the incoming wave packets which decrease quicker than r^{-1} at constant u cannot be expressed in U's and U^*'s (they are zero at $\mathcal{I}^+$). Let us denote a

Klein-Gordon-orthonormal system of such packets, complete at the horizon, by $\{V_k, V_k^*\}$. Then, the whole outgoing basis is $\{U_{\omega\ell m}, V_k, U^*_{\omega\ell m}, V^*_k\}$.

To verify whether the condition (b) of the Theorem is satisfied, we have to calculate Klein-Gordon scalar product of two solutions of which we know the past asymptotic data only, or the future asymptotic and horizon data only. The most elegant way to do this is based on the observation that the conformally deformed fields

$$\tilde{X} = \Omega^{-1} X$$

satisfy the Klein-Gordon equation

$$\tilde{g}^{ij} \tilde{\nabla}_i \tilde{\nabla}_j \tilde{X} = 0$$

(because $R_{ij} = 0$) in the conformally deformed spacetime $(M, \tilde{g})$,

$$\tilde{g}_{ij} = \Omega^2 g_{ij} ,$$

where Ω is an arbitrary smooth non-zero function [17]. Moreover, the integral (2) is invariant under the conformal deformation, that is to say

$$(F,G) = \frac{i}{2}\int_S \left(\tilde{F}^* \frac{\partial \tilde{G}}{\partial x^a} - \frac{\partial \tilde{F}^*}{\partial x^a}\tilde{G}\right) \sqrt{-\tilde{g}}\ \tilde{g}^{ab} \varepsilon_{cdeb} \frac{\partial x^c}{\partial u^1} \frac{\partial x^d}{\partial u^2} \frac{\partial x^e}{\partial u^3} du^1 du^2 du^3 . \tag{21}$$

On the other hand, we can bring an asymptotic region to a finite distance by such a deformation [18]. Let us show this for T^-. In the line element (13), perform the coordinate transformation $\xi = \frac{1}{r}$. On M, $\xi > 0$, at T^-, $\xi = 0$. The new line element reads

$$ds^2 = (1-R_g\xi)c^2dv^2 + 2cdv\,\frac{d\xi}{\xi^2} - \frac{1}{\xi^2}\,(d\theta^2 + \sin^2\theta d\phi^2)\ .$$

Choose $\Omega = \xi\,(= r^{-1})$. Then,

$$d\tilde{s}^2 = (1-R_g\xi)\xi^2c^2dv^2 + 2cdvd\xi - d\theta^2 - \sin^2\theta d\phi^2\ .$$

The conformally deformed metric in regular at $\xi = 0$, the points with $\xi = 0$ forming a null hypersurface. Using (21), we can calculate (F,G) by integrating the conformally deformed quantities over the hypersurface $\xi = 0$:

$$\sqrt{-\tilde{g}} = c\sin\theta,\quad \tilde{g}^{ij} = \begin{bmatrix} 0 & \frac{1}{c} & 0 & 0 \\ \frac{1}{c} & 0 & 0 & 0 \\ 0 & 0 & -1 & 0 \\ 0 & 0 & 0 & -\sin^{-2}\theta \end{bmatrix}$$

and the parametrical definition of S is

$$v = u^1,\ \theta = u^2,\ \phi = u^3,\ \xi = 0\ .$$

Setting this in (21), we obtain

$$(F,G) = \frac{i}{2}\int_{-\infty}^{\infty}dv\int_0^{\pi}d\theta\int_0^{2\pi}d\phi\ \left(\tilde{F}^*\frac{\partial\tilde{G}}{\partial v} - \frac{\partial\tilde{F}^*}{\partial v}\tilde{G}\right)\sin\theta\ , \qquad (22)$$

where

$$\tilde{F} = \lim_{r\to\infty} (rF), \quad \tilde{G} = \lim_{r\to\infty} (rG) \quad , \tag{23}$$

the limit to be taken at constant v, θ, ϕ.

Similar results hold for future asymptotic data:

$$(F,G) = \frac{i}{2} \int_0^\infty dv \int_0^\pi d\theta \int_0^{2\pi} d\phi \, \left(F^* \frac{\partial G}{\partial v} - \frac{\partial F^*}{\partial v} G\right)\Big|_{r=R_g} \cdot R_g^2 \sin\theta \, + \tag{24}$$

$$+ \frac{i}{2} \int_{-\infty}^{\infty} du \int_0^\pi d\theta \int_0^{2\pi} d\phi \, \left(\tilde{F}^* \frac{\partial \tilde{G}}{\partial u} - \frac{\partial \tilde{F}^*}{\partial u} \tilde{G}\right) \sin\theta \quad ,$$

where, now, the limit (23) is taken at constant u, θ, ϕ. The first integral in (24) is the contribution of the horizon data, the second of the asymptotic data.

The relations (22) and (24) imply that our "bases" are "normalized" to δ-function. However, one can always form wave packets from them ([1]). With this proviso, the condition b) of the Theorem is satisfied.

The condition c) is clearly satisfied at $S_i = T^-$. It also is satisfied at $T^+ \subset S_f$. However, it is not clear at all, whether $\{V_k\}$ defines positive frequencies at the horizon; we have even no idea what such a "positive frequency" means. One could choose another system of wave packets as well, e.g. $\{V'_h\}$:

$$V'_h = \sum_k (\lambda_{hk} V_k + \kappa_{hk} V_k^*) \quad , \tag{25}$$

where λ_{hk} and κ_{hk} are some constants satisfying the Bogoljubov conditions, but otherwise completely arbitrary. Fortunately, the expectation values of measurable quantities are invariant under (25). Proof:

The transformation (3) reads, in our case:

$$U_r = \sum_n (\alpha_{rn} F_n + \beta_{rn} F_n^*) ,$$

$$V_k = \sum_n (\alpha_{kn} F_n + \beta_{kn} F_n^*) . \tag{26}$$

Inserting the second relation of (26) into (25), we have

$$U_r = \sum_n (\alpha_{rn} F_n + \beta_{rn} F_n^*) ,$$

$$V'_n = \sum_n [\sum_k (\lambda_{hk}\alpha_{kn}+\kappa_{hk}\beta^*_{kn}) F_n + \sum_k (\lambda_{hk}\beta_{kn}+\kappa_{hk}\alpha^*_{kn}) F^*_n] . \tag{27}$$

Denoting the creation operator of U_r-mode by b_r^+, V_k-mode by d_k^+, V'_n-mode by d'^+_n and F_n-mode by a_n^+, we obtain from (4) and (26)

$$b_r = \sum_n (\alpha^*_{rn} a_n - \beta^*_{rn} a^+_n)$$

$$d_k = \sum_n (\alpha^*_{kn} a_n - \beta^*_{kn} a^+_n) ,$$

and from (4) and (27)

$$b_r = \sum_n (\alpha^*_{rn} a_n - \beta^*_{rn} a^+_n) ,$$

$$d'_h = \sum_n [\sum_k (\lambda^*_{hk}\alpha^*_{kn}+\kappa^*_{hk}\beta_{kn})a_n - \sum_k (\lambda^*_{hk}\beta^*_{kn}+\kappa^*_{hk}\alpha_{kn})a^+_n] \ .$$

Hence, the creation and anihilation operators of U-modes are invariant under (25). On the other hand, the local algebra of field operators at T^+ is generated by polynomials in b's and b^+'s, so the expectation values of this algebra elements are not influenced by (25).

This is an extremely important result: the quantities measurable at T^+ are not influenced by what is chosen for positive frequency near horizon. Not only we need not worry about how the particles near horizon are to be defined; we even can use the freedom in the choice of V's to simplify the mathematics.

On the other hand, the choice (20) of the horizon boundary conditions on U's is unique and forced to us by the requirements that

1) the horizon modes V_k do not influence the measurements at T^+ and $\{V_k, V^*_k\}$ form a complete system of function at the horizon;

2) $\{U_n, V_k, U^*_n, V^*_k\}$ is a basis satisfying the conditions a) and b) of the Theorem (in particular, all U's are orthogonal to all V's).

Proceeding to the condition c^*) of the Theorem, we observe at once that it can never be satisfied: the initial and the final Cauchy surfaces are simply not isometrical. Thus, one expects that the notion of development in the sense of the conventional quantum field theory cannot be maintained in curved spacetimes. This may be related to the difficulties with the definition of the energy density of gravitational field even in the classical theory.

Calculation of α's and β's

The simplest method is to propagate the functions U_r back in the spacetime up to T^- so that we find $U_r|_{T^-}$, and then to expand these in $F_n|_{T^-}$,

$$U_{\omega\ell m}|_{T^-} = \sum_{\omega'} (\alpha_{\omega\omega'} F_{\omega'\ell m}|_{T^-} + \beta_{\omega\omega'} F^{\star}_{\omega'\ell m}|_{T^-}) \tag{28}$$

(it is clear that the angular part of α's and β's is diagonal, because the spacetime is spherically symmetrical). Hawking's estimate of the high-frequency (in ω') part of $\alpha_{\omega\omega'}$ and $\beta_{\omega\omega'}$ is based on the observation that the frequency of $U_{\omega\ell m}$ is very high in a strip along the horizon, at the horizon even infinity (see (19) and (20)). Here, instead of solving (1) exactly, we can use the geometrical optics approximation. According to it and to (19), the hypersurfaces of constant phase of U's coincide with our null hypersurfaces u = const, which we have studied at the beginning. In such a way, near the point v = 0 at T^-, $U_{\omega\ell m}$ has the form (see the relation (16)):

$$\begin{aligned} U_{\omega\ell m}|_{T^-} &\propto e^{i\omega\frac{2R_g}{c}\lg(-v)} & v < 0\,, \\ U_{\omega\ell m}|_{T^-} &\propto 0 & v > 0\,. \end{aligned} \tag{29}$$

A part of the original mode $U_{\omega \ell m}$ will be scattered by the static Schwarzschild metric outside the star and comes to T^- with the same frequency ω. If we neglect this and all low frequency contributions to $U|_{T^-}$, we can use (29) at the whole of T^-. The function $U_{\omega \ell m}|_{T^-}$ is determined by the requirement that it should be normalized by (22):

$$v < 0 : U_{\omega \ell m}|_{T^-} = \frac{e^{i\omega \frac{2R_g}{c} \lg(-v)}}{\sqrt{2\pi\omega \frac{2R_g}{c}}} \cdot \frac{1}{r} \cdot Y_{\ell m}(\theta,\phi) ,$$
(30)
$$v > 0 : U_{\omega \ell m}|_{T^-} = 0 .$$

Such U's form a Klein-Gordon orthonormal system for the $(v < 0)$-part of T^-. They are naturally completed by V's defined by

$$v < 0 : V_{\omega \ell m}|_{T^-} = 0 ,$$
(31)
$$v > 0 : V_{\omega \ell m}|_{T^-} = \frac{e^{-i\omega \frac{2R_g}{c} \lg v}}{\sqrt{2\pi\omega \frac{2R_g}{c}}} \cdot \frac{1}{r} \cdot Y_{\ell m}(\theta,\phi) .$$

Then, $\{U_{\omega \ell m}, V_{\omega \ell m}, U^*_{\omega \ell m}, V^*_{\omega \ell m}\}$ is a complete basis for T^-, and the system $\{U_{\omega \ell m}, V_{\omega \ell m}\}$ is Klein-Gordon orthonormal.

To obtain α's and β's, we should expand the functions (30) into (17). Instead, we make use of the following trick. Let us continue the function $U_{\omega\ell m}|_{T^-}$ analytically around the singularity at v = 0 through the lower half of the complex v-plane to the positive values of v and similarly for $U^*_{\omega\ell m}|_{T^-}$. The following functions result,

$$U_{\omega\ell m}|_{T^-} + e^{-\frac{2\pi R_g}{c}\omega} V^*_{\omega\ell m}|_{T^-} \quad ,$$

$$U^*_{\omega\ell m}|_{T^-} + e^{\frac{2\pi R_g}{c}\omega} V_{\omega\ell m}|_{T^-} \quad .$$

We normalize them, using the orthonormality of U's and V's and the well-known property of the Klein-Gordon product

$$(F^*, G^*) = - (G,F) \quad .$$

In this manner, we obtain two functions called $X_{\omega\ell m}|_{T^-}$ and $Y_{\omega\ell M}|_{T^-}$:

$$X_{\omega\ell m}|_{T^-} = \sqrt{\frac{e^{\frac{2\pi R_g}{c}\omega}}{2\mathrm{sh}\frac{2\pi R_g}{c}\omega}}\, U_{\omega\ell m}|_{T^-} + \sqrt{\frac{e^{-\frac{2\pi R_g}{c}\omega}}{2\mathrm{sh}\frac{2\pi R_g}{c}\omega}}\, V^*_{\omega\ell m}|_{T^-} \quad ,$$

$$Y_{\omega\ell m}|_{T^-} = \sqrt{\frac{e^{-\frac{2\pi R_g}{c}\omega}}{2\mathrm{sh}\frac{2\pi R_g}{c}\omega}}\, U^*_{\omega\ell m}|_{T^-} + \sqrt{\frac{e^{\frac{2\pi R_g}{c}\omega}}{2\mathrm{sh}\frac{2\pi R_g}{c}\omega}}\, V_{\omega\ell m}|_{T^-} \quad . \qquad (32)$$

The inverse relation is

$$U_{\omega\ell m}|_{T^-} = \sqrt{\frac{e^{\frac{2\pi R}{c}g_\omega}}{2\mathrm{sh}\frac{2\pi R}{c}g_\omega}}\, X_{\omega\ell m}|_{T^-} - \sqrt{\frac{e^{-\frac{2\pi R}{c}g_\omega}}{2\mathrm{sh}\frac{2\pi R}{c}g_\omega}}\, Y^*_{\omega\ell m}|_{T^-}\,,$$

$$V_{\omega\ell m}|_{T^-} = -\sqrt{\frac{e^{-\frac{2\pi R}{c}g_\omega}}{2\mathrm{sh}\frac{2\pi R}{c}g_\omega}}\, X^*_{\omega\ell m}|_{T^-} + \sqrt{\frac{e^{\frac{2\pi R}{c}g_\omega}}{2\mathrm{sh}\frac{2\pi R}{c}g_\omega}}\, Y_{\omega\ell m}|_{T^-}\,. \tag{33}$$

The relations (32) and (33) imply that $\{X_{\omega\ell m}, Y_{\omega\ell m}, X^*_{\omega\ell m}, Y^*_{\omega\ell m}\}$ is a basis at T^- satisfying the condition b) of the Theorem. Moreover, from the construction it follows that X's and Y's contain only positive frequencies with respect to v, because they are analytic in the lower half of the complex v-plane. We can, therefore, use them instead of F's (they are analogous to F's in (5)).

On the other hand because V's are orthogonal to all U's they can be chosen as our complete system for the horizon.

The transformation (33) has the special form (6) with

$$\mathrm{tgh}\,\chi_{\omega\ell m} = e^{-\frac{2\pi R}{c}g_\omega}\,;$$

taking $E_{\omega\ell m} = \hbar\omega$, the relation (12) implies

$$T = -\frac{1}{k}\,\frac{\hbar\omega}{\dfrac{4\pi R}{\left(-\dfrac{g}{c}\omega\right)}} = \frac{\hbar c^3}{8\pi G k}\cdot\frac{1}{M}\text{, q.e.d..}$$

Observe that the series (9) does not converge for two reasons with these χ's:

1) The energy E_ω is allowed to be arbitarily low (massless field). It follows from the relation (11) that χ_ω can, then, be arbitarily large. $sh^2\,\chi_\omega$ is the expectation value of the particle number in the mode ω, so this is just the infrared divergence.

2) For any ω, an infinite number of equal $\chi_{\omega\ell m}$ exists. Thus, the average number of particles with a given energy diverges. This is due to the infinite radiation time of a static black hole, because in the semiclassical approximation we are using, the radiation does not influence the background.

Whereas the first divergence is more or less insignificant, the second is more serious. It is also present in the quantum field approach to the Klein paradox [19]. As a remedy, one can, e.g. attempt to calculate the effective energy-momentum tensor of the radiation and to couple it back to the metric by Einstein's equations. Such an approach is proposed by B. De Witt [16] and forms the current program of several groups (London). Another possibility is to consider an equilibrium of the hole with thermal radiation. This is mostly done in Cambridge.

Summarizing: we have proved that an observer far away from the black hole will see a thermal flux of particles at late times, under the condition that no particle went inwards at the past null infinity.

It should be stressed that we are able only to interpret our calculation far away from the star and the hole where the spacetime is flat. Thus, we cannot answer the question, from where the particles come out. If their source is the star, then no horizon seems to be able to form during the collapse (Boulware, Gerlach). On the other hand, Hawking's interpretation is that the particles are sent out at a constant rate from the neighbourhood of the horizon, as described at the beginning of this lecture.

In the literature, one finds two different attempts to obtain a more localized picture: regularizing the expectation value of the energy-momentum tensor in the incoming vacuum (B. De Witt, Fulling, Davies, Unruh, Christensen) and constructing models of local particle detectors (Unruh). Both methods seem not to contradict Hawking's conjecture. However, localizing the energy and momentum fluxes by taking seriously some regularized value of $\langle 0i|T^{ab}|0i\rangle$ may be misleading, if the fluctuations of T^{ab} are greater than this value. On the other hand, local particle detectors need not determine all of the energy flux.

REFERENCES

1. S.W. Hawking, Commun. Math. Phys. 43, 199 (1975).

2. D.N. Page, Phys. Rev. D14, 3260 (1976).

3. D.N. Page, Phys. Rev. D13, 198 (1976).

4. D.N. Page, S.W. Hawking, Ap. J. 206, 1 (1976).

5. S.W. Hawking, Phys. Rev. D13, 191 (1976).

6. S.W. Hawking, Mon.Not. R. Astron. Soc. 152, 75 (1971), B.J. Carr, S.W. Hawking, Mon. Not. R. Astron. Soc. 168, 399 (1974).
B. J. Carr, Ap. J. 206, 5 (1976).

7. S.W. Hawking, Phys. Rev. D14, 2460 (1976).

8. Ya.B. Zeldovich, Phys. Letters 59A, 254 (1976).

9. G.W. Gibbons, S.W. Hawking, Action Integrals and Partition Functions in Quantum Gravity, Phys. Rev. D, to be published.
G.W. Gibbons, Thermal Zeta Function, Preprint, München, 1977.
S.W. Hawking, Zeta Function Regularization of Path Integrals in Curved Spacetime, Preprint, Cambridge, 1977.

10. V. Urbantke, Nuovo Cim. 63B, 203 (1969).

11. A.Z. Capri, J. Math. Phys. 10, 575 (1969).
B. Schroer, R.Seiler, J.A. Swieca, Phys. Rev. D2, 2927 (1970),
P.J.M. Bongaarts, Linear Fields According to E.I. Segal. In: Mathematics of Contemporary Physics, Ed. by R.F. Streater, Acad. Press, London 1972.

12. R.M. Wald, Commun. Math. Phys. 45, 9 (1975).

13. L. Parber, Phys. Rev. D12, 1519 (1975).

14. P. Hájicek, Theory of Particle Detection in Curved Spacetimes, Preprint, München, 1976.

15. J.D. Jackson, Classical Electrodynamics, Wiley, New York, 1962.

16. B. De Witt, Phys. Reports 19, 295 (1975).

17. R. Penrose, Proc. Roy. Soc. A 284, 186 (1965).

18. R. Penrose, Phys. Rev. Letters 10, 66 (1963).

19. P.J.M. Bongaarts, S.N.M. Ruijsenaars, Ann. Phys. 101, 289 (1976).

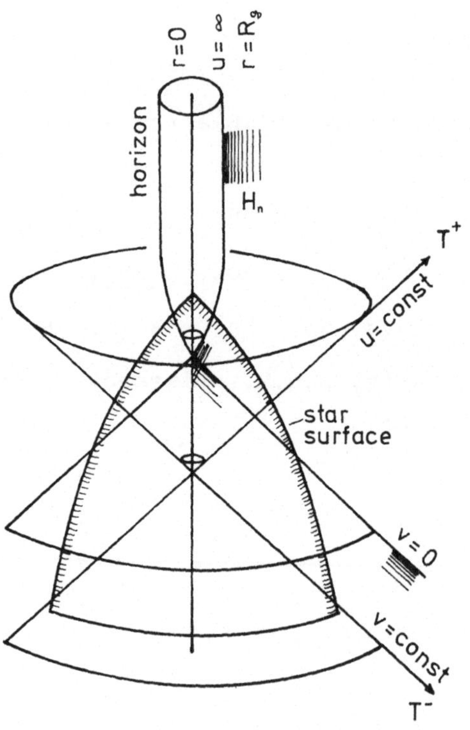
r=0
u=∞
r=R_g
horizon
H_n
T^+
u=const
star
surface
v=0
v=const
T^-

Fig.1

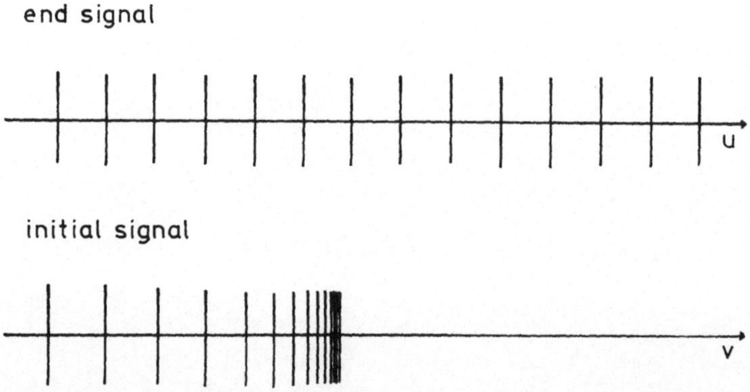
end signal
u
initial signal
v

Fig. 2

Acta Physica Austriaca, Suppl. XVIII, 873–883 (1977)

COVARIANT TREATMENT OF PARTICLE CREATION IN EXTERNAL FIELDS+

by

H. RUMPF
Institute for Theoretical Physics
University of Vienna, Austria

1. INTRODUCTION AND MOTIVATION

Quantum field theory in an external background poses a fundamental problem. In simple terms, it may be formulated as: "What is a particle in a generic external, e.g. electromagnetic or gravitational, field?" Here the word "generic" has to be given special emphasis. It means that the external field is not assumed to vanish asymptotically in space or time or to admit a symmetry group. In this seminar I want to propose a particle definition that makes use of none of these features. Rather, the assumptions that I will make arise naturally from the mathematical structure of (scalar) field theory and seem to be so general that one might accept them as the indispensible ingredients for the notion of "particle" to make sense at all.

+Seminar given at XVI. Internationale Universitätswochen für Kernphysik, Schladming, Austria, February 24 - March 5, 1977.

Although generalizations are possible, I will consider only a charged scalar field $\Phi(x)$ coupled to an external electromagnetic potential $A^{\mu}(x)$ and metric $g_{\mu\nu}(x)$. Let me recall some basic mathematical features:

The field equation

$$(\Box + \xi R + m^2)\Phi(x) = 0 \ ,$$

$$\Box : = \frac{1}{\sqrt{-g}} (\partial_{\mu} + ieA_{\mu})\sqrt{-g}\, g^{\mu\nu}(\partial_{\nu} + ieA_{\nu}) \ , \qquad (1)$$

R ... curvature scalar ,

is conformally covariant for $m^2 = 0$, if we choose $\xi = 1/6$, but we will leave ξ unspecified. We assume the initial-value problem for eq.(1) to be well-posed and its solution to be given by

$$\psi(x) = - \int_{\Sigma} d\sigma^{\mu}(y) G(x,y) (\frac{\overleftrightarrow{\partial}}{\partial y^{\mu}} + 2ieA_{\mu}(y))\psi(y) =$$

$$= : (G * \psi)_{\Sigma}(x) \ , \qquad (2)$$

where Σ is any space-like hypersurface and $G(x,y)$ the Cauchy propagator. Then the commutation relations

$$[\Phi(x), \Phi^{+}(y)] = iG(x,y) \qquad (3)$$

are consistent and together with eq.(1) constitute the "field algebra". The problem now is to complete the physical description by introducing "physical" states. If one wants to retain the particle aspect of the field, this amounts to select a certain Fock representation of the field algebra. The general construction of such a representation goes as follows:

Current conservation defines on the linear space H of classical solutions of (1) the "charge form" (,) by

$$(\phi,\psi) := \int_{\Sigma} d\sigma^{\mu}\ \phi^{*}(i\overleftrightarrow{\partial}_{\mu} - 2eA_{\mu})\psi \ . \tag{4}$$

Assume that there exists a decomposition

$$H = H^{+} \oplus H^{-} , \qquad (H^{+}, H^{-}) = 0 ,$$

$$(\ , \ )\Big|_{H^{+}} \geq 0 , \qquad (\ , \ )\Big|_{H^{-}} \leq 0 \tag{5}$$

and introduce orthonormal bases $\{{}^{+}\psi_i\}$, $\{{}^{-}\psi_k\}$ in H^{+} and H^{-}:

$$({}^{+}\psi_i, {}^{+}\psi_j) = \delta_{ij}, \ ({}^{-}\psi_k, {}^{-}\psi_\ell) = -\delta_{k\ell}, \ ({}^{+}\psi_i, {}^{-}\psi_k) = 0. \tag{6}$$

Then because of (3) the coefficients defined by the expansion

$$\phi(x) = \sum_i {}^{+}a_i \ {}^{+}\psi_i(x) + \sum_k {}^{-}a_k^{+} \ {}^{-}\psi_k(x) \tag{7}$$

obey

$$[{}^{+}a_i, {}^{+}a_j^{+}] = \delta_{ij}, \ \ [{}^{-}a_k, {}^{-}a_\ell^{+}] = \delta_{k\ell} \tag{8}$$

(all other commutators vanish). As is well known, these relations allow to interprete the various coefficients as annihilation or creation operators of "quase-particles" and "quase-antiparticles" and to define a "quasi-vacuum" $|0\rangle$ by

$$ {}^{+}a_i |0\rangle = {}^{-}a_k |0\rangle = 0 \quad . \tag{9} $$

But which of the infinitely many inequivalent Fock representations that exist in general is the right one? In order to prepare an answer to this question let us consider an example where the answer is well known. We take the case of an external field of finite duration. There we have two physical representations: one with vacuum $|0\ in\rangle$ for the time before the field is switched on (let the corresponding decomposition be $H = H_+ \oplus H_-$), and one with vacuum $|0\ out\rangle$ for the time after the field has been switched off (based on a decomposition $H = H^+ \oplus H^-$). $H_+(H^+)$ consists of solutions of (1) that go over into positive frequency modes at early (late) times. Orthonormal bases corresponding to the two decompositions will be related by

$$ {}^{+}\psi_r = \alpha_{rk}\ {}_{+}\psi_k + \beta_{r\ell}\ {}_{-}\psi_\ell $$

$$ {}^{-}\psi_s = \gamma_{sk}\ {}_{+}\psi_k + \varepsilon_{s\ell}\ {}_{-}\psi_\ell \quad . \tag{10} $$

Nonvanishing β's and γ's give account of particle creation by the external field.

Now consider the Green's function

$$ K_\infty(x,y) := -i\ \langle 0\ out|T(\phi(x)\phi^+(y))|0\ in\rangle / \langle 0\ out|0\ in\rangle \quad . \tag{11} $$

With some algebra one finds

$$K_{\infty}(x,y) = -i[\theta(x,\Sigma(y))(\alpha^{-1})_{kr}\ {}^{+}\psi_r(x)\ {}_{+}\psi_k^{*}(y) +$$

$$+ \theta(\Sigma(y),x)(\varepsilon^{-1})^{*}_{\ell s}\ {}_{-}\psi_{\ell}(x)\ {}^{-}\psi_s^{*}(y)] , \tag{12}$$

where we have introduced the notation

$$\theta(x,\Sigma) = 1 - \theta(\Sigma,x) = \begin{cases} 1 & \text{if} \quad x > \Sigma \\ 0 & \text{if} \quad x < \Sigma \end{cases} \tag{13}$$

and $\Sigma(y)$ is any spacelike hypersurface containing y. (It can be shown quite generally that α and ε are invertible.) From (13), (10) and the orthonormality relations we conclude

$$(K_{\infty} * {}^{+}\psi_r)_{\Sigma}(x) = \theta(x,\Sigma)\,{}^{+}\psi_r(x) ,$$

$$(K_{\infty} * {}_{-}\psi_{\ell})_{\Sigma}(x) = -\theta(\Sigma,x)\ {}_{-}\psi_{\ell}(x) . \tag{14}$$

Similarly, if we define

$$\tilde{K}_{\infty}(x,y) = K^{*}_{\infty}(y,x) , \tag{15}$$

we find

$$(\tilde{K}_{\infty} * {}^{-}\psi_s)_{\Sigma}(x) = \theta(x,\Sigma)\,{}^{-}\psi_s(x) ,$$

$$(\tilde{K}_{\infty} * {}_{+}\psi_k)_{\Sigma}(x) = -\theta(\Sigma,x)\,{}_{+}\psi_k(x) . \tag{16}$$

Thus the 4 subspaces H^{+}, H^{-}, H_{+}, H_{-} can be reconstructed out of the single Green's function K_{∞}.

Now it was recognized by Schwinger already in 1951 [1] that for the present example the Green's function K is the one that corresponds to adding an infinitessimal negative imaginary part to the mass parameter in the Green's function equation:

$$(\Box_x + \xi R(x) + m^2 - i0)K(x,y) = -\frac{1}{\sqrt{-g}}\,\delta^{(4)}(x-y) \ . \qquad (17)$$

For the case of a vanishing external field, the solution of this equation is the well-known "Feynman propagator", which propagates positive frequency solutions into the future and negative frequency solutions into the past. However, as we have seen, in the more general case of our example the "propagation properties" (14), (16) introduce in addition a distinction between "ingoing" and "outgoing" solutions.

2. PARTICLE DEFINITION

It is important to observe that the "propagation properties" and the m^2-i0-prescription are concepts that can be maintaned even in much more general cases than the special scattering situation alluded to above. First, however, we have to give eq. (17) a more precise meaning. The operator $\Box + \xi R$ is formally self-adjoint with respect to the scalar product

$$\langle f,g\rangle := \int_M d^4x \sqrt{-g}\, f^*(x)g(x) \qquad (18)$$

on the Hilbert space $L^2(M)$ of square-integrable functions on the space-time manifold M. If this operator possesses a domain of self-adjointness, the resolvents $(\Box + \xi R + m^2)^{-1}$ exist for <u>complex</u> m^2, and we may define K, $\tilde{K}$ as the integral kernels of their boundary values as the real m^2-axis is approached from below and above, respectively.

Furthermore we define linear operators $\overset{(\sim)}{P}\uparrow$, $\overset{(\sim)}{P}\downarrow$ on the space H by

$$(\overset{(\sim)}{P}\uparrow\psi)(x) := (\overset{(\sim)}{K} * \psi)_\Sigma (x) , \qquad x > \Sigma ,$$

$$(\overset{(\sim)}{P}\downarrow\psi)(x) := -(\overset{(\sim)}{K} * \psi)_\Sigma (x) , \qquad x < \Sigma , \tag{19}$$

and finally the 4 subspaces

$$H^+ := P\uparrow H, \; H^- := \tilde{P}\uparrow H, \; H_+ := \tilde{P}\downarrow H, \; H_- := P\downarrow H . \tag{20}$$

This is our particle definition for the generic case. But is it consistent? We know that a Fock space construction is possible only if the formal properties (5) are fulfilled. They imply that the operators $P\uparrow$, $P\downarrow$ have to be projectors (which is just the propagation property). This property as well as the definiteness of charge in the subspaces considered can be shown by considering solutions of the Klein-Gordon-equation with <u>complex</u> m^2 and then applying an analyticity argument. In this seminar, I only want to mention a surprising "by-product" of this argument. It allows an alternative formulation of the definition which does not use the propagators: "A solution of (1) is an $\genfrac{}{}{0pt}{}{\text{outgoing}}{\text{ingoing}}$ particle (antiparticle) mode if there exists an analytical continuation of ψ into the $\genfrac{}{}{0pt}{}{\text{lower (upper)}}{\text{upper (lower)}}$ complex m^2-plane which is "regular" near the $\genfrac{}{}{0pt}{}{\text{future}}{\text{past}}$ boundary of M." The term "regular" can be made more precise by the explicit construction of a space of "allowed" distributions.

Summarizing,our definition singles out a unique Green's function by introducing a new structure which is neglected in conventional field theory. This new structure is the Hilbert space $L^2(M)$, resp. the "scalar product" (18). Since this scalar product as well as the charge form is gauge invariant and invariant under the identity component of any symmetry group admitted by the external field, the same is true for the particle definition.

Another interesting feature of the definition is the following: (18) is the natural scalar product of the "five-dimensional Schrödinger equation"

$$(\Box + \xi R)\Psi(x,s) = i \frac{\partial}{\partial s}\Psi(x,s) \ , \qquad (21)$$

which has been discussed by some outhors [2], although the physical meaning of the "fifth parameter" s is still unclear. That it actually might have a meaning is suggested by the following fact: There exist canonical bases in the subspaces H_+, H_-, H^+, H^- with the property that the relative probabilities for scattering and pair creation in these modes evaluated with the scalar product (18) coincide with the relative probabilities of quantum field theory. E.g. the relative probability for pair creation in the canonical modes ${}^+\psi_i$, ${}^-\psi_k$ is

$$w_{ik} = \frac{|\langle {}^+\psi_i, {}^-\psi_k \rangle|^2}{\langle {}^+\psi_i, {}^+\psi_i \rangle \langle {}^-\psi_k, {}^-\psi_k \rangle} = \frac{|\langle 0 \text{ out}| \, {}^+a_i \, {}^-a_k | 0 \text{ in} \rangle|^2}{|\langle 0 \text{ out}| \, 0 \text{ in} \rangle|^2} \ .$$

3. RESULTS

After this formal exposition I want to report about some results obtained with this definition in special electromagnetic and gravitational fields. Although the problem I started with in the beginning is more urgent in the gravitational context, where fields cannot be switched, also the electromagnetic application is interesting, since at least it may serve as a testing ground for the physical relevance of the formalism proposed.

a) Electromagnetic fields

1. Homogeneous, time-dependent electric field that becomes constant as $t \to \pm\infty$. With the gauge $A^\mu = (0,\vec{A}(t))$ the physical modes turn out to behave as

$$^{\pm}\psi \xrightarrow{t\to\infty} \frac{1}{\sqrt{2\dot{S}}} \exp\,(\mp iS)\,,\quad S = \vec{k}\vec{x} + \int^{t} dt' \sqrt{m^2+(\vec{k}-e\vec{A}(t'))^2}$$

where S is the classical action. The ingoing modes behave correspondingly.

2. $A^\mu = (0,\vec{A}(t))$ with $\vec{A}(t)$ periodic in time. Here no creation takes place, since the physical modes turn out to be temporal analogs of the Bloch waves of solid state physics. For the same reason one would expect no particles to be created in a plane electromagnetic wave.

3. Static potential step $A^\mu = (V\theta(x^3),\vec{0})$. If $eV > 2m$, Klein's paradox occurs, and the definition predicts a splitting $H^+ \neq H_+$, $H^- \neq H_-$, which takes place exactly in the "critical" energy region $m < E < eV - m$. The

creation rate is proportional to the transmission coefficient as expected intuitively.

4. Deep square well. The complex frequency modes reported in [3] turn out to belong to $H^+ \cap H^-$ for Im $\omega < 0$ and to $H_+ \cap H_-$ for Im $\omega > 0$. A more detailed analysis suggests that the particle creation takes place at an ever increasing rate. The phenomenon is qualitatively the same in a Coulomb potential which is "cut off" at short distances.

b) Gravitational fields

1. There is no creation in Gaussian metrics of the type $ds^2 = dt^2 + g_{\alpha\beta}(x^\gamma)\, dx^\alpha dx^\beta$ (Greek indices run from 1 to 3), e.g. the static closed Einstein universe (the situation here is quite analogous to a static magnetic field). The vacuum is also stable in a spherically symmetric, globally static, asymptotically flat space-time, i.e. around a star which is not too dense. However, if the metric is not globally stationary, particle creation will occur, as simple examples show.

2. de Sitter space. This "universe" can be visualized as the hyperboloid $(x^0)^2-(x^1)^2-(x^2)^2-(x^3)^2-(x^4)^2=-R^2$ embedded in 5-dimensional Minkowski space. In suitable coordinates the metric is $ds^2 = dt^2 - R^2 \cosh^2(t/R)d\sigma^2$, where $d\sigma^2$ is the line element on the 3-dimensional unit sphere. There exists a family of geodesics for which t is the proper time. The distance x between 2 such geodesics obeys the law $d^2x/dt^2 = x/R^2$. Therefore locally the relative motion of these freely falling objects can be described by the Newtonian potential $V(x) = -x^2/(2R^2)$.

This potential exhibits a gravitational analog of Klein's paradox: Virtual particle pairs must tunnel apart a distance $x = 2R$ to be "allowed" to become real (because then $\Delta V = -2$). From this heuristic reasoning one would expect the creation probability to be proportional to $\exp(-4mR)$. As a matter of fact, our formalism yields a total number of $\sim 4 \exp(-2\pi mR)$ particles created in every mode defined by the 3-dimensional spherical harmonics, if $m^2 >> 1/4\ R^2$. (For $m^2 < 1/4\ R^2$ the mass spectrum becomes discrete, the corresponding eigenfunctions might be called "linear instantons".)

3. If space-time has an initial singularity, as in the most cosmological models and the Kruskal manifold, the dfinition predicts a divergent number of particles created in every mode. However, in our opinion this pathological result only confirms that the concept of an "externally prescribed" singularity is not a very realistic one.

4. Last but not least we mention that the definition reproduces Hawking's effect for the case of a spherically symmetric gravitational collapse starting from regular initial conditions, which has been described in much detail by Dr. Hajicek during this school.

REFERENCES

1. J. Schwinger, Phys. Rev. 82 (1951) 664.
2. e.g. R.P. Feynman, Phys. Rev. 80 (1950) 440.
   Y. Nambu, Progr. Theor. Phys. 5 (1950) 82.
   C. Garrod, Rev. Mod. Phys. 38 (1966) 483.
3. L.H. Schiff, H.Snyder and J.Weinberg, Phys. Rev. 57 (1940) 315.

Acta Physica Austriaca, Suppl. XVIII, 885–897 (1977)
© by Springer-Verlag 1977

PHOTON CONDENSATION IN AN EINSTEIN UNIVERSE[+]

by

J.D. BECKER and L. CASTELL
Max-Planck-Institut zur Erforschung der Lebensbedingungen
der wissenschaftlich-technischen Welt
Starnberg, West Germany

INTRODUCTION

In many discussions with the authors, C.F. von Weizsäcker has suggested that matter arises from a condensation phenomenon. In this paper we give a model for such a mechanism.

We start with an Einstein universe with a metric arising from an initially unspecified isotropic matter density. We then assume that a large number of free photons exists in this universe. We treat this system of photons with the usual methods of Bose-Einstein statistical mechanics. The result is a Bose-Einstein condensation, i.e. below a certain temperature T_c there

---

[+]Seminar given at XVI.Internationale Universitätswochen für Kernphysik,Schladming,Austria,February 24–March 5,1977.

is a photon "liquid" besides the usual photon "gas". (Since we have a finite universe the energy of the lowest photon energy level is greater than zero.)

We now identify the matter in the universe with this photon liquid. Then we are able to calculate T_c from the radius of the universe. We obtain $T_c = 3.5 \cdot 10^{11}$ K. The mass $m_c = \frac{kT_c}{c^2}$ associated with this temperature is $m_c = 30$ MeV.

PHOTONS IN AN EINSTEIN UNIVERSE

The metric of the (static) Einstein universe $S_3 \times T$ is

$$R^2\, du^2 - c^2\, dt^2$$

$$du^2 = d\rho^2 + \sin^2\rho\, (d\theta^2 + \sin^2\theta\, d\phi^2) \qquad (1)$$

$$0 \leq \rho, \theta \leq \pi;\ 0 \leq \phi \leq 2\pi;\ -\infty < t < +\infty.$$

R is the radius of the sphere S_3; t is the cosmic time; c is the velocity of light. R is connected with the pressure p and the matter density ρ by [1]

$$R^2 = \frac{2}{\kappa(p + \rho)} \qquad (2)$$

where $\kappa = 1.865 \cdot 10^{-27}$ cm g^{-1} is the gravitational constant. We shall assume that one can neglect the pressure (indeed, this can be done in our universe the pressure of which arises from the 2.7 K background radiation). Thus, with $\rho = 10^{-30}$ g cm^{-3}, we obtain $R = 3\cdot 10^{28}$ cm which is of the order of the inverse Hubble constant, $R_H = 1.8\cdot 10^{28}$ cm. In this stationary closed universe we consider particles of mass 0 and integer helicity s. For photons (s = 1), e.g. the equations of motion read

$$g^{\mu\nu} \nabla_\mu F_{\nu\rho} = 0 , \tag{3}$$

with ∇_μ the covariant derivative. However, for the statistical treatment we don't need the solutions of these equations but only the eigenvalues of the energy operator $i\hbar \frac{\partial}{\partial t}$ and the degeneracies g_n of the corresponding stationary solutions. These are given by [2]

$$\varepsilon_n = \frac{\hbar c}{R} (n + s + 1) ;$$

$$g_n = (n + 2s + 1)(n + 1) ; \tag{4}$$

$$n = 0, 1, 2, \ldots .$$

<u>We note:</u>

1. The spectrum is discrete and integer spaced. Hence, the eigenfunctions are periodic in time (with the period $T = \frac{2\pi R}{c} = 6\cdot 10^{18}$ s $= 2\cdot 10^{11}$ years). Thus, instead of the universe $S_3 \times T$, we could consider $S_3 \times S_1$.

2. The lowest energy level is proportional to R^{-1}.

For finite R this is greater than O. (This is the same situation as for photons in a box where $\varepsilon_o \sim L^{-1}$ where L is the length of the box. In the usual treatment one sets $\varepsilon_o = 0$. However, we shall not use this approximation.) For photons (s = 1), we get $\varepsilon_o = \frac{2\hbar c}{R} = 1.3 \cdot 10^{-39}$ MeV.

THE THERMODYNAMICAL QUANTITIES

The thermodynamical quantities depend on three parameters: the volume $V = 2\pi^2 R^3$ of the S_3, the number of photons N (which is finite because $\varepsilon_o > 0$), and the temperature T. The grand canonical partition function Ξ, or rather the grand canonical potential $J = \ell n\, \Xi$ is given by [3], [4]

$$J = \ell n\, \Xi = - \sum_{n=o}^{\infty} g_n\, \ell n\, (1 - \Lambda\, e^{-\beta\varepsilon_n}) \; . \qquad (5)$$

The absolute activity Λ is to be determined from

$$N = \Lambda \frac{\partial}{\partial \Lambda} J ; \quad 0 < \Lambda < 1 . \qquad (6)$$

In the sequel we shall use the abbreviations

$$z := \exp\left(- \frac{T_R}{T}\right) ; \quad T_R = \frac{\hbar c}{kR} = \frac{\varepsilon_o}{2k} \; . \qquad (7)$$

The characteristic temperature T_R is $0.75 \cdot 10^{-29}$ K. Now we split the sum in (5) into the part J_o arising from the contribution of the lowest state (n = 0) and into the rest J^* arising from the contributions of all other states ($n \geq 1$). In the expression for J^*, we expand the logarithm, perform the sum over n, and use the identity $1-z^j = (1-z)\cdot\sum_{\nu=0}^{j-1} z^\nu$. Thus, we may write

$$J = J_0 + J^* ;$$

$$J_0 = -(2s+1)\,\ell n\,(1-\Lambda) ; \tag{8}$$

$$J^* = \frac{2}{(1-z)^3}\sum_{j=1}^{\infty}\frac{\Lambda^j z^j(3-3z^j+z^{2j})}{j\cdot\left(\sum_{\nu=0}^{j-1} z^\nu\right)^3} + \frac{2s-1}{(1-z)^2}\sum_{j=1}^{\infty}\frac{\Lambda^j z^j(2-z^j)}{j\cdot\left(\sum_{\nu=0}^{j-1} z^\nu\right)^2} .$$

The functions defined by the sums are bounded and well behaved in the whole domain $0 \leq z, \Lambda \leq 1$.
The form of J^*, in particular, the powers 2 and 3 of the denominators, arises from the degeneracies g_n. From J, we can derive all other thermodynamical quantities, as the particle number N, energy U, and the specific heat C_V. We split all these quantities in two parts (e.g., $N = N_0 + N^*$), according to the split in J. We note that for $T \ll T_R$, i.e., $z \ll 1$, J^* vanishes exponentially. N^*, U^*, and C_V^* show the same behaviour.

Now we assume $T \gg T_R$; since T_R is very small, all temperatures of interest are much bigger. Then we may write approximately

$$1 - z = \frac{T_R}{T} . \tag{9}$$

Thus, for $T \gg T_R$, we may write

$$J^* = 2 \left(\frac{T}{T_R}\right)^3 \sum_{j=1}^{\infty} \frac{\Lambda^j}{j^4} . \tag{10}$$

Hence,

$$N_o = \frac{(2s + 1)\Lambda}{1-\Lambda} ; \tag{11}$$

$$N^* = 2 \left(\frac{T}{T_R}\right)^3 \sum_{j=1}^{\infty} \frac{\Lambda^j}{j^3} ;$$

$$U_o = \frac{\hbar c}{R} (s+1) N_o = k T_R (s+1) N_o ;$$

$$U^* = 6k \frac{T^4}{T_R^3} \sum_{j=1}^{\infty} \frac{\Lambda^j}{j^4} + k T_R (s+1) N^* . \tag{12}$$

For $T >> T_R$ we may neglect the term $kT_R (s+1) N^*$, as well.

THE BOSE-EINSTEIN CONDENSATION

From (11) we see that N^* is bounded by T^3. Thus, for large N, most particles must fall into the ground state as T decreases. This is the Bose-Einstein condensation. To make it more transparent, we now determine $\Lambda = \Lambda(T)$ from (11) for a given $N = N_o + N^*$.

With $\tau := 2 \left(\frac{T}{T_R}\right)^3 \sum_{j=1}^{\infty} \frac{\Lambda^{j-i}}{j^3}$ we may re-write eq. (11) in the form

$$\Lambda^2\tau - \Lambda(N + (2s + 1) + \tau) + N = 0 . \tag{13}$$

Since $N_3(\Lambda) := \sum_{j=1}^{\infty} \frac{\Lambda^{j-1}}{j^3}$ is monotonous in Λ, and $N_3(0) = 1$, $N_3(1) = \zeta(3) = 1.202$, ζ the Riemann ζ-function, we disregard the Λ dependence of τ for the moment. Then we can solve (13) for Λ. We obtain

$$\Lambda = \frac{1}{2\tau} \left(N+(2s+1)+\tau - \sqrt{(N+(2s+1)+\tau)^2 - 4\tau N}\right) . \tag{14}$$

For $\tau = 0$, Λ has its maximal value $1 - \frac{2s+1}{N+2s+1}$. It then decreases extremely slowly. For $\tau = N$, $\Lambda \gtrsim 1 - \sqrt{\frac{2s+1}{N}}$ which is still very close to 1. For $\tau > N$, however, Λ decreases like $\frac{N}{\tau}$.
(Remark: If for $\tau > N$ a higher accuracy is required, (14) can be solved iteratively for Λ.) The critical temperature T_c for the Bose condensation is therefore given by $\tau = N$, i.e.

$$\left(\frac{T_c}{T_R}\right)^3 = \frac{N}{2\zeta(3)} . \tag{15}$$

Thus, for $T \leq T_c$, we have

$$\begin{aligned} N^* &= 2\left(\frac{T}{T_R}\right)^3 \cdot \zeta(3) = \left(\frac{T}{T_c}\right)^3 \cdot N ; \\ U^* &= 6k \frac{T^4}{T_R^3} \cdot \zeta(4) = 3\, kT\, N^* \frac{\zeta(4)}{\zeta(3)} . \end{aligned} \tag{16}$$

For $T >> T_c$, we have

$$\begin{aligned} N^* &= N ; \\ U^* &= 3\, kT\, N . \end{aligned} \tag{17}$$

At T_c, the specific heat C_V has a discontinuity. C_V is given by

$$C_V = \left(\frac{\partial U}{\partial T}\right)_\Lambda + \left(\frac{\partial U}{\partial \Lambda}\right)_T \frac{\partial \Lambda}{\partial T} . \tag{18}$$

The discontinuity arises from the change in the slope of Λ. We obtain

$$C_V\ (T=T_c-0) = 6k\left(\frac{T_c}{T_R}\right)^3 \cdot 4\zeta(4) = 3kN \cdot 4\,\frac{\zeta(4)}{\zeta(3)} ;$$

$$C_V\ (T=T_c+0) = 6k\left(\frac{T_c}{T_R}\right)^3 \cdot \left(4\zeta(4) - 3\,\frac{\zeta(3)}{\zeta(2)} \cdot \zeta(3)\right) \tag{19}$$

$$= 3\ kN\ \left(4\,\frac{\zeta(4)}{\zeta(3)} - 3\,\frac{\zeta(3)}{\zeta(2)}\right) .$$

We note that $C_V^o << C_V^*$ for $T >> T_R$.

<u>Remarks:</u>

1. If we express T_R in terms of the volume $V = 2\pi^2 R^3$ and insert $\zeta(4) = \frac{\pi^4}{90}$, we obtain from (16)

$$U^* = \frac{4}{15} \cdot \frac{\pi^5}{c^3} \cdot \frac{k^4}{h^3} VT^4 \tag{20}$$

which is the usual formula for the black body radiation (for one helicity state).

2. From (17) we see that for $T >> T_c$ the photons behave like a gas with 6 degrees of freedom. This is due to the fact that photons are oscillators rather than classical particles.

Figures 1a,b,c and d show the T dependence of Λ, N^*, U^*, and C_V for fixed N and R (qualitatively).

CONCLUSIONS

As already mentioned, our approach differs from the usual treatment of photon statistics by keeping the volume finite. As a consequence, there is a characteristic temperature T_R, the Bose-Einstein condensation temperature T_c is finite, and the ground state of the photons differs from the vacuum because it has a finite energy. Thus, there exists a photon "liquid". Now we identify the photon condensate with the matter of the universe. The total energy of this matter is

$$U_M = \rho \; V \; c^2 = \rho \cdot 2\pi^2 R^3 \cdot c^2 \; . \qquad (21)$$

We equate this energy with the energy of the photon condensate (eq. 12), i.e. we set $U_o := U_M$. Apparently the temperature of the universe is far below T_c, so that we may write $N_o = N$ in (12). Thus, from (12) and (21), we can calculate N. We obtain

$$N = \frac{\rho \cdot 2\pi^2 R^4 \cdot c^2}{\hbar c(s+1)} \; . \qquad (22)$$

Inserting this into (15), we get the critical temperature:

$$T_c^3 = \frac{2\pi^2}{2\zeta(3)(s+1)} \cdot \frac{\hbar^2 c^4}{k^3} \cdot R \cdot \rho . \tag{23}$$

(We observe that, for ρ fixed, T_c tends to infinity for $R \to \infty$. However, because of (2), this would imply that the gravitational constant vanishes. We shall keep R finite anyway.) With the aid of eq. (2) we can eliminate the matter density ρ from (23). We obtain

$$T_c^3 = \frac{2\pi^2}{\zeta(3)\cdot(s+1)} \cdot \frac{\hbar^2 c^4}{k^3} \cdot \frac{1}{\kappa R} . \tag{24}$$

Thus, we have expressed the condensation temperature T_c in terms of the radius of the universe. We note that the exponents in $T_c^3 \cdot R \sim$ const. are not trivial but occur as a consequence of the fact that the degeneracy is a polynomial of degree 2 in n(cf.eq. (4)).

Inserting the numbers into (22) and (24) we finally obtain

$$N = 2.4 \cdot 10^{122} \; ; \qquad T_c = 3.5 \cdot 10^{11}\ \mathrm{K}. \tag{25}$$

Remarks:

1. Unlike the critical temperature T_H in Hagedorn's programme [4], which has the value $1.2 \cdot 10^{12}$ K, our temperature T_c is not the highest possible temperature. However, above T_c all matter has undergone the transition to radiation.

2. Formally one can associate a critical mass m_c with T_c by $m_c = k\, T_c\, c^{-2}$. We obtain m_c = 30 MeV. It is hoped that this correspondence has some deeper significance and that it is possible to establish a particle structure of the photon condensate in our model.

3. The first statistical treatment of mass 0 particles in an Einstein universe was done by P. Roman [5].

ACKNOWLEDGEMENTS

We thank our colleagues in the physics group of our institute for many valuable discussions.

In particular, we express our warmest thanks to Prof. C. F. v. Weizsäcker for many discussions and suggestions.

REFERENCES

1. R.U. Sexl and H.K. Urbantke, Gravitation und Kosmologie B.I. - Wissenschaftsverlag, Zürich 1975.
   D.W. Sciama, Modern Cosmology, Cambridge University Press 1971.

2. L. Castell, Nucl. Phys. B13 (1969) 231.

3. E.A.Guggenheim in "Handbuch der Physik", Vol. III/2 Springer, Heidelberg 1959.

4. W. Weidlich, Thermodynamik und statistische Mechanik, Akademische Verlagsgesellschaft, Wiesbaden 1976.

5. P. Roman in "Quantum Theory and the Structures of Time and Space 2" ed. L. Castell, M. Drieschner, C. F. v. Weizsäcker, Hanser München 1977.

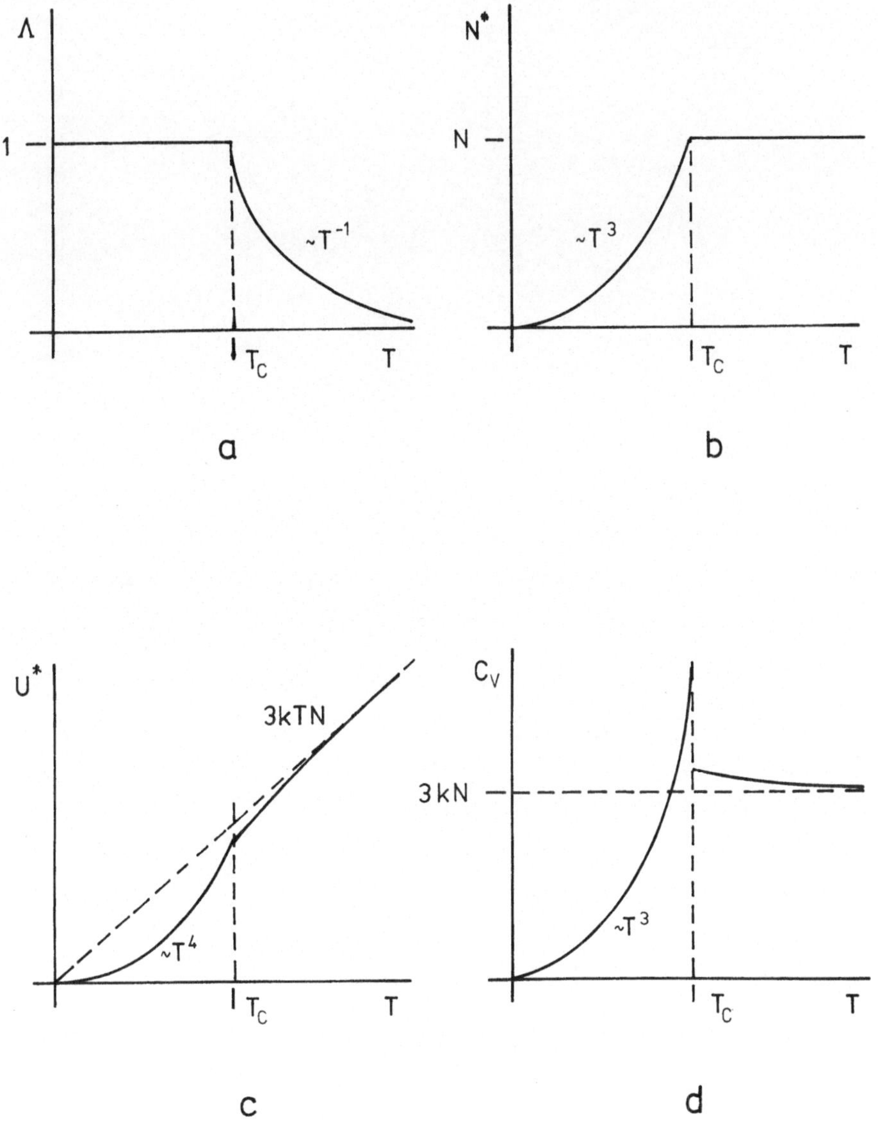

Fig. 1